Geodynamics of the Mexican Pacific Margin

Edited by
William L. Bandy
Yuri Taran
Carlos Mortera Gutiérrez
Vladimir Kostoglodov

Previously published in *Pure and Applied Geophysics*
(PAGEOPH), Volume 168, Nos. 8–9, 2011

Editors

William L. Bandy
Departamento de Geomagnetismo y Exploración
Instituto de Geofísica
Universidad Nacional Autónoma de México
Ciudad Universitaria
Coyoacán, 04510 México, DF
México
bandy@geofisica.unam.mx

Yuri Taran
Departamento de Vulcanología
Instituto de Geofísica
Universidad Nacional Autónoma de México
Ciudad Universitaria
Coyoacán, 04510, México, DF
México
taran@geofisica.unam.mx

Carlos Mortera Gutiérrez
Departamento de Sismología
Instituto de Geofísica
Universidad Nacional Autónoma de México
Ciudad Universitaria
Coyoacán, 04510 México, DF
México
carlosm@ollin.igeofcu.unam.mx

Vladimir Kostoglodov
Departamento de Sismología
Instituto de Geofísica
Universidad Nacional Autónoma de México
Ciudad Universitaria
Coyoacán, 04510 México, DF
México
vladi@servidor.unam.mx

ISBN 978-3-0348-0196-6   ISBN 978-3-0348-0197-3 (eBook)
DOI 10.1007/978-3-0348-0197-3

Library of Congress Control Number: 2011934864

Cover illustration: Based on a photograph provided by Yuri Taran.

Cover design: deblik, Berlin.

Printed on acid-free paper

Springer Basel AG is part of Springer Science+Business Media

www.birkhauser-science.com

Contents

Pure Appl. Geophys. 168 (2011), 1251–1253
© 2010 Springer Basel AG
DOI 10.1007/s00024-010-0241-7

Introduction

WILLIAM L. BANDY,[1] YURI TARAN,[2] CARLOS MORTERA GUTIÉRREZ,[3] and VLADIMIR KOSTOGLODOV[3]

The Mexican Pacific Margin contains a diverse but interdependent array of geodynamic settings providing for a wealth of research opportunities. Ancient ridge-trench collisions, cessation of subduction, continental rifting and the initiation of seafloor spreading are present within the peninsular province of Baja California. Between Puerto Vallarta and Acapulco one can find recent ridge-trench collisions, plate fragmentation and plate motion reorganizations, ridge propagation events, slab windowing, flat-slab subduction, continental rifting, subduction erosion, newly initiated subduction zones, great earthquakes and slow slip events, non-volcanic tremors, and magmatism that has produced a puzzling assortment of volcanic series as well as an unusual non-parallel alignment of the volcanic arc and the trench. The process of plate margin truncation by forearc translation and subduction erosion has significantly altered the tectonic landscape of the southern part of the Mexican Pacific Margin.

To understand these complex geodynamic settings and their interrelationships is challenging, and the solutions of many of the outstanding problems will require multidisciplinary investigations. Many such investigations are presently underway, and the aim of this special volume is to provide a means to facilitate the dissemination of the results of these studies amongst the various disciplines.

Seventeen papers are included in this volume. They include contributions from the fields of seismology, marine geophysics, geomorphology, petrology, geodesy and mantle/subduction dynamics.

The contributions from the field of seismology better define the active plate boundary, seismotectonics and seismic hazard potential occurring within the Gulf of California and its northern extension as well as at the intersection of the Rivera–Cocos plate boundary and the Middle America Trench off Manzanillo. Egill et al. discuss the implications of the El Mayor-Cucapah earthquake sequence on the seismotectonics at the north end of the Gulf of California and propose that this earthquake, as well as the 1992 Landers and the 1999 Hector Mine earthquakes, may have been controlled by bends in the plate boundary. Castro et al. make significant improvements in the epicentral locations in the Gulf of California by incorporating data from local broadband stations and conclude that the spatial distribution seismicity is quite complex in the northern Gulf of California, whereas to the south it is confined to narrow zones. Three papers address the debate concerning the interplate versus intra-plate origin of the Tecoman earthquake (Mw 7.4) of 22 January 2003. Núñez-Cornú et al. incorporate new data into a reassessment of the source region of this event and again conclude that this event was an intra-plate event. Andrews et al. perform a double-difference relocation of the aftershocks of this event and conclude that the main event occurred along the plate interface and was not an intra-plate event. Quintanar Robles et al. re-analyze the aftershock locations, source parameters and slip distribution using new additional near-field strong motion data and find that the aftershocks lie mainly along the coupling zone of the plates and inside the subducted slab, favouring an inter-plate event. They also found a more complicated slip pattern than previously determined. Zobin analyzes the earthquake intensity patterns of the 9 October 1995 (Mw

---

[1] Departamento de Geomagnetismo y Exploración, Instituto de Geofísica, Universidad Nacional Autónoma de México, Mexico D.F., Mexico. E-mail: bandy@geofisica.unam.mx
[2] Departamento de Volcanología, Instituto de Geofísica, Universidad Nacional Autónoma de México, Mexico D.F., Mexico.
[3] Departamento de Sismología, Instituto de Geofísica, Universidad Nacional Autónoma de México, Mexico D.F., Mexico.

8.0) and the 22 January 2003 (Mw 7.5) events and find that, for the City of Colima located at roughly the same distance from the two events, the intensity was greater for the smaller 2003 event. He postulates that this may be the result of the tectonic settings of the epicentral zones and/or the directivity of the rupture processes.

Marine geophysical investigations reveal the detailed morphology, structure and tectonics occurring along the Pacific margin of southern Baja California, the Jalisco subduction zone and along the Rivera Transform. Michaud et al. re-analyze the existing multi-beam bathymetric data off southern Baja California at a higher spatial resolution (60 m grid spacing) and find convincing evidence for several transcurrent right-lateral faults, consistent with the proposal that Baja California has yet to fully attach itself to the Pacific Plate. Bartolomé et al. use multichannel seismic reflection data collected along the offshore area of the Jalisco subduction zone off Puerto Vallarta during the 1996 CORTES survey to image for the first time the deep crustal structure of this area. Bandy et al. use seismic reflection and multi-beam data collected along the Rivera Transform during the 2002 FAMEX and BART surveys to characterize in more detail the morphotectonic elements of the transform. They propose that the transform is currently in the process of adjusting to changes in the relative motion between the Rivera and Pacific Plates.

Ramírez-Herrera et al. present an overview and synthesis of the geomorphologic and stratigraphic evidence and possible causes of coastal uplift/subsidence along the Middle America Trench from Puerto Vallarta to the Gulf of Tehuantepec. They find that, although spatial and temporal variability exists, coastal subsidence is occurring at the southern Colima Graben and the Guerrero seismic gap near Acapulco, whereas coastal uplift is occurring between Puerto Vallarta and Manzanillo and along the coast south of the Colima/El Gordo Graben until Lázaro Cárdenas.

Petrologic and geodetic investigations have focussed on better defining the continental rifting process in the area of the Gulf of California, the Trans-Mexican Volcanic Belt, and the SE boundaries of the Jalisco Block. Calmus et al. analyze the geochemical composition and ages of ca. 350 lava samples in the Gulf of California Province to define three successive stages in the evolution of the gulf and show that the spatial and temporal distribution of these lavas is consistent with the development of a slab tear beneath the present-day Gulf of California. Verma et al. use recently created discrimination diagrams to infer the tectonic setting of the Trans-Mexican Volcanic Belt. They find that the Trans-Mexican Volcanic Belt shows a dominantly continental rift setting, with the west-central part being consistent with dual tectonics of arc and rift. In contrast, the Central American Volcanic Arc shows an arc setting. Selvans et al. discuss the implications of their recent global positioning system (GPS) campaign near the triple junction near Guadalajara on the motion of the Jalisco Block relative to North America. They find that a motion of ∼2 mm/year to the southwest may be attributed to this tectonic motion.

Several papers focus on better defining the geometry of the crust and subducted slab, the thermal structure and subduction processes occurring beneath central and southern Mexico. Melgar and Pérez–Campos analyze receiver functions along a transect of broadband stations across the Isthmus of Tehuantepec. They show that the Moho in this area is not a simple horizontal interface and that the subducted oceanic crust has a dip of 26° between 140 and 310 km from the trench axis. Skinner and Clayton undertake a critical evaluation of the proposed mechanisms of slab flattening for central Mexico and find no obvious explanation for the shallow subduction with the possible exception of a change in wedge viscosity produced by the introduction of water into the wedge by the subducting slab. V. Manea and M. Manea use the results of recent studies, which better constrain the shape and state of the subducting slab and continental crust beneath central Mexico, to derive a new model for the thermal structure beneath central Mexico. The new thermal model is then used to estimate the amount and location of dehydration pulses along the slab interface, and a correlation is noted between the location of these dehydration pulses and areas of non-volcanic tremors in the overlying continental crust. M. Marina and V. Marina use Curie point depths estimated from published magnetic data to study the regional thermal structure of the crust in Mexico.

The diversity of the papers presented in this volume clearly illustrates the complex nature of the geodynamic processes occurring along the Mexican Pacific Margin and the need for a multidisciplinary approach to unraveling these processes. The Mexican Pacific Margin remains fertile ground for investigating a wide variety of geodynamic processes and will continue to attract an increasingly diverse group of investigators.

Acknowledgements

We thank all the authors for their contributions to this special volume, Renata Dmowska for her help during the preparation of this volume and Brian Mitchell for handling the reviews of our contributions. We are most grateful to the many reviewers for their efforts to improve the quality of this special volume. These people include, in alphabetical order, Samuele Agostini, Salil Agrawal, Aydin Büyüksarac, Thierry Calmus, Raúl Ramón Castro, Ina Cecic, Elizabeth Cochran, Claire Currie, Luca Ferrari, Gary Fuis, Antonio González Fernández, Stephen Grand, Jeanne Hardebeck, Rebecca Lange, Hervé Martin, Marina Menea, Vlad Contantin Manea, Carlos Mendoza, Alejandro Nava Pichardo, Mike Oskin, Javier Pacheco, Nikolai Shapiro, Francisco Suárez, Gerardo Suárez, Akiko Tanaka, Carlos Valdés González, Jesús Vidal Solano, Paul Wallace, Dante Morán Zenteno, Ramón Zúñiga Dávila and several anonymous reviewers who reviewed our contributions.

(Received October 19, 2010, revised October 21, 2010, accepted October 22, 2010, Published online December 21, 2010)

Pure Appl. Geophys. 168 (2011), 1255–1277
© 2010 Springer Basel AG
DOI 10.1007/s00024-010-0209-7

The 2010 M_w 7.2 El Mayor-Cucapah Earthquake Sequence, Baja California, Mexico and Southernmost California, USA: Active Seismotectonics along the Mexican Pacific Margin

Egill Hauksson,[1] Joann Stock,[1] Kate Hutton,[1] Wenzheng Yang,[1] J. Antonio Vidal-Villegas,[2] and Hiroo Kanamori[1]

Abstract—The El Mayor-Cucapah earthquake sequence started with a few foreshocks in March 2010, and a second sequence of 15 foreshocks of $M > 2$ (up to M4.4) that occurred during the 24 h preceding the mainshock. The foreshocks occurred along a north–south trend near the mainshock epicenter. The M_w 7.2 mainshock on April 4 exhibited complex faulting, possibly starting with a ~M6 normal faulting event, followed ~15 s later by the main event, which included simultaneous normal and right-lateral strike-slip faulting. The aftershock zone extends for 120 km from the south end of the Elsinore fault zone north of the US–Mexico border almost to the northern tip of the Gulf of California. The waveform-relocated aftershocks form two abutting clusters, each about 50 km long, as well as a 10 km north–south aftershock zone just north of the epicenter of the mainshock. Even though the Baja California data are included, the magnitude of completeness and the hypocentral errors increase gradually with distance south of the international border. The spatial distribution of large aftershocks is asymmetric with five M5+ aftershocks located to the south of the mainshock, and only one M5.7 aftershock, but numerous smaller aftershocks to the north. Further, the northwest aftershock cluster exhibits complex faulting on both northwest and northeast planes. Thus, the aftershocks also express a complex pattern of stress release along strike. The overall rate of decay of the aftershocks is similar to the rate of decay of a generic California aftershock sequence. In addition, some triggered seismicity was recorded along the Elsinore and San Jacinto faults to the north, but significant northward migration of aftershocks has not occurred. The synthesis of the El Mayor-Cucapah sequence reveals transtensional regional tectonics, including the westward growth of the Mexicali Valley and the transfer of Pacific–North America plate motion from the Gulf of California in the south into the southernmost San Andreas fault system to the north. We propose that the location of the 2010 El Mayor-Cucapah, as well as the 1992 Landers and 1999 Hector Mine earthquakes, may have been controlled by the bends in the plate boundary.

Key words: Seismicity, El Mayor-Cucapah earthquake, Mexican Pacific Margin, Baja California, Southern California, faults, aftershocks, seismotectonics, margin processes, Pacific–North America plate boundary, tomography.

1. Introduction

The 2010 M_w 7.2 El Mayor-Cucapah earthquake sequence occurred within the Mexican Pacific margin in northern Baja California (BC), a region of high seismicity straddling the complex Pacific–North America plate boundary. The seismically active parts of the northeast Baja California region follow the Pacific–North America plate boundary from the northern shores of the Gulf of California, crossing major sedimentary basins and connecting up to the faults of the plate boundary zone in southern California, including the Elsinore and San Jacinto faults, and the southernmost San Andreas fault in the Salton sea to the north (Fig. 1). Because sediments cover many of the faults, mapping their exact locations and estimating slip rates is difficult (Pacheco et al., 2006). Nonetheless, this earthquake sequence has released some of the right-lateral shear that had accumulated adjacent to the plate boundary for more than a century. This earthquake sequence also sheds light on the structural connections between the principal plate boundary fault zone in the northern Gulf of California/Salton trough region and the active fault systems found farther west on both sides of the US–Mexico border.

The El Mayor mainshock rupture can be divided into two rupture zones. The northern zone extends from the mainshock epicenter to the northwest, terminating near the US–Mexico international border. The southern zone extends from the epicenter to the

---
[1] Seismological Laboratory, Division of Geological and Planetary Sciences, California Institute of Technology, Pasadena, CA 91125, USA. E-mail: Hauksson@caltech.edu
[2] CICESE, Carretera Ensenada-Tijuana No. 3918 Zona Playitas Código Postal 22860 Apdo. Postal 360, Ensenada, BC, México.

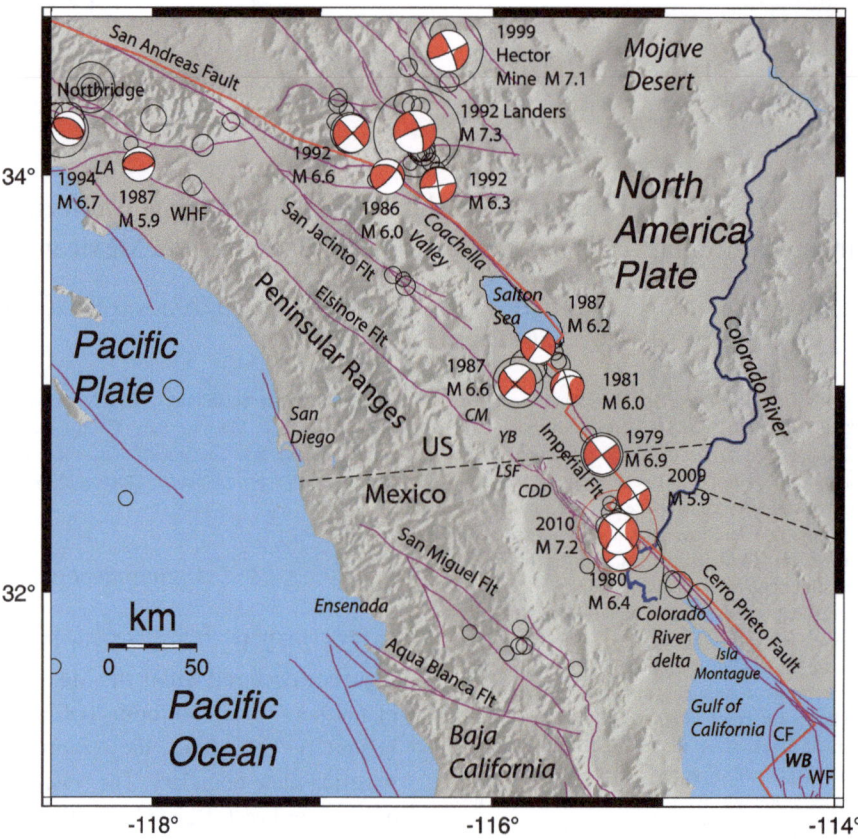

Figure 1
Tectonic overview of Southern California and Baja California, including the Pacific–North America plate boundary (*red*) and global moment tensor solutions from the www.gcmt.org catalog. Earthquakes in the ANSS catalog from 1970 to 2010 of $M \geq 5.0$ are shown as *open circles*. Fault traces from JENNINGS (1994), GONZÁLEZ-ESCOBAR *et al.* (2010), FENBY and GASTIL (1991), and FLETCHER and SPELZ (2009). Approximate plate boundary trace from BIRD (2003). *CDD* Cañada David detachment, *CF* Consag fault, *CM* Coyote mountains, *LA* Los Angeles, *LSF* Laguna Salada fault, *WB* Wagner basin, *WF* Wagner fault, *WHF* Whittier fault, *YB* Yuha basin

southeast, apparently terminating at the shoreline of the Gulf of California, although the detection capability of the joint network is limited in this latter region. It was surprising that the northwestern half of the El Mayor sequence occurred within the Sierra Cucapah mountain range as opposed to along its western side, which is controlled by a major active steeply west-dipping dextral fault with a small normal component, the Laguna Salada fault (MUELLER and ROCKWELL, 1995). The faults that ruptured in the northwestern zone during the mainshock had been mapped by BARNARD (1968) and FLETCHER and SPELZ (2009). In contrast, the region of the southeastern half of the rupture consists of deep basins, including the Mexicali Valley, the Colorado River delta, and the Altar basin (PACHECO *et al.*, 2006). In addition to

dextral motion, the east-side-down displacement in the El Mayor sequence also contributed to the collapse of the range, and thus to widening of the adjacent sedimentary basin. Because the southeastern half of the rupture zone occurred beneath the Colorado River delta basin sediments, surface rupture was, in part, masked by disturbance of the surficial sedimentary layers, possibly caused by liquefaction at depth. Furthermore, there are no nearby seismic stations and the hypocenters are of progressively poor quality to the southeast of the mainshock epicenter, making the mapping of the mainshock rupture surface difficult at best.

The El Mayor-Cucapah sequence is the largest sequence recorded in the plate boundary zone since the 1992 M_w 7.3 Landers and 1999 M_w 7.1 Hector

Mine sequences in the eastern Mojave Desert to the north in California. The El Mayor-Cucapah sequence shares some of the same seismotectonic characteristics (Fig. 1). First, none of these sequences occurred on the main plate boundary faults, such as the San Andreas, Imperial or the Cerro Prieto faults. Second, they all ruptured multiple fault strands with oblique normal-dextral sense of shear, causing similar aftershock sequences. Third, the Landers sequence occurred on a series of faults adjacent to the San Andreas fault. The El Major-Cucapah sequence occurred on a series of faults that are also adjacent to the Cerro Prieto plate-boundary fault. The occurrence of the El Mayor-Cucapah earthquake to the west of the Cerro Prieto fault is consistent with the westward shifting of strain localization in the Gulf area (PACHECO et al., 2006). The mainshock also caused minor slip on the Laguna Salada fault near the international border and the Cañada David detachment, a low-angle normal fault mapped for 60 km distance along the western side of the Sierra El Mayor (FLETCHER and SPELZ, 2009). A previous large earthquake in the region occurred in 1892 (e.g., HOUGH and ELLIOTT, 2004), and may have ruptured the northwestern part of the Laguna Salada fault and parts of the Cañada David detachment (FLETCHER and SPELZ, 2009; MUNGUÍA et al., 2010). The Laguna Salada fault, with a slip rate of ~2–3 mm/yr, is the southeast extension of the Elsinore fault (MUELLER and ROCKWELL, 1995). These authors showed that at least 22 km of the fault ruptured in the 1892 earthquake.

The mainshock on April 4, 2010 was felt across Baja California, the southwestern US, and northwestern Mexico. After the mainshock, SUÁREZ-VIDAL et al. (2010) searched for damage in Mexicali Valley. Small villages and agricultural areas had significant structural damage caused by soil liquefaction and ground fracturing. The damage consisted of flooded areas, subsurface zones of liquefaction, fissures across paved roads, cracking of water canals, tilting of power line towers, and damage or collapse of residential houses. A total of 25,000–35,000 people were evacuated from the earthquake zone. Geotechnical instruments and visual inspection reports identified up to 30 cm of triggered slip on the faults around the Cerro Prieto pull-apart (GLOWACKA et al.,

2010). MUNGUÍA et al. (2010) showed that recorded peak accelerations were up to ten times larger at sediment sites as compared with rock sites, with peak acceleration of 0.81 g at a sediment site located 12 km away from the epicenter.

We provide synthesis of the seismicity following the El Mayor-Cucapah mainshock and the decade preceding it. We also provide insight into the regional seismotectonics along the plate boundary, as well as a Coulomb failure model to explain the occurrence of large earthquakes adjacent to the plate boundary. The El Mayor-Cucapah earthquake sequence probably added tectonic stress to the Elsinore and San Jacinto faults to the north. It caused some minor triggered seismicity within hours of the mainshock, an M5.0 aftershock at Ocotillo, California, with its own aftershock sequence on June 15, and a M5.4 mainshock aftershock sequence along the San Jacinto fault on July 7.

2. Data and Methods

The Caltech/USGS Southern California Seismic Network (SCSN) and the CICESE Baja California, Mexico Seismic Network (RESNOM [Red Sísmica del Noroeste de México]) have recorded more than 10,000 earthquakes in the El Mayor-Cucapah earthquake sequence. We relocated the foreshocks, mainshock and aftershocks using a 3-D velocity model determined with SIMULPS (THURBER, 1993). We made differential travel times from the picks and applied HypoDD (WALDHAUSER and ELLSWORTH, 2000) to refine the hypocenters. Using the same approach, we also relocated the preceding decade of seismicity. The event relocations benefited from arrival time picks determined from real-time waveform data streams of several stations exchanged between the two networks. In addition, arrival time picks from several other stations were provided on the CICESE Baja California, Mexico Seismic Network web site (Fig. 2).

Real-time SCSN moment tensor solutions (CLINTON et al., 2006) were derived for the mainshock and $M \geq 4$ aftershocks using an automated analysis of waveforms from up to six broadband stations and the inversion method of DREGER and HELMBERGER (1993).

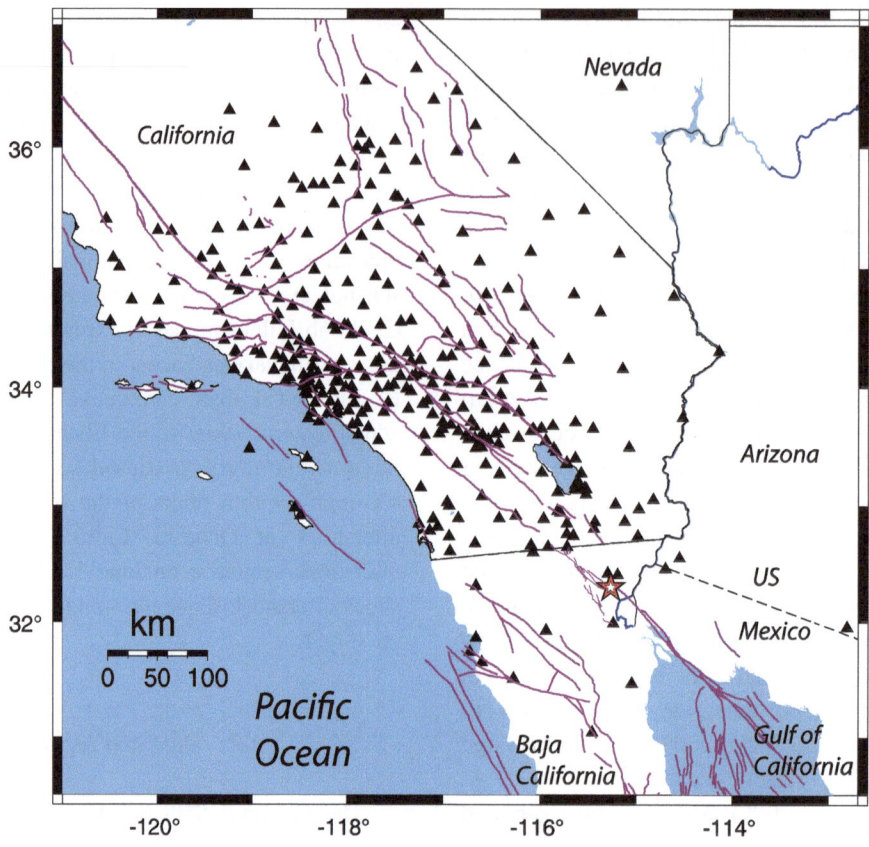

Figure 2
Seismic stations that recorded data from the El Mayor-Cucapah earthquake sequence. Stations located in Mexico are part of the RESNOM network. Stations in the US are part of the Southern California Seismic Network (SCSN) or contributed stations from partner networks

Because of the large azimuthal gap, with almost all of the broadband stations located to the north, the large non-double-couple components shown for the M4.4 foreshock and some of the aftershocks are considered to be artifacts.

Using the available arrival time data, we determined a refined 3-D V_p velocity model. The starting model was based on a 1-D model from HAUKSSON (2000). This new 3-D model has somewhat limited resolution south of the US–Mexico international border, but is essential for constraining the focal depths of events in this sequence.

3. Results

The 2010 M_w 7.2 El Mayor-Cucapah earthquake sequence occurred in northern Baja California, along the Mexican Pacific margin. The sequence ruptured along a number of fault segments trending from the northernmost tip of the Gulf of California, in the southeast, to Ocotillo in the northwest, just north of the US Mexico international border, for a distance of more than 120 km. The crustal deformation caused by the sequence reflects the tectonics of the region, which is characterized by complex interactions of strike-slip faults and normal faults leading to right-lateral transtensional deformation, including basin extension.

3.1. Background Seismicity

The rate of background seismicity in the Baja California region during the previous decade (2001–2010) has remained high and has exhibited complex temporal and spatial evolution (Fig. 3). Numerous swarms and mainshock-aftershock sequences have

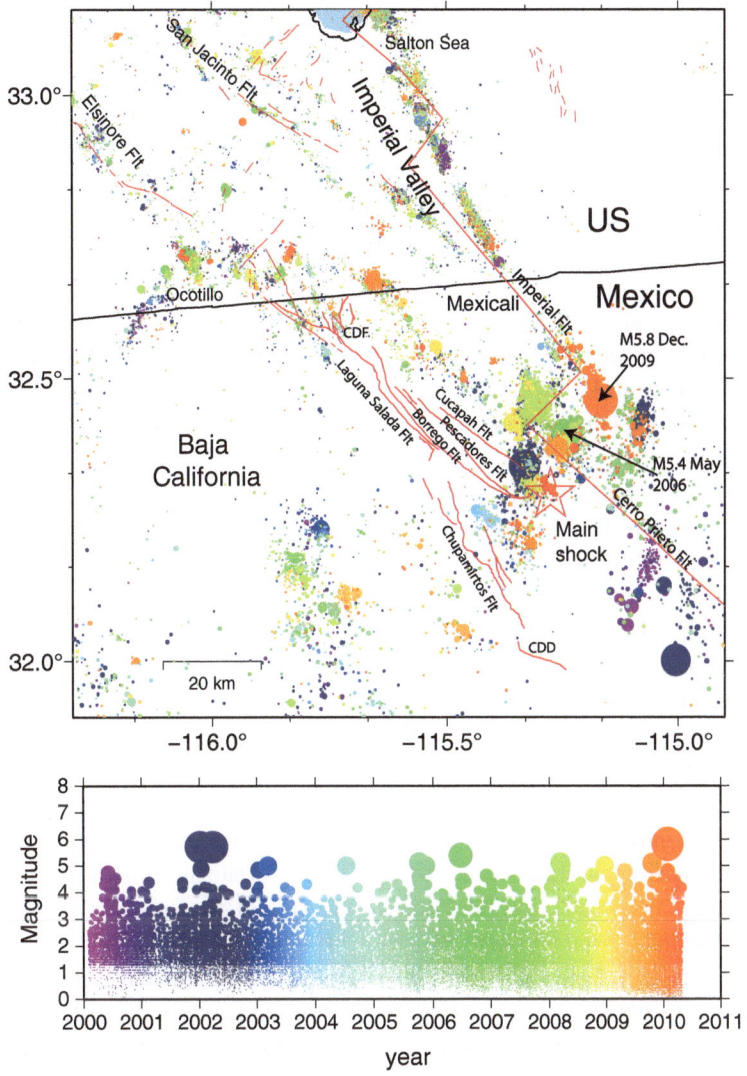

Figure 3

The seismicity in the study area from 2000 to April 3, 2010. The seismicity surrounds the rupture zone of the M7.2 El Mayor-Cucapah mainshock but only the north-northeast trend near the mainshock epicenter is within the rupture zone. Seismicity was recorded by the SCSN and RESNOM seismic networks. The approximate plate boundary trace from BIRD (2003) is also shown. *CDF* Centinela Detachment fault, *CDD* Cañada David detachment

occurred, with nine earthquakes of $M \geq 5.0$ culminating in the El Mayor-Cucapah earthquake sequence. The swarms lasted for weeks to months, and were sometimes associated with geothermal areas (SUÁREZ-VIDAL *et al.*, 2007, 2008).

The background seismicity shows prominent geographical trends in the northeast and northwest directions. The longest bands of seismicity (>20 km long) trend northwest and include bands along the Cerro Prieto and Imperial faults, regarded as the Pacific–North America plate boundary (PACHECO

et al., 2006). The coexisting conjugate northwest and north to northeast faulting produces both right-lateral strike-slip and normal focal mechanisms. In addition, there is a long band that extends northwest from the northwest end of the Cerro Prieto fault to the international border, on the northeast side of the Sierra Cucapah. There is a more poorly defined northwest-trending band along the northern part of the Laguna Salada fault, on the west side of Sierra Cucapah, on either side of the international border. Numerous shorter bands (~20 km long) trend north-northeast to northeast, including prominent bands east and southeast of the El Mayor-Cucapah mainshock and also through Ocotillo, straddling the international border.

A sequence of earthquakes in May 2006 culminated with a M_w 5.4 mainshock causing extensional ground failure, with surface rupture for a distance of more than 5 km, ground acceleration up to 0.5 g (4.9 m/s^2), and water level changes up to 6 m (SUÁREZ-VIDAL et al., 2007; MUNGUIA et al.,2009; SARYCHIKHINA et al., 2009). The finite fault modeled determined by SARYCHIKHINA et al. (2009), has a strike, rake, and dip of (48°, 89°, 45°) and a length of 5.2 km, width of 6.7 km, and 34 cm of uniform slip.

The latest foreshock–mainshock–aftershock sequence to occur prior to the El Mayor-Cucapah earthquake had a M_w 5.8 mainshock on December 30, 2009, near Ejido Jalapa in the Municipality of Mexicali, Baja California. This sequence had a strike-slip faulting mainshock that caused some ground failure in the immediate vicinity of the epicenter. Both the northwest-striking focal plane of the mainshock and the north–northwest alignment of the aftershocks were consistent with faulting along the Imperial fault. Subsequently, about three months later, the El Mayor-Cucapah mainshock occurred ~30 km to the southwest.

Although the background seismicity had been high during the last decade in the general region of Baja California, it did not coincide geographically with the future rupture zone of the El Mayor-Cucapah mainshock, but formed a halo surrounding the Sierra Cucapah. This geographical separation between the mainshock and the background seismicity illustrates the complexities of earthquake processes and the tectonics in the region.

3.2. Foreshocks

Foreshocks with magnitudes ranging from M1.5 to M4.4 occurred mostly in two temporal clusters (Fig. 4). The foreshocks were located within 2–3 km distance of the mainshock, and formed an approximate north–south trend. The first occurred on March 21 and 22 and the second occurred on April 3 and 4, during the 24 h preceding the mainshock. The largest foreshock had a magnitude of M_L 4.4 or M_w 4.2, and

exhibited left-lateral strike-slip faulting on a vertical nodal plane striking almost north–south (N3°E). The non-double-couple component of the foreshock moment tensor is not considered to be real because of the lack of azimuthal coverage. This mechanism has a P-axis trending N38°W, in contrast to most of the other aftershock focal mechanisms, whose P-axes trend more northerly. The fault orientation is also rotated ~35°–45° clockwise away from the strike of other faults in the region. A similar orientation of the foreshock faults and the seismicity trends is also seen in some of the background seismicity recorded during the past decade (Fig. 3). As discussed below, the mainshock may have started with similar fault strike, but different dip as the M_L 4.4 foreshock, and quickly evolved into an extended northwest-striking rupture.

3.3. Mainshock Moment Tensor

The El Mayor-Cucapah mainshock epicenter was located at the southeast corner of the Sierra Cucapah. The global moment tensor solution (GCMT www. gcmt.org) exhibited a large non-double-couple (N-DC) component as shown in Fig. 5a(1). Frequently, an N-DC component is considered to be an artifact of an inappropriate modeling procedure (e.g., ADAMOVA and SILENY, 2010). However, as the quality and quantity of teleseismic data and modeling methods have improved significantly, it has been shown that for some earthquakes the N-DC mechanisms reflect the complexity of the source mechanism (e.g., the 2000 Sumatra earthquake (ABERCROMBIE et al., 2003); the 2010 Samoa Islands Earthquake (LAY et al., 2010)).

Since the source region of the El Mayor earthquake has very complex tectonic structures, the GCMT N-DC source mechanism most likely reflects a real feature rather than an artifact of modeling. To explore this possibility further, we examined the mechanism of this earthquake at even longer periods using the W phase recorded at the Global Seismographic Network stations (KANAMORI and RIVERA, 2008). At longer periods, the assumption of a point source is more justified. We calculated moment tensor solutions using various initial locations and frequency bands. Although, the solutions vary with the chosen initial location and frequency bands, the solutions always have a significant N-DC component

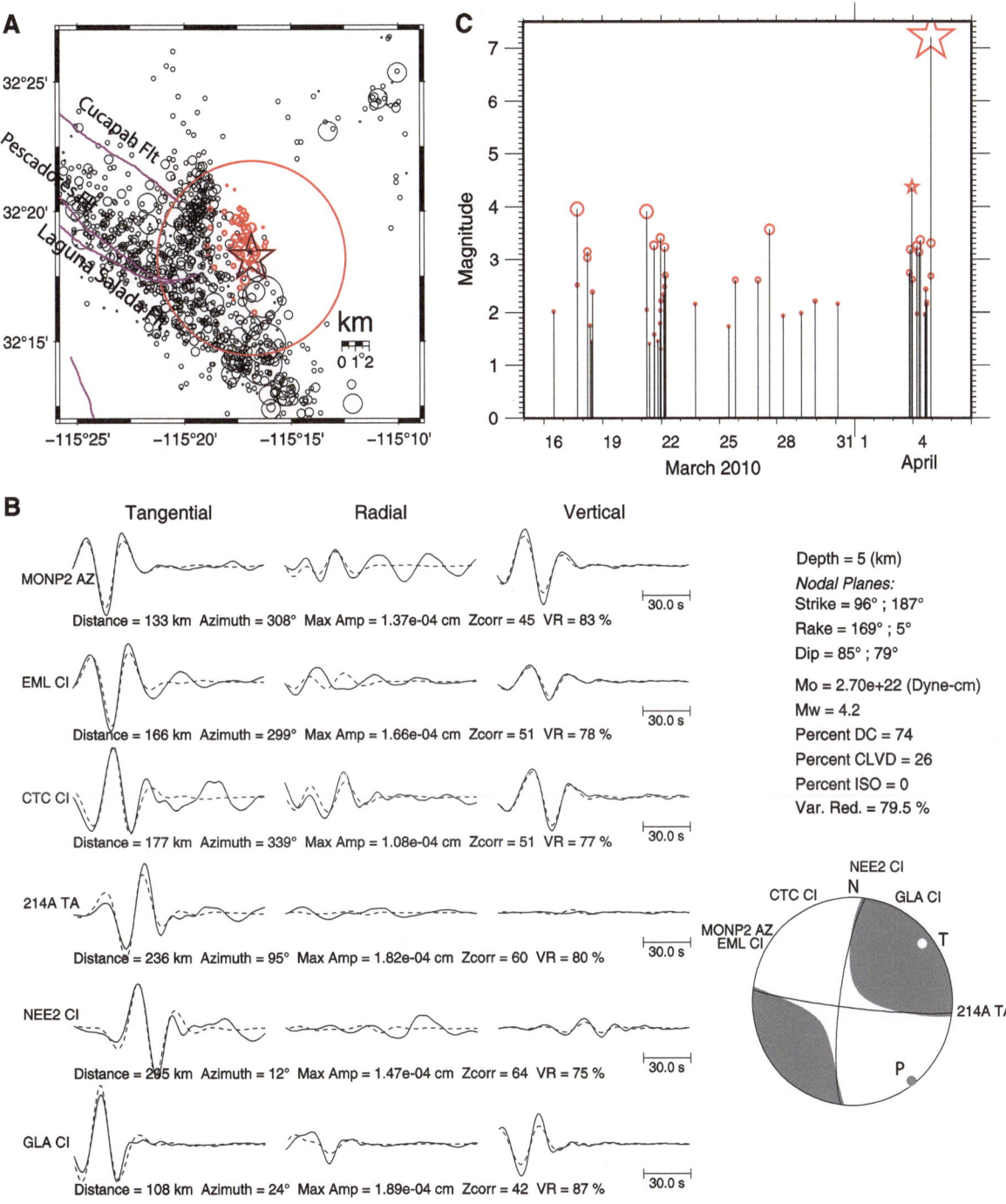

Figure 4

a Map view, the El Mayor-Cucapah foreshocks (*red open circles*) and aftershocks (*black open circles*). **b** Waveform fits with data shown as solid curves and modeled waveforms shown as *dashed*, and moment tensor of the M_L 4.3 foreshock (origin time: 23 h, 3 min, 47 s, 3 April, 2010). Station and network codes are shown next to waveforms and the lower hemisphere mechanism. *CLVD* compensated linear vector dipole, *DC* double couple, *ISO* isotropic component, *Mo* moment magnitude, *Mw* moment magnitude, *Var. Red.* variance reduction. The non-double-couple component of the moment tensor is not considered to be real because of the limited azimuthal coverage. **c** Magnitude versus time stick plot. The $M \geq 4.3$ events are designated by a *star*

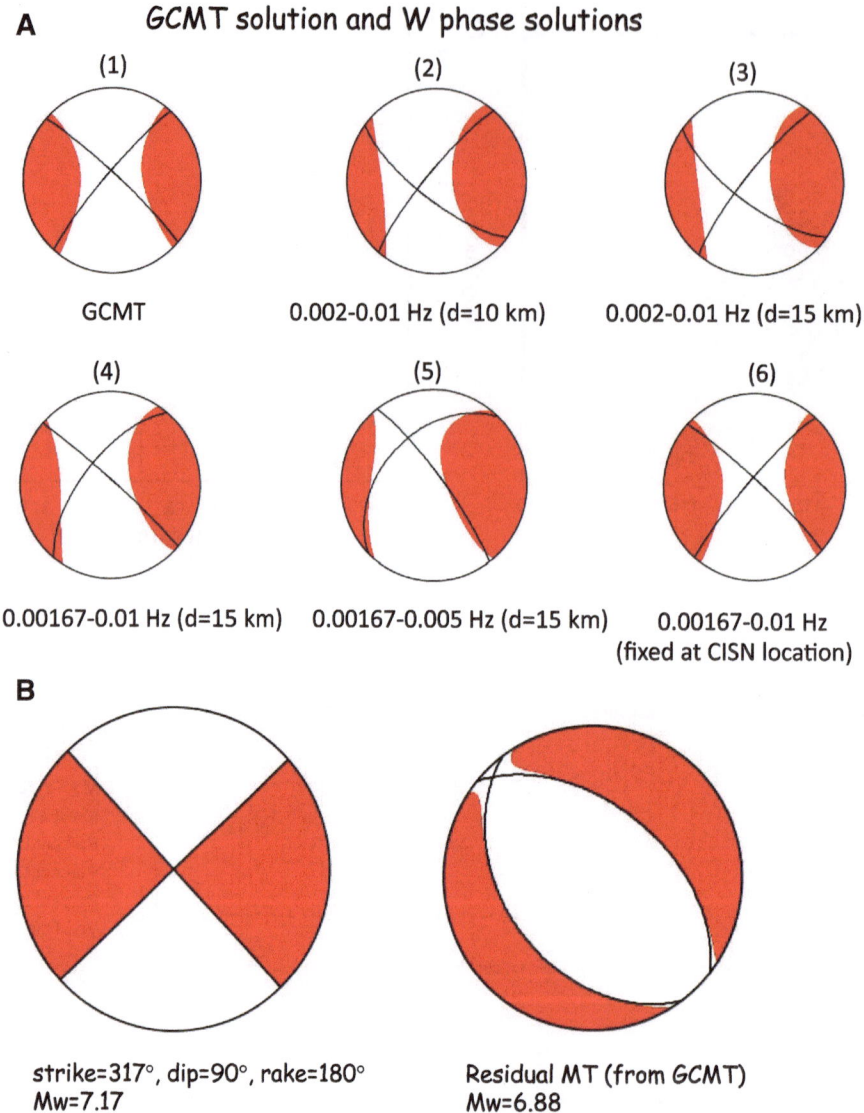

Figure 5
a Mechanism diagrams from the Global Centroid Moment Tensor Project (1), and those from W phase inversions with various initial locations and frequency bands. (2) 0.002–0.01 Hz ($d = 10$ km), (3) 0.002–0.01 Hz ($d = 15$ km), (4) 0.00167–0.01 Hz ($d = 15$ km), (5) 0.00167–0.005 Hz ($d = 15$ km), (6) 0.00167–0.01 Hz (location fixed at the CISN location, $d = 10$ km). **b** One decomposition of GCMT non-double-couple mechanism: vertical strike-slip mechanism (*left*) and the residual moment tensor

as shown in Fig. 5a, HAYES *et al.* (2009) showed that for events in this magnitude range, a frequency band from 0.00167 Hz (600 s) to 0.01 Hz (100 s) yields the most stable solutions. Shown in Fig. 5a, solutions (4) and (6) obtained with this frequency band are almost identical to the GCMT solution. This modeling of the waveforms suggests that the N-DC component is a real feature for this event, and the

GCMT solution is a good representation of the overall mechanism.

A long-period teleseismic solution cannot constrain the details of the mainshock source. However, it represents the overall source deformation pattern and provides an important constraint for the source mechanism. Any coseismic source model, either from field studies, local data, high-frequency teleseismic body wave data or static (GPS and InSAR) data, must satisfy this constraint.

The decomposition of a N-DC event to multiple mechanism events (e.g., double-couple, and compensated linear vector dipole (CLVD), etc.) is non-unique. We show one example of decomposition of the GCMT mechanism in Fig. 5b. As a representative big tectonic picture of the region, we assume a simple vertical strike-slip fault with a strike of 317° (i.e., approximate trend of the aftershock zone) is the principal mechanism of the mainshock. Then, we subtract the moment tensor of an earthquake with this mechanism having a magnitude M_w from the GCMT moment tensor such that the residual moment tensor measured with a chosen norm is minimized. Here the norm is the sum of the square of all the elements of the residual moment tensor. This process determines the M_w of the principal source (i.e., strike-slip) and the residual source. More specifically, let b_{ij} be the moment tensor elements of the observed moment tensor shown in Fig. 5a(1), and let a_{ij} be the moment tensor of the assumed strike-slip event with a unit moment. Then, the moment tensor of the residual event is given by $c_{ij} \equiv b_{ij} - M_0 a_{ij}$ where M_0 is the scalar moment of the principal event. We determine M_0 by minimizing $S = \sum (b_{ij} - M_0 a_{ij})^2$.

Figure 5b shows the mechanisms of the principal event and the residual event thus determined. As mentioned earlier, the decomposition is not unique, and this figure is not meant to show that the El Mayor-Cucapah mainshock consists of the two events shown. What we intend to show is that if the principal fault pattern of the El Mayor-Cucapah mainshock is given by the right-lateral vertical strike-slip earthquake with $M_w = 7.17$ ($M_0 = 7.2 \times 10^{19}$ N m) (left figure on Fig. 5b), then the sum of all the secondary deformations, taken together, can be represented by a normal fault earthquake with $M_w = 6.88$ ($M_0 = 2.6 \times 10^{19}$ N m). In other words, the mechanism diagram shown on the right of Fig. 5b represents the residual deformation distributed near the fault zone, in the sense as discussed by KOSTROV (1974).

3.4. Finite Source of the Mainshock

The mainshock occurred along the western edge of the plate boundary deformation zone and ruptured through both the Sierra Cucapah to the northwest and the Colorado River delta region to the southeast, extending to the northern shorelines of the Gulf of California. Ground ruptures were identified both from satellite images and field mapping by various investigative teams (e.g., FLETCHER et al., 2010).

FIELDING et al. (2010) applied interferometric analysis of synthetic aperture radar images (InSAR) and pixel correlation to map the surface rupture and ground deformation. They identified a system of mapped faults through the Sierra Cucapah up to the international border and a set of unmapped faults extending 60 km to the southeast beneath the Colorado River delta. Similarly, SANDWELL et al. (2010) used InSAR data to identify the same major zones of faulting with right-lateral and east-side down normal faulting. To the southeast of the mainshock epicenter they identified a deformation zone that was 18 km wide and 60 km long, with extensive liquefaction superimposed on a zone of deep slip. This zone, bracketed to the east by the Cerro Prieto fault, includes the sub-surface fault that may have accommodated slip in the mainshock and was named the Indiviso fault (J. GONZALEZ, written communication, 2010).

On the ground and using aerial flights, FLETCHER et al. (2010) mapped a ~60 km long complex rupture along multiple fault strands from the mainshock epicenter to the international border, just south of Ocotillo, California. Some scarps associated with the 1892 Laguna Salada earthquake were re-ruptured. The rupture southeast of the epicenter consisted of distributed fracturing superimposed on liquefaction. The rupture to the northwest extended along the Laguna Salada and Pescadores faults. Their field mapping showed that the rupture was discontinuous in the high Sierra Cucapah and jumped ~10 km to the north, forming a left step to resume on the Borrego fault with slip up to 4 m. The rupture finally terminated 30 km to the north on the Paso Superior fault. The results of FIELDING et al. (2010), SANDWELL et al. (2010), and FLETCHER et al. (2010) showed that faulting was very complex, with slip partitioning between dip–slip faults, including detachment faults, and right-lateral strike-slip motion.

To determine a finite source model, WEI et al. (2010) used teleseismic waveforms and the sub-pixel correlation of optical SPOT 2.5 m panchromatic images from FIELDING *et al.* (2010) to determine that the mainshock rupture extended 55 km to the northwest. The average right-lateral slip was 2.5 m. The data are less clear to the southeast but suggest a meter of offset over a distance of more than 60 km. They found a total seismic moment of $M_0 = 7.9 \times 10^{19}$ N m during the initial 40 s of rupture. They interpreted the first 15 s of mainshock waveforms as a M_w6.3 subevent with a focal mechanism with strike of N5°W and dip of 45° towards east. They also inferred that, following a few seconds of rupture hiatus, the rupture resumed with a N50°W strike exhibiting transtensional bilateral rupture. The slip to the northwest was coseismic, but the slip to the southeast is still being investigated as possibly having been aseismic (WEI *et al.*, 2010).

Similarly, XU *et al.* (2010) inverted high sample rate GPS waveforms, teleseismic body waves, and surface waves from the mainshock. They determined a total seismic moment of $M_0 = 5 \times 10^{19}$ N m. In their model, the rupture to the northwest started as pure strike-slip with peak slip of 5 m and evolved into oblique motion as it traveled farther north. The southeastern rupture extended N133°E with, possibly, two large asperities, one at the epicenter and the other located about 45 km to the southeast with maximum slip of 4 m. The results of the inversion by XU *et al.* (2010) also required a large subevent with normal faulting motion on a 65° dipping plane. Adding both the strike-slip and normal faulting subevents, they obtained a total seismic moment of 1.2×10^{20} N m (M_w 7.3). Their model showed gradual onset of slip, with the slip reaching high levels from 18 to 40 s after the origin time.

3.5. *Aftershocks*

The ability to detect and determine accurate hypocenters decreases with distance from the international border to the south, because the majority of the seismic stations are located in the US, with only a few stations recording south of the border (Fig. 2). As an example, two measures of quality, the calculated horizontal and vertical errors, become larger with distance from the border (Fig. 6a, b). Similarly, the distance to the nearest station influences the quality of the depth determination (Fig. 6c). If the distance to

the nearest station is greater than twice the focal depth, the depth is poorly determined. The accuracy of focal depths progressively decreases as distance to the nearest station increases. Similarly, magnitude $M > 1.0$ events are well recorded at the border while, at the northern tip of the Gulf of California, the combined networks are detecting only $M \geq 3.0$ events (Fig. 6d).

The foreshock, mainshock, and aftershock magnitude versus time diagram shows the temporal evolution of the sequence (Fig. 7). Most of the $M \geq 4.4$ aftershocks occurred within days of the mainshock, with a late M5.7 aftershock on June 14. The lack of seismic stations south of the border also leads to heterogeneity in the earthquake statistics and requires more stringent assumptions about magnitude of completeness (M_c) for the whole sequence (Fig. 7). We used the approach of WIEMER (2001) to determine a p-value for Omori's law (decay of the sequence with time), assuming $M_c = 3.0$ for the whole sequence and a c-value of 0.5 days, which is the duration of the most incomplete part of the catalog following the mainshock. The p-value of 1.0 ± 0.03 is similar to what has been observed for other aftershock sequences in California (REASENBERG and JONES, 1989). The b-value of 0.96 ± 0.03 is also similar to what is observed for the Southern California catalog (HUTTON *et al.*, 2010).

Most of the aftershocks occurred within the latitude box of 31.5° to 32.75° and longitude −116.20° to −114.7°. The joint seismic networks have recorded and processed data for the following aftershocks from April 4 to June 15, 2010 (Table 1).

We estimate that the SCSN has unprocessed data for about 30% more aftershocks during this time interval. This would give a total number of ~14,000 aftershocks. If this sequence had occurred within the more densely instrumented Southern California Seismic Network, as did the Landers and Northridge earthquakes, the number of located aftershocks would likely have been in the range of 20,000–30,000.

The aftershocks extended from the northern tip of Gulf of California, across the US–Mexico international border, to Ocotillo, California (Fig. 8). The aftershocks formed three separate spatial clusters, as indicated in Fig. 8. The first extended from Ocotillo to the step-over in slip (mapped by FLETCHER *et al.*,

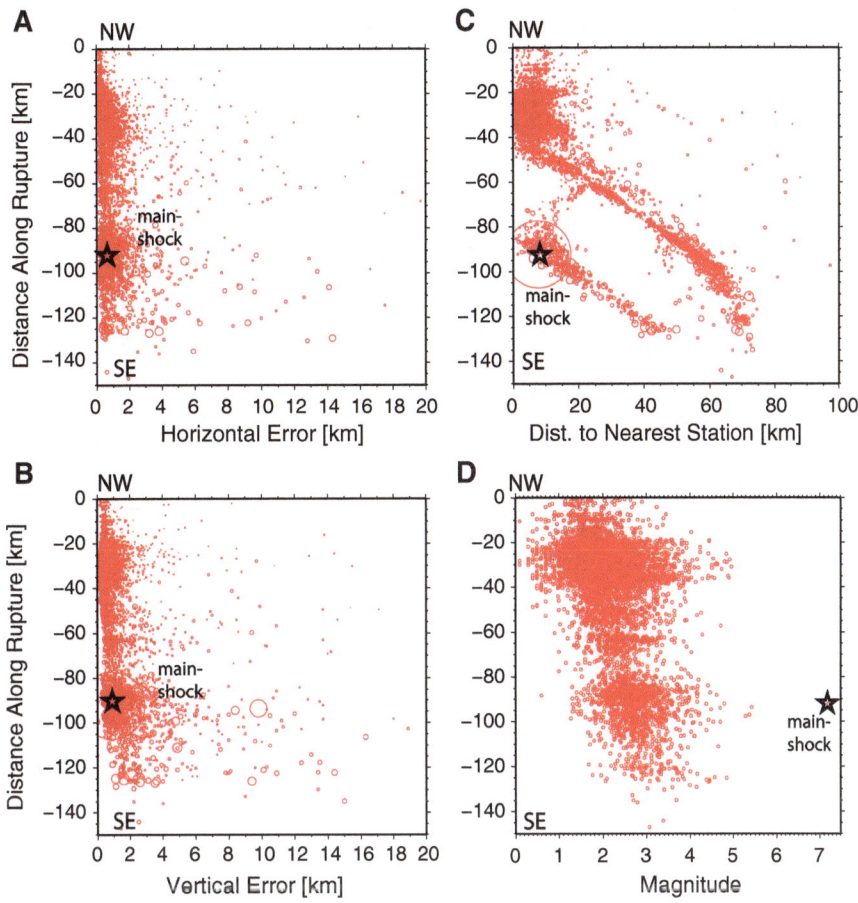

Figure 6

a Horizontal, and **b** vertical one-sigma errors calculated with Simulps (THURBER, 1993). The northwest and southeast end points coincide with the A–A' cross-section shown in Fig. 8. **c** The distance to the closest station versus distance along the rupture. Note that, this distance increases southward. **d** Distance along the rupture versus magnitude, showing that fewer small earthquakes are detected to the south

2010) between the Borrego and Pescadores faults, about mid-way between the mainshock epicenter and the international border. The second extended from 10 km northwest of the mainshock epicenter southeastward, to the northern tip of Gulf of California where it is much less well-defined. The third cluster formed a 10 km north-trending aftershock zone just north of the epicenter of the mainshock. It defined a trend similar to the foreshock sequence discussed above.

The first and largest cluster of aftershocks exhibited unusually high activity straddling the international border in the Yuha basin desert area. TREIMAN *et al.* (2010) mapped triggered slip and found up to 10–20 mm of triggered slip along the Laguna Salada and other faults in the Yuha Desert

area. They also identified a previously unknown fault in the Yuha basin striking northeast with 45 mm of left-lateral slip. Similarly, SANDWELL *et al.* (2010) used ENVISAT and field mapping to identify triggered slip on the northeast striking faults as well as the Superstition Hills, Imperial, Coyote Creek, and San Andreas faults. Thus, both the overlapping northwest and northeast trends of aftershocks and the mapped triggered slip confirm the complex crustal deformation at the north end of the aftershock zone.

The M5.7 aftershock to the El Mayor-Cucapah mainshock occurred on June 14 in the Yuha basin, about 8 km southeast of Ocotillo, CA. This aftershock was the largest so far of the M7.2 El Mayor-Cucapah sequence. It was located within the large

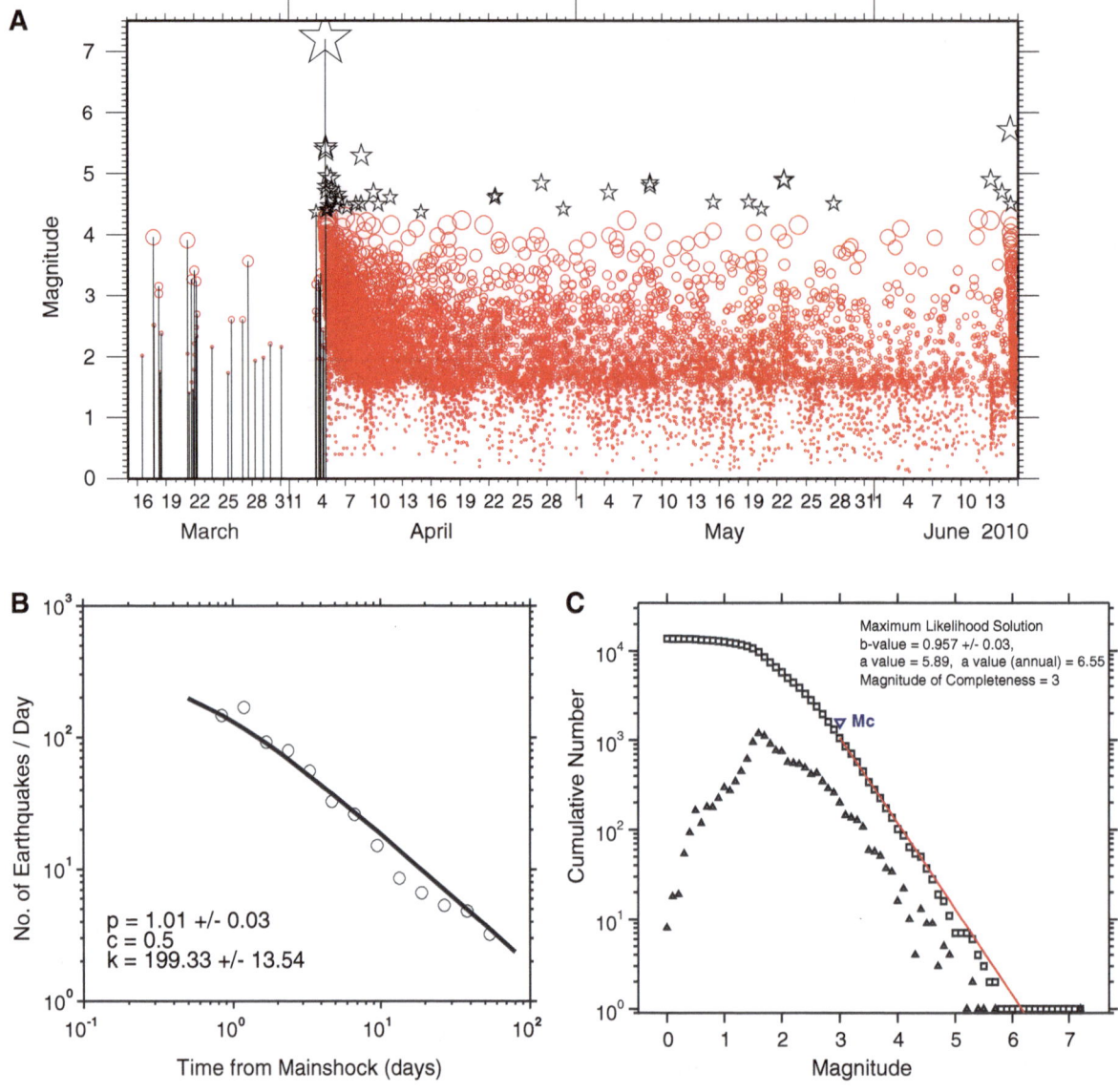

Figure 7

a Magnitude versus time plot of foreshocks, mainshock, and aftershocks for the events shown in Fig. 8. Foreshocks and mainshock are shown with a *vertical bar* and a symbol. Aftershocks of $M \geq 4.3$ are shown as *stars*, and smaller events are shown as *red circles* (*the vertical bars* are excluded for clarity). **b** The decay of El Mayor-Cucapah aftershocks showing number of events per day versus time, and including only $M \geq 3.0$ events. The values of the parameters of Omori's law are also shown; and **c** A *b*-value plot. The curve formed by *triangles* is the number of earthquakes per magnitude bin of 0.1, and the curve formed by the *squares* is the cumulative number of earthquakes. The *red line* is the maximum likelihood fit with the *b*-value as the slope and the *a*-value as the *y*-axis intercept. The *inverted triangle* labeled M_c indicates the selected magnitude of completeness of 3.0

cluster of aftershocks at the northwest end of the aftershock sequence and exhibited a strike-slip focal mechanism. The M5.7 event was followed by its own vigorous aftershock sequence, with four M4+ and 35 M3 to M4 events, during the first 12 h. Rupture associated with this event was identified on a 1 km

long NNW-striking break, with up to 5 cm motion down to the west (M. Rymer, written communication, August 2010).

The second aftershock cluster to the southeast, which straddles the mainshock epicenter along the Indiviso fault, increased in depth to the south and was

Table 1

Number of mainshock and aftershocks recorded in the 2010 El Mayor-Cucapah sequence

Minimum magnitude	Maximum magnitude	Number of events
Mainshock	7.2	1
5.00	5.70	6
4.00	4.99	91
3.00	3.99	865
2.0	2.99	4,153
1.0	1.99	4,935
0.00	0.99	207
Total		10,293

scattered over a broad region (Fig. 8). This depth increase could be an artifact or it could indicate a real change in the tectonic style. On one hand, the availability of arrival time data decreases to the south, thus providing almost no depth constraints. On the other hand, the deeper focal depths may be an artifact of the slower crustal velocities. However, the finite source modeling of the mainshock suggests that some of the slip was deeper to the southeast than to the northwest (WEI *et al.*, 2010; and XU *et al.*, 2010).

3.6. 3-D V_p Velocity Model

Many factors, including starting model, parameterization, data errors, and uneven ray coverage, contribute to errors in the final 3-D velocity model. We use values of the derivative weighted sum (DWS), which is a measure of the ray density next to each node in the 3-D velocity model, to evaluate the quality of the model (THURBER, 1993; HAUKSSON, 2000). The model is mostly well resolved except for the near surface, where nodes are progressively more poorly resolved to the south. In Fig. 8 the cross-sectional areas of the model with DWS values of less than 1,000 are crossed out with white lines.

The 3-D V_p model of the region images series of en-echelon rift basin structures such as the Laguna Salada, Mexicali Valley, and the Altar basin, at shallow depth ($V_p < 4.5$ km/s at depths of less than 5 km). These basins flank both sides of the mainshock rupture. At greater crustal depths, the 3-D V_p model images a high velocity zone ($V_p > 6.7$ km/s) in the depth range of 15–25 km, which is located mostly to

the east of the aftershocks and decreases in width to the southeast (see cross-sections b, c, and d in Fig. 8). The high V_p lower crust, which presumably is the high V_p lid below the basins, extends north–south from the Salton Sea to the Gulf of California. The rift zone is oriented more northerly than the aftershock zone, and thus the cross-sections slice into the high V_p lid more to the southeast than near the international border. In cross-section a, the 3-D model shows a strong change in the V_p structure about 15 km north of the mainshock epicenter, with lower crustal velocities and deeper aftershock focal depths to the south (Fig. 8). This simultaneous change in focal depths and decrease in the crustal velocities may be affected by lack of seismic stations in the area.

4. Discussion

The El Mayor-Cucapah sequence represents deformation in the Pacific plate as the plate boundary comes ashore in northern Baja California, Mexico. Early models of such deformation by LOMNITZ *et al.* (1970), and ELDERS *et al.* (1972) explain the Elsinore and San Jacinto faults as fracture-zone extensions of the Cerro Prieto and Imperial transform faults, respectively. These interpreted fracture zones do not appear to form perfect extensions of the plate-boundary transform faults, however, but seem to merge obliquely with them. The causative faults of the El Major-Cucapah earthquake and the southern part of the San Jacinto fault zone (Superstition Mountain and Superstition Hills faults) trend obliquely towards the Cerro Prieto and Imperial faults, respectively. We provide a model suggesting why the El Mayor-Cucapah, as well as the 1992 Landers and 1999 Hector Mine earthquakes, did not occur on the plate boundary itself but within the broader plate boundary deformation zone. Eventually, more large earthquakes will happen to the north along some of the major fault zones, but it is impossible to tell which faults will rupture first and how soon.

4.1. Relation to Mexican Pacific Margin

One of the intriguing questions raised by the El Mayor-Cucapah earthquake is how it is related to the

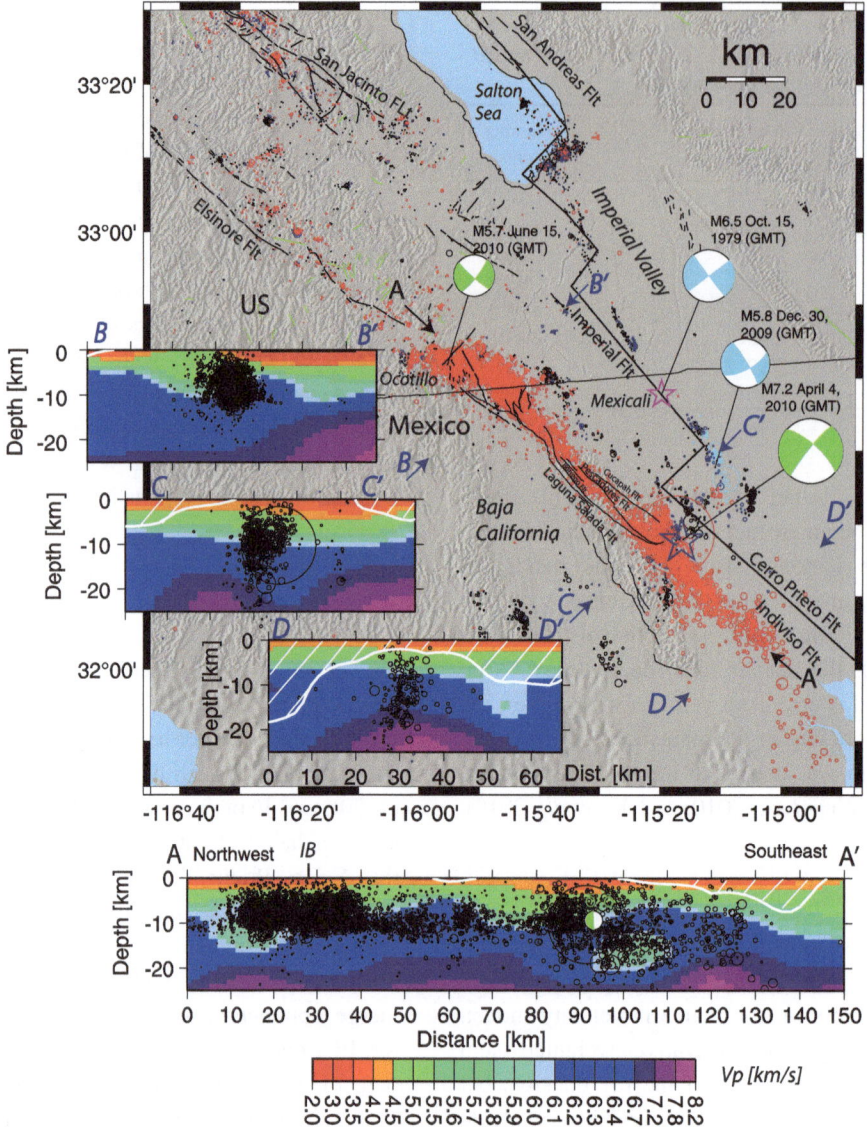

Figure 8
Map of the relocated 2010 hypocenters of background seismicity, foreshocks, mainshock, and aftershocks and 2009 background seismicity.
30th Dec. 2009 sequence shown in *light blue*. The mainshock epicenter is indicated by a *star*. Foreshocks are shown as *blue open circles*
beneath the mainshock *star*. The locations of the V_p cross-sections are indicated by the A–A' (includes focal mechanism of mainshock, and
IB–US–Mexico international border), B–B', C–C', and D–D' end points. The model areas that are crossed out with *white lines* are poorly
resolved are poorly resolved; aftershock hypocenters are plotted as *black open circles*

plate boundary, and whether the southeastern end of the fault corresponds to a major buried structure that may have significant cumulative offset from previous earthquakes. Because this part of the rupture lies within the Colorado River delta and corresponding tidal flat, periodic floods have removed any evidence of past fault slip at the surface. Geophysical potential field data may shed some light on possible structures that are buried here. Also, because the zone of displacement in the El Mayor-Cucapah earthquake extended nearly to the shoreline of the Gulf of California, where it was still west of the Cerro Prieto fault, the configuration of faults offshore in the northern Gulf of California can be examined to address the long-term structural importance of this fault zone.

In the delta region, there is a Bouguer gravity low southwest of the Cerro Prieto Fault (PACHECO *et al.*, 2006) with a curved southeastern boundary along the southernmost channel of the Colorado River, continuing out under Isla Montague in the delta region. PEMEX well W-3 within this gravity low contains 4.5 km of sedimentary rock overlying granitic basement (PACHECO *et al.*, 2006). The southwestern margin of the gravity low may bound this greater depth of sediments and approximately coincide with the April 4, 2010 rupture.

Faults and sedimentary strata underwater in the Gulf of California have been imaged in several marine multichannel seismic surveys close to the region (GONZÁLEZ-ESCOBAR *et al.*, 2009, 2010, and references therein). These studies show that the major structure along the plate boundary in this zone is the Cerro Prieto fault, which bounds the NE side of the Wagner basin (Fig. 1). The Wagner basin is an active extensional zone characterized by numerous minor faults and a pronounced accumulation of young sediments. Its major bounding normal faults, the Consag fault (on the west) and the Wagner fault (on the SE) trend N–S to NE ,where they intersect the Cerro Prieto fault (Fig. 1). No other major faults have been identified parallel to but southwest of the Cerro Prieto fault. Nevertheless, a narrow bathymetric channel extends NW from the WSW margin of the Wagner Basin, in a zone characterized by faults striking N25°W–N5°E, dipping eastward 50–60 degrees (HURTADO-ARTUNDAGA, 2002). This channel has a scarp on the NE side, along which the bathymetric contours (30–100 m water depth) are deflected about 10 km dextrally. The origin of this channel and scarp is not known, but its linearity and location suggest that it may be fault-controlled and project along strike to the fault zone of the April 4, 2010 rupture, perhaps serving as the structural connection between the Sierra Cucapah fault systems and the southern Wagner basin in the Gulf of California. This region needs to be studied in more detail, but it is clear that the April 4, 2010 rupture projects along strike into the Wagner basin segment of the Pacific–North America plate boundary, and does not appear to merge with the Cerro Prieto fault.

The plate boundary in the northern Gulf of California is generally simplified into a set of extensional basins connected by transform faults parallel to plate motion (e.g., LOMNITZ *et al.*, 1970). Under this paradigm, one can divide the plate boundary zone into segments usually named after the major extensional basin within the segment (ARAGÓN-ARREOLA and MARTÍN-BARAJAS, 2007). This simplification breaks down in the northernmost Gulf of California, where multiple active basins are found within the same basin segment (ARAGÓN-ARREOLA and MARTÍN-BARAJAS, 2007). In the Wagner basin segment, southwest of the Cerro Prieto fault, two extensional basins were known to be active: the Wagner basin and the Laguna Salada basin. From the April 4, 2010 earthquake we also see the clear involvement of normal faults in a third area: the basin east of the southern segment of the April 4 fault rupture, just east of the Sierra El Mayor/Sierra Cucapah.

Simultaneous activity in three basins within a single plate boundary segment of the Mexican Pacific margin highlights the fact that fault systems here have not yet localized into a single, well-defined plate boundary fault system, as in the simple model of LOMNITZ *et al.* (1970); rather, multiple processes in different locations are accommodating crustal separation. Although one process in the Wagner basin is thought to be magmatism, involving intrusion into a thick pile of sedimentary rocks, akin to that documented on land in the Imperial Valley (SCHMITT and VAZQUEZ, 2006), tectonic extension is still playing an important role. In addition to the extension seen on high-angle normal faults during the April 4, 2010 earthquake, low-angle detachment faults (AXEN *et al.*, 1999; FLETCHER and SPELZ, 2009) are also important within this plate boundary segment.

4.2. Seismotectonics

The main Pacific–North America plate boundary in southernmost California and Baja California includes the major plate boundary faults, the southern San Andreas, Imperial and Cerro Prieto faults. These faults connect in a complex geometrical pattern that sometimes does not follow the optimum angle of plate motion, on either large or small scales (LUNDGREN *et al.*, 2009). The San Andreas fault accommodates nearly pure strike-slip motion. South of the Salton Sea, the boundary contains a number of extensional step-overs that accommodate crustal thinning. The El Mayor-Cucapah mainshock consisted of simultaneous strike-slip and normal faulting, accommodating both plate motion and basin formation.

The El Mayor-Cucapah earthquake sequence differed from past sequences in Southern California with a style of faulting more complex than had been previously observed. The El Mayor-Cucapah, Landers, and Hector Mine earthquakes all ruptured multiple fault strands with varying degrees of overlap. Both the Landers and Hector Mine mainshocks ruptured, successively, four major fault segments each, which previously were thought to be unlikely to rupture in one earthquake. Similarly, the El Mayor-Cucapah earthquake ruptured at least four segments, with two different styles of faulting, accommodating complementary seismotectonic processes. Prior to the occurrence of the earthquake, FLETCHER and SPELZ (2009) had mapped these two styles of faulting in the field but it was unclear if they happened simultaneously. Further, the El Mayor-Cucapah mainshock did not rupture along seismicity trends that had developed over the last decade. Rather, it ruptured along the length of the Sierra Cucapah causing partial down-dropping of the eastern side of the mountain range. Thus, the mainshock contributed, also, to the westward widening of the Mexicali Valley through the extensional part of its focal mechanism. Such an extension, that contributes to East–West widening of basins in the northern Gulf extensional province, has not been documented in detail previously in the Baja California region.

The focal mechanisms of the mainshock and M4+ aftershocks exhibit mostly northwest to west-north-west-striking dextral strike-slip faulting (red mechanisms in Fig. 9). The prominent cluster of strike-slip mechanisms in the Yuha basin coincides with the region of mixed northwest and northeast striking faults. A few mechanisms show normal faulting on mostly north-northwest to north striking planes (green mechanisms in Fig. 9). Three events exhibit predominantly thrust faulting, probably indicating geometrical complexities along the fault rupture (blue mechanisms in Fig. 9). One possible interpretation of the normal faulting events near the

mainshock is an accommodation zone coincident with a change in dip, as suggested by the finite source models for the mainshock. The dips to the northwest of the mainshock are steep, while the dips to the southeast are shallower (WEI et al., 2010).

4.3. Migration of Seismicity to the Northwest

Within hours following the El Mayor-Cucapah mainshock, triggered earthquakes occurred farther north along the Elsinore and San Jacinto faults (Fig. 10). Each of these faults is capable of a major earthquake, which would significantly affect the large metropolitan areas of southern California.

The triggered seismicity along the Elsinore fault extended about 60 km to the north. Many of the aftershocks near Ocotillo occurred on northwest-striking faults that trend subparallel to the Elsinore fault in this region. The Elsinore fault is more than 170 km long, and extends into the Los Angeles area as the Whittier fault. MAGISTRALE and ROCKWELL (1996) used seismicity to map the double strands of the Elsinore fault with a second set of right-lateral faults located 7–12 km east of the main strand. They argued that the Elsinore fault ends in the Yuha basin, which is a stepover basin to the Laguna Salada fault. The presence of the basin with different strain release mechanisms may have contributed to the termination of the mainshock rupture and the abundant aftershock activity.

The triggered seismicity along the San Jacinto fault extended 80 km to the north (Fig. 10). In this paper, we discuss the seismicity that occurred through the end of June, 2010; however, we note that as of mid-July 2010, the largest triggered earthquake on the San Jacinto fault was M_w 5.4 on July 7, 2010. During the twentieth century, the San Jacinto fault has been the most active fault in Southern California with more than a dozen earthquakes of $M > 6$ (SANDERS and KANAMORI, 1984). In particular, the southern San Jacinto fault has accommodated major earthquakes in the past (GURROLA and ROCKWELL, 1996). These authors showed evidence for at least three events during the past 1,200 years, and argued that the most recent major earthquake on the Superstition Mountain fault occurred more than 400 years ago.

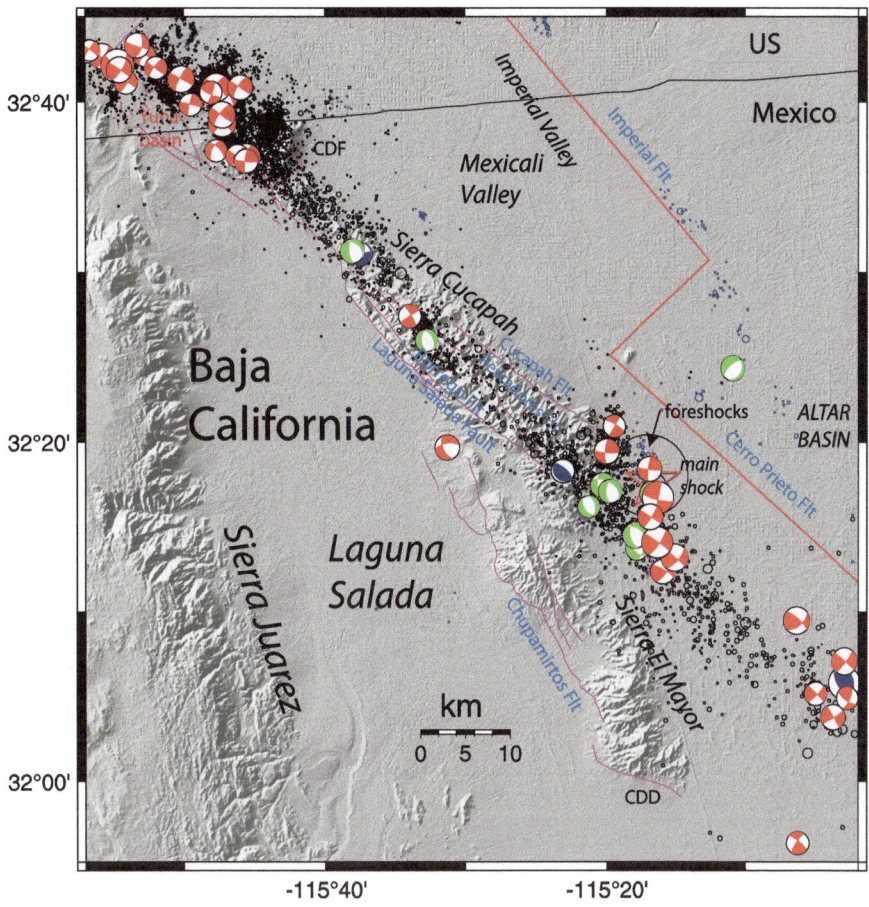

Figure 9

Map showing the locations of the El Mayor-Cucapah foreshock, mainshock, and aftershocks. The aftershock hypocenters plotted in *black* illustrate how the El Mayor-Cucapah sequence cuts across the Sierra Cucapah range. Moment tensors for selected M4+ aftershocks are shown; *red* strike-slip, *blue* thrust, and *green* normal faulting. The Pacific–North America plate boundary is shown in *red*, and marked as Imperial and Cerro Prieto faults. Topography is from: http://www.gdem.aster.ersdac.or.jp/. *CDF* Centinela Detachment fault, *CDD* Cañada David detachment

Both geodetic and geological slip rates for the southern San Andreas, San Jacinto, and Elsinore faults decrease from east to west (LUNDGREN *et al.*, 2009). To explain the sharp velocity gradient across the faults and the elapsed time since the last major earthquake, their model required a high-viscosity lower crust and a low-viscosity upper mantle. They argued that the Borrego Mountain and Superstition Hills segments of the San Jacinto fault had completed only about a third of their earthquake cycle since the last large earthquake. This may explain why the southernmost part of the San Jacinto fault has less triggered seismicity than the segments farther to the north.

No triggered seismicity was associated with the southern San Andreas fault, although the fault is close

to reaching the end of its interseismic loading phase (FIALKO, 2006). However, surficial triggered slip was documented along the fault (SANDWELL *et al.*, 2010; TREIMAN *et al.*, 2010).

There are only a few previously documented cases of aftershock migration and subsequent triggering of a major earthquake. In one case, HAUKSSON *et al.* (1993) reported that the 1992 M6.1 Joshua Tree aftershocks migrated over a time period of two months for a distance of ~10 km to the northnorthwest, towards the future epicenter of the 1992 M_w 7.3 Landers earthquake. However, HELMSTETTER *et al.* (2003), who used data from 20 California earthquake sequences, argued that aftershock diffusion is very limited and in most cases does not

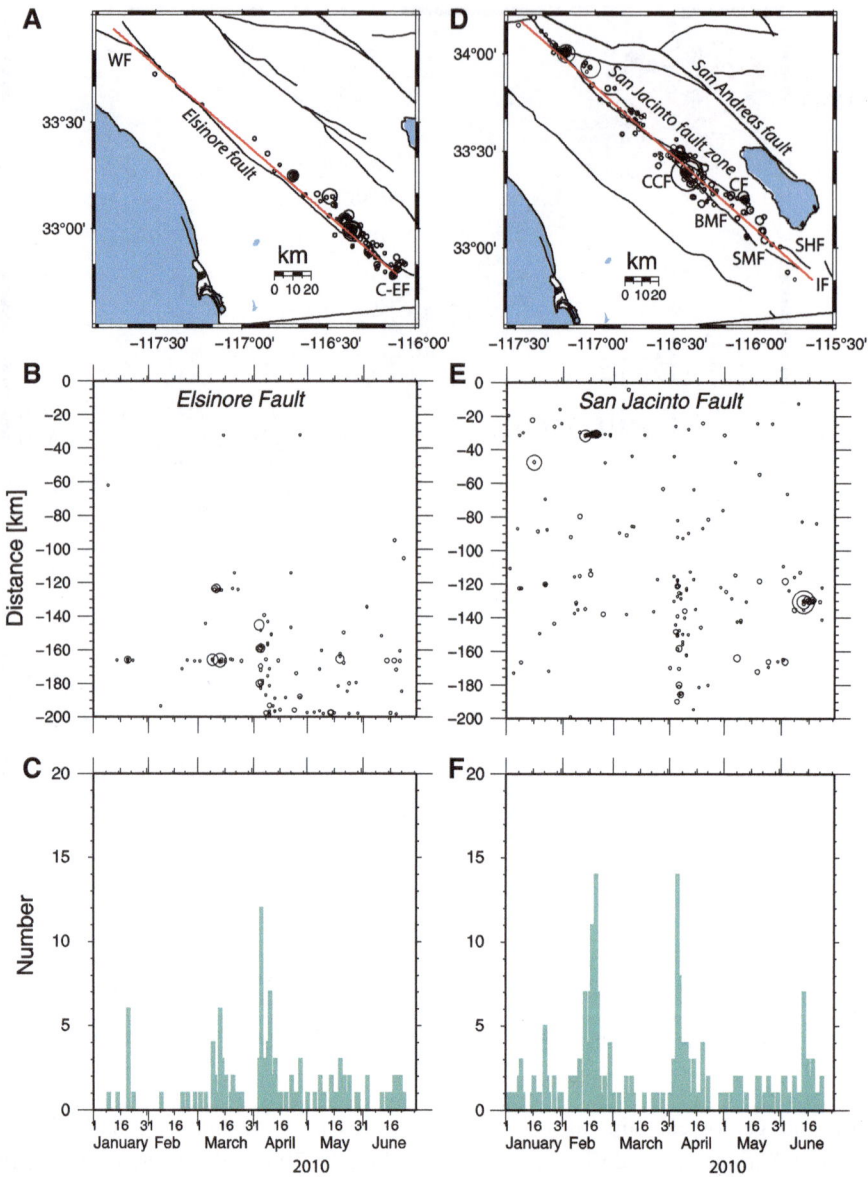

Figure 10

a Map view of triggered seismicity along the Elsinore fault. *C-EF* Coyote segment of the Elsinore fault. **b** Distance measured from north along the Elsinore fault (approx. by the *red line* in (**a**)) from northwest to southeast versus date showing seismicity of $M \geq 1.8$ that has occurred in 2010. **c** Histogram of number of events. **d** Map view of triggered seismicity along the San Jacinto fault. Segments of the San Jacinto fault: *BMF* Borrego Mountain fault, *CF* Clark fault, *CCF* Coyote Creek fault, *SHF* Superstition Hill fault, *SMF* Superstition Mountain fault, *IF* Imperial fault. **e** Distance measured from north along the San Jacinto fault (approx. by the *red line* in (**d**)) from northwest to southeast versus date showing seismicity of $M \geq 1.8$ that has occurred in 2010. **f** Histogram of number of events

occur. Thus, when future earthquakes happen along the Elsinore and San Jacinto faults, their causative relation to the El Mayor-Cucapah earthquake will probably be stated in terms of triggered seismicity.

4.4. Coulomb Stress Transfer from Deep Slip

Although the plate boundary zone is broad, the 1992 Landers, 1999 Hector Mine, and the 2010 El Mayor-Cucapah sequences occurred on faults that are not considered to be part of the Pacific–North America plate boundary (Fig. 1). Their locations suggest some tectonic relationship to the plate boundary itself. We speculate that the aseismic deep slip below the seismogenic zone along the plate boundary and nearby faults may cause stress concentrations adjacent to the plate boundary which are sufficient to cause these types of events on secondary faults. Such stress concentrations may occur preferentially at bends in the plate boundary faults.

To illustrate this model we use the computer program Coulomb 3 (http://earthquake.usgs.gov/research/software/) developed by TODA et al. (2005) and LIN and STEIN (2004) to determine the Coulomb stress change associated with continuous slip below 12 km depth along the plate boundary (Fig. 11). We use a friction coefficient of 0.4. We model, as source faults, three fault segments, one along the southernmost segment of the San Andreas fault, a second along the Brawley seismic zone, and a third along the Imperial fault, as being the main high-slip-rate plate boundary at depth below the seismogenic zone. Our schematic representation of the Imperial fault follows the seismicity in the region and ignores small ridge-type step-overs in the region. Adding such local features to the model does not affect the large-scale distribution of Coulomb stress. We also included the San Jacinto and Elsinore faults, as well as the rupture segments of the Landers, Hector Mine and El Mayor-Cucapah mainshocks to represent the earthquakes themselves. Assuming a 40 year period (1970–2010), with slip rates of 35 mm/yr for southern San Andreas and Imperial Valley faults, 15 mm/yr for San Jacinto fault, 2.7 mm/yr for Elsinore fault, we calculate cumulative slip on the San Andreas to be 1.4 m, on the San Jacinto to be 0.6 m, and Elsinore to be 0.1 m below 12 km depth or below the seismogenic zone.

This model is similar to the model published by Wdowinski (2009). He proposed a mechanism of continuous deep creep in the depth range of 10–17 km along the San Jacinto fault to explain the apparent relatively high rate of seismicity as compared to the low rate of seismicity along the southern San Andreas fault. We treat other faults in the region as receiver faults with zero cumulative slip. Receiver faults are defined as planes with specified strike, dip, and rake, on which the stress changes caused by the source faults are resolved. As the representative receiver fault, we use the average of the Landers and the El Mayor focal mechanisms (335° strike, 90° dip, −170° rake).

Using these parameters, the Coulomb stress change calculation shows primary loading of the seismogenic parts of the San Andreas, Imperial, and San Jacinto faults, which is consistent with the analysis of geodetic data by FIALKO (2006). The loading is somewhat less along the Elsinore fault and in the Landers and Hector Mine and El Mayor-Cucapah epicentral areas (Fig. 11). Some of the $M \geq 4.0$ background seismicity also occurs within the region of increased stress. Stress decreases across a broader region, mostly in between the San Andreas and San Jacinto faults. The El Mayor-Cucapah and Landers earthquakes occurred near geometrical bends in the plate boundary, and thus could be interpreted as secondary crustal deformation. In contrast, the continuous slip below the seismogenic zone of the plate boundary faults (FIALKO, 2006) loads the seismogenic part of the plate boundary faults themselves and brings them close to failure, although providing timing of possible future major earthquakes or pinpointing the most likely segment for future rupture is not possible.

5. Conclusions

The M_w 7.2 El Mayor-Cucapah earthquake sequence ruptured bilaterally for a distance of ∼ 120 km, from the northern tip of the Gulf of California to the US–Mexico international border. The mainshock was preceded by a high seismicity rate during the previous decade occurring on adjacent faults, and two temporally separated clusters of foreshocks. The overall earthquake statistics of the sequence, such as productivity and rate of decay, are similar to previous earthquake sequences in the region. The mainshock rupture involved complex transtensional rupture on several subparallel fault strands, with a duration twice as long as expected for this size earthquake. The overall effects of this sequence were to accommodate right-lateral Pacific–North America plate motion, as well as collapse and associated crustal extension in the Sierra Cucapah.

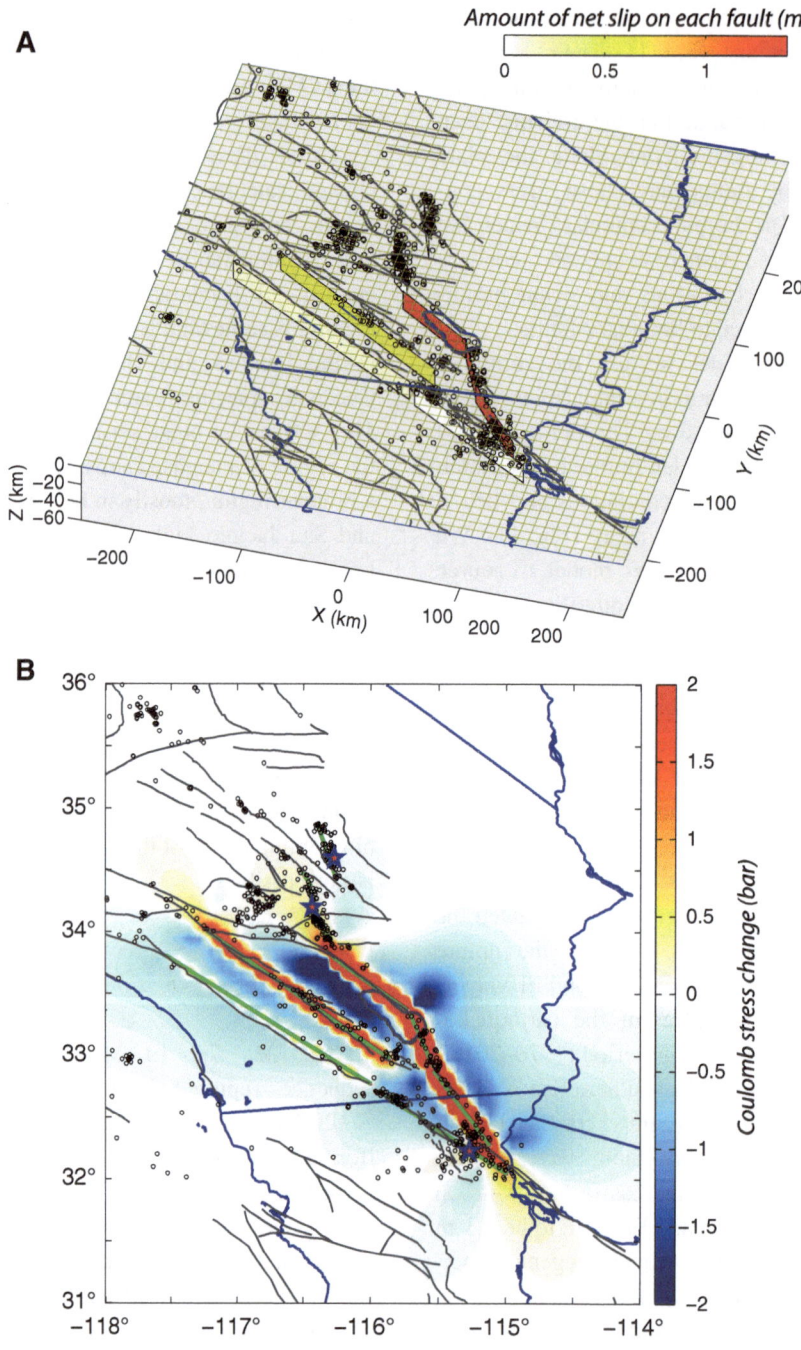

Figure 11

a Faults that are assumed to have steady state slip-rate below the seismogenic zone are shown in colors, *red, yellow,* and *light yellow*. The color shows the accumulated slip over a 40-year time period on each fault segment, assuming an average slip rate appropriate for each fault. **b** The Coulomb stress change mapped at 7.5 km depth from deep slip along the faults shown in (**a**), and as *green lines* in (**b**). Fault friction is assumed to be 0.40. The *black circles* are $M \geq 4.0$ from SCSN catalog between 1970 and 2010. The geographical boundaries are bold *blue curves*, and the plate boundary, San Andreas and Imperial faults, is a thin *blue* curve. The three major earthquakes with magnitude above 7.0: the Landers, the Hector Mine and the El Mayor-Cucapah earthquakes, are marked by *blue* and *red star* symbols at epicenters, respectively. Receiver faults are selected to have strike 335°, dip 90°, and rake −170°

Thus, the sequence contributes to the westward migration of the plate boundary zone. The very active aftershock cluster in the Yuha basin near Ocotillo suggests a change in deformation style to a mixed northwest and northeast strike-slip faulting, which may transfer slip to the more westerly trending Coyote segment of the Elsinore fault to the north. The deep slip, below the seismogenic zone, along the plate boundary faults and stress concentrations associated with geometrical bends, could be the driving force causing off-plate-boundary earthquakes, such as El Mayor-Cucapah and Landers and Hector Mine.

Acknowledgments

This research was supported by the US Geological Survey Grant G10AP00017; NSF grants EAR-0911761 and OCE-0742253 to Caltech, and by the Southern California Earthquake Center. SCEC is funded by NSF Cooperative Agreement EAR-0529922 and USGS Cooperative Agreement 07HQAG0008. The Incorporated Research Institutions for Seismology (IRIS) Data Management System (DMS) was used to access the Global Seismographic Network data. Funds for maintenance of RESNOM network are provided by CICESE. We thank S. Wei, S. Skinner, and E. Glowacka for feedback and discussions. We thank L. Munguía, G. Diaz, F. Farfan, I. Mendez, L. Orozco, O. Galvez, and S. Arregui, Department of Seismology, CICESE, Baja California, Mexico, and N. Scheckel, A. Guarino, and B. Wu of the SCSN for help with data collection and processing. Most figures were done using GMT (WESSEL and SMITH, 1998). SCEC contribution number 1,439. Contribution number 10,047, Seismological Laboratory, Division of Geological and Planetary Sciences, California Institute of Technology, Pasadena.

REFERENCES

ABERCROMBIE, R.E., ANTOLIK, M., and EKSTROM, G. (2003), *The June 2000 M_w 7.9 earthquakes south of Sumatra: deformation in the India–Australia plate* J. Geophys. Res., 108, B1, 2018. doi: 2010.1029/2001JB000674.

ADAMOVA, P., and SILENY, J. (2010), *Non-double-couple earthquake mechanism as an artifact of the point-source approach applied to a finite-extent focus,* Bull. Seismol. Soc. Am., 100, 447–457. doi: 410.1785/0120090097.

ARAGÓN-ARREOLA, M., and MARTÍN-BARAJAS, A. (2007), *Westward migration of extension in the northern Gulf of California, Mexico,* Geology, v. 35, 571–574. doi:10.1130/G23360A.1.

AXEN, G. J, FLETCHER, J. M., COWGILL, E., MURPHY, M., KAPP, P., MACMILLAN, I., RAMOS-VELAZQUEZ, E., and ARANDA-GOMEZ, J. (1999), *Range-front fault scarps of the Sierra El Mayor, Baja California: formed above a low-angle normal fault?* Geology, v. 27, 247–250.

BARNARD, F. L. (1968), *Structural geology of the Sierra de los Cucapahs, northeastern Baja California, Mexico, and Imperial County, California,* Thesis, University of Colorado, Boulder, Colorado, 157p.

BIRD, P. (2003), *An updated digital model of plate boundaries,* Geochem. Geophys. Geosyst., 4(3), 1027. doi:10.1029/2001 GC000252.

CLINTON, J. F., HAUKSSON, E., and SOLANKI, K. (2006), *An evaluation of the SCSN moment tensor solutions: robustness of the M_W magnitude scale, style of faulting, and automation of the method,* Bull. Seismol. Soc. Am., 96(5). doi:10.1785/0120050241.

DREGER, D.S., and HELMBERGER, D. V. (1993), *Determination of source parameters at regional distances with three-component sparse network data,* J. Geophys. Res., 98, 8107–8125.

ELDERS, W.A., REX, R.W., MEIDAV, T., ROBINSON, P.T., and BIEHLER, S. (1972), *Crustal spreading in southern California,* Science, 178, 15–24.

FENBY, S. S., and GASTIL, R. G. (1991), Geologic-Tectonic map of the Gulf of California and surrounding areas, in, J. P. Dauphin and B. R. Simoneit, Eds., *The Gulf and Peninsular Provinces of the Californias, AAPG Memoir 47,* Tulsa, Oklahoma, p. 79–83.

FIALKO, Y. (2006), *Interseismic strain accumulation and the earthquake potential on the southern San Andreas fault system,* Nature, v. 441. doi:10.1038/nature04797.

FIELDING, E. J., LEPRINCE, S., WEI, S., SLADEN, A., SIMONS, M., AVOUAC, J.-P., LOHMAN, R., BRIGGS, R., HUDNUT, K.,and HELMBERGER, D. (2010), *InSAR and subpixel-correlation pixel-tracking measurements of the 2010 El Mayor-Cucapah earthquake,* Geological Society of America, Cordilleran Section, Abstract LB2-1, Anaheim, California.

FLETCHER, J. M., and SPELZ, R. M. (2009), *Pattterns of quaternary deformation and rupture propagation associated with an active low-angle normal fault, Laguna Salada, Mexico: evidence of a rolling hinge,* Geosphere, v. 5, no. 4, 385–407. doi:10.1130/ GES00206.1.

FLETCHER, J., ROCKWELL, T. K., TERAN, O., MASANA, E., FANEROS, G., HUDNUT, K., GONZALEZ, J., GONZALEZ, A., SPELZ, R., and MUELLER, K. (2010), *The surface ruptures associated with the El Mayor-Borrego earthquake sequence,* Geological Society of America, Cordilleran section, Abstract LB1-5, Anaheim, California.

GLOWACKA, E., VAZQUEZ GONZALEZ, R., SARYCHIKHINA, O. V., NAVA PICHARDO, F. A., FARFAN SANCHEZ, F. J., and DIAZ DE COSSIO BATANI, G. E. (2010), *Deformations triggered by the M = 7.2, April 4, 2010, El Mayor earthquake on faults in the Mexicali Valley.* Geological Society of America, Cordilleran section, Abstract LB2-3, Anaheim, California.

GONZÁLEZ-ESCOBAR, M., AGUILAR-CAMPOS, C., SUÁREZ-VIDAL, F., and MARTÍN-BARAJAS, A. (2009), *Geometry of the Wagner basin, upper Gulf of California based on seismic reflections,* Int. Geol.Rev., v. 51, no. 2, p. 133–144.

GONZÁLEZ-ESCOBAR, M., SUÁREZ-VIDAL, F., HERNÁNDEZ-PÉREZ, J. A., and MARTÍN-BARAJAS, A. (2010), *Seismic reflection-based evidence of a transfer zone between the Wagner and Consag basins: implications for defining the structural geometry of the northern Gulf of California*, Geo-Marine Lett. doi:10.1007/s00367-010-0204-0, 10 pages.

GURROLA, L. D., and ROCKWELL T. K. (1996), *Timing and slip for prehistoric earthquakes on the Superstition Mountain fault, Imperial Valley, southern California*, J. Geophys. Res., v. 101, no. B3, 5977–5985.

HAUKSSON, E. (2000), *Crustal structure and seismicity distributions adjacent to the Pacific-North America plate boundary in southern California*, J. Geophys. Res., 105, 13,875–13,903.

HAUKSSON, E., JONES, L. M., HUTTON, K., and EBERHART-PHILLIPS, D. (1993), *The 1992 Landers Earthquake Sequence: Seismological observations*, J. Geophys. Res., 98, 19,835–19,858.

HAYES, G.P., RIVERA, L., and KANAMORI, H. (2009), *Source Inversion of the W-Phase: real-time implementation and extension to low magnitudes*, Seismol. Res. Lett., 80, 817–822.

HELMSTETTER, A., QUILLON, G., and SORNETTE, D. (2003), *Are aftershocks of large Californian earthquakes diffusing?* J. Geophys. Res., v. 108, no. B10. doi:10.1029/2003JB002503.

HOUGH, S. E., and ELLIOTT, A. (2004), *Revisiting the February 23, 1892 Laguna Salada Earthquake*, Bull. Seismol. Soc. Am., v. 94, no. 4, p. 1571–1578.

HURTADO-ARTUNDUAGA, A. D. (2002), *Modelo estructural de la cuenca Wagner en el Golfo de California basado en sísmica de reflexión multicanal*, M.S. Thesis, Universidad Nacional Autónoma de México, México, D. F., México. 93p.

HUTTON, L. K., WOESSNER, J., and HAUKSSON, E. (2010), *Seventy-seven years (1932–2009) of Earthquake Monitoring in Southern California*, Bull. Seismol. Soc. Am., v. 100, no. 2, p. 423–446. doi:10.1785/0120090130.

JENNINGS, C.W. (1994), Fault activity map of California and adjacent areas: California Department of Conservation, Division of Mines and Geology, Geologic Data Map No. 6, scale 1:750,000.

KANAMORI, H., and RIVERA, L. (2008), *Source inversion of W phase: speeding up seismic tsunami warning*: Geophys. J. Int. 175, p. 222–238.

KOSTROV, B.V. (1974), *Seismic moment and energy of earthquakes, and seismic flow of rock*, Izv. Acad. Sci. USSR Phys. Solid Earth, 1, 23–44.

LAY, T., AMMON, C.J., KANAMORI, H., RIVERA, L., KOPER, K.D., and HUTKO, A.R. (2010), *The 2009 Samoa-Tonga great earthquake triggered doublet*, Nature, Letter, 466, 964–968. doi:10.1038/nature09214.

LIN, J., and STEIN R.S. (2004), *Stress triggering in thrust and subduction earthquakes, and stress interaction between the southern San Andreas and nearby thrust and strike-slip faults*, J. Geophys. Res., 109, B02303. doi:10.1029/2003JB002607.

LOMNITZ, C., MOOSER, C.R., ALLEN, C.R., BRUNE, J.N., and THATCHER, W. (1970), *Seismicity and tectonics of the northern Gulf of California region, Mexico, Preliminary results*, Geofis. Int. 10, 37–48.

LUNDGREN, P., HETLAND, E. A., LIU, Z., and FIELDING, E. J. (2009), *Southern San Andreas–San Jacinto fault system slip rates estimated from earthquake cycle models constrained by GPS and interferometric synthetic aperture radar observations*, J. Geophys. Res., v. 114, B02403. doi:10.1029/2008JB005996.

MAGISTRALE, H., and ROCKWELL, T. K. (1996), *The Central and Southern Elsinore Fault Zone, Southern California*, Bull. Seismo. Soc. Am., v. 86, no. 6, pp. 1793–1803.

MUELLER, K.J., and ROCKWELL, T.K. (1995), *Late quaternary activity of the Laguna Salada fault in northern Baja California, Mexico*. Geol. Soc. Am. Bull., v. 107, p. 8–18.

MUNGUÍA, L., GLOWACKA, E., SUÁREZ,F., SARYCHIKHINA, O., and LIRA, H. (2009), *Near-Fault Strong Ground Motions Recorded during the Morelia Normal-Fault Earthquakes of May 2006 in Mexicali Valley, B. C., México*, Bull. Seismol. Soc. Am., v. 99, no. 3, 1538–1551.

MUNGUÍA, L., NAVARRO, M., VALDEZ, T., and LUNA, M. (2010), *Strong-motion data collected in Baja California during the El Mayor-Cucapah earthquake of April 4, 2010 (M_w7.2): Preliminary results*, Geological Society of America Cordilleran Section, Abstract LB1-2, Anaheim, California.

PACHECO, M., MARTÍN-BARAJAS, A., ELDERS, W., ESPINOSA-CARDENA, J. M., HELENES, J., and SEGURA, A. (2006), *Stratigraphy and structure of the Altar basin of NW Sonora: implications for the history of the Colorado River delta and the Salton Trough, Revista Mexicana de Ciencias*, Geológicas, v. 23, no. 1, 1–22.

REASENBERG, P.A., and JONES, L.M. (1989), *Earthquake hazard after a mainshock in California*, Science 265, 1173–1176.

SANDERS, C. O., and KANAMORI, H. (1984), *A seismotectonic analysis of the Anza Seismic Gap, San Jacinto Fault Zone, southern California*, J. Geophys. Res., v. 89, 5873–5890.

SANDWELL, D. T., WEI, M., GONZALES, J., GONZALES, A., LIPOVSKI, B., FUNNING, G., FIALKO, Y., MELLORS, R., AGNEW, D., and PETERSON, R. (2010), *InSAR and GPS measurements of crustal deformation from the El Mayor earthquake: liquefaction and triggered slip*, Cordilleran section, Abstract LB1-8, Anaheim, California.

SARYCHIKHINA, O., GLOWACKA, E., MELLORS, R., VÁZQUEZ, R., MUNGUÍA, L. and GUZMÁN, M. (2009), *Surface displacement and groundwater level changes associated with the May 24, 2006 M_w 5.4 Morelia Fault Earthquake, Mexicali Valley, Baja California, Mexico*. Bull. Seismol. Soc. Am. v. 99, no. 4, 2180–2189. doi: 10.1785/0120080228.

SCHMITT, A. K., and VAZQUEZ, J. A. (2006), *Alteration and remelting of nascent oceanic crust during continental rupture: evidence from zircon geochemistry of rhyolites and xenoliths from the Salton Trough, California*, Earth plan. Sci. Lett., v. 252, p. 260–274. doi:10.1016/j.epsl.2006.09.41.

SUÁREZ-VIDAL, F., MENDOZA-BORUNDA R., NAFARRETE-ZAMARRIPA L.M., RÁMIREZ J., and GLOWACKA E. (2008), *Shape and dimensions of the Cerro Prieto pull-apart basin, Mexicali, Baja California, Mexico, based on the regional seismic record and surface structures*. Int. Geol. Rev., v. 50 (7), p. 636–649 (PA: 68049). doi:10.2747/0020-6814.50.7.636.

SUÁREZ-VIDAL, F., MENDOZA-BORUNDA, R., VAZQUEZ, S.,and MENDOZA, L. (2010), *Preliminary report of the geological effects and damage distribution of the April 4, 2010 Cucapah-El Mayor earthquake, Mexicali Valley, BC, Mexico*, Geological Society of America, Cordilleran Section, Abstract LB2-2, Anaheim, California.

SUÁREZ-VIDAL, F., MUNGUÍA-OROZCO, L., GONZÁLEZ-ESCOBAR, M., GONZÁLEZ-GARCÍA, J., and GLOWACKA, E. (2007), *Surface rupture of the Morelia fault near the Cerro Prieto Geothermal field, Mexicali, Baja California, Mexico, during the M_w 5.4 Earthquake of May 24, 2006*, Seismol. Res. Lett., v. 78, no. 3, p. 394–399. doi:10.1785/gssrl.78.3.394.

THURBER, C. H. (1993), Local earthquake tomography: velocities and VP=VS-theory, in *Seismic Tomography: Theory and Practice* H. M. Iyer and K. Hirahara (Eds.), Chapman and Hall, London, pp 563–583.

TODA, S., STEIN, R. S., RICHARDS-DINGER, K., and BOZKURT, S. (2005), *Forecasting the evolution of seismicity in southern California: animations built on earthquake stress transfer,* J. Geophys. Res., B05S16. doi:10.1029/2004JB003415.

TREIMAN, J.A., RYMER, M.J., KENDRICK, K.J., LIENKAMPER, J.J., WELDON, R.J. II, HERNANDEZ, J.L., IRVINE, P.J., KNEPPRATH, N., OLSON, B.P.E., and SICKLER, R.R. (2010), *Triggered slip in southern California as a result of the April 5, 2010 El Mayor-Cucapah earthquake,* Geological Society of America, Cordilleran Section, Abstract LB2-5, Anaheim, California.

WALDHAUSER, F and ELLSWORTH W.L., (2000), *A double-difference earthquake location algorithm: method and application to the northern Hayward Fault, California,* Bull. Seismol. Soc. Am., 90, pp 1353–1368.

WDOWINSKI, S. (2009), *Deep creep as a cause for the excess seismicity along the San Jacinto fault,* Nat. Geosci., v. 2. doi: 10.1038/NGEO684.

WEI, S., SLADEN, A., LEPRINCE, S., AVOUAC, J.-P., FIELDING, E. J., CHU, R., SIMONS, M., HELMBERGER, D., HAUKSSON, E., and LOHMAN, R. (2010), *Joint inversion of geodetic and seismic slip models for the April 4, 2010 El Mayor-Cucapah earthquake,* Geological Society of America, Cordilleran Section, Abstract LB1-4, Anaheim, California.

WESSEL, P., and SMITH W. H. F. (1998). *New version of the generic mapping tools released,* EOS, 79, p.579.

WIEMER, S. (2001), *A software package to analyze seismicity: ZMAP,* Seismol. Res. Lett., 373–382.

XU, Z., SHAO, G., JI, C., and LARSON, K. (2010), *Slip distribution of April 4, 2010 Baja California, Mexico earthquake constrained using teleseismic body and surface waves and high rate GPS,* Geological Society of America, Cordilleran Section, Abstract LB1-7, Anaheim, California.

(Received July 10, 2010, revised September 1, 2010, accepted September 6, 2010, Published online November 16, 2010)

Pure Appl. Geophys. 168 (2011), 1279–1292
© 2010 Springer Basel AG
DOI 10.1007/s00024-010-0177-y

Location of Moderate-Sized Earthquakes Recorded by the NARS–Baja Array in the Gulf of California Region Between 2002 and 2006

RAUL R. CASTRO,[1] ARTURO PEREZ-VERTTI,[1] IGNACIO MENDEZ,[1] ANTONIO MENDOZA,[1] and LUIS INZUNZA[1]

Abstract—We relocated the hypocentral coordinates of small to moderate-sized earthquakes reported by the National Earthquake Information Center (NEIC) between April 2002 and August 2006 in the Gulf of California region and recorded by the broadband stations of the network of autonomously recording seismographs (NARS–Baja array). The NARS–Baja array consists of 19 stations installed in the Baja California peninsula, Sonora and Sinaloa, Mexico. The events reported by the preliminary determinations of epicenters (PDE) catalog within the period of interest have moment magnitudes (M_w) ranging between 1.1 and 6.7. We estimated the hypocentral location of these events using P and S wave arrivals recorded by the regional broadband stations of the NARS–Baja and the RESBAN (*Red Sismológica de Banda Ancha*) arrays and using a standard location procedure with the HYPOCENTER code (LIENERT and HAVSKOV in Seism Res Lett 66:26–36, 1995) as a preliminary step. To refine the location of the initial hypocenters, we used the shrinking box source-specific station term method of LIN and SHEARER (J Geophys Res 110, B04304, 2005). We found that most of the seismicity is distributed in the NW–SE direction along the axis of the Gulf of California, following a linear trend that, from north to south, steps southward near the main basins (Wagner, Delfin, Guaymas, Carmen, Farallon, Pescadero and Alarcon) and spreading centers. We compared the epicentral locations reported in the PDE with the locations obtained using regional arrival times, and we found that earthquakes with magnitudes in the range 3.2–5.0 mb differ on the average by as much as 43 km. For the M_w magnitude range between 5 and 6.7 the discrepancy is less, differing on the average by about 25 km. We found that the relocated epicenters correlate well with the main bathymetric features of the Gulf.

Key words: Seismicity, Gulf of California, Mexico.

1. Introduction

The transform faults and spreading centers of the Gulf of California (GoC hereafter) form a transform-rift plate boundary between the Pacific and the North American plates. The seismicity in this region consists of right lateral strike-slip events located near transform faults and normal fault earthquakes that occur mostly on the spreading centers (GOFF et al., 1987). In the northern GoC, there are three active basins, from north to south: the Consag and Wagner basins; the upper Delfin basin; and the lower Delfin and Salsipuedes basins (Fig. 1). These basins are all distributed within a shallow depression. The lower Delfin basin has a high fault density and is magmatically the most active of the three (PERSAUD et al., 2003). In general, the northern GoC has a more complex fault system than the southern GoC, where transform fault-spreading center geometry is dominant. The crust in the southern region of the Gulf is oceanic and thinner than in the northern GoC (ZHANG et al., 2007) where the crust is not typically oceanic. In the southern GoC, the Guaymas basin seems to have a transitional crust, since it is different from the typical continental or oceanic crusts (EINSELE 1982; FABRIOL et al., 1999). Farther south, the crust of the Alarcon basin is oceanic, based on magnetic anomaly lineations (DEMETS, 1995).

The GoC is characterized for having fast spreading centers (REICHLE and REID, 1977). Although earthquakes are not commonly observed from fast spreading centers (ISACKS et al., 1968), these types of boundaries are seismically active in the GoC (REICHLE and REID, 1977). Previous studies of seismicity in this region have been done based on worldwide networks (e.g. SYKES, 1968, 1970; MOLNAR, 1973; GOFF et al.,

---

[1] División Ciencias de la Tierra, Departamento de Sismología, Centro de Investigación Científica y de Educación Superior de Ensenada (CICESE), Apartado Postal 2732, 22860 Ensenada, Baja California, Mexico. E-mail: raul@cicese.mx

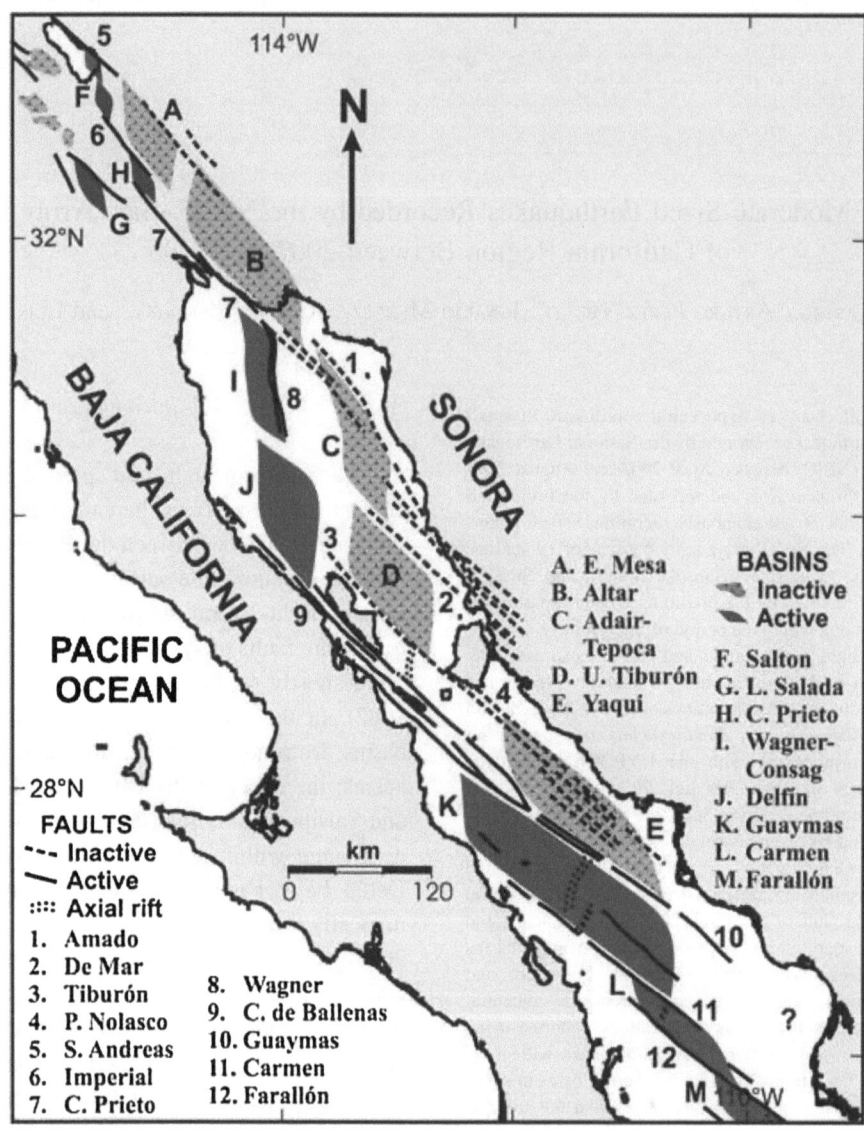

Figure 1
Tectonic map of the region, modified from Aragón-Arreola and Martin-Barajas (2007)

1987) and a few using regional stations (LOMNITZ *et al.*, 1970) and sonobuoys data (THATCHER and BRUNE, 1971; REID *et al.*, 1973; REICHLE and REID, 1977).

In this paper, we present accurate located epicenters, from moderate magnitude earthquakes, obtained using regional data and modern location techniques. We will show that the epicentral distribution of the earthquakes analyzed correlates well with the main bathymetric features of the GoC region. We also evaluate the epicentral location

deviations expected from the preliminary determinations of epicenters (PDE) catalog.

2. Data

We picked arrival times from earthquakes reported by the PDE from April 2002 to August 2006 and recorded by the broadband stations of the NARS–Baja array (Network of Autonomously Recording Seismographs) (TRAMPERT *et al.*, 2003; CLAYTON *et al.*, 2004) and the RESBAN array (Red Sism-ológica de Banda Ancha). The NARS–Baja array operated from the spring of 2002 to the fall of 2008 and consisted of 14 broadband seismic stations, pro-vided by Utrecht University, with STS2 sensors, a global positioning system (GPS) and a 24-bit data logger. The RESBAN array started operating in 1995 with two broadbad stations, one in Bahia de los Angeles, Baja California, and the other in Guaymas, Sonora (REBOLLAR *et al.*, 2001). During the period 2002–2006 three additional stations were added to the network. This array is operated by CICESE (Centro de Investigación Científica y de Educación Superior de Ensenada). The stations of RESBAN consist of Guralp CMG-40T or CMG-3ESP sensors, 24-bit

Guralp digitizers, a CMG-SAM2 acquisition module and GPS for time control. The stations of both arrays record continuously at 20 samples per second. The triangles in Fig. 2 show the distribution of the seismic stations of these arrays. Stations NE70–NE83 belong to the NARS–Baja array and the rest to RESBAN.

The data set consists of *P* and *S* wave arrival times handpicked, with an approximate accuracy of one tenth of a second, from the recordings of both broadband arrays from earthquakes that occurred between April 2002 and August 2006 and reported by PDE. The earthquakes analyzed range in M_w mag-nitude between 1.1 and 6.7, and are located between 22°N–31°N and 108°W–117°W.

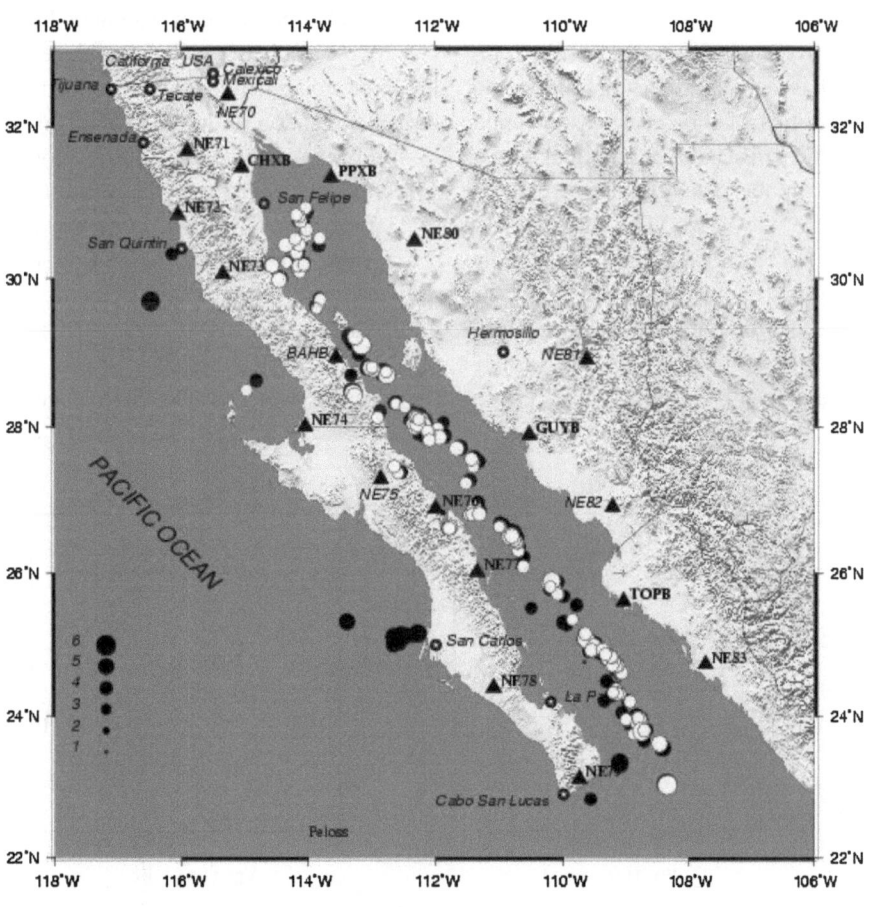

Figure 2

Earthquake epicenter location map. The *black dots* are epicenters reported by NEIC and relocated using arrival times of the broadband stations (*triangles*) of the NARS–Baja (NE70-NE83) and RESBAN (EXCB, BAHB, PPXB, GUAY, TOPB) arrays, and the HYPOCENTER code of LIENERT and HAVSKOV (1995). The *white dots* are the epicenters of the same data set but relocated using the SSST method of LIN and SHEARER (2005). The size of the *circles* is proportional to the magnitude of the earthquakes

3. Method

We calculated an initial hypocentral location with the HYPOCENTER code of LIENERT and HAVSKOV (1995) using at least four P and S wave arrival times from the stations shown in Fig. 2. For events reported by PDE in the northern GoC, between 30°N and 31.5°N, we used the velocity model obtained by NAVA and BRUNE (1982). This model is based on an approximate reverse refraction line obtained using a blast from Corona, Southern California ($\sim$33.8°N, 117.5°W) and the well-located Pino Solo earthquake (5.1M_L) of July 1975 (31.82°N, 115.85°W). This model consists of three layers with P velocities of 6.57–6.95 km/s, an upper mantle velocity of 8.0 km/s and a crustal thickness of approximately 42 km. For earthquakes reported in the lower Delfin basin, between 28°N and 30°N, we used a velocity model based on surface wave analysis (LÓPEZ-PINEDA et al., 2007). This model is composed by four horizontal layers with P velocities of 4.6–7.9 km/s, an upper mantle velocity of 8.2 km/s and a crustal thickness of 40 km. For events in the Guaymas basin and the southern end of the GoC, we used the velocity structure proposed by FABRIOL et al. (1999) for the Tortuga rift, on the western margin of the GoC. This velocity model consists of four homogenous layers with P wave velocities of 4.0–6.9 km/s over a half-space with a P velocity of 7.6 km/s and a crustal

thickness of 24 km. The comparison of these velocity models is shown in Fig. 3a. The black dots in Fig. 2 show the distribution of the epicenters located using these models and the body wave arrivals recorded in the regional stations of the NARS–Baja and the RESBAN arrays.

To account for lateral heterogeneities of the velocity structure the HYPOCENTER code uses static station terms (ST). The ST correction is determined by the mean of the travel-time residuals from all the events recorded at a given station. This approach is adequate if the source-station paths are the same for all the events. However, in the GoC the seismicity is distributed over a large region (Fig. 2) and the ST do not account completely for the travel-time perturbation introduced by lateral velocity heterogeneities. The use of source-specific station terms (SSST) as proposed by RICHARDS-DINGER and SHEARER (2000), is a better technique, since each station has a correction function that varies with source position. A generalization of the SSST method was implemented by LIN and SHEARER (2005) in their COMPLOC earthquake location code. We used a modified version of the COMPLOC code that permits the use of regional phases (Pn, Pg, Sn, Sg) and weights the phase arrival picks according to the source-station distance (e.g. CASTRO et al., 2010).

To minimize regional variations of the velocity structure, we relocated the hypocenters dividing the

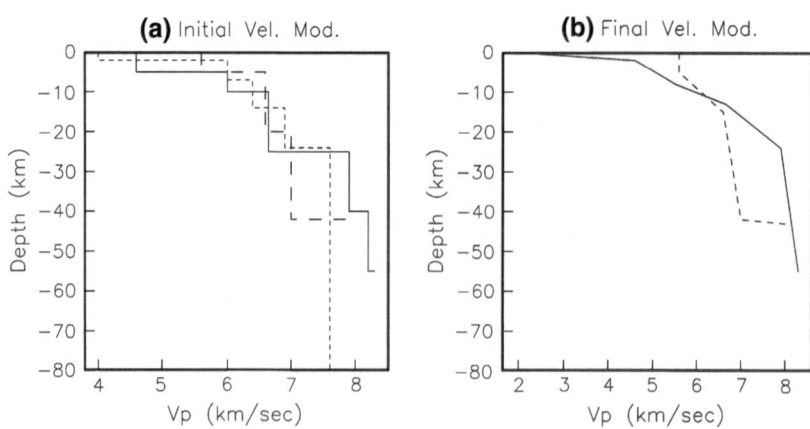

Figure 3
P wave velocity models tested. a Initial models used to locate the events with HYPOCENTER. *Solid lines* correspond to the model of LOPEZ-PINEDA et al. (2007), *dashed lines* to the model of NAVA and BRUNE (1982) and *dotted lines* to the model of FABRIOL et al. (1999). b Additional models tested. *Dashed line* is a model based on NAVA and BRUNE's (1982) but with vertical gradients. The model obtained by GONZALEZ-FERNANDEZ et al. (2005) is represented with *solid lines*

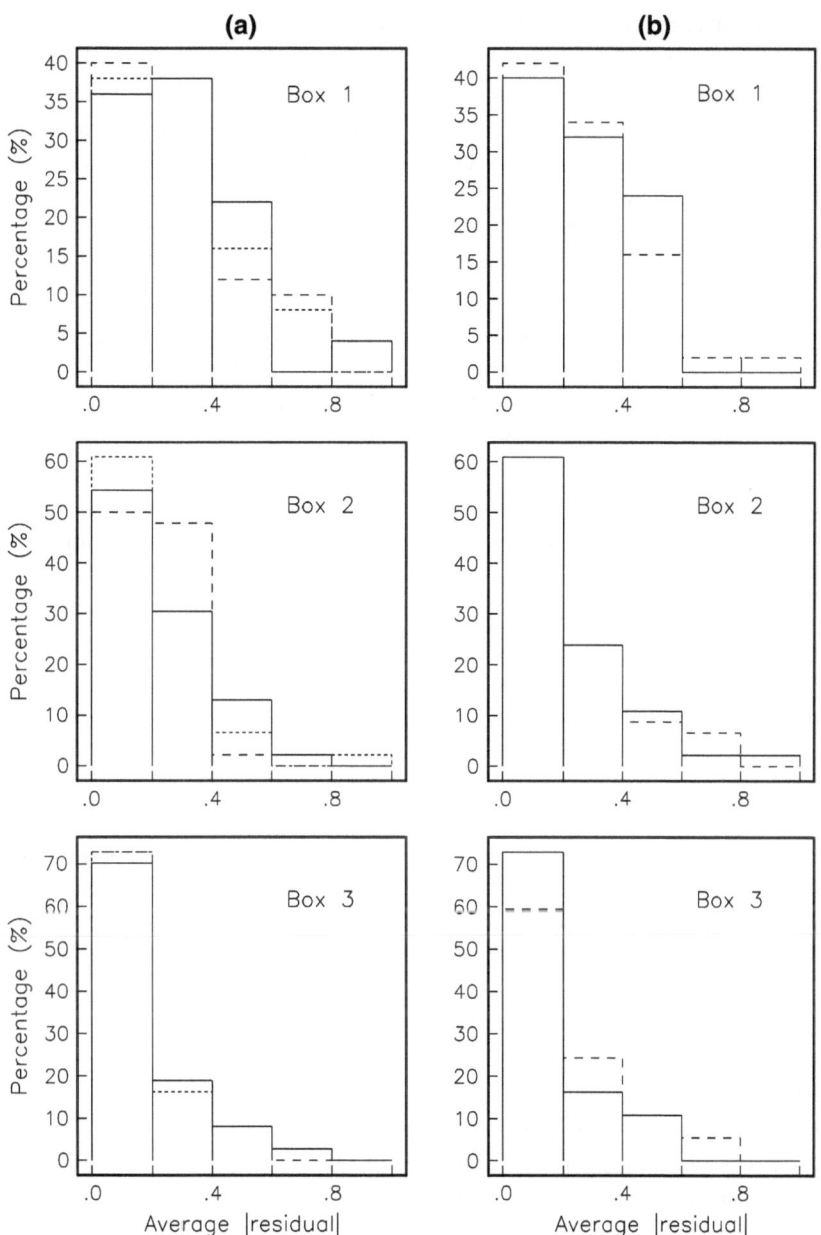

Figure 4

Average of the absolute value of the travel time residuals obtained using different velocity models. *Column a* shows histograms of residuals for the three rectangles used to divide the GoC region. *Solid lines* were used to represent residuals obtained with the velocity model of LOPEZ-PINEDA *et al.* (2007), *dashed lines* correspond to residuals obtained with the model proposed by NAVA and BRUNE (1982) and for the velocity model obtained by FABRIOL *et al.* (1999) we used *dotted lines*. *Column b* shows with *dashed lines* the residuals obtained using a model similar to that of NAVA and BRUNE (1982) but with vertical gradients. The *solid lines* correspond to the residuals obtained with the model of GONZALEZ-FERNANDEZ *et al.* (2005)

GoC region into three rectangles based on the distribution of the epicenters located with HYPOCENTER (Fig. 2). One of the rectangles covered from the San Pedro Martir basin (28°N) to the Wagner basin (31.5°N), another from Farallon basin (25.5°N) to the northern Guaymas basin (28°N) and the last one from the southern end of the GoC (22°N) to the Farallon basin. We relocated the hypocenters using records from all the stations available regardless of the rectangle but only the events within the corresponding rectangle were used to calculate the SSST. We tested the three velocity models described above (Fig. 3a) and two additional models (Fig. 3b) to relocate the earthquakes in each rectangle. One of these models, represented with dashed lines in Fig. 3b, is similar to that proposed by NAVA and BRUNE (1982), but we introduced vertical velocity gradients. The other velocity model (solid line in Fig. 3b) was proposed by GONZALEZ-FERNANDEZ et al. (2005) for the northern GoC and is based on a 280-km-long profile that included deep multi-channel seismic reflection data, densely sampled refraction data and wide-angle reflection information. This model is consistent with that obtained by PHILLIPS (1964) using active sources and sonobouys. The model of GONZALEZ-FERNANDEZ et al. (2005) consists of four layers. The first layer represents the uppermost sediments, and it has P wave velocities between 1.77 and 2.15 km/s. The second layer, corresponding to lower sediments, has velocities between 4.11 and 5.09 km/s. The middle crust is represented by a layer with velocities of 5.37–5.67 km/s. The lower crust has velocities of 6.58–6.73 km/s, and below the Moho (at 24–25 km depth) is the upper mantle with a velocity of 7.9 km/s.

We calculated average travel time residuals using the five velocity models tested. Column a of Fig. 4 shows histograms calculated every 0.2 s for each of the rectangles, from north to south, used to divide the GoC region. This column shows the average residuals obtained with the SSST method and the three initial velocity models used to locate the earthquakes. For the northern GoC (Box 1) the model of NAVA and BRUNE (1982) gives the smaller average residuals and for the southern GoC (Box 3) the model of LOPEZ-PINEDA et al. (2007). In general, the velocity models that have a vertical velocity gradient (column b in Fig. 4) give the smaller residuals, particularly the model of GONZALEZ-FERNANDEZ et al. (2005), shown with solid lines in Fig. 4b.

4. Results

We used the velocity model obtained by GONZALEZ-FERNANDEZ et al. (2005) for the final

estimates of the hypocentral coordinates. Figure 5 shows the histogram of focal depths obtained with the final locations. Most earthquakes have focal depths of less than 10 km and 52% have depths of less than 5 km. The lower maps of Figs. 6, 7, 8 and 9 show the final distribution of epicenters for different regions of the GoC and Fig. 10 shows the global distribution of the bigger earthquakes (M_w 4.8–6.6). We also compared in Fig. 2 the initial locations obtained with HYPOCENTER (black dots) and the relocated epicenters using the SSST method (white dots). In many cases the epicentral locations obtained with these two methods are not very different because, as described above, when we located the epicenters with the HYPOCENTER code we used different velocity models depending on the initial location reported by PDE and according with the region. By doing this the static station corrections calculated with HYPO-CENTER became space dependent. When we relocated the hypocenters using the SSST method, we used the same velocity model for all regions but we calculated specific stations terms that depend on the spatial distribution of the events surrounding the target event. In contrast, the epicenters reported in the PDE catalog show important differences with respect to the relocated epicenters using the SSST method (Fig. 11).

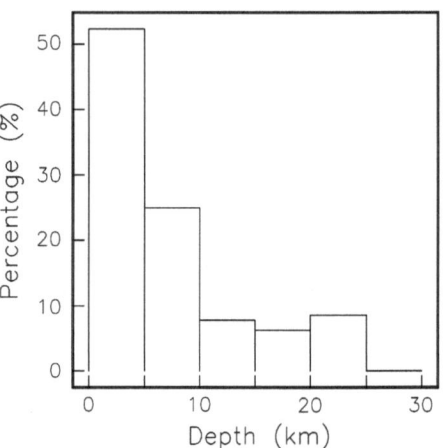

Figure 5

Histogram of focal depths of the relocated hypocenters (using regional arrival times, the SSST method and the velocity model of GONZALEZ-FERNANDEZ et al., 2005)

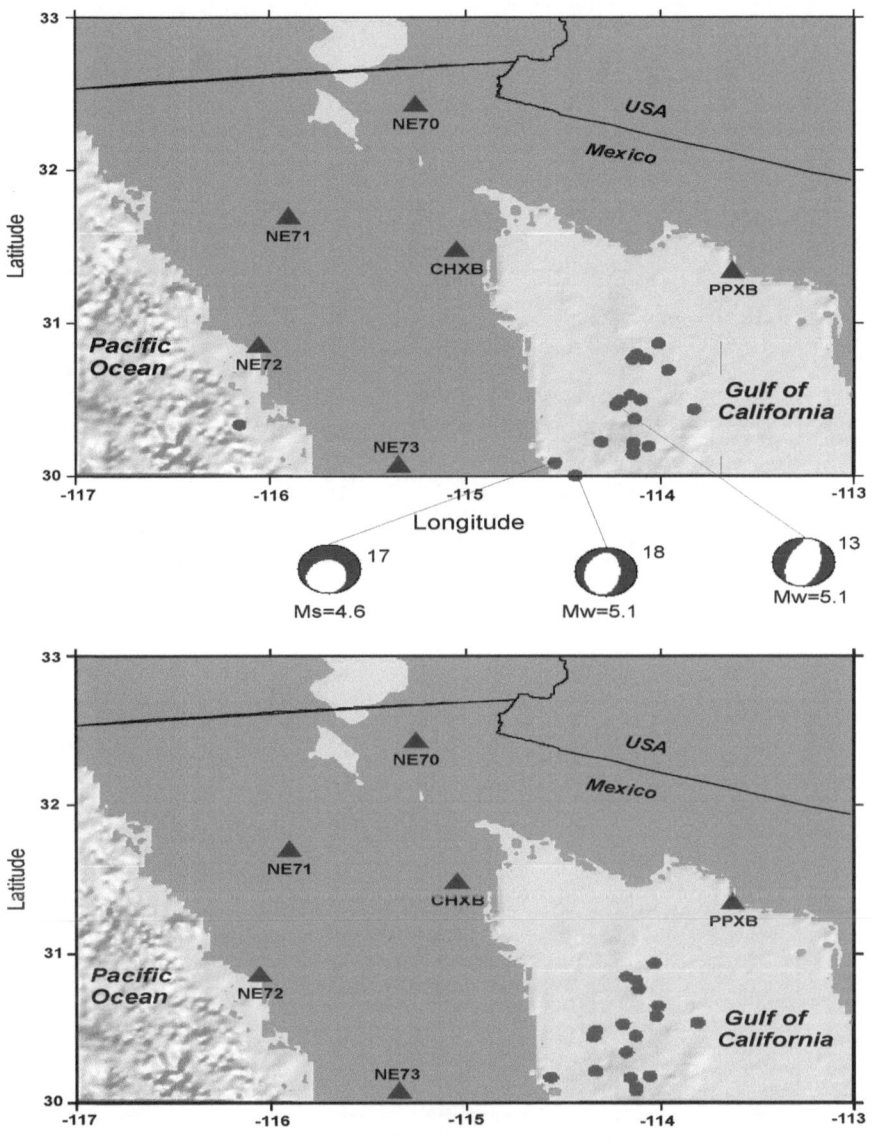

Figure 6

Upper map shows epicenters located with HYPOCENTER and the focal mechanisms available in the Harvard CMT catalog. The *lower map* shows the epicentral locations obtained using the SSST method of LIN and SHEARER (2005)

Comparing the epicentral locations reported by PDE with the relocated epicentral coordinates obtained with the regional arrival times and the SSST method (Fig. 11), we found that earthquakes with *mb* magnitudes in the range 3.2–5.0 differ on the average by as much as 43 km. For earthquakes with M_w magnitude between 5.0 and 6.7 the discrepancy is less, differing on the average by about 25 km. In general, the epicenters reported by PDE are shifted towards the northeast, where most seismic stations in North America are located.

4.1. The Consag, Wagner and Upper Delfin Basins

On the northern end of this region (30.0°N–31.0°N) the epicenters are disperse in the middle of the GoC (Fig. 6). However, the most northern events (30.4°N–31.0°N) tend to align in the NW–SE direction. This part of the Gulf is characterized for having a complex fault system where multiple oblique-slip faults interact with each other (PERSAUD *et al.*, 2003). South of 30.5°N the earthquakes scatter toward the western margin of the GoC and for the period

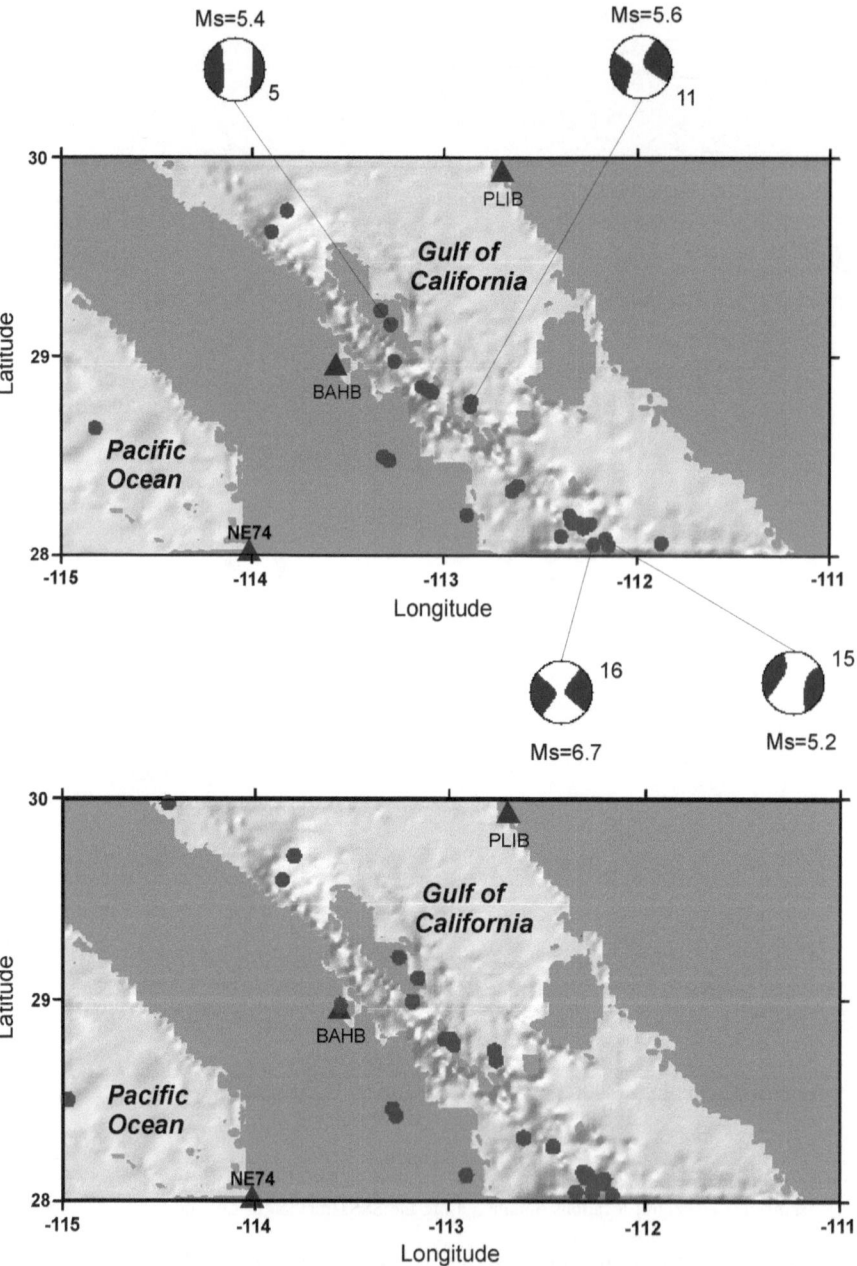

Figure 7
Same as in Fig. 6 but for the San Pedro Martir–Delfin basins

analyzed the bigger events have normal fault mechanisms (upper map of Fig. 6).

4.2. The Lower Delfin Basin and Canal de Ballenas

We relocated two events in the lower Delfin basin, distributed in the NE–SW direction, approximately

perpendicular to the GoC axis. However, most events align near the Canal de Ballenas fault, which is located between the station BAHB and the Angel de la Guarda island (Fig. 7). This transform fault is oriented in the NW–SE direction and runs from the northern end of Angel de la Guarda (~29.5°N) to the south, where most of the epicenters are located

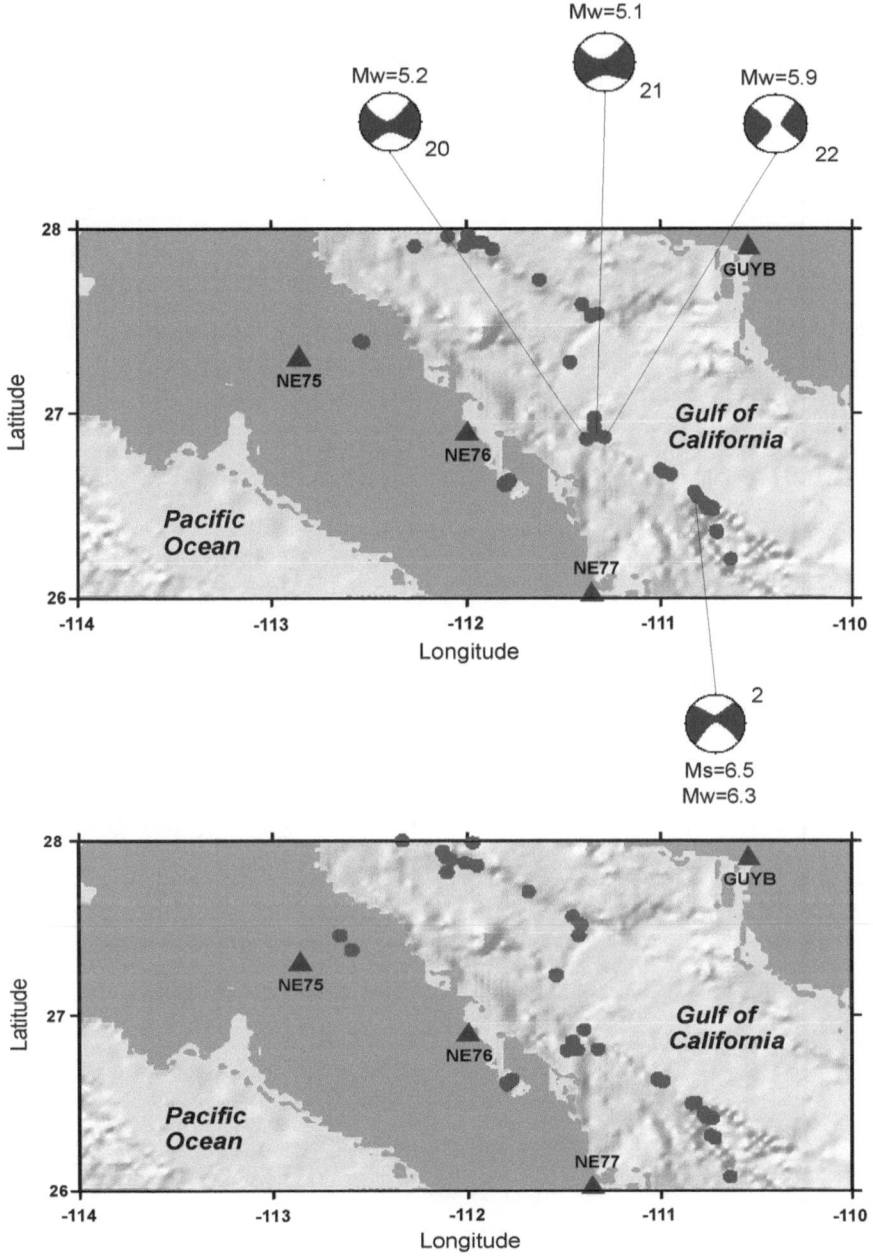

Figure 8
Same as in Fig. 6 but for the Carmen–Guaymas basins

(∼28.1°N). In this region the events have strike-slip focal mechanisms (upper map of Fig. 7). RODRIGUEZ-LOZOYA et al. (2008) studied events 5 and 11 (see Table 1 and upper map of Fig. 7). The earthquake located near the island Angel de la Guarda (event 5) occurred on November 12, 2003, they estimated a $M_w = 5.6$ and a focal mechanism solution corresponding to a normal fault striking 301°. Event 11 occurred on 24 September 2004 northwest of island San Lorenzo. RODRIGUEZ-LOZOYA et al. (2008) estimated an $M_w = 5.8$ and a right-lateral strike-slip fault mechanism (strike = 117°, dip = 75°, rake = 175°). Notice also in Fig. 7 that a few smaller events were located in the stable, central part of the Baja Peninsula

37

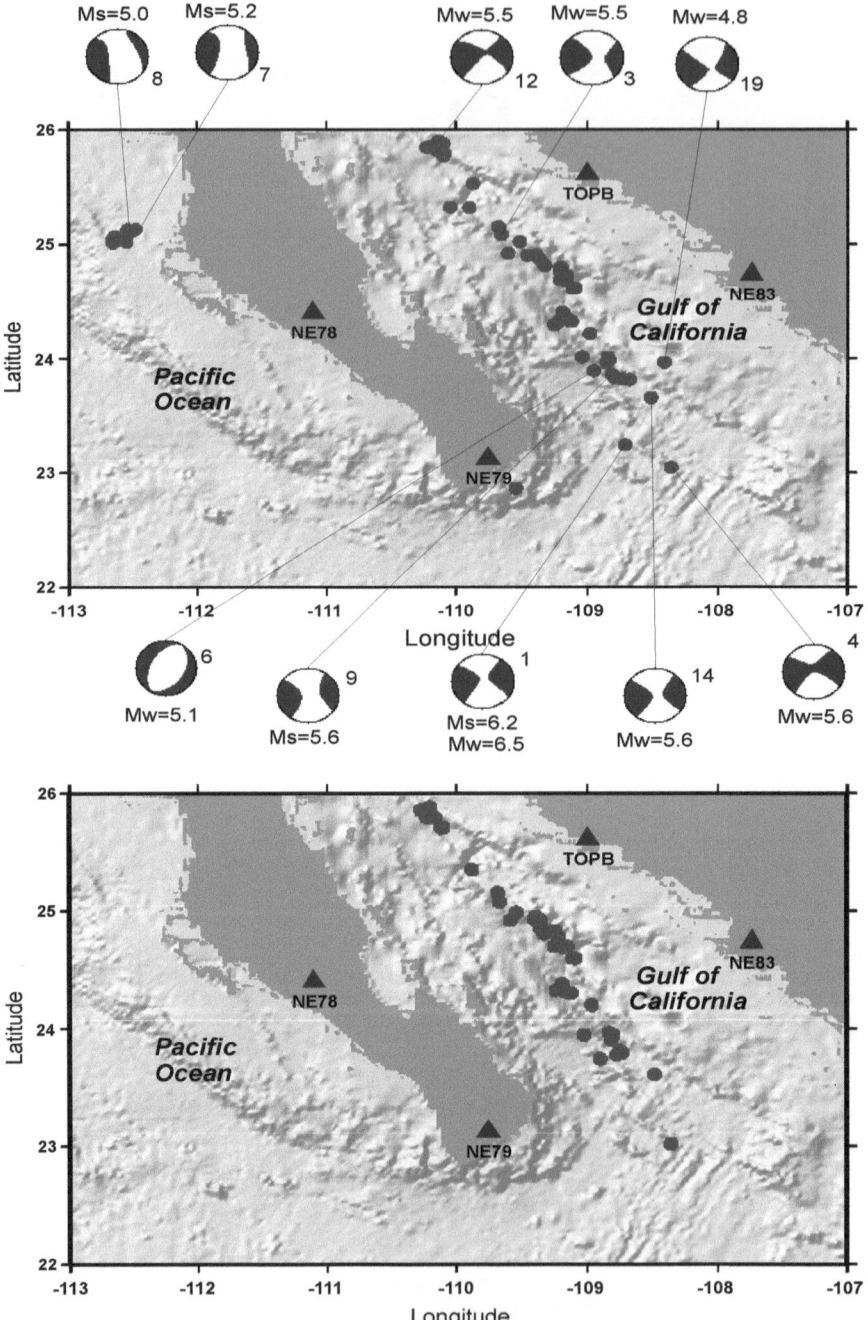

Figure 9
Same as in Fig. 6 but from the southern end of the Gulf of California to the Farallon basins

where the unextended and relatively unfaulted crust is not expected to have seismic activity. Thus, these events may result as the response of the Pacific plate to the stress accumulated due to extensional forces from the GoC.

4.3. Guaymas and Carmen Basins

This is a region of the GoC where some of the bigger events have occurred, particularly on the transform faults that connect the Guaymas and Carmen basins (Fig. 1). The earthquake of 12 March

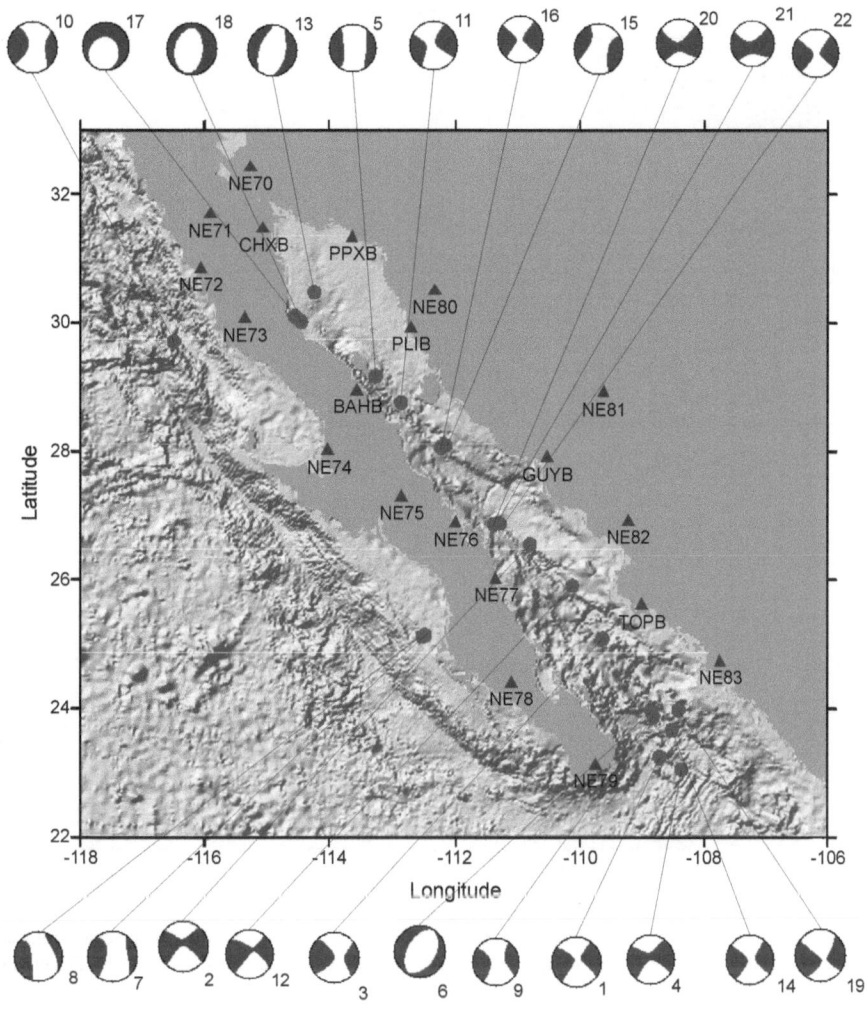

Figure 10
Distribution of main events (M_w 4.8–6.6) and focal mechanisms taken from the Global CMT Catalog

2003, $M_w = 6.3$ (event 2 in Table 1 and upper map of Fig. 8) and a foreshock of $M_w = 4.2$ were studied by LÓPEZ-PINEDA and REBOLLAR (2005). They calculated an $M_w = 6.2$ and a focal mechanism consistent with a right-lateral strike-slip event (strike = 117°, dip = 79°, rake = 168°). Other important events in this region have occurred in 1971 ($M_w = 6.5$), 1974 ($M_w = 6.3$), 1988 ($M_w = 6.6$) and 1995 ($M_w = 6.0$) (LÓPEZ-PINEDA and REBOLLAR 2005). South of Carmen basin (26°N, 110°W) PACHECO and SYKES (1992) also reported an earthquake with $M_w = 7.0$ that occurred on 07 January 1901. The relocated epicenters for the 2002–2006 interval (lower map of Fig. 8) tend to align in the NW–SE direction between 27.6°N and

28°N, where the transform fault that connects the San Pedro Martir basin with the Guaymas basin can be inferred from the bathymetry (LONSDALE 1989). Farther south ($\sim$27°N), in the Guaymas basin, the seismicity jumps towards the western margin of the Gulf and aligns again in the NW–SE direction along the transform fault that connects the Guaymas and Carmen basins.

4.4. Farallon, Pescadero and Alarcon Basins

This region is the most active of the four. Figure 9 shows the distribution of the relocated epicenters and the focal mechanism reported by the Global CMT

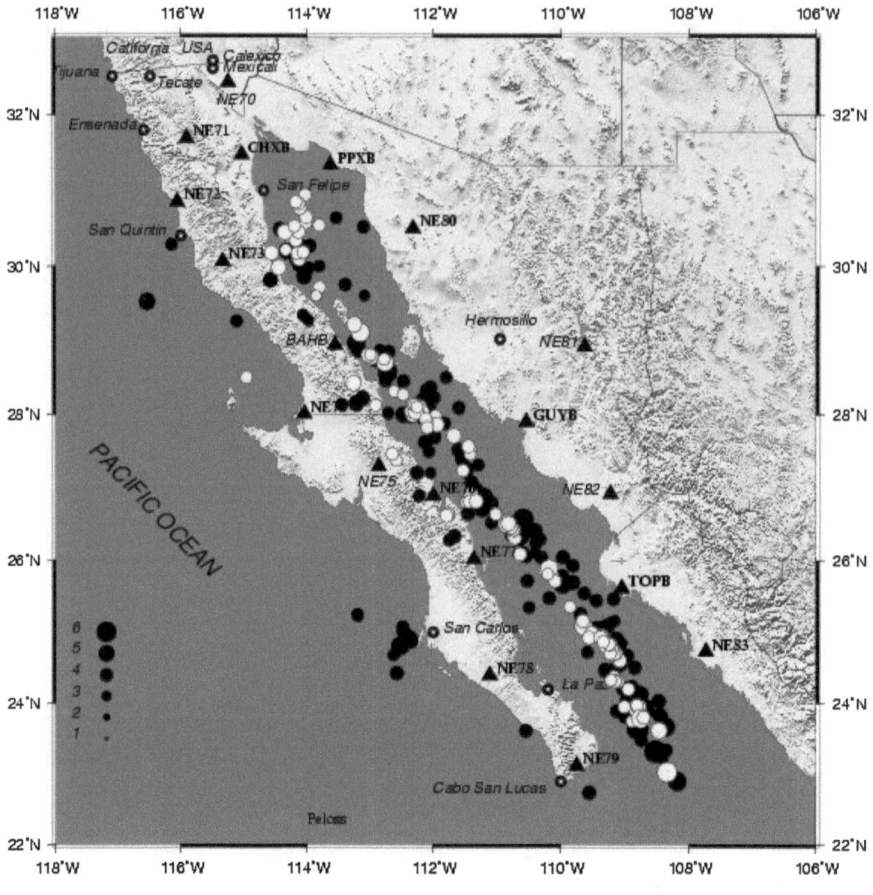

Figure 11
Earthquake epicenter location map. The *black dots* are epicenters reported by NEIC. *White dots* are the epicenters relocated using the SSST method of LIN and SHEARER (2005) and arrival times of the regional stations of NARS–Baja and RESBAN seismic networks. The size of the *circles* is proportional to the magnitude of the earthquakes

Catalog for the bigger events. Most of these earthquakes are strike-slip events, although normal fault events also occur in this region. Event 6 ($M_w = 5.1$) of Table 1, upper map of Fig. 9, is a normal fault earthquake. Northwest of Farallon basin ($\sim 25.9°$N, $110.1°$W) we relocated a cluster of earthquakes, where on February 22, 2005 occurred event 12 ($M_w = 5.5$). This cluster is located on the transform fault that connects the Carmen and Farallon basins. RODRIGUEZ-LOZOYA *et al.* (2008) calculated the focal mechanism of event 12 and found that the best solution is consistent with a right-lateral strike-slip fault (strike = 309°, dip = 65°, rake = 159°). Note that southwest of the cluster, there are some events located on the spreading center, near the Farallon basin, that connects to another transform fault to the

south. This transform fault extends toward the southeast, where we relocated another group of earthquakes aligning in the NW–SE direction. Another group of earthquakes that jump towards the south, between 108.5°W and 109°W, are located on the transform fault that connects Pescadero and Alarcon basins.

Figure 10 summarizes the location and focal mechanisms of the main events relocated in the 2002–2006 period. Most of the events (68%) are strike-slip events associated to the transform faults of the GoC. Only seven events are normal fault earthquakes and have magnitudes $M_w \leq 5.6$, indicating that most of the seismic energy is relieved along the transform faults of the GoC, where the bigger strike-slip events occur.

Table 1

Coordinates of events shown in Fig. 10, taken from the Global CMT catalog

	Date	Time	Lat.	Lon.	H	M_w
1	2002-10-03	16:08:41	23.22	−108.27	15.0	6.5
2	2003-03-12	23:41:39	26.63	−110.91	15.0	6.3
3	2003-04-15	08:21:21	25.51	−109.60	15.0	5.5
4	2003-07-02	05:11:40	23.11	−108.33	15.0	5.6
5	2003-11-12	04:55:01	29.34	−113.45	15.0	5.6
6	2004-02-09	00:01:48	24.04	−108.93	12.0	5.1
7	2004-02-09	01:24:40	25.06	−112.60	12.0	5.4
8	2004-02-09	09:03:50	25.13	−112.56	12.0	5.3
9	2004-02-18	10:59:25	23.91	−108.75	12.6	5.8
10	2004-08-30	05:35:18	29.61	−116.59	12.0	5.1
11	2004-09-24	14:43:14	28.82	−112.99	12.0	5.9
12	2005-02-22	19:15:52	25.82	−110.09	18.9	5.5
13	2005-04-27	00:32:60	30.45	−114.27	12.0	5.1
14	2005-06-05	08:28:52	23.70	−108.51	12.0	5.6
15	2006-01-04	01:05:12	28.32	−112.28	12.0	5.2
16	2006-01-04	08:32:37	28.38	−112.51	15.0	6.6
17	2006-05-01	20:53:22	30.41	−114.46	13.8	4.9
18	2006-05-01	21:04:41	30.13	−114.49	12.0	5.1
19	2006-05-11	11:02:52	24.16	−108.83	12.0	4.8
20	2006-05-28	14:02:59	26.96	−111.34	19.1	5.2
21	2006-05-28	14:18:08	27.00	−111.32	17.3	5.1
22	2006-07-30	01:21:01	26.87	−111.36	22.9	5.9

5. Conclusions

We relocated 128 hypocenters of earthquakes with M_w magnitudes 1.1–6.7 reported by PDE between April 2002 and August 2006 using regional arrival times recorded by the NARS–Baja and RES-BAN arrays. We found that earthquakes with magnitudes in the range 3.2–5.0 *mb* differ on the average by as much as 43 km. Epicenters from events with magnitudes 5.0–6.7 M_w have a discrepancy of about 25 km on the average. The epicenters relocated using the SSST method of LIN and SHEARER (2005) correlate well with the bathymetry of the GoC and permit to infer the active transform faults. We found that the main events occur on or near transform faults and tend to have right-lateral strike slip focal mechanisms. Since epicenters located with regional stations have a much better azimuthal coverage and since we have accounted for 3D velocity variations by using source specific station term corrections, our results confirm, with a higher degree of confidence than previous studies, that the spatial distribution of seismicity is complex in the northern GoC whereas in the southern GoC, it is confined to narrow zones near the most active faults.

Acknowledgments

The operation of the RESBAN array (*Red Sismológica de Banda Ancha*) has been possible thanks to the financial support of the Mexican National Council for Science and Technology (CONACYT) by means of the projects 48852 and COC022. The NARS–Baja array was the result of a collaboration project among Utrecht University, California Institute of Technology and CICESE. The operation and data acquisition was possible thanks to Arie van Wettum, Robert Clayton, Jeannot Trampert and Hanneke Paulssen. Part of the analysis of this paper was made while one of the authors (RRC) was a UC MEXUS-CONACYT Visiting Scholar at the University of California, San Diego, IGPP-SIO. Peter Shearer kindly provided the modified version of the COMPLOC code. The useful comments and suggestions of the editor William Lee Bandy help us to improve the manuscript. We also thank the two anonymous reviewers.

REFERENCES

ARAGÓN-ARREOLA, M., and MARTIN-BARAJAS, A. (2007), *Westward migration of extension in the Northern Gulf of California, Mexico*, Geology 35, 571–574

CASTRO, R. R., SHEARER, P. M., ASTIZ, L., SUTER, M., JACQUES-AYALA, C., and VERNON, F. (2010), *The Long-lasting aftershock series of the 3 May 1887 M_W 7.5 Sonora Earthquake in the Mexican Basin and Range Province*, Bull Seism Soc Am 100, 1153–1164

CLAYTON, R. W., TRAMPERT, J., REBOLLAR, C. J., RITSEMA, J., PERSAUD, P., PAULSSEN, H., PÉREZ-CAMPOS, X., VAN WETTUM, A., PÉREZ-VERTTI, A., and DI LUCCIO, F. (2004), *The NARS-Baja array in the Gulf of California R ift zone*, Margins Newslett 13, 1–4

DEMETS, C. (1995), *A reappraisal of seafloor spreading lineations in the Gulf of California: implications for the transfer of Baja California to the Pacific Plate and estimates of Pacific-North America Motion*, Geophys Res Lett 22, 3545–3548

EINSELE, G. (1982), *Mechanism of sill intrusion into soft sediment and expulsion of pore water*, Initial Rep Deep Sea Drill Proj 64, 1169–1176

FABRIOL, H., DELGADO-ARGOTE, L.A., DAÑOBEITIA, J.J., CÓRDOBA, D., GONZÁLEZ, A., GARCÍA-ABDESLEM, J., BARTOLOMÉ, R., and MARTÍN-ATIENZA, B. (1999), *Backscattering and geophysical features of volcanic rifts offshore Santa Rosalía, Baja California Sur, Gulf of California, México*, J Volcano Geotherm Res 93, 75–92

GOFF, J.A., BERGMAN, E.A., and SOLOMON, S.C. (1987), *Earthquake source mechanism and transform fault tectonics in the Gulf of California*, J Geophys Res *92*, 10485–10510

GONZALEZ-FERNANDEZ, A., DAÑOBEITIA, J. J., DELGADO-ARGOTE, L. A., MICHAUD, F., CORDOBA D., and BARTOLOMÉ, R. (2005), *Mode of extensión and rifting history of upper Tiburón and upper Delfín basins, Northern Gulf of California*, J Geophys Res *110*, B01313. doi:10.1029/2003JB002941

ISACKS, B., OLIVER, J., and SYKES, L. (1968), *Seismology and the new global tectonics*, J Geophys Res *73*, 5855–5899

LIENERT, B. R. E., and HAVSKOV, J. (1995), *A computer program for locating earthquakes both locally and globally*, Seism Res Lett *66*, 26–36

LIN, G., and SHEARER, P. M. (2005), *Test of relative earthquake location techniques using synthetic data*, J Geophys Res *110*, B04304. doi:10.1029/2004JB003380.

LOMNITZ, C., MOOSER, F., ALLEN, C. R., BRUNE, J. N., and THATCHER, W. (1970), *Seismicity and tectonics of the Northern Gulf of California region, Mexico. Preliminary Results*, Geof Int *10*, 37–48

LONSDALE, P. L., *Geology and Tectonics History of the Gulf of California*, In: The Geology of North America: The Eastern Pacific Ocean and Hawaii (The Geological Society of America 1989).

LÓPEZ-PINEDA, L., and REBOLLAR, C.J. (2005), *Source characteristics of the Mw 6.2 Loreto earthquake of 12 March 2003 that occurred in a transform fault in the middle of the Gulf of California, Mexico*, Bull Seism Soc Am *95*, 419–430

LÓPEZ-PINEDA, L., REBOLLAR, C. J., and QUINTANAR, L. (2007), *Crustal Thickness estimates for Baja California, Sonora, and Sinaloa, Mexico, using disperse surface waves*, J Geophys Res, *112*, B04308. doi:10.1029/2005JB003899

MOLNAR, P. (1973), *Fault plane solutions of earthquakes and direction of motion in the Gulf of California and on the rivera fracture zone*, Geol Soc Am Bull *84*, 1651–1658

NAVA, F. A., and BRUNE, J. N. (1982), *An earthquake-explosion reversed refraction line in the peninsular ranges of southern California and Baja California Norte*, Bull Seism Soc Am *72*, 1195–1206

PACHECO, J. F., and SYKES, L. R. (1992), *Seismic moment catalog of large shallow earthquakes, 1900 to 1989*, Bull Seism Soc Am *82*, 1306–1349

PERSAUD, P., STOCK, J. M., STECKLER, M. S., MARTIN-BARAJAS, A., DIEBOLD, J. B., GONZALEZ-FERNANDEZ, A., and MOUNTAIN, G. S.

(2003), *Active deformation and shallow structure of the Wagner, Consag, and Delfin basins, northern Gulf of California, Mexico*, J Geophys Res *108*(B7), 2355. doi:10.1029/2002JB001937.

PHILLIPS, R.P. (1964), *Seismic refraction Studies in the Gulf of California, in Marine Geology of the Gulf of California*, In Van Andel T. H., and Shor G.G., AAPG Mem *3*, 90–121

REBOLLAR, C. J., QUINTANAR, L., CASTRO, R. R., DAY, S. M., MADRID, J., BRUNE, J. N., ASTIZ, L., and VERNON, F. (2001), *Source characteristics of a 5.5 magnitude earthquake that occurred in the transform fault system of the Delfin basin in the Gulf of California*, Bull Seism Soc Am *91*, 781–791

REICHLE, M., and REID, I. (1977), *Detailed study of earthquake swarms from the Gulf of California*, Bull Seism Soc Am *67*, 159–171

REID, I., REICHLE, M., BRUNE, J., and BRADNER, H. (1973), *Micro-earthquake studies using sonobuoys, preliminary results from the Gulf of California*, J R Astronom Soc *34*, 365–379

RICHARDS-DINGER, K. and SHEARER P. (2000), *Earthquake locations in southern California obtained using source-specific station terms*, J Geophys Res, *105*, 10939–10960

RODRIGUEZ-LOZOYA, H. E., QUINTANAR, L., ORTEGA, R., REBOLLAR, C. J., and YAGI, Y. (2008), *Ruptura process of four medium-size earthquakes that occurred in the gulf of california*, J Geophys Res, *113*, B10301. doi:10.1029/2007JB005323.

SYKES, L. R. (1968), *Seismological Evidence for Transform Faults, Sea-floor Spreading and Continental Drift*, in the History of the Earth's Crust, In (ed. Phinney), pp 120–150, Princeton Univ. Press, Princeton

SYKES, L. R., (1970), *Focal mechanism solutions for earthquakes along the world rift system*, Bull Seismol Soc Am *60*, 1749–1752

THATCHER, W., and BRUNE, J. (1971), *Seismic study of an oceanic ridge earthquake swarm in the Gulf of California*, Geophys J R Astron Soc *22*, 473–489

TRAMPERT, J., PAULSEN, H., VAN WETTUM, A., RITSEMA, J., CLAYTON, R., CASTRO, R., REBOLLAR, C., and PÉREZ-VERTTI, A. (2003), *New array monitors seismic activity near the Gulf of California in México*, EOS, Trans Am Geoph Union *84*, 29–32

ZHANG, X., PAULSEN, H., LEBEDEV, S. and MEIER, T. (2007), *Surface Wave Tomography of the Gulf of California*, Geophys Res Lett, *34*, L15305. doi:10.1029/2007GL030631

(Received December 21, 2009, revised June 8, 2010, accepted June 16, 2010, Published online July 30, 2010)

Pure Appl. Geophys. 168 (2011), 1293–1302
© 2010 The Author(s)
This article is published with open access at Springerlink.com
DOI 10.1007/s00024-010-0178-x

Using an Enhanced Dataset for Reassessing the Source Region of the 2003 Armería, Mexico Earthquake

Francisco J. Núñez-Cornú,[1] Marta Rutz-López,[1] Víctor Márquez-Ramírez,[2]
Carlos Suárez-Plascencia,[1,2] and Elizabeth Trejo-Gómez[1]

Abstract—We present a fresh look at the source region of the 22 January 2003 M_w 7.4 Armería earthquake, which occurred off the Pacific coast of the state of Colima, Mexico, near the town of Armería. The effects of this earthquake in the neighboring states of Colima and Jalisco were different and stronger than those of previous recent major earthquakes in the region. This earthquake and its aftershocks were recorded by two local telemetered seismograph networks (RESCO and RESJAL). From 22 January to 24 January 2003, no important seismicity was located on the plates interface, or within the Rivera Plate, and most epicenters were located west of the Armería River, which is the western border of the Colima Graben, and is located outside of the Colima Gap region. From 24 January to 31 January, the seismicity recorded by both networks showed a migration in depth, with an almost vertical offshore distribution between 4 and 24 km in depth. For this period, a seven-station portable digital seismograph network, equipped with three-component seismometers, was deployed in the epicentral area to study the aftershock sequence in detail. With this denser network more than 200 $M_L > 2.0$ aftershocks were recorded. The aftershock foci were deeper than those recorded during the early period and most of them locate on a hypothetical 12° dipping interface between the Rivera and North American Plates. Composite focal mechanism solutions for the aftershocks located during both periods indicate a reverse fault character that changes with time. Analysis of the new dataset still indicates that the earthquake was a shallow intraplate event.

Key words: Jalisco Block, Colima Graben, intraplate earthquake, aftershocks, subduction.

1. Introduction

On 22 January 2003 a shallow M_w 7.4 earthquake (18.6658°W, 104.0895°N, depth = 5.0 km) occurred off the Pacific coast of the state of Colima, Mexico, near the town of Armería. This earthquake is the largest during the early twenty-first century in the area. Both the damage pattern and surface effects of this earthquake in the neighboring states of Colima and Jalisco were stronger than those caused by recent large earthquakes in these regions and affected different areas; especially in Colima City and Zapotitlan de Vadillo, Jalisco, on the western flank of the Colima volcano. This earthquake and its aftershocks were recorded by two local telemetered seismograph networks: The Red Sísmica Telemétrica de Colima (RESCO) and the Red Sísmica Digital Telemétrica de Jalisco (RESJAL). Additionally, to have detailed data of the aftershock sequence, a portable digital seismograph network (PN) was deployed from 24 to 31 January 2003. The PN recorded more than 200 $M_L > 2.0$ earthquakes. In this paper we present the results of the analysis of the newly gathered dataset, and compare them with the results reported for the initial 72 h of seismicity (Núñez-Cornú *et al.*, 2004).

2. Historical Seismicity and Previous Studies

The largest ($M = 8.2$) earthquake of the twentieth century in Mexico occurred in 1932 in the Jalisco Block (JB) region (Fig. 1), a tectonically complex, seismically active zone. This earthquake was followed 15 days later by another large event ($M = 7.8$). Singh *et al.* (1985) studied this sequence

---
[1] Centro de Sismología y Volcanología de Occidente, Universidad de Guadalajara, 48280 Puerto Vallarta, JAL, Mexico. E-mail: pacornu77@gmail.com
[2] Posgrado en Ciencias de la Tierra, Centro de Investigación Científica y Educación Superior de Ensenada, Ensenada, B.C., Mexico.

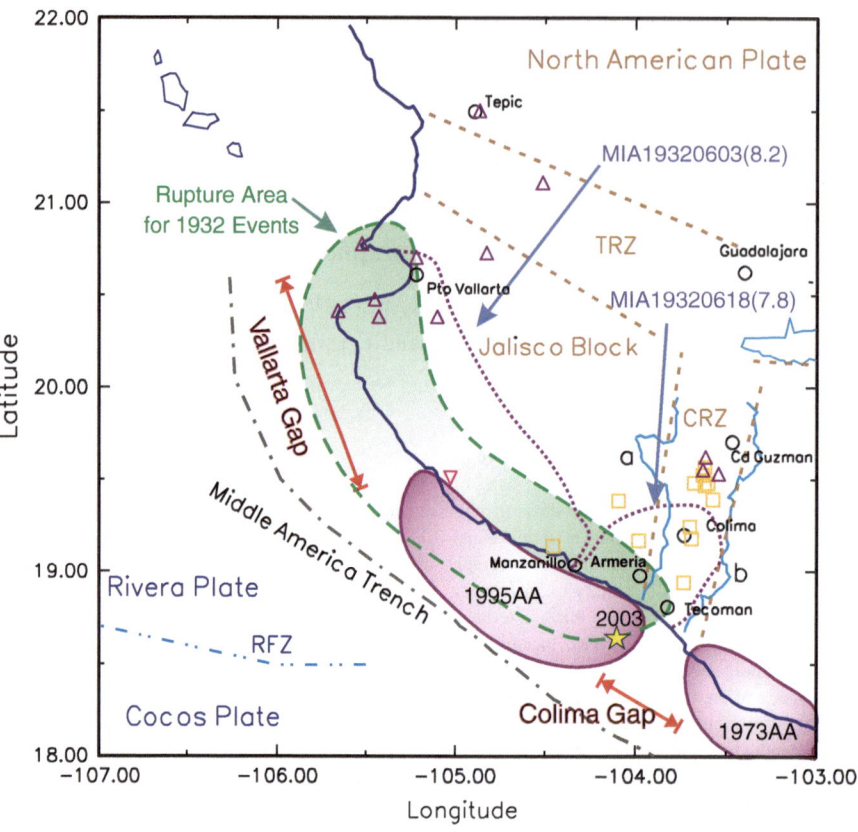

Figure 1

Map of a portion of western Mexico showing the seismotectonic features of the Jalisco region. *RFZ* Rivera Fault Zone, *CRZ* Colima Rift Zone, *TRZ* Tepic–Zacoalco Rift Zone, *a* Armería River, *b* Cohuayana River, *MIA* maximum intensity areas for earthquakes in 1932 (dates and magnitudes indicated), *AA* aftershocks areas. Seismograph stations are marked with squares (RESCO network), triangles (RESJAL network), and inverted triangle (CJIG station of the SSN). *Circles* cities. *Solid star* epicenter of the Armería earthquake (NÚÑEZ–CORNÚ *et al.*, 2004)

and concluded that the composite rupture area straddles the entire coast of the states of Jalisco and Colima, they also suggested a recurrence time of 77 years for major earthquakes in this region. In 1995 a $M_w = 8.0$ earthquake (E95) near Manzanillo involved part of the 1932 rupture area but left unruptured patches to the north and south, termed the Vallarta and Colima gaps, respectively (Fig. 1). In spite of the occurrence of large earthquakes and the associated high seismic hazard in the region, only one permanent broadband seismograph station at Chamela [station CJIG, from the Mexican Seismograph Network (SSN)] and a short-period, analog-telemetry network (RESCO) were operating on the Colima Rift Zone to monitor seismicity in the area until 2001 (Fig. 1). In late 2001, a new seismograph network, RESJAL, began to be deployed

by the Jalisco Civil Defense and the Centro de Sismología y Volcanología de Occidente, SisVOc (NÚÑEZ-CORNÚ *et al.*, 2001). Currently, this network features 11 stations (six of them telemetered); all equipped with three component-1 Hz sensors (LE3D type) and 24-bit data loggers (Fig. 1). From studies of microearthquakes in the JB during 1996–1998, NÚÑEZ-CORNÚ *et al.* (2002) found that the subduction of the Rivera Plate under the North American Plate exhibits a double seismic zone, with a subduction angle smaller than 15° up to 160 km from the trench; and identified various seismogenic zones, showing that data from the SSN alone are not enough to accurately determine the seismic patterns in the JB region. The seismicity recorded throughout 2002 by RESJAL and RESCO confirmed these results (NÚÑEZ-CORNÚ *et al.*, 2003).

3. Tectonic Setting

The North American, Pacific, Cocos, and Rivera lithospheric plates interact in western Mexico (Fig. 1), yet the seismotectonic setting is poorly understood. A tectonic unit known as the Jalisco Block has been proposed in this region (LUHR et al., 1985; BURGOIS et al., 1988). The Jalisco Block is limited to the east by the Colima Rift Zone (CRZ) and to the north by the Tepic–Zacoalco Rift Zone (TRZ). There is general agreement that the Jalisco Block is actively separating from the continent, moving in a WNW direction away from mainland Mexico (BURGOIS et al., 1988; GARDUÑO and TIBALDI, 1991; ALLAN et al., 1991), although alternative models have also been proposed (FERRARI et al., 1994; ROSAS-ELGUERA et al., 1996).

To date, the geometry of Rivera Plate under the JB is also unclear. EISSLER and McNALLY (1984) and SINGH et al. (1985) suggested a dip angle of 20° for the interface between the Rivera and North American Plates. NÚÑEZ–CORNÚ and SÁNCHEZ–MORA (1999), reported a dip angle of 12° as estimated from local seismicity data, and similar values were obtained from reflection–refraction studies carried out in the central and northern regions of the Jalisco coast (DAÑOBEITIA et al., 1997). More controversial are the geometry of the interface and the relative motion between the Rivera and Cocos Plates. Does the interface lie under the CRZ (NIXON, 1982) or further north (EISSLER and McNALLY, 1984)? Is it a convergent (DeMETS and STEIN, 1990) or divergent (BANDY, 1992) boundary? The CRZ, which is roughly delimited at surface by the Armería and Cohuayana rivers (Figs. 1, 2), has been related to this boundary (BANDY, 1992; FERRARI et al., 1994; BANDY et al., 1995), or alternatively, to a spreading center location (LUHR et al., 1985). More recently GRAND et al. (2007) reported on results from teleseismic tomography that indicates a tear between the subducting Rivera and Cocos Plates along a trend that is at west of the CRZ, consistent with the proposal of BANDY et al. (1995). Thus, it is not clear what underlies the CRZ: a contact zone between the subducted Rivera and Cocos Plates, a tear zone, or a spreading center.

4. Data and Results

During the first 3 days following the mainshock (22–24 January, Period I) a group of 72 aftershocks recorded by RESJAL and RESCO were processed for accurate locations. For details of the analysis of this initial dataset the reader is referred to NÚÑEZ–CORNÚ et al. (2004). Here we present the analysis of the aftershocks recorded during 24–31 January (Period II) and compare it with that of Period I. Also for this period, SisVOc and RESCO, with the support of Colima Civil Defense, deployed a portable network (PN) with seven digital 3D seismographs to study the aftershocks; an analysis of these data is presented and compared with results from RESJAL and RESCO data.

For period II, 40 aftershocks could be located using RESJAL and RESCO data, employing the same criteria as for Period I. Figure 2a and b show maps of epicenters located during Period I and Period II, respectively. We observe that during Period II seismicity persisted in the area of the Armería cluster (AC), beneath the epicentral area of the mainshock of 22 January with a different depth distribution, while it almost disappeared from the 95 earthquake cluster (95C) (NÚÑEZ-CORNÚ et al., 2004). Figure 2c and d show the depth distribution, along a profile perpendicular to the trench; here we observe the change in depth distribution for the period II, Many of the aftershocks are deeper than during period I, and are vertically distributed 3 and 24 km depth. In Fig. 2e and f we present the distribution of these aftershocks along a profile parallel to the trench, we observe clearly that most of the seismicity is consistent with the AC cluster and shows an almost vertical distribution.

An analysis of the frequency-magnitude distribution (FMD) for periods I and II results in $b = 0.66$, which is consistent with average values estimated for crustal seismicity (FROLICH and DAVIS, 1993), the magnitude threshold for this data is 3.0 (Fig. 3).

With the deployed portable stations the detection capabilities were enhanced, and 200 aftershocks with magnitude ≥ 2.0 were located during Period II (this magnitude threshold allows us to include readings from RESCO and nearby RESJAL stations), the

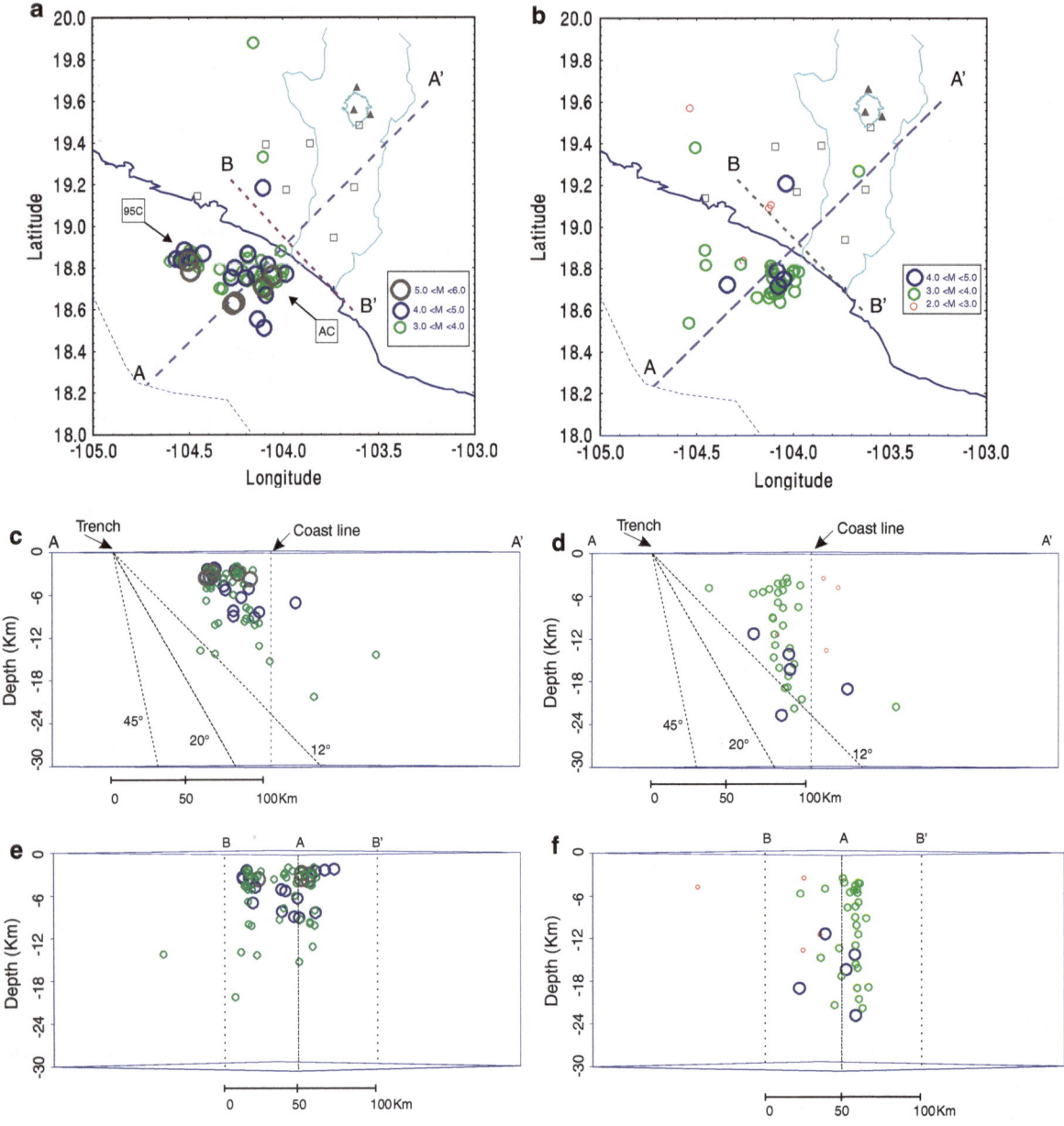

Figure 2

Locations of earthquakes recorded from 24 January to January 31 using different sets of data. *Top row* map views, *center row* profile perpendicular to the trench (*A–A'*), *bottom row* profile parallel to the trench (*B–B'*). *Circles* mark earthquakes locations. *Dashed line* outline of the Middle America trench. *Solid thin gray lines* Armería and Cohuayana rivers. **a, c, e** Locations obtained using arrival phase data from RESCO and RESJAL stations as in Núñez-Cornú *et al.* (2004). **b, d, f** locations obtained using arrival phase data from RESCO and RESJAL stations for Period II (June 24–31). *Open squares* RESCO stations, *solid triangles* RESJAL stations. *AC* Armería cluster; *95C* Earthquake 1995 cluster

velocity model used is the one proposed in Núñez-Cornú *et al.* (2004). To estimate the quality of the hypocentral locations, we show the aftershocks located with data from the RESCO and RESJAL networks (Fig. 2b, d, f) and the results with data from the PN, RESCO and RESJAL networks (Fig. 4a–c) using the same residual criteria (Núñez-Cornú *et al.*, 2004). No significant differences are observed in the

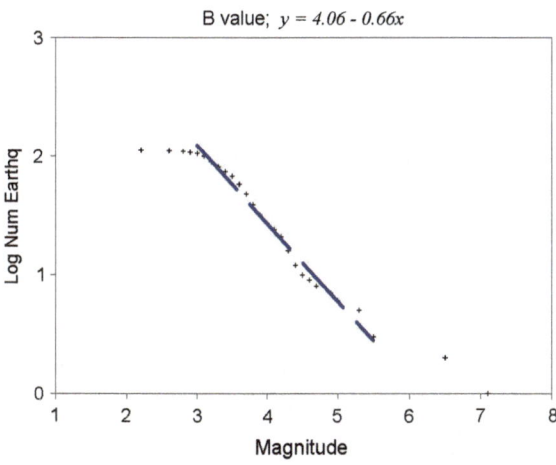

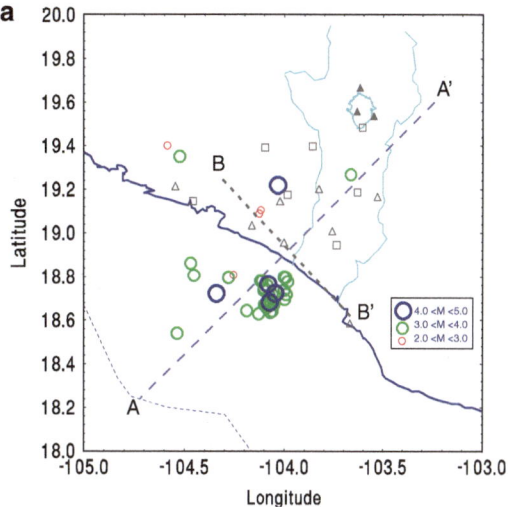

Figure 3

Frequency–magnitude distribution (FMD) for aftershocks recorded during periods I and II using RESJAL and RESCO data

epicentral distribution (Figs. 2b, 4a). However, the inclusion of the local stations shows a difference in the depth distribution which is observed among Fig. 4b and c, the depth range in Fig. 4b and c is between 6 and 21 km, instead of 3–24 km in Fig. 2d and f, we see that shallower events are 3–5 km deeper. Two hypocenter clusters can be observed, a shallower one between 6 and 10 km depth which corresponds to the zone of the AC cluster of the first 72 h aftershocks and a second, deeper one which locates between 12 and 21 km depth, along a hypothetical slab interface dipping 12°. The vertical alignment observed in Fig. 2c is also observed in Fig. 4c.

Figure 5a shows the epicentral distribution of the 200 aftershocks located with the portable array data, which is, in general, similar to Fig. 4a. In Fig. 5b, which shows a section perpendicular to the trench, we observe a hypocenter group that agrees with the AC cluster, a second larger group of the hypocenters locates between 12 and 21 km depth, along a hypothetical slab interface dipping 12°, another group locates roughly between 21 and 31 km depth with an almost vertical distribution that appears to be inside the slab, or perhaps delineating the slab tip. The hypocentral distribution in Fig. 5c is roughly the same that Fig. 4c.

To gain insight into the source properties of the Armería seismic sequence, we compute composite focal mechanism solutions using the computer

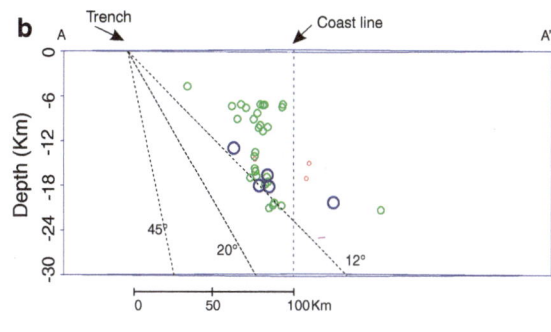

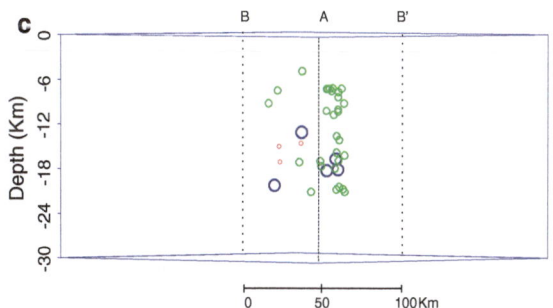

Figure 4

Locations of the same aftershocks in Fig. 2b, d, f, including Portable Network data. **a** Epicentral distribution. **b** Profile perpendicular to the trench (*A–A'*). **c** Profile parallel to the trench (*B–B'*). *Open squares* RESCO stations, *solid triangles* RESJAL stations, *open triangles* temporal network

program MEC93, a modified version of the computer program MECSTA (UDÍAS *et al.*, 1982), which use polarities of first arrivals and is based on the probabilistic algorithm proposed by BRILLINGER *et al.* (1980), we use the output data (azimuth and takeoff angle) from HYPO71 (LEE and LAHR, 1975). From a total of 272 events 115 were selected and grouped by

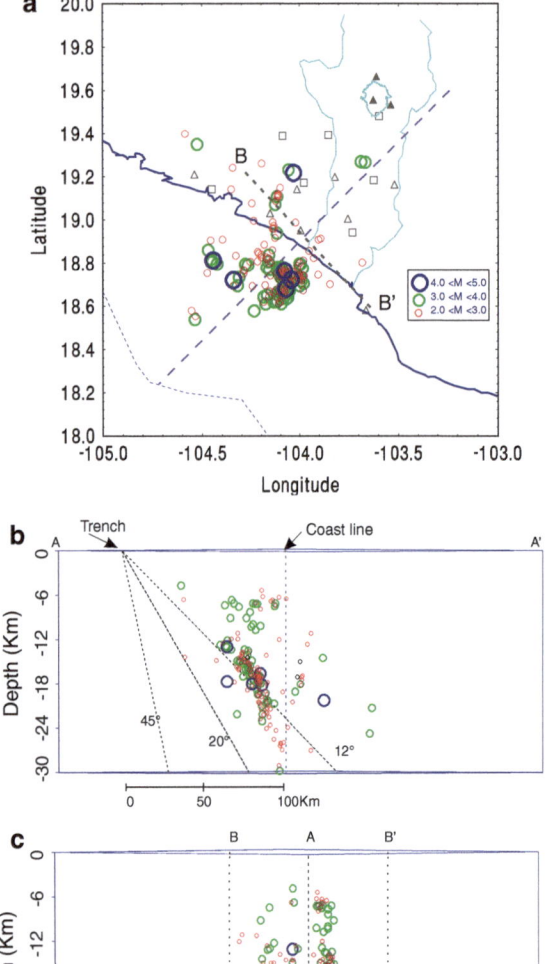

Figure 5

Aftershocks located using portable network seismic data for period 24–31 January **a** epicentral distribution. **b** Profile perpendicular to the trench (*A–A'*). **c** Profile parallel to the trench (*B–B'*). *Open squares* RESCO stations, *solid triangles* RESJAL stations, *open triangles* temporal network

the program into four solutions (Fig. 6): Group 1 composed of 25 earthquakes, Group 2 composed of 46 earthquakes (this solution is similar to one reported for the two principal aftershocks by Núñez-Cornú *et al.*, 2004), Group 3 composed of 35 earthquakes (this solution is similar to one reported for the Main Event by Núñez-Cornú *et al.*, 2004), and these groups are consistent with reverse fault solutions. The

fourth group with nine earthquakes is consistent with a normal fault solution.

The spatio-temporal distribution of aftershocks according to their focal mechanisms is shown in Fig. 7a–c, although no clear spatial pattern is observable (Fig. 7a, b), a possible variation with time of the composite reverse focal mechanism solutions is apparent from Fig. 7c. From this figure it is possible to observe that aftershocks of Group 1 took place between 22 January and 26 January; aftershocks of Group 2 between 26 January and 29 January and the events included in Group 3 between January 29 and January 31. Meanwhile shocks of Group 4 occurred sparsely from 22 January to 31 January.

5. Discussion and Conclusions

In a previous work (Núñez-Cornú *et al.*, 2004) it was found that during the early days (22–24 January) following the M_w 7.4 Armería mainshock, no important seismicity was located on the hypothetical interface between the Rivera and North American Plates that could support the suggestion of the Armería shock being a subduction earthquake breaking the Colima Gap. In this study, using the data from RESCO and RESJAL, one can observe a migration in depth of the aftershocks after 72 h.

The same conditions of station coverage and location procedure for the earthquakes in Period I were applied to a set of earthquakes recorded during the Period II. We found that the hypocentral distribution is different from one period to another with earthquakes during Period II locating at deeper levels within the crust and being aligned almost vertically, reaching the plate interface [assuming a dip of 12°] (Fig. 2d, f). The inclusion of the data from the PN stations improves the quality of the hypocentral depth of the locations; however, the location of AC cluster remains almost at the same depth. This fact supports that the main event was an intraplate continental earthquake.

We also used an expanded dataset to analyze the aftershock activity from 22 January to 31 January 2003, and found that: (1) Most of the aftershock activity ($M < 3.0$) is located on the plate interface (assuming a dip of 12°), and possibly delineates the

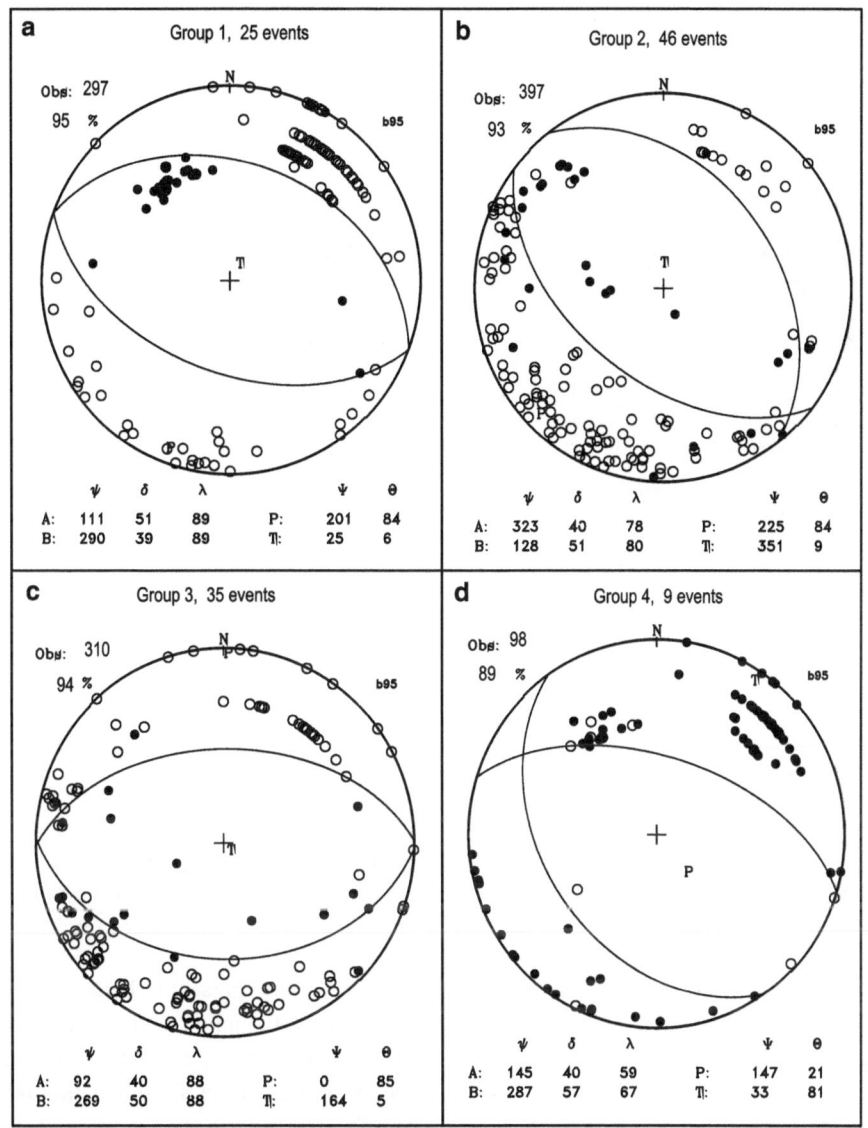

Figure 6
Focal mechanism composite solutions for aftershocks groups. **a** Group 1: earthquakes from 22 January to 26 January, **b** Group 2: earthquakes from 26 January to 29 January, **c** Group 3: earthquakes from 29 January to 31 January, and **d** Group 4: earthquakes from 22 January to 31 January. The group and the number of earthquakes used to compute each solution are indicated above the plots. Lower hemisphere projections are used. *Solid black circles* compressions, *hollow circles* dilatations. *T* tension axis, *P* pressure axis. *A* and *B* labels beneath each solution indicate the two planes and their orientations in degrees (ψ strike, δ dip, λ rake, Ψ azimuth, Θ plunge). *Obs* number of first motion polarities used. Percentages indicate values of *p* scores, which measure the fit of a set of observations with respect to the joint solution (BRILLINGER *et al.*, 1980)

tip of the slab, or a fault in it; (2) the vertical aftershock alignment observed in Figs. 2f, 4c and 5c suggests a feature coinciding with the mouth of Armería River, that could be related with the western border of Colima Graben; (3) the composite focal mechanisms for temporarily grouped subsets of earthquakes are mostly consistent with steeply dipping reverse faulting; and (4) there is an difference in the orientation of faulting as evidenced by the rotated fault plane solution for Group 3. This difference could indicate that the rupture at shallower levels during the mainshock and early aftershocks

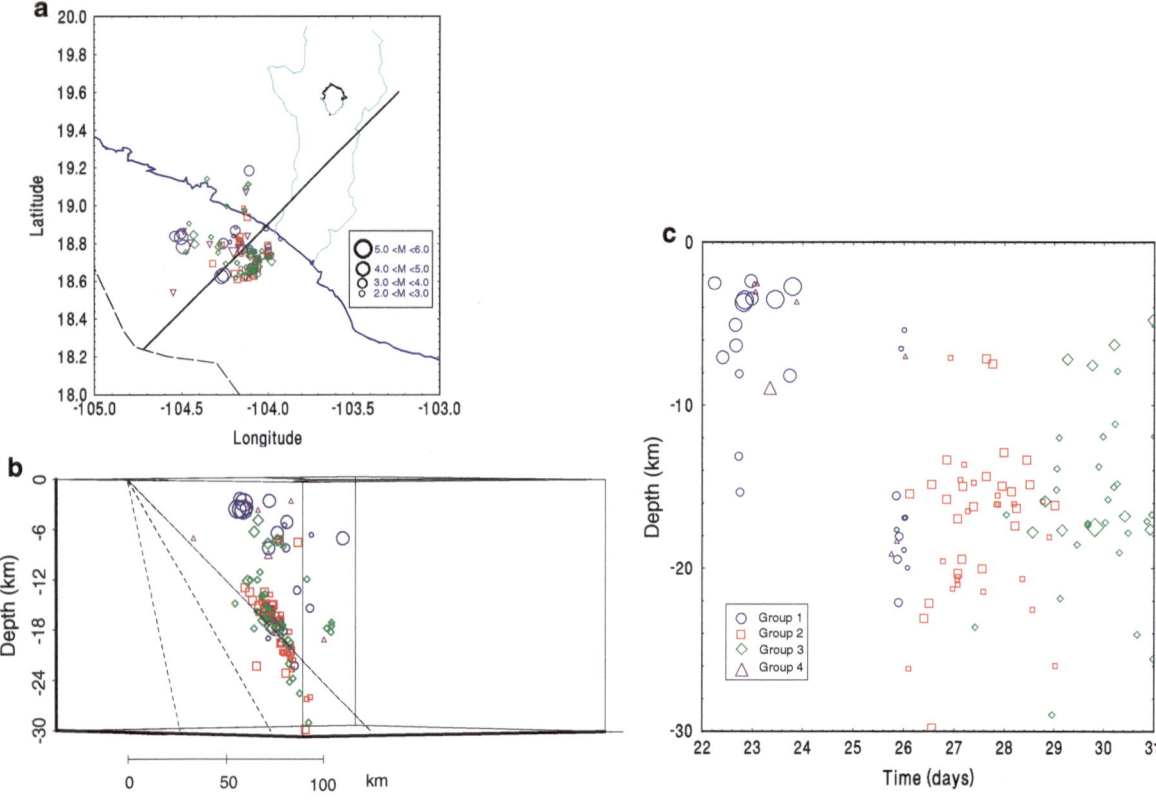

Figure 7
Spatial distribution for hypocenters of aftershocks according to their similar focal mechanism solutions. *Circles* Group 1, *squares* Group 2, *diamonds* Group 3, *triangles* Group 4. Sizes of symbols are proportional to magnitudes. **a** Epicentral distribution, **b** distribution along a profile perpendicular to the trench. **c** Temporal against depth distribution of aftershocks. Conventions as in **a**

proceeded, and possibly contributed to, faulting at deeper levels which may have reached the plate interface and penetrated into the subducted Rivera Plate.

The composite fault plane solutions computed for groups of aftershocks indicate mostly reverse faulting on likely steep planes (dipping between 39° and 51°) and oriented NW–SE or roughly E–W. Given that no surface rupture, cracks, or liquefaction could be documented in the epicentral area during the main-shock, we cannot ascertain the precise orientation of the associated fault. The apparent switch in orienta-tion and location of faulting with time (Fig. 7c) could indicate highly heterogeneous conditions in the upper 20 km of the crust in the region surrounding the epicentral area of the Armería earthquake, as well as short-term temporal variability of frictional resistance and stress loading on the faults (DEICHMANN and GARCÍA-FERNANDEZ, 1992).

The results presented here indicate that the 2003 Armería earthquake occurred on a continental intra-plate reverse fault, in contrast to other works that favor faulting along the subduction interface between the Rivera and North American Plates (YAGI *et al.*, 2004; SINGH *et al.*, 2003). We suggest that the Arm-ería shock and its aftershocks could represent partial accommodation of deformation in the continental crust caused by oblique subduction.

The 2003 Armería, Mexico earthquake is the largest earthquake in the Jalisco region during the last 14 years, and the damage pattern was quite different to the damage due to the 1995 Earthquake. This earthquake has been the focus of attention because there is the open question as to what it could repre-sent in terms of the seismotectonics of the region and the associated seismic hazard. In this work we show results from an extended dataset with accurate loca-tion of aftershocks (recorded during the first 9 days following the mainshock), and fault plane modeling from local P wave arrivals and find that the observations are consistent with a rupture along a steeply dipping plane that locates mainly above any projection of a subducting interface within a wide range of dip angles. We suggest that the Armería earthquake represents slip on a fault in the continental crust. Our observations do not support rupture of the Colima Gap, but point to disruption of a region within the 1995 earthquake rupture zone. This has implications for seismic hazard because the Armería earthquake could have caused stress changes and possibly loading of a region within the largely unaffected Colima Gap. This is a matter for future research. We note that the epicentral region of the 2003 Armería earthquake shows activity that is consistent with shallow crustal seismicity. No triggered seismicity effects were reported by RESCO and RESJAL outside the epicentral region, including the active Colima Volcano.

Acknowledgments

The authors wish to thank comments and revisions by anonymous reviewers which greatly improved the manuscript. Support from Congreso del Estado de Jalisco (C. Dip. Jesus García) for field work is acknowledged. Digital Seismograms used in this study were collected as part of joint projects of University of Guadalajara and PCJal. Additional RESCO data was kindly provided by G. Reyes-Davila.

References

ALLAN, J. F., NELSON, J., LUHR, J., CARMICHAEL, J., WOPAT, M., and WALLACE P., Pliocene-Holocene rifting and associated volcanism in southwest Mexico: An exotic terrain in the making. In *The Gulf and Peninsular Provinces of the Californias*, vol 47, (ed. J. P. Dauphin and R. R. T. Simoneit) (American Association Petroleum Geologists, Tulsa, Oklahoma, 425–445, 1991).

BANDY, W. L. (1992), *Geological and geophysical investigation of the Rivera-Cocos Plate boundary: implications for plate fragmentation*, Ph.D. Thesis, Texas A&M University, College Station, Texas.

BANDY, W. L., MORTERA-GUTIERREZ, C., URRUTIA-FUCUGAUCHI, J., and HILDE, T. W. C. (1995), *The subducted Rivera-Cocos plate boundary: Where is it, what is it, and what is its relationship to the Colima rift?* Geophys. Res. Lett. *22*, 3075–3078.

BRILLINGER, D., UDÍAS, A., and BOLT, B. (1980), *A probability model for regional focal mechanism solutions*, Bull. Seism. Soc. Am. *70*, 1121–1133.

BOURGOIS J., RENARD, D., AUBOIN, J., BANDY, W., BARRIER, E., CALMUS, T., CARFANTAN, J. C., GUERRERO, J., MAMMERICKX, J., MERCIER DE LEPINAY, B., MICHAUD, F., and SOSSON, R. (1988). *Fragmentation en cours du bord continental nord américain: les frontières sous marines du bloc de Jalisco (Mexique)*. Campagne Seamat du N/O Jean Charcot, Juin–Juillet 1987, C. R. Acad. Sci. Paris *307*(II), 1121–1130.

DAÑOBEITIA, J. J., CORDOBA, D., DELGADO-ARGOTE, L. A., MICHAUD, F., BARTOLOMÉ, R., FARRAN, M., NÚÑEZ-CORNÚ, F., and CARBONELL, R. (1997), *Expedition gathers new data on crust beneath Mexican west coast*, EOS Trans. AGU 78, 565.

DEICHMANN, N. and GARCÍA-FERNANDEZ, M. (1992), *Rupture geometry from high-precision relative hypocenter locations of microearthquake clusters*, Geophys. J. Int. *110*, 501–517.

DEMETS, C. and STEIN, S. (1990), *Present-day kinematics of the Rivera plate and implications for tectonics in southwestern Mexico*, J. Geophys. Res. 95, 21931–21948.

EISSLER H. and MCNALLY, K. C. (1984), *Seismicity and tectonics of the Rivera Plate and implications for the 1932 Jalisco, Mexico, earthquake*, J. Geophys. Res. 89, 4520–4530.

FERRARI, L., PASQUARÉ, G., VENEGAS, S., CASTILLO, D., and ROMERO, F. (1994). *Regional tectonics of western Mexico and its implications for the northern boundary of Jalisco Block*. Geofís Inter *33*, 139–151.

FROLICH, C. and DAVIS, S. (1993), *Teleseismic b values: or much ado about 1.0*, J. Geophys. Res. 98, 631–634.

GARDUÑO V. H. and TIBALDI, A. (1991), *Kinematic evolution of the continental active triple junction of the western Mexican Volcanic Belt*, C. R. Acad. Sci. Paris *307*, 135–142.

GRAND, S. P., YANG, T., WILSON, D., GUZMÁN SPEZIALE, M., GÓMEZ GONZÁLEZ, J., DOMÍNGUEZ-REYES, T., and NI, J. (2007), *Seismic structure of the Rivera subduction zone*, EOS Trans. AGU 88 (Fall Meet. Suppl.), abstract T41C-0702.

LEE, W. H. K and LAHR, J. C. (1975), *HYPO71 (revised): a computer program for determining hypocenter, magnitude, and first motion pattern of local earthquakes*, U.S. Geol. Surv. Open-File Rept. 75–311, 59 pp.

LUHR, J., NELSON, S., ALLAN, J., and CARMICHAEL, I. (1985), *Active rifting in southwestern Mexico: Manifestations of an incipient eastward spreading-ridge jump*, Geology *13*, 54–57.

NIXON, G. T. (1982), *The relationship between Quaternary volcanism in central Mexico and the seismicity and structure of subducted ocean lithosphere*, Geol. Soc. Am. Bull. *98*, 514–523.

NÚÑEZ-CORNÚ, F. J. and SÁNCHEZ-MORA, C. (1999), *Stress field estimations for Colima Volcano, Mexico based on Seismic data*, Bull. Volc. *60*, 568–580.

NÚÑEZ-CORNÚ, F., REYES-DÁVILA, G., SUÁREZ-PLASCENCIA, C., GONZÁLEZ-LEDEZMA, M., and GARCÍA-PUGA, J. (2001), *The Jalisco Seismic Telemetric Network*, EOS Trans. AGU *82*, F821.

NÚÑEZ-CORNÚ, F. J., RUTZ, M., NAVA, F. A., REYES-DAVILA, G., and SUÁREZ-PLASCENCIA, C. (2002), *Characteristics of Seismicity in* the Coast and North of Jalisco Block, Mexico, Phys. Earth Planet. Int. *132*, 141–155.

NÚÑEZ-CORNÚ, F. J., RUTZ, M., SUÁREZ-PLASCENCIA, C., REYES-DÁVILA, G., and NAVA, F. A. (2003), Seismotectonics of Jalisco Block, Mexico, Conference proceedings of The Geological Society of America annual meeting, Cordilleran Section, *35*(4), 71–72.

NÚÑEZ-CORNÚ, F. J., REYES-DÁVILA, G. A., RUTZ, M., TREJO-GÓMEZ, E., CAMARENA-GARCÍA, M. A., and RAMÍREZ-VAZQUEZ, C. A. (2004), *The 2003 Armería, Mexico earthquake (Mw 7.4): Mainshock and early aftershocks*, Seism. Res. Lett. *75*, 734–743.

ROSAS-ELGUERA, J., FERRARI, L., GARDUÑO-MONROY, V., and URRUTIA-FUCUGAUCHI, J. (1996), *Continental boundaries of the Jalisco Block in the Pliocene-Quaternary kinematics of western Mexico*, Geology, *24*, 921–924.

SINGH, S. K., PONCE, L., and NINSHENKO, S. P. (1985), *The great Jalisco, Mexico earthquakes of 1932: Subduction of the Rivera plate*, Bull. Seism. Soc. Am. *75*, 1301–1313.

SINGH, S. K., PACHECO, J. F., ALCÁNTARA, L., REYES, G., ORDÁZ, M., IGLESIAS, A., ALCOCER, S. M., GUTIÉRREZ, C., VALDÉS, C., and KOSTOGLODOV, V. (2003), *A preliminary report on the Tecomán, Mexico earthquake of 22 January 2003 (Mw 7.4) and its effects*, Seism. Res. Lett. *74*(3), 279–289.

UDÍAS, A., BUFFORN, E., BRILLINGER, D., and BOLT, B. (1982), *Joint statistical determination of fault-plane parameters*, Phys. Earth Planet. Int. *30*, 178–184.

YAGI, Y., MIKUMO, T., PACHECO, J., and REYES, G. (2004), *Source Rupture Process of the Tecoman, Colima, Mexico Earthquake of 22 January 2003, determined by Joint Inversion of Teleseismic Body-Wave and Near Source Data*, Bull. Seism. Soc. Am. *94*(5), 1795–1807.

(Received January 6, 2010, revised June 16, 2010, accepted June 25, 2010, Published online July 30, 2010)

Pure Appl. Geophys. 168 (2011), 1303–1330
© 2010 Springer Basel AG
DOI 10.1007/s00024-010-0204-z

Volcanic Markers of the Post-Subduction Evolution of Baja California and Sonora, Mexico: Slab Tearing Versus Lithospheric Rupture of the Gulf of California

THIERRY CALMUS,[1] CARLOS PALLARES,[1,2,3] RENÉ C. MAURY,[2,3] ALFREDO AGUILLÓN-ROBLES,[4] HERVÉ BELLON,[2,3] MATHIEU BENOIT,[5] and FRANÇOIS MICHAUD[6]

Abstract—The study of the geochemical compositions and K-Ar or Ar-Ar ages of ca. 350 Neogene and Quaternary lavas from Baja California, the Gulf of California and Sonora allows us to discuss the nature of their mantle or crustal sources, the conditions of their melting and the tectonic regime prevailing during their genesis and emplacement. Nine petrographic/geochemical groups are distinguished: "regular" calc-alkaline lavas; adakites; magnesian andesites and related basalts and basaltic andesites; niobium-enriched basalts; alkali basalts and trachybasalts; oceanic (MORB-type) basalts; tholeiitic/transitional basalts and basaltic andesites; peralkaline rhyolites (comendites); and icelandites. We show that the spatial and temporal distribution of these lava types provides constraints on their sources and the geodynamic setting controlling their partial melting. Three successive stages are distinguished. Between 23 and 13 Ma, calc-alkaline lavas linked to the subduction of the Pacific-Farallon plate formed the Comondú and central coast of the Sonora volcanic arc. In the extensional domain of western Sonora, lithospheric mantle-derived tholeiitic to transitional basalts and basaltic andesites were emplaced within the southern extension of the Basin and Range province. The end of the Farallon subduction was marked by the emplacement of much more complex Middle to Late Miocene volcanic associations, between 13 and 7 Ma. Calc-alkaline activity became sporadic and was replaced by unusual post-subduction magma types including adakites, niobium-enriched basalts, magnesian andesites, comendites and icelandites. The spatial and temporal distribution of these lavas is consistent with the development of a slab tear, evolving into a 200-km-wide slab window sub-parallel to the trench, and extending from the Pacific coast of Baja California to coastal Sonora. Tholeiitic, transitional and alkali basalts of subslab origin ascended through this window, and adakites derived from the partial melting of its upper lip, relatively close to the trench. Calc-alkaline lavas, magnesian andesites and niobium-enriched basalts formed from hydrous melting of the supraslab mantle triggered by the uprise of hot Pacific asthenosphere through the window. During the Plio-Quaternary, the "no-slab" regime following the sinking of the old part of the Farallon plate within the deep mantle allowed the emplacement of alkali and tholeiitic/transitional basalts of deep asthenospheric origin in Baja California and Sonora. The lithospheric rupture connected with the opening of the Gulf of California generated a high thermal regime associated to asthenospheric uprise and emplaced Quaternary depleted MORB-type tholeiites. This thermal regime also induced partial melting of the thinned lithospheric mantle of the Gulf area, generating calc-alkaline lavas as well as adakites derived from slivers of oceanic crust incorporated within this mantle.

Key words: Slab tearing, slab melting, ridge-trench collision, adakite, basalt, comendite, magnesian andesite, asthenospheric window, basin and range, Gulf of California, Baja California, Sonora, México.

[1] Estación Regional del Noroeste, Instituto de Geología, Universidad Nacional Autónoma de México, C.P 83000 Hermosillo, Sonora, Mexico. E-mail: tcalmus@servidor.unam.mx

[2] Université Européenne de Bretagne, Université de Brest, Brest Cedex 3, France.

[3] CNRS, UMR 6538 Domaines Océaniques, Institut Universitaire Européen de la Mer, Place N. Copernic, 29280 Plouzané, France.

[4] Instituto de Geología, UASLP, Av. Dr. Manuel Nava no. 5, Zona Universitaria, San Luis Potosí, S.L.P, C.P 78250 Mexico, Mexico.

[5] UMR 5562, OMP, Université Paul Sabatier, 14 Avenue Edouard Belin, 31400 Toulouse, France.

[6] UMR 6526, Géosciences Azur, Université Pierre et Marie Curie, 06235 Villefranche sur Mer, France.

1. Introduction

The geochemical (major, trace elements and isotopic) compositions of fresh magmatic rocks are mostly inherited from those of their source materials during partial melting, although they may have been modified later by intracrustal petrogenetic processes such as fractional crystallization coupled or not with assimilation of host rocks, or magma mixing. On one hand, experimental studies allow the petrologist to take into account the geochemical effects linked to variable source mineralogy, temperature, pressure and melting rate on the composition of the melts. On the other hand, the presence of a given source at depth and the physical conditions governing its

partial melting are controlled by the regional geodynamic setting. Magmatic rocks are thus potential
markers of the tectonic regime prevailing during their
emplacement.

The Neogene and Quaternary geological history of
Baja California, Sonora and the Gulf of California has
been marked by the almost continuous emplacement
of volcanic rocks showing an exceptional geochemical
diversity (Gastil et al., 1979; Sawlan, 1991; Benoit
et al., 2002). Mafic lavas encompass the whole range
of basaltic compositions, from depleted mid-oceanic
ridge basalts (MORB) to plume-type alkali basalts,
through various kinds of tholeiitic, transitional and
calc-alkaline basalts and the very rare niobium-enriched basalts (NEB: Aguillón-Robles et al., 2001).
Intermediate and evolved lavas are also highly diver-
sified. In addition to the types commonly found in
calc-alkaline series, they include unusual rocks such as
magnesian andesites (Saunders et al., 1987; Calmus
et al., 2003), adakites (Aguillón-Robles et al., 2001;
Calmus et al., 2008), icelandites and peralkaline rhyolites (Vidal-Solano et al., 2008a, b).

A majority of authors have considered this geochemical diversity as resulting from the partial melting
of contrasted mantle and crustal sources, during the
complex tectonic evolution of the Pacific margin,
which followed the end of the subduction of the
Farallon oceanic plate around 12.5 Ma. In Baja
California, the wide range of erupted magmas is
generally attributed to the opening of an asthenospheric window, although the details of the process are
debated: for instance, the source of adakites is thought
to be either the subducted Farallon crust or the mafic
base of the continental crust (see Pallares et al., 2007,
2008; Castillo, 2008, 2009; Maury et al., 2009, and
references therein). In Sonora, the association of tholeiitic to transitional basalts (temporally evolving
towards alkali basalts) with icelandites and peralkaline
rhyolites is linked to the transition from a typical Basin
and Range regime to rift opening in the nearby Gulf of
California (Vidal-Solano et al., 2008a, b).

However, a rather different point of view has been
developed in two recent articles. Negrete-Aranda
and Cañón-Tapia (2008) consider that a stalled
Farallon slab is still present beneath Baja California,
and that the post-subduction magmas originated from
sources located in the mantle wedge or the overlying

continental crust. These authors claim that the partial
melting of these sources was due to the thermal
rebound following the end of subduction, and that the
temporal and spatial distribution of post-subduction
lavas resulted from local tectonic features like the
stress field and the tensile strength of the Baja
California crustal rocks. Till et al. (2009) consider all the
Miocene volcanism in Sonora as subduction-related
(continental arc type), and find only subtle geochemical changes (slight variations of incompatible
element ratios, e.g., La/Nb) concomitant with the
ridge-trench collision off Baja California at 12.5 Ma.
They suggest that the subduction signature of the sub-
arc Sonoran mantle was not erased 4 m.y. after the
end of subduction, and thus that what they call
"petrotectonic modeling" is a perilous exercise.

Numerous good quality geochemical analyses of
K-Ar and/or Ar-Ar dated lavas from Baja California
Peninsula, the Gulf of California islands and coastal
Sonora have been published during the last 10 years.
The purpose of this paper is to review them, to dis-
cuss their implications on the nature of the magmatic
sources at depth and finally to examine critically the
constraints that they may provide on the tectonic
evolution of the Pacific margin of northwestern
México.

2. Tectonic Framework

The Middle Miocene to Recent tectonic and
magmatic evolution of northwestern Mexico is closely
related to the transition between subduction regime
and the opening of the Gulf of California. After the
Pacific-Farallon ridge entered the trench at the latitude
of present-day Los Angeles, the Rivera triple junction
migrated progressively to the south, until the eastwards subduction of the Farallon plate and subsequent
microplates below the North America plate ended
between 12.5 and 12.3 Ma with the capture of microplates by the Pacific plate (Lonsdale, 1991). This is
the case north of the Shirley transform fault, when the
seafloor spreading between the Pacific and Farallon
plates stopped after Chron 5AB (Lonsdale, 1991) and
more precisely during the younger part of Chron 5A
(Dyment, 2003) along the Guadalupe ridge. South of
the Shirley transform fault, this capture was

progressive until 8–7 Ma, the period during which the Pacific-Magdalena ridge experienced a break into several segments, together with a ~50° clockwise rotation (MICHAUD *et al.*, 2006). After that rotation, the direction of demising seafloor spreading along the ridge segments was closely parallel to the margin, which suggests (1) that the spreading centers' segments accommodated the main part of the transcurrent motion between the Pacific and North America plates before 8–7 Ma (MICHAUD *et al.*, 2006), and (2) that the onset of the activity of the San Benito-Tosco Abreojos fault zone (SPENCER and NORMARK, 1989) probably occurred at that time.

The limit between the Pacific and North America plates was located along the Tosco-Abreojos and San Benito fault zones from 8 to 7 Ma until ca. 6 Ma, when the transtensional regime in the Gulf of California became established. The displacement along the Tosco-Abreojos-San Benito fault system is evaluated to 350–400 km, which has been the offset necessary to complete the 650–700 km of the northwest relative motion of Pacific plate with respect to North America since 12.3 Ma (ATWATER and STOCK, 1998), after restoring the offset of 276 km accumulated within the Gulf of California since 6.3 Ma (OSKIN and STOCK, 2003). Based on the provenance data of detrital zircons, FLETCHER *et al.* (2007) concluded that the dextral slip along the Tosco-Abreojos fault was less that 150 km, and that the main transform boundary between Pacific and North America plates was the Gulf of California, between 12.5 and 6 Ma. In both cases, west of the Main Gulf Escarpement (MGE), Baja California Peninsula is considered to be stable between the Upper Miocene and Present. No major fault is known between Tosco-Abreojos fault and the MGE. East verging normal faults of the western Los Cabos block belong to the rifting structures of the Gulf Extensional Province (GEP). Minor faults are associated with some volcanic fields of Baja California Sur such as La Purisima (BELLON *et al.*, 2006). The adakitic domes and dykes of Santa Clara volcanic field are aligned along a NW-SE direction, subparallel to the paleotrench off Vizcaino Peninsula.

On the continent, prior to the opening of the Gulf of California, we can distinguish two morpho-tectonic regions that were inherited from the Paleogene and Early Miocene geologic evolution, marked by the subduction of the Farallon plate: (1) To the east, a large Basin and Range province, which is part of the Southern Basin and Range province (SONDER and JONES, 1999). It is divided into three NNW-SSE trending areas (e.g., HENRY, 1989); an eastern Basin and Range subprovince that extends from the Rio Grande Rift to the Trans-Mexican Volcanic Belt; a western Basin and Range province extending between the western limit of the Sierra Madre Occidental (SMO) and the Main Gulf Escarpment (MGE); and finally the Sierra Madre Occidental, a huge Oligocene volcanic belt, corresponding to a relatively unextended domain located between the western and eastern Basin and Range subprovinces, respectively. (2) West of the MGE is the Peninsular Ranges belt whose backbone is composed mainly by plutons of the Cretaceous magmatic arc, together with some accreted terranes along the western margin. This zone is not affected by Basin and Range extension. Based on the previous morphotectonic distinction, we conclude that the MGE coincides with the western limit of the Basin and Range province. Henry (1989, his Fig. 7) reported that the region surrounding the Gulf of California experienced intense faulting during Basin and Range extension, before the opening of the Gulf of California. He also considered that normal faults aligned with the MGE represent the western limit of the Basin and Range Province.

A majority of authors agree to consider that the NE-SW extension observed along and east of the MGE is related the opening of the Gulf of California (e.g., STOCK and HODGES, 1989). Nevertheless, in the southern Sierra Juarez, along the present eastern coast of Baja California, LEE *et al.* (1996) interpreted west-dipping normal faulting, which occurred between 15.98 ± 0.13 and 10.96 ± 0.05 Ma, as the first east-west extension of the Gulf Extensional Province (GEP). That age is coeval with some ages of Basin and Range extension determined in Sonora, and it is thus necessary to distinguish the extension related to Basin and Range from the extension due to lithosphere breakup at the beginning of the opening of the Gulf of California. That distinction has been also questioned by DOKKA and MERRIAM (1982) for the region of Puertecitos, and STOCK and HODGES (1989) for the whole Gulf region. These latter authors propose that the eastern limit of the GEP coincides with the western limit of the SMO, but recognize at the same time that the limit of the GEP is not well defined in Sonora, due to the lack of reliable data. In the work of STOCK and HODGES (1989) as well as in many others papers, the use of the acronym GEP is accompanied by a reference to the classical study of the geology of the state of Baja California, Mexico, by GASTIL et al. (1975). Nevertheless, a detailed lecture of that exhaustive work does not show any evidence for a precise description of such a morphologic or tectonic province. GASTIL et al. (1975) present the Gulf of California depression (p. 76 and 131) as a structural province limited by the coast of Sonora and the MGE in Baja California, probably reported for the first time by GABB (1882) as "an enormous fault" along the coastline between La Paz and Mulege. Later maps of the GEP (LEE et al., 1996; TILL et al., 2009 and many others) refer to the Fig. 1 of GASTIL et al. (1975) where the authors presented the Basin and Range province (and not a hypothetical GEP) as extending in northwestern Mexico from the Sierra Madre Occidental western escarpment to the east to the MGE to the west. HENRY and ARANDA GÓMEZ (2000) presented a new evaluation of the 12-6-Ma extension around the Gulf of California, and concluded that it occurred probably throughout the southern Basin and Range province, including the eastern Basin and Range, east of the SMO. In the case of Sonora, they reported tilted volcanic rocks younger than 12-10 Ma, but the highest angles of dip were observed in Sierra Santa Ursula, close to the Gulf of California. Following ROLDÁN-QUINTANA et al. (2004) and CALMUS et al. (1997), we will consider that the eastern limit of the GEP in Sonora might correspond to the Empalme graben and its continuation toward the north along the Hermosillo graben. To the east of that limit, minor tilting and extension could be associated to Late Miocene waning Basin and Range extension within a Miocene Extensional Arc Province as suggested by GANS (1997).

3. Lava Types: Their Occurrences, Specific Geochemical Features and Mantle/Crustal Sources

3.1. Database and Classification

We have compiled ca. 350 chemical analyses of ^{40}K-^{40}Ar or ^{40}Ar-^{39}Ar dated Neogene and Quaternary lavas for which a large set of major and trace element data, mostly obtained by Inductively Coupled Plasma-Atomic Emission Spectrometry (ICP-AES) and Inductively Coupled Plasma-Mass Spectrometry (ICP-MS), is available. Our main source of data for Baja California is a set of papers (AGUILLÓN-ROBLES et al., 2001; BENOIT et al., 2002; CALMUS et al., 2003; BELLON et al., 2006; PALLARES et al., 2007, 2008; CALMUS et al., 2008) for which the analytical techniques are described by COTTEN et al. (1995). For coastal Sonora, we have mostly used the set of data of VIDAL-SOLANO et al. (2005, 2008a, b) and TILL et al. (2009). Selected analyses of dated lavas are reported in Table 1.

Calc-alkaline lavas showing a typical subduction-related geochemical signatures have been classified according to their K_2O and SiO_2 contents (PECCERILLO AND TAYLOR, 1976), and other lavas according to the TAS diagram (LE BAS et al., 1986). Transitional basalts are defined according to MIDDLEMOST'S (1975) criteria, and adakites according to the Sr/Y versus Y plot of DEFANT and DRUMMOND (1990). Niobium-enriched basalts (NEB) are named after their original description in Santa Clara volcanic field, Baja California (AGUILLÓN-ROBLES et al., 2001), although their definition and relationships with alkali basalts are still a matter of debate (CASTILLO, 2008, 2009; MAURY et al., 2009).

3.2. "Regular" Calc-Alkaline Lavas

Calc-alkaline lavas form the bulk of the Comondú Late Oligocene to Middle Miocene calc-alkaline belt (GASTIL et al., 1979; HAUSBACK, 1984; UMHOEFER et al., 2001). It extends all along the eastern part of the Baja California Peninsula from 32°N to 24°N (Fig. 1), mostly as coalescent stratovolcanoes along its main ridge, and also forms the bulk of volcanic cover of Isla Tiburón and a number of sierras close to the coast of central Sonora. Medium-K to high-K andesites are prominent, with subordinate dacites, basaltic andesites and basalts. These lavas are highly porphyritic, with pheocrysts of plagioclase (labradorite), clinopyroxene (augite, diopside) and olivine altered to iddingsite, together with occasional crystals of hornblende, titanomagnetite and orthopyroxene. They display a characteristic "subduction-related"

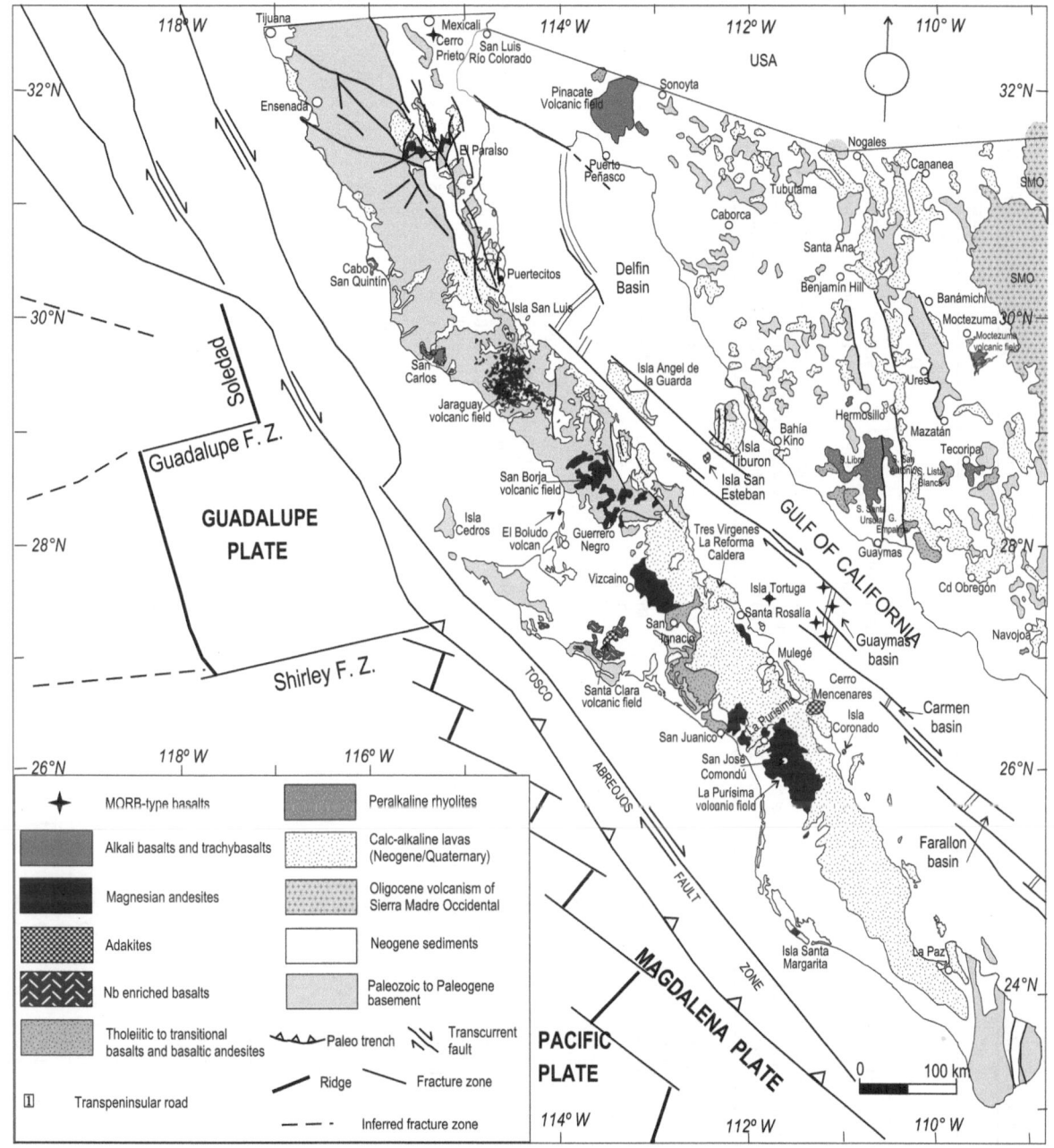

Figure 1

Geological sketch map of Baja California and Sonora, showing the major Neogene and Quaternary volcanic fields and the distribution of the main lava types. The regional geology is modified from ORTEGA-GUTIÉRREZ et al. (1992), CALMUS et al. (2003, 2008), PALLARES et al. (2007) and CONSEJO DE RECURSOS MINERALES (CRM) (1992)

incompatible element signature (an. 1–4, Table 1), with high ratios of large ion lithophile elements (LILE) such as Ba and Sr over high field strength elements (HFSE) and typical depletion of the latter, resulting in negative Nb and Ti anomalies and positive Nb spikes in their multi-element patterns (Fig. 2a). The fractionated character of their rare earth element (REE) patterns is mostly due to their high contents of light REE (LREE; e.g., La, Ce). Unlike adakites and related rocks, they are not

Table 1

Selected major and trace element analyses and K-Ar ages of Neogene and Quaternary lavas from Baja California, the Gulf of California and coastal Sonora

Lava type	Calc-alkaline lavas				Adakites				Magnesian basalts,	
Number	1	2	3	4	5	6	7	8	9	10
Rock type	BA	B	B	BA	D	D	A	D	B	A
Sample	BC05-42	BC05-40	SE96-10	TV96-14	BC99-122	BC05-29A	SE9601	SE9627	BC0539	BC99-65
Volc. field	S. Borja	S. Borja	S. Esteban	Tres Virgenes	S. Clara	Jaraguay	S. Esteban	S. Esteban	S. Borja	La Purisima
Age (Ma)	14.64 ± 0.46	10.93 ± 0.32	4.01 ± 0.18	0.00	10.50 ± 0.40	9.40 ± 0.33	3.44 ± 0.19	2.70 ± 0.14	10.93 ± 0.34	9.67 ± 0.25
Reference	PALLARES et al., 2007	PALLARES et al., 2007	CALMUS et al., 2008	AGUILLÓN-ROBLES, 2002	AGUILLÓN-ROBLES et al., 2001	PALLARES et al., 2007	CALMUS et al., 2008	CALMUS et al., 2008	PALLARES et al., 2008	BENOIT et al., 2002
Major elements (wt. %)										
SiO_2	54.50	50.50	52.90	52.60	69.00	62.80	61.50	64.70	48.20	56.70
TiO_2	0.84	1.53	1.38	1.30	0.31	0.52	0.85	0.58	1.7	1.52
Al_2O_3	15.20	14.90	17.45	17.35	15.65	16.00	16.70	16.50	14.5	14.62
$Fe_2O_3^a$	7.22	9.65	7.44	7.77	1.95	3.50	5.07	4.11	9.13	6.37
MnO	0.12	0.15	0.13	0.12	0.03	0.05	0.10	0.07	0.14	0.08
MgO	6.64	7.76	4.60	6.05	1.53	3.48	2.50	1.82	8.55	5.68
CaO	8.62	9.75	8.60	9.75	3.15	5.17	5.46	4.85	9.2	6.52
Na_2O	3.37	3.28	4.24	3.86	5.58	4.21	4.50	4.57	3.97	4.15
K_2O	1.38	1.21	0.68	0.73	0.92	1.68	2.16	1.67	1.36	3.01
P_2O_5	0.26	0.44	0.26	0.31	0.10	0.16	0.37	0.20	0.75	0.67
LOI	1.40	0.71	2.33	0.01	1.32	1.78	1.05	0.64	1.65	0.49
Total	99.55	99.88	100.01	99.85	99.54	99.35	100.26	99.71	99.15	99.81
Normative minerals (CIPW)										
Q	3.87	0.00	1.34	0.00	24.06	17.55	12.15	18.61	0.00	2.07
ne	0.00	0.00	0.00	0.00	0.00	0.00	0.00	0.00	3.21	0.00
hy	14.67	8.80	12.48	11.89	5.33	7.13	8.20	6.60	0.00	12.33
Mg^b	68.18	65.20	60.87	66.87	67.87	69.85	56.22	54.40	64.99	70.60
Trace elements (ppm)										
Rb	21.30	14.50	5.40	5.70	10.60	22.00	23.00	22.30	20.00	21.50
Ba	770.00	688.00	316.00	315.00	214.00	770.00	1150.00	716.00	1150.00	1395.00
Th	2.75	4.85	1.80	0.60	1.20	4.15	3.45	2.90	9.90	3.80
Nb	7.00	9.70	5.60	4.40	1.80	2.10	4.90	4.40	11.60	13.50
La	19.50	26.00	14.80	11.30	8.00	13.80	24.50	16.50	53.00	56.00
Ce	41.00	58.00	35.00	28.00	18.50	29.00	55.00	33.00	120.00	110.00
Sr	570.00	670.00	565.00	1030.00	705.00	982.00	1153.00	640.00	1170.00	2070.00
Nd	20.50	32.00	20.00	16.80	10.60	15.00	27.00	18.50	60.00	48.00
Sm	4.20	6.40	4.75	3.60	2.50	2.85	5.35	3.15	10.70	7.70
Zr	160.00	181.00	147.00	91.00	28.00	124.00	156.00	132.00	246.00	232.00
Eu	1.25	1.80	1.50	1.23	0.66	0.83	1.31	0.88	2.80	1.87
Gd	4.10	5.70	4.80	3.50	1.95	2.40	3.50	2.50		

Table 1 *continued*

	Calc-alkaline lavas				Adakites				Magnesian basalts,	
Number	1	2	3	4	5	6	7	8	9	10
Rock type	BA	B	B	BA	E	D	A	D	B	A
Sample	BC05-42	BC05-40	SE96-10	TV96-14	BC99-122	BC05-29A	SE9601	SE9627	BC0539	BC99-65
Volc. field	S. Borja	S. Borja	S. Esteban	Tres Virgenes	S. Clara	Jaraguay	S. Esteban	S. Esteban	S. Borja	La Purisima
Age (Ma)	14.64 ± 0.46	10.93 ± 0.32	4.01 ± 0.18	0.00	10.50 ± 0.40	9.40 ± 0.33	3.44 ± 0.19	2.70 ± 0.14	10.93 ± 0.34	9.67 ± 0.25
Reference	PALLARES et al., 2007	PALLARES et al., 2007	CALMUS et al., 2008	AGUILLÓN-ROBLES, 2002	AGUILLÓN-ROBLES et al., 2001	PALLARES et al., 2007	CALMUS et al., 2008	CALMUS et al., 2008	PALLARES et al., 2008	BENOIT et al., 2002
Dy	3.45	4.85	4.40	3.05	0.95	1.30	2.45	2.30	5.30	2.55
Y	21.00	28.00	27.00	16.60	4.30	7.40	14.20	13.00	28.50	12.50
Er	2.00	2.60	2.50	1.65		0.70	1.30	1.25	2.50	1.05
Yb	1.93	2.52	2.48	1.39	0.25	0.59	1.24	1.22	2.25	0.91
Sc	20.00	27.00	28.00	23.00	4.00	8.40	10.60	8.50	24.00	12.70
V	138.00	238.00	217.00	200.00	36.00	71.00	132.00	80.00	222.00	170.00
Cr	160.00	274.00	90.00	110.00	44.00	120.00	30.00	13.00	274.00	211.00
Co	33.00	40.00	24.00	30.00	6.00	14.00	13.00	9.00	37.00	28.00
Ni	162.00	154.00	37.00	55.00	24.00	98.00	17.00	7.00	186.00	160.00

	Bas. and., andesites		NEB	Alkali bas., trachybas.		MORB	Tholeiites		Comendite	Icelandite
Number	11	12	13	14	15	16	17	18	19	20
Rock type	B	B	B	TB	B	B	BA	B	R	I
Sample	JA96-23	BC99-30	BC99-131	BC05-103	SQ96-19	T02-20	BC00-35	M28	JR0435A	JR03-1
Volc. field	Jaraguay	La Purisima	S. Clara	S. Ignacio	S. Quintin	Is. Tortuga	S. Ignacio	Sra S. Ursula	C. Sarpullido	C. Chapala
Age (Ma)	3.14	0.29		7.63 ± 0.19		<1.70	10.60 ± 0.50	10.10 ± 0.30	12.10 ± 0.10	10.90 ± 0.40
Reference	CALMUS et al., 2003	BELLON et al., 2006	AGUILLÓN-ROBLES et al., 2001	PALLARES et al., 2007	AGUILLÓN-ROBLES, 2002	BATIZA et al., 1979	BENOIT et al., 2002	MORA-KLEPEIS AND MCDOWELL, 2004	VIDAL-SOLANO et al., 2008b	VIDAL-SOLANO et al., 2008b
Major elements (wt. %)										
SiO_2	52.00	49.60	49.00	49.10	47.80	47.00	54.40	51.78	74.18	61.41
TiO_2	0.70	2.86	1.67	2.74	2.20	1.42	1.70	1.68	0.14	0.85
Al_2O_3	16.10	13.80	14.86	14.90	15.60	16.40	14.25	16.56	11.75	13.34
$Fe_2O_3^a$	7.05	9.30	11.50	10.35	11.35	8.66	11.95	4.76	2.43	9.97
MnO	0.12	0.10	0.15	0.14	0.17	0.15	0.15	0.16	0.03	0.11
MgO	7.85	5.08	9.10	6.92	8.68	8.50	5.95	5.96		1.00
CaO	9.35	9.70	7.80	7.75	9.00	11.79	8.80	9.02	0.58	2.22
Na_2O	3.62	3.50	3.60	4.17	3.60	3.50	2.75	3.41	3.12	3.12
K_2O	0.98	3.77	1.15	2.01	1.51	0.07	0.20	1.02	5.08	3.48
P_2O_5	0.31	1.10	0.35	1.03	0.51	0.16	0.14	0.36	0.22	0.22
LOI	0.80	0.69	0.63	0.25	−0.60		−0.01	1.29	3.42	1.97

Table 1 *continued*

Lava type	Bas. and, andesites		NEB	Alkali bas., trachybas.		MORB	Tholeiites		Comendite	Icelandite
Number	11	12	13	14	15	16	17	18	19	20
Rock type	B	B	B	TB	B	B	BA	B	R	I
Sample	JA96-23	BC99-30	BC99-131	BC05-103	SQ96-19	T02-20	BC00-35	M28	JR0435A	JR03-1
Volc. field	Jaraguay	La Purisima	S. Clara	S. Ignacio	S. Quintin	Is. Tortuga	S. Ignacio	Sra S. Ursula	C. Sarpullido	C. Chapala
Age (Ma)	3.14	0.29		7.63 ± 0.19		<1.70	10.60 ± 0.50	10.10 ± 0.30	12.10 ± 0.10	10.90 ± 0.40
Reference	CALMUS et al., 2003	BELLON et al., 2006	AGUILLÓN-ROBLES et al., 2001	PALLARES et al., 2007	AGUILLÓN-ROBLES 2002	BATIZA et al., 1979	BENOIT et al., 2002	MORA-KLEPEIS AND McDOWELL, 2004	VIDAL-SOLANO et al., 2008b	VIDAL-SOLANO et al., 2008b
Total	100.06	99.50	99.86	99.36	99.82		100.28	100.7	100.85	100.59
Normative minerals (CIPW)										
Q	0.00	0.00	0.00	0.00	0.00	0.00	9.02	3.06	35.47	21.41
ne	0.00	5.55	0.00	2.03	3.91	4.54	0.00	0.00		0.00
hy	9.87	0.00	0.14	0.00	0.00	0.00	19.12	12.60	2.05	11.81
Mg^b	73.69	58.72	64.90	60.91	65.80	69.64	55.60	75.90	8.61	21.27
Trace elements (ppm)										
Rb	8.00	21.00	14.60	28.50	29.00	1.33	2.70	21.18	188.00	124.00
Ba	820.00	2300.00	255.00	480.00	338.00	16.70	70.00	542.00	170.00	1032.00
Th	1.65	2.00	2.35	3.80	3.70		0.20	2.00	20.80	13.45
Nb	4.00	17.70	26.50	56.00	43.00		4.40	12.00	22.40	16.00
La	16.80	54.00	19.50	43.00	28.00	13.60	3.90	18.60	55.30	42.80
Ce	36.00	125.00	41.00	100.00	55.00		10.00	39.80	119.00	96.00
Sr	835.00	2970.00	495.00	865.00	552.00	245.00	229.00	451.00	10.00	186.00
Nd	18.40	68.00	22.00	61.00	28.50	10.90	8.70	23.80	54.00	44.82
Sm	3.50	10.40	5.30	12.70	6.60	3.40	3.30	6.10	10.90	9.53
Zr	103.00	230.00	149.00	460.00	222.00		82.00	182.00	272.00	424.00
Eu	1.02	2.24	1.60	3.90	1.94	1.29	1.31	1.85	0.55	1.82
Gd	3.00	5.50	4.60	10.40					9.32	8.57
Dy	2.55	2.65	3.70	6.95	4.90		4.20	6.20	9.93	
Y	14.50	11.70	18.50	32.50	26.50	5.10	21.50	32.00	58.40	46.00
Er	1.35	1.00	1.80	2.60	2.60	3.50	1.95	3.46	5.89	5.10
Yb	1.40	0.77	1.42	2.06	2.26	3.70	1.70	2.86	5.70	4.72
Sc	24.00	19.00	21.00	17.50	27.00		24.00			
V	210.00	345.00	180.00	195.00	234.00		160.00			89.00
Cr	270.00	128.00	333.00	230.00	250.00	229.00	240.00			8.00
Co	32.00	30.00	48.00	39.00	43.00	49.00	40.00			34.00
Ni	170.00	41.00	208.00	210.00	185.00	91.00	110.00			10.00

B Basalts, *TB* trachybasalts, *BA* basaltic andesites, *A* andesites, *D* dacites, *I* icelandites, *R* rhyolites

^{a} Total iron as Fe_2O_3

^{b} Mg number (100 mg/mg +Fe)

selectively depleted in yttrium and heavy REE (HREE), with Y > 10 ppm and Yb > 1 ppm (average values and related standard deviations: 18.8 ± 3.8 ppm for Y and 1.6 ± 0.4 ppm for Yb; CASTILLO, 2008). Therefore, they display low Sr/Y (<20) and La/Yb (<10) ratios. In the following discussion they will be referred to as "regular" calc-alkaline lavas (opposed to adakites and related rocks).

The Comondú volcanic belt was active until 15-14.5 Ma in northern Baja California (MARTÍN et al., 2000; PALLARES et al., 2007; Fig. 3a); until 11 Ma in Baja California Sur (SAWLAN and SMITH, 1984; SAWLAN, 1991; BELLON et al., 2006); and until 12-11 Ma in Sonora (MORA-KLEPEIS and MCDOWELL, 2004; VIDAL-SOLANO et al., 2008a; TILL et al., 2009). However, limited calc-alkaline activity persisted during the Late Miocene in various volcanic fields of Baja California and Sonora (Fig. 3b), until 5.8 Ma in Puertecitos (MARTÍN-BARAJAS et al., 1995), 7.3 Ma in Jaraguay (PALLARES et al., 2007) and 8.3 Ma in Sierra El Aguaje (TILL et al., 2009). Then, it resumed during the Plio-Quaternary along the eastern coast of Baja California and within the Gulf of California (Figs. 1 and 3c), emplacing several volcanic edifices, most of them of medium-K composition. These include from north to south: the youngest lavas of the Puertecitos volcanic field (3.2-2.7 Ma; MARTÍN-BARAJAS et al., 1995), Isla San Luis (Pleistocene; PAZ-MORENO and DEMANT, 1999), the Cerro Starship center in the SW of Isla Tiburón (5.7-3.7 Ma; OSKIN and STOCK, 2003), Isla San Esteban (4.5-2.5 Ma; DESONIE, 1992; CALMUS et al., 2008), Tres Virgenes young volcano (160-36 ka; SCHMITT et al., 2006), La Reforma and El Aguajito calderas (1.4-1.2 Ma; DEMANT, 1984; GARDUÑO-MONROY et al., 1993; SCHMITT et al., 2006), Cerro Los Mencenares volcanic center (4.3-3.8 Ma; BIGIOGGERO et al., 1995; AGUILLÓN-ROBLES, 2002) and Isla Coronado (0.69 Ma-Holocene; BIGIOGGERO et al., 1987). Some of these volcanic edifices contain, in addition to "regular" calc-alkaline lavas, adakites (Isla San Esteban; CALMUS et al., 2008) or lavas plotting within the adakite field in most geochemical diagrams (Isla Coronado, Cerro Mencenares, Tres Virgenes; see below).

Although no specific geochemical modeling of the origin of the Comondú "regular" calc-alkaline lavas

has been attempted, all authors have assumed that these andesite-dominated suites derive from the usual arc lava source, i.e., the supraslab mantle metasomatized by hydrous fluids transferred from the downgoing Pacific-Farallon plate (ARCULUS, 1994; STERN, 2002). The origin of calc-alkaline lavas younger than the postulated end of the subduction event (ca. 12.5 Ma) has also been attributed to the delayed partial melting of this previously metasomatized source (BELLON et al., 2006; TILL et al. 2009). This melting may have been triggered by the heat supply from the Pacific asthenosphere during the Late Miocene development of an asthenospheric window (PALLARES et al., 2007, 2008), and later by the high thermal regime linked to the opening of the Gulf of California (CALMUS et al., 2008).

3.3. Adakites and Lavas Intermediate Between Adakites and Calc-Alkaline Lavas

Adakites (DEFANT and DRUMMOND, 1990) or high-silica adakites (MARTIN et al., 2005) are low-K to medium-K andesitic and dacitic rocks (SiO_2 = 56–70 wt%), usually amphibole-rich, the geochemical signature of which shows the high LILE/HFSE ratios and relative depletion in Nb typical of calc-alkaline magmas (Fig. 2b). They have highly fractionated REE patterns (La/Yb > 20), and very low Y and HREE contents, e.g., Y < 12 ppm and Yb < 1 ppm (average values and related standard deviations: 8.8 ± 3.4 ppm for Y and 0.65 ± 0.32 ppm for Yb; CASTILLO, 2008). In addition, they display positive anomalies in Sr, very high Sr/Y (>50) and equivalent ratios, and isotopic Sr, Nd, Pb signatures similar to that of oceanic MORB-type basalts. Their characteristic Y and HREE depletion (Fig. 2b) is thought to traduce the selective incorporation of these elements in garnet (either residual garnet in their source or deep fractionation of garnet from the early melts). Adakites are generally considered as derived either (1) from the partial melting of subducted oceanic crust metamorphosed into the garnet amphibolite or eclogite facies (DEFANT and DRUMMOND, 1990; SEN and DUNN, 1994a; MARTIN, 1999; DEFANT and KEPEZHINSKAS, 2001; MARTIN et al., 2005), or (2) from the partial melting of the mafic base of thickened Andean-type crust (ATHERTON and PETFORD, 1993; PETFORD and ATHERTON, 1996; ARCULUS

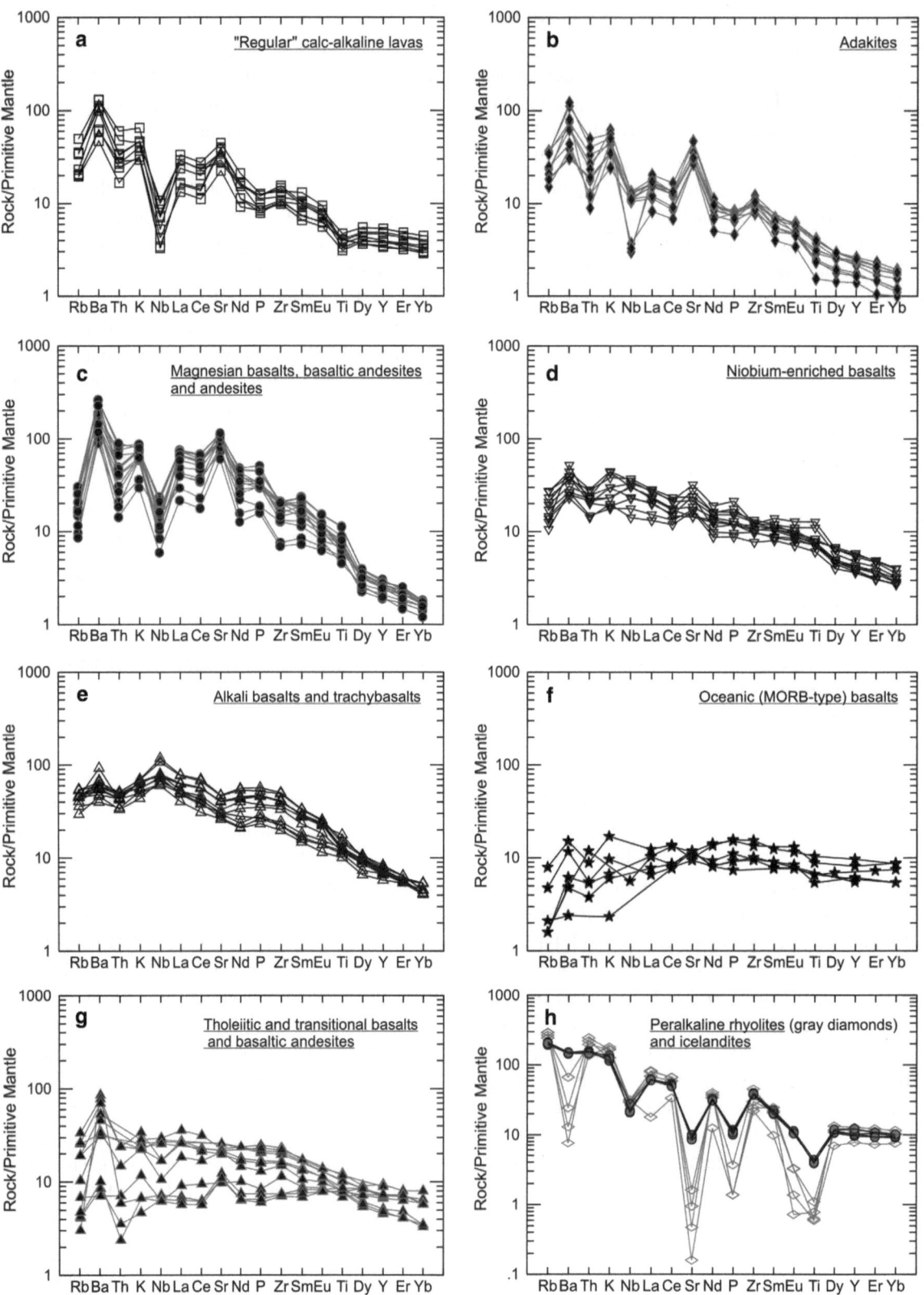

Figure 2

Incompatible multi-element patterns of selected lava types from the northwestern Mexican margin normalized to the Primitive Mantle of SUN and McDONOUGH (1989). **a** "Regular" calc-alkaline lavas from Baja California volcanic fields (MARTÍN-BARAJAS et al., 1995; AGUILLÓN-ROBLES et al., 2001; BENOIT et al., 2002; PALLARES et al., 2007, 2008;) and Sonora volcanic fields (VIDAL-SOLANO et al., 2005, 2007, 2008a; TILL et al., 2009). **b** Adakites from Santa Clara Volcanic Field (AGUILLÓN-ROBLES et al., 2001) and Jaraguay Volcanic Field (PALLARES et al., 2007, 2008). **c** Magnesian basalts, basaltic andesites and andesites from Baja California volcanics fields (CALMUS et al., 2003; BELLON et al., 2006; PALLARES et al., 2007, 2008). **d** Niobium-enriched basalts from Santa Clara Volcanic Field (AGUILLÓN-ROBLES et al., 2001). **e** Alkali basalts and trachybasalts from San Quintín Volcanic Field (LUHR et al., 1995), Moctezuma Volcanic Field (PAZ-MORENO et al., 2003), San Carlos Volcanic Field (PALLARES et al., 2007, 2008), and Pinacate Volcanic Field (VIDAL-SOLANO et al. 2008a). **f** Oceanic (MORB-type) basalts from Isla Tortuga (BATIZA, 1978) and Guaymas Basin (SAUNDERS et al., 1982b). **g** Tholeiitic and transitional basalts and basaltic andesites from San Ignacio-San Juanico Volcanic Field (BENOIT et al., 2002), Las Trincheras Volcanic Field (MORA-KLEPEIS and McDOWELL, 2004), Moctezuma Volcanic Field (PAZ-MORENO et al., 2003), Central Sonora (VIDAL-SOLANO et al., 2005, 2007, 2008a) and Coastal and Eastern Sonora (TILL et al., 2009). **h** Peralkaline rhyolites and icelandites from Sonora volcanic fields (VIDAL-SOLANO et al., 2005, 2007, 2008a)

et al., 1999), or finally from (3) high-pressure fractionation (involving separation of garnet) of basaltic to andesitic liquids in mantle conditions (PROUTEAU and SCAILLET, 2003; MÜNTENER and ULMER, 2006; ALONSO-PEREZ et al., 2009). In addition to typical adakites, the compositions of which match those of experimental garnet amphibolite or garnet eclogite partial melts, many adakite associations contain rocks displaying Sr/Y and La/Yb ratios intermediate between the former and calc-alkaline melts. These rocks might result from mixing involving the two kinds of melts, as shown for their occurrences in the Philippines (JÉGO et al., 2005) and Ecuador (SAMANIEGO et al., 2005; HIDALGO et al., 2007; SCHIANO et al., 2010). Alternatively, they might derive from variable fractionation of garnet or garnet + amphibole in calc-alkaline magmas (e.g., MACPHERSON et al., 2006; CHIARADIA et al., 2009).

Adakites have never been identified in Sonora, but several occurrences are described (and others suspected) in Baja California and the Gulf of California islands (an. 5–8, Table 1). The largest is the Late Miocene Santa Clara volcanic field (Vizcaino Peninsula), the volume of which is estimated to 25 km^3. In this area, ca. 20 dacitic domes (an. 5, Table 1) and up

to 250-m-thick associated lava flow and pyroclastic flow sequences were emplaced between 11 and 8.7 Ma (AGUILLÓN-ROBLES et al., 2001; BENOIT et al., 2002), in close spatial and temporal association with niobium-enriched basalts (NEB). In addition, adakites dated between 6.2 and 4.9 Ma have been reported by BONINI and BALDWIN (1998) from Isla Santa Margarita, and an adakitic dyke 9.4 Ma old (an. 6, Table 1) from the Jaraguay volcanic field (PALLARES et al., 2007). In the Santa Rosalía basin, adakitic flows (the Santa Rosalía dacites) were emplaced between 12.5 and 12.3 Ma (CONLY et al., 2005).

Pliocene (4.5-2.6 Ma old) adakitic andesites and dacites (an. 7–8, Table 1), associated with contemporaneous calc-alkaline lavas (an. 3, Table 1), have been reported from Isla San Esteban in the Gulf of California (CALMUS et al., 2008). Although they were not identified as such by former authors, Pliocene-Quaternary adakites or adakite-related lavas might also occur in several other volcanic centers. Indeed, analyses of dacites or andesites displaying low heavy rare earth elements (HREE) and Y contents together with high Sr/Y and La/Yb ratios have been recorded from the Pliocene Cerro Mencenares complex (BIG-IOGGERO et al., 1995), and the Quaternary Isla Coronado (BIGIOGGERO et al., 1987) and Tres Virgenes volcanoes (CAMERON and CAMERON, 1985).

The origin of Late Miocene Baja California adakites has been attributed to the melting of the subducting Pacific-Farallon oceanic crust along the edges of an asthenospheric window (AGUILLÓN-RO-BLES et al., 2001; BENOIT et al., 2002; PALLARES et al., 2007). This interpretation has been challenged by CASTILLO (2008), who prefers that of the partial melting of metabasites from the base of the Baja crust. However, the thickness of this crust (less than 33 km) may not allow garnet to be stable in such conditions (MAURY et al., 2009). In addition, the occurrence of mantle-derived ultramafic xenocrysts and xenoliths in Rancho San Lucas adakite (Santa Clara) suggests that adakitic melts ascended through the upper Baja mantle, and thus derived either from the downgoing slab (MAURY et al., 2009) or alternatively from delaminated Baja crust slivers (CASTILLO, 2009). The source of the Pliocene San Esteban adakites, which overlie a still thinner continental

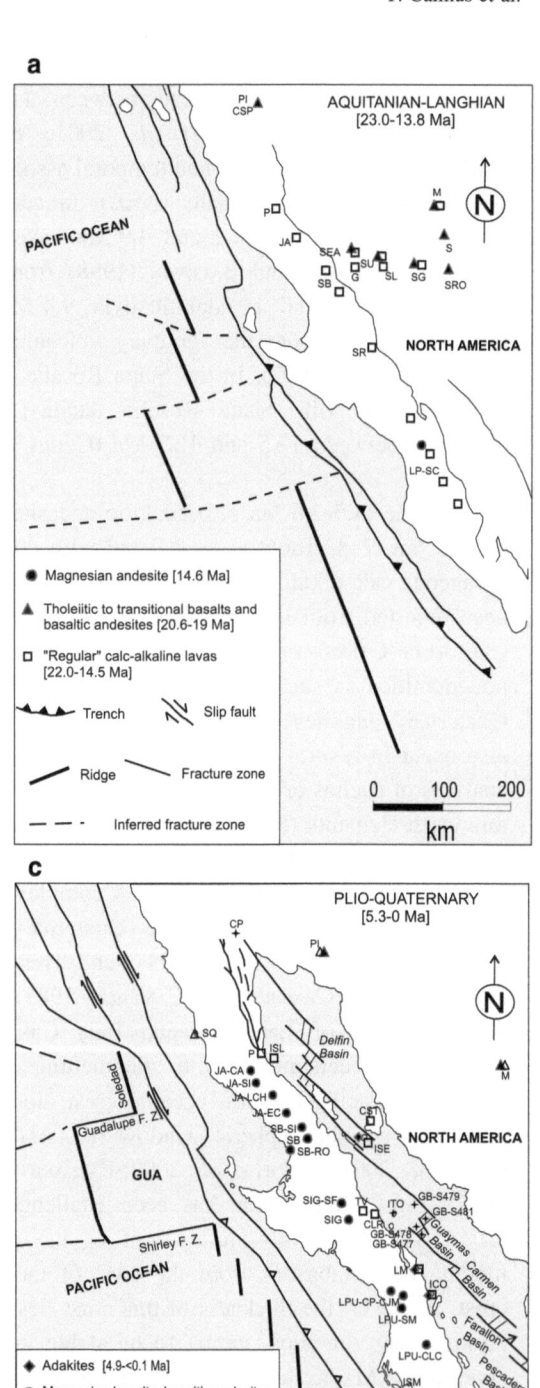

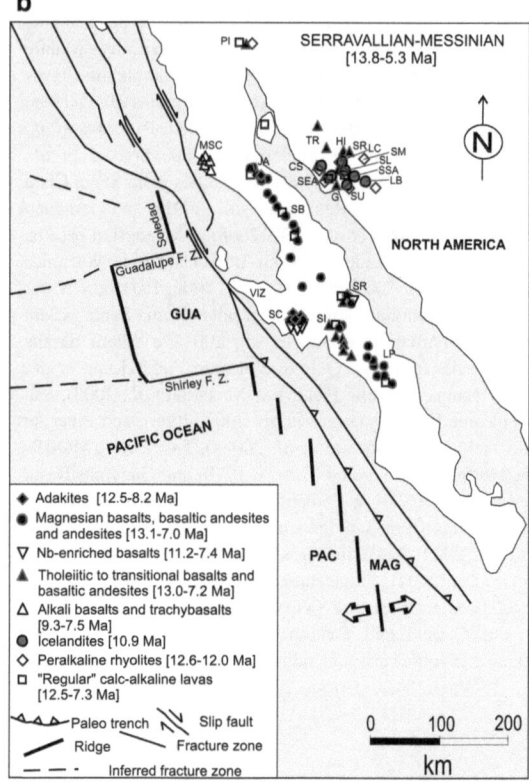

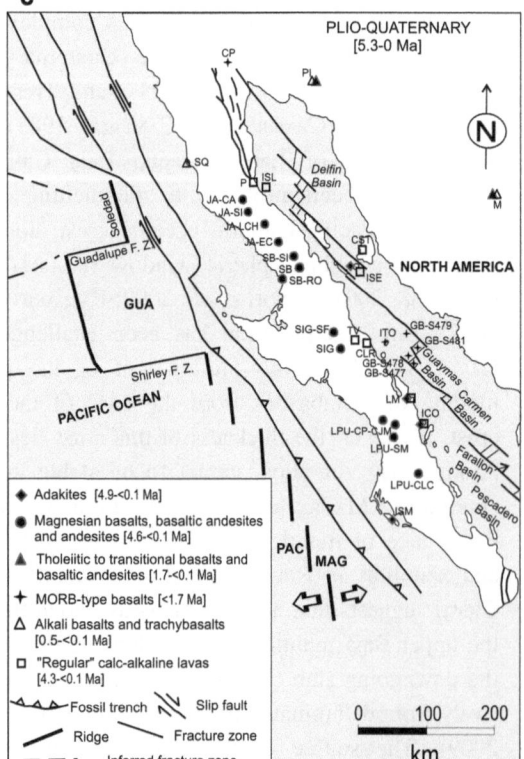

Figure 3
Sketches of Baja California and Sonora at various periods showing the locations and types of lavas together with structural elements of the adjacent Pacific Ocean and the Gulf of California. **a** Early-Middle Miocene (Aquitanian to Langhian): tectonic setting drawn at 16 Ma from WILSON *et al.* (2005). **b** Middle-Upper Miocene (Serravallian to Messinian): tectonic setting at 11 Ma (WILSON *et al.*, 2005; PALLARES *et al.*, 2007; MAURY *et al.*, 2009), during the peak of adakitic volcanism in Santa Clara. **c** Pliocene-Quaternary: present-day tectonic setting from CALMUS *et al.* (2008). The present-day coastal lines are depicted on this reconstruction, but the Gulf of California did not exist at 16 and 11 Ma. Abbreviated plate names: *PAC* Pacific, *MAG* Magdalena, *GUA* Guadalupe. The distribution of Aquitanian-Langhian, Serravallian-Messinian and Plio-Quaternary volcanic rocks is from BATIZA (1978), SAUNDERS *et al.* (1982b), LUHR *et al.* (1995), MARTÍN-BARAJAS *et al.* (1995), AGUILLÓN-ROBLES *et al.* (2001), BENOIT *et al.* (2002), CALMUS *et al.* (2003, 2008), OSKIN and STOCK (2003), PAZ-MORENO *et al.* (2003), MORA-KLEPEIS and McDOWELL (2004), CONLY *et al.*, (2005), BELLON *et al.* (2006), PALLARES *et al.* (2007), PALLARES *et al.* (2007, 2008), VIDAL-SOLANO *et al.* (2005, 2007, 2008a,b) qnd TILL *et al.* (2009). Abbreviations for volcanic fields/localities discussed in the text are as follows: *CLR* Caldera La Reforma, *CP* Cerro Prieto, *CSP* Cerro San Pedro, *CS* Coastal Sonora, *CST* Cerro Starship, Isla Tiburón, *EP* El Paraiso, *G* Guaymas, *GB-S477* Guaymas Basin DSDP site 477, *GB-S478* Guaymas Basin DSDP site 478, *GB-S479* Guaymas Basin DSDP site 479, *GB-S481* Guaymas Basin DSDP site 481, *HI* Hilarenos, *ICO* Isla Coronado, *ISE* Isla San Esteban, *ISL* Isla San Luís, *ISM* Isla Santa Margarita, *JA* Jaraguay, *JA-CA* Jaraguay-Cataviña, *JA-EC* Jaraguay-El Crucero, *JA-LCH* Jaraguay-Laguna de Chapala, *JA-SI* Jaraguay-Santa Inés, *LB* Sierra Lista Blanca, *LM* Los Mencenares, *LP* La Purísima, *LP-SC* La Purísima-San José de Comondú, *LPU-CLC* La Purísima-Cerro Los Cerritos, *LPU-CP-CJM* La Purísima-Cerro Pabellón-Cerro Jesús del Monte, *LPU-SM* La Purísima-San Miguel, *LT* Las Trincheras, *M* Moctezuma, *MSC* Mesas San Carlos and Santa Catarina, *P* Puertecitos Volcanic Province, *PI* Pinacate, *S* Sahuaripa, *SB* San Borja, *SB-RO* San Borja-Rosarito, *SB-SI* San Borja-San Ignacito, *SC* Santa Clara, *SEA* Sierra El Aguaje, *SF* San Felipe, *SG* Suaqui Grande, *SI* San Ignacio, *SIG-SF* San Ignacio-San Francisco, *SL* Sierra Libre, *SM* Sierra de Mazatán, *SQ* San Quintin, *SR* Santa Rosalía, *SRO* Santa Rosa, *SSA* Sierra San Antonio, *SU* Sierra Santa Ursula, *TV* Tres Virgenes, *VIZ* Vizcaíno

crust and clearly post-date the subduction event, is thought to be isolated slivers of oceanic crust left within the Gulf mantle. Their melting might have been triggered by the hot thermal regime linked to its opening, especially during the spreading stage of the Lower Tiburón basin, concomitant with volcanic activity in San Esteban (CALMUS *et al.*, 2008).

3.4. Magnesian Basalts, Basaltic Andesites and Andesites: The "Bajaite" Suite

Magnesian andesite suites are characterized by high contents in incompatible elements, although they display the relative depletion in Nb typical of calc-alkaline magmas. They include basalts, basaltic andesites and andesites ($SiO_2 < 60$ wt%) showing MgO and compatible transition elements (Cr, Co, Ni) contents higher than expected from their silica contents (when compared to regular calc-alkaline series). A majority of them are rich in Sr ($>1,000$ ppm), and their multielement patterns (Fig. 2c) show positive anomalies for this element, together with variable but often marked depletions in Y and HREE (average values and related standard deviations: 13 ± 3 ppm for Y and 0.92 ± 0.24 ppm for Yb; CASTILLO, 2008). These chemical features have been interpreted as indicative of an adakitic imprint (DEFANT AND DRUMMOND, 1990; "low-silica adakites" of MARTIN *et al.*, 2005) and are often thought to be inherited from mantle metasomatism of garnet-bearing peridotites by adakitic melts (CALMUS *et al.*, 2003; PALLARES *et al.*, 2007, 2008; MOYEN, 2009). In Baja California, these unusual lavas (an. 9–12, Table 1), which were termed "bajaites" by ROGERS *et al.* (1985), form six volcanic fields delineating a 500-km-long array parallel to the Gulf of California from Jaraguay to La Purísima (Fig. 1), and their emplacement was largely controlled by the regional orientation of tectonic stress (NEGRETE-ARANDA *et al.*, 2010). They cover a total surface of ca. 9,000 km^2 and range in age from Middle Miocene (14.6 Ma; Calmus et al., 2003; PALLARES *et al.*, 2008) to Holocene in La Purísima (BELLON *et al.*, 2006). They are moderately porphyritic, with 5–20 modal% phenocrysts which include, by order of decreasing abundances, olivine, diopsidic clinopyroxene, orthopyroxene, labradorite, titanomagnetite, and occasional phlogopite and sanidine. They display easily recognizable geochemical characteristics (an. 9–12, Table 1): SiO_2 up to 57%, MgO up to 8%, low FeO*/MgO ratios usually less than 2, high Na/K, low Rb/Sr (<0.01), very high contents in Sr (up to 3,000 ppm) and Ba ($>1,000$ ppm), highly fractionated REE patterns (Fig. 2c) and relatively low $^{87}Sr/^{86}Sr < 0.7048$ (ROGERS *et al.*, 1985; SAUNDERS *et al.*, 1987; ROGERS and SAUNDERS, 1989; CALMUS *et al.*, 2003).

SAUNDERS *et al.* (1987) and ROGERS and SAUNDERS (1989) concluded that the genesis of the Baja California magnesian andesite suite was a two-stage process, involving mantle metasomatism by silicic melts during subduction followed by melting of this metasomatized

65

mantle during a post-subduction (extensional) event. CALMUS *et al.* (2003) and PALLARES *et al.* (2008) proposed a detailed petrogenetic model involving melting of supraslab mantle having interacted with slab melts. In this model, reactions between adakitic melts and the deep supraslab mantle led to metasomatized pargasite-rich peridotites. This process has been documented experimentally by numerous authors (SEN AND DUNN 1994; YAXLEY and GREEN, 1998; RAPP *et al.*, 1999, 2006; PROUTEAU *et al.*, 2001). Then, incongruent dehydration melting of pargasitic amphibole (FRANCIS and LUDDEN, 1995; DALPÉ and BAKER, 2000), at depths of ca. 80 km (NIIDA and GREEN, 1999), triggered the genesis of bajaitic melts, and left a garnet-rich residue. This melting occurred at minimal temperatures of 1,050–1,075°C, consistent with a high thermal flux in the mantle wedge during the opening of an asthenospheric window following ridge-trench collision, as well as during the subsequent "no-slab" regime, which followed the sinking of the Farallon plate into the deep mantle (PALLARES *et al.*, 2008). The model proposed by CASTILLO (2008) is rather similar to the former one, except that metasomatic mantle minerals are thought to result from the percolation of hydrous fluids rather than slab melts.

3.5. Niobium-Enriched Basalts

Niobium-enriched basalts (NEB) are an extremely unusual rock type, only found in close spatial and temporal association with Baja California adakites in Santa Clara, Vizcaíno Peninsula (11.7-8.7 Ma; AGUILLÓN-ROBLES *et al.*, 2001) and possibly in Santa Rosalía (11-9 Ma; CONLY *et al.*, 2005). In Santa Clara, they occur as fluidal lava flows of olivine-plagioclase-phyric basalts that form horizontal mesas (AGUILLÓN-ROBLES *et al.*, 2001). They overlie adakitic pyroclastic flow deposits, some of which contain NEB blocks (MAURY *et al.*, 2009). These NEBs are silica-saturated or -oversaturated, highly sodic (an. 13, Table 1), and differ from the vast majority of arc basalts by their higher Nb (10–30 ppm) and TiO_2 (1.3–1.8 wt%) contents. They display rather smooth enriched incompatible element patterns culminating at the level of Nb (Fig. 2d), with variable positive anomalies in Ba, Sr and Ti that are almost identical to those of the Philippine NEB (SAJONA *et al.*, 1996).

There is an ongoing debate regarding the genesis of these unusual basalts, which differ from the alkali basalts described below by their silica-saturated character, their generally lower enrichment in incompatible elements and their lower Sr isotopic ratios (AGUILLÓN-ROBLES *et al.*, 2001; BENOIT *et al.*, 2002). Unlike the magnesian andesite suite, they show neither negative Nb anomalies nor strong depletion in Y and HREE (average values and related standard deviations: 19.1 ± 10.4 ppm for Nb, 20.2 ± 3.5 ppm for Y and 4.76 ± 1.18 ppm for Yb; CASTILLO, 2008). AGUILLÓN-ROBLES *et al.* (2001) and BENOIT *et al.* (2002) proposed that these NEB derived from the melting at relatively low pressures (depths of 40–60 km), i.e., in the spinel stability field, of amphibole-rich supraslab mantle having interacted with adakitic melts (DEFANT *et al.*, 1992; KEPEZHINSKAS *et al.*, 1996; SAJONA *et al.*, 1996). A major difference with the petrogenetic history of the magnesian andesite suite is that, for the latter, interactions between slab melts and the supraslab mantle occurred at greater depths (ca. 80 km), i.e., in the garnet stability field. CASTILLO (2008, 2009) proposed an entirely different model, in which NEBs of Baja California are genetically unrelated to adakites despite ample field evidence for their association (MAURY *et al.*, 2009). They are thought to result from the fractional crystallization of San Quintín-type melts coupled with their contamination by tholeiitic mantle materials and/or the Baja California continental crust. The San Quintín-type melts would have ascended through a slab window located beneath the Proto-Gulf of California and should therefore have traveled ca. 150 km towards the fossil trench within the Baja lithospheric mantle. In addition, Santa Clara NEBs are less radiogenic in Sr than San Quintín basalts, a feature hardly consistent with the contamination hypothesis (MAURY *et al.*, 2009).

3.6. Alkali Basalts and Trachybasalts

Alkali basalts in Baja California and Sonora are mostly Quaternary in age. In Baja California, the well-preserved San Quintín strombolian cones and associated flows, which contain peridotitic and granulitic xenoliths (GASTIL *et al.*, 1979), have been extensively studied (LUHR *et al.*, 1995). Available ages range from 126 to 90 ka, but the eruptive activity probably continued into the Holocene. In Sonora, Quaternary alkali basalts and trachybasalts occur in two volcanic fields, where they are associated with tholeiitic basalts. In the Pinacate volcanic field, alkali basaltic to trachytic activity emplaced the Santa Clara volcanic shield between 1.7 and 1.1 Ma. Then, the 1,500 km^2 Pinacate series magmas erupted until the Holocene (13 ka; TURRIN *et al.*, 2008), forming hundreds of strombolian cones, maars and tuff rings (GUTMANN, 2002) and emplacing tholeiitic, transitional and mildly alkali basalts together with trachybasalts and minor trachyandesites. The ca. 300 km^2 Quaternary Moctezuma volcanic field (PAZ-MORENO *et al.*, 2003) is located at the foothills of the Sierra Madre Occidental. It includes tholeiitic lava flows emitted during an early fissural event (1.7 Ma), overlain by younger (0.53-0.44 Ma) alkali trachybasaltic lava flows erupted from small monogenetic cones.

Late Miocene alkali basalt-related lavas have only been described in northern Baja California. The two large plateaus (ca. 600 km^2 each) of Mesa San Carlos and Mesa Santa Catarina, located along the west coast of Peninsula 100 km south of El Rosario, are capped by trachybasaltic (hawaiitic) flows dated between 9.3 and 7.5 Ma (PALLARES *et al.*, 2007).

Quaternary alkali basalts and trachybasalts from San Quintín, Pinacate and Moctezuma volcanic fields and Late Miocene San Carlos-Santa Catarina trachybasalts (an. 14–15, Table 1) contain olivine, clinopyroxene, plagioclase, titanomagnetite and ilmenite. They are mildly silica-undersaturated (less than 5% normative nepheline). Their multielement patterns (Fig. 2e) display a considerable enrichment in most incompatible elements, culminating at the level of Nb, although they show a relative depletion in Rb, Ba and K. They are typical of an ocean island basalt (OIB)-type source with little or no crustal contamination (LUHR *et al.*, 1995). Isotopic (Sr, Nd, Pb) data are only available for the Quaternary San Quintín (LUHR *et al.*, 1995), Pinacate (ASMEROM and EDWARDS, 1995; GOSS *et al.*, 2008) and Moctezuma (PAZ-MORENO *et al.*, 2003) volcanic fields. They plot consistently within the OIB field, and evidence for crustal contamination is limited to two San Quintín cones. These alkali basaltic magmas are thought to derive from plume-type asthenospheric mantle (ASMEROM and EDWARDS,

1995; PAZ-MORENO *et al.*, 2003), at depths increasing from the spinel lherzolite field below San Quintín (LUHR *et al.*, 1995) to the garnet lherzolite field below the Pinacate (GOSS *et al.*, 2008).

3.7. Oceanic (MORB-type) Basalts and Related Rocks

Oceanic basalts in the studied area are exclusively Quaternary. Their occurrence is restricted to the Gulf of California, although Holocene tholeiitic dacites from the Cerro Prieto geothermal field, south of Mexicali, have been considered as MORB-related (HERZIG, 1990). These oceanic basalts form the small Isla Tortuga (an. 16, Table 1), made of tholeiitic basaltic and ferrobasaltic flows (<1.70 Ma; BATIZA, 1978; BATIZA *et al.*, 1979) overlain by possibly Holocene hyaloclastic tuffs (MEDINA *et al.*, 1989). They have also been found in several holes drilled during DSDP Leg 64 (SAUNDERS *et al.*, 1982a, b; PERFIT *et al.*, 1982; FORNARI *et al.*, 1982) along the Gulf rise (DSDP site 479) and in the Guaymas Basin (<0.4 Ma; DSDP sites 477, 478) and Yaqui Basin (DSDP site 481). These basalts range from subaphyric to sparsely porphyritic lavas containing phenocrysts of calcic plagioclase or olivine + plagioclase. Their major element compositions are broadly similar to those of normal MORB from the East Pacific Rise (EPR). Like the latter, they are poor in incompatible elements, and their flat to depleted multielement patterns (Fig. 2f) are consistent with their derivation from depleted asthenospheric sources similar to the EPR ones. However, the Sr contents of Guaymas Basin and Tortuga basalts are higher than those from basalts from the Gulf mouth area, and the former also display higher La/Yb, Sr/Zr, Zr/Ti and Th/Hf ratios, which suggest that their mantle source contained a "residual calc-alkaline component" (SAUNDERS *et al.* 1982a).

3.8. Tholeiitic or Transitional Basalts and Basaltic Andesites

Late Miocene tholeiitic basaltic andesites (11.3-7.2 Ma; BENOIT *et al.*, 2002; BELLON *et al.* 2006) crop out in Baja California, between San Juanico and San Ignacio, as very fluid flows overlying tilted Tertiary sedimentary rocks (an. 17, Table 1). They cap large sub-horizontal plateaus (mesas) and have probably been emitted from fissures. Similar tholeiitic lavas dated to 6 Ma have been collected in the northern Baja California peninsula near 31°22′N (El Paraiso, AGUILLÓN-ROBLES, 2002). These tholeiitic rocks contain sparse plagioclase and olivine phenocrysts, are silica-oversaturated, and characterized by low K_2O (<0.6 wt%) and rather high TiO_2 (1.6–1.9 wt%) contents. They are also richer in Nb and other HFSE than their calc-alkaline equivalents, and their rather flat REE and multielement patterns (Fig. 2g, bottom patterns) do not display any subduction imprint. Melting of depleted subslab mantle accounts satisfactorily for these flat patterns and depleted Sr and Nd isotopic signatures of these tholeiites (BENOIT et al., 2002). However, a small sediment contribution is required to explain their enriched Pb isotopic features.

Three episodes of Neogene-Quaternary tholeiitic to transitional basaltic volcanism can be recognized in Sonora. During the Early Miocene (20.6–19 Ma) tholeiitic lava flows, which were later affected by extensional tectonics, were emplaced in the substratum of the Pinacate Volcanic Field (VIDAL-SOLANO et al., 2008a) and in several sierras. These lavas contain less than 5% phenocrysts (olivine and plagioclase), and are olivine + hypersthene to quartz-normative basalts and basaltic andesites. Their multielement patterns (Fig. 2g, bottom patterns) are flat to slightly enriched. They display small negative Nb anomalies, which, together with their Sr and Nd isotopic signatures, suggest the involvement of lithospheric materials, possibly from a Precambrian lithospheric mantle source (Vidal-Solano et al., 2008a). Indeed, these rocks are rather similar to the lithospheric mantle-derived Early Miocene basalts of southern Nevada and westernmost Arizona, and the basaltic andesites from the Mojave Desert, California (MILLER et al., 2000).

The second episode of tholeiitic to transitional magmatism in Sonora occurred during the Middle and Upper Miocene (Serravallian-Tortonian, 13.0-7.2 Ma). It emplaced lava flows in the Pinacate substratum (VIDAL-SOLANO et al., 2008a), Las Trincheras (MORA-KLEPEIS and MCDOWELL, 2004), and possibly, according to TILL et al. (2009)'s analyses, in Sierra Libre, Sierra El Aguaje, Coastal Sonora and Guaymas. This event was concomitant with the eruption in the same areas of peralkaline rhyolites and icelandites (see below), which are considered as derived from the open system fractional crystallization of such tholeiitic/transitional magmas (VIDAL-SOLANO et al., 2005, 2007, 2008a, b). The mafic lavas from this episode (an. 18, Table 1) display multielement patterns (Fig. 2g, upper group) more enriched than the older ones, with weaker Nb anomalies indicative of lithospheric contribution. Their Sr and Nd isotopic compositions are also closer to the MORB field (VIDAL-SOLANO et al., 2008a). These features might traduce a temporal change in the composition of the mantle sources of the tholeiites, from shallow lithospheric to deeper asthenospheric mantle, as the result of convective thinning and extension of the Basin and Range lithosphere (FITTON et al., 1991; DEPAOLO and DALEY, 2000; VIDAL-SOLANO et al., 2008a).

Finally, tholeiitic basalts and basaltic andesites were also emplaced in Sonora during the Quaternary, in association with alkali basalts, in the Pinacate and Moctezuma volcanic fields (see Sect. 3.6). They contain less than 10% olivine phenocrysts, and are either olivine + hypersthene or quartz-normative. Their main mineralogical difference with respect to the alkali basalts lies in the composition of clinopyroxenes. These are Ca-rich in the alkaline lavas and subcalcic with orthopyroxene or pigeonite in the tholeiitic lavas (PAZ-MORENO et al., 2003). The multielement patterns of the tholeiitic lavas are subparallel to (although less enriched in the most incompatible elements) those of the associated alkali basalts and trachybasalts, and typical of their OIB affinity. Isotopic (Sr, Nd, Pb) data on the Pinacate (ASMEROM and EDWARDS, 1995; GOSS et al., 2008) and Moctezuma (PAZ-MORENO et al., 2003) tholeiites plot within the OIB field and do not indicate the occurrence of lithospheric contamination processes. Like the associated alkali basalts, these Quaternary tholeiites are thought to derive from plume-type asthenospheric mantle (ASMEROM and EDWARDS, 1995; PAZ-MORENO et al., 2003; GOSS et al., 2008).

3.9. Peralkaline Rhyolites and Icelandites

VIDAL-SOLANO (2005) and VIDAL-SOLANO et al. (2005, 2007, 2008a, b) demonstrated that two rock types outcropping in central Sonora and the

Puertecitos area in Baja California, and considered as calc-alkaline by other authors (SAWLAN, 1991; MARTÍN-BARAJAS *et al.*, 1995; TILL *et al.*, 2009), are respectively peralkaline rhyolites (comendites) and icelandites. Both types commonly occur in tholeiitic to transitional basalt series erupted in extensional settings, and are generally thought to derive from the closed- or open-system fractional crystallization of the associated basaltic magmas. The Sonora and Puertecitos peralkaline rhyolites occur as ignimbrite deposits and less commonly as rhyolitic domes, and have been dated to 12.6-12 Ma (VIDAL-SOLANO *et al.*, 2008a, b). Most ignimbrite deposits have suffered some weathering, and thus their major element analyses (an. 19, Table 1) do not show anymore the (Na+K)/Al ratios higher than unity which are characteristic of peralkaline rhyolites. Their Sr and Nd isotopic ratios indicate that they have also experienced limited contamination by the Precambrian substratum of Sonora (VIDAL-SOLANO *et al.*, 2008a, b). However, they have retained many typical features (VIDAL-SOLANO *et al.*, 2005, 2007), including (1) their unmistakable mineralogical association, with phenocrysts of fayalite, Fe-rich augite, alkali feldspar and zircon, (2) major element compositions characterized by high silica (>70 wt%), low alumina (~ 12 wt%) and high alkalies, and (3) strong enrichment in the most incompatible elements, including the LREE, and marked depletion in Ba, Sr and Eu typical of feldspar fractionation (Fig. 2h). TILL *et al.* (2009) have published rhyolite analyses displaying similar characteristics from the Miocene volcanic fields of Suaqui Grande, Sierra Libre, Santa Ursula, Sierra El Aguaje, Sierra Mazatan and coastal Sonora, together with Ar-Ar ages (12.5–10.1 Ma) close to those of VIDAL-SOLANO *et al.* (2008a). These rhyolitic sequences, more than 1,000 m-thick in Sierra Santa Ursula, are interpreted as the first volcanic manifestation of the continental breakup of the Gulf of California. The main emission center of the 12.5 Ma old San Felipe Tuff is supposed to be located in the Sierra Kunkaak between Bahía de Kino and Punta Chueca (OSKIN, 2002). However, the very large distribution of its deposits from San Felipe area in Baja California until Guaymas region and Central Sonora, as well as their variable thickness, may be due to the occurrence of several emission centers.

The Lista Blanca Formation of La Colorada, Mazatán region and Tecoripa in Central Sonora may correspond to the most distal pyroclastic deposits to the east.

In the same areas of central Sonora (e.g., in Sierra Lista Blanca and Sierra San Antonio), peralkaline comenditic ignimbrites are often capped by black porphyritic lava flows containing andesine, augite, pigeonite and Fe-Ti oxide phenocrysts set in a glassy groundmass (VIDAL-SOLANO *et al.*, 2008b). One of these flows yielded an Ar-Ar age of 10.9 ± 0.4 Ma. These rocks show major and trace element features (an. 20, Table 1) typical of intermediate lavas from the tholeiitic basalt series (icelandites): intermediate silica contents (60–65 wt%), high total iron oxide (5–13 wt%) and FeO/MgO ratios, together with TiO_2 contents (>1 wt%) higher than those of calc-alkaline lavas. Their enriched incompatible element patterns (Fig. 2h) show negative anomalies in Ba, Sr, Eu and Ti weaker than those of the comendites (VIDAL-SOLANO, 2005). Like the comendites, these icelandites also display small negative Nb anomalies and Sr, Nd and Pb isotopic compositions consistent with minor assimilation of upper crustal Precambrian materials (VIDAL-SOLANO *et al.*, 2008a, b). Therefore, they are thought to derive from open-system fractional crystallization of tholeiitic to transitional magmas. TILL *et al.* (2009) have published andesite and dacite analyses with similar characteristics from the Miocene volcanic fields of Sierra Libre, Santa Ursula, Sierra El Aguaje and coastal Sonora, together with one Ar-Ar age (11.41 ± 0.04 Ma) and several ages estimated by stratigraphic correlation for these rocks (12.2-11.1 Ma), which are close to that of the former authors.

4. Discussion

4.1. Volcanism and Tectonic Reconstructions: Spatial and Temporal Constraints

There are two prerequisite conditions for reconstructing past tectonic regimes from the spatial and temporal patterns of volcanic activity described above. First, the geographic position of the volcanics should mark those of their sources at the time of their emplacement: in other words, most authors implicitly or explicitly assume that magmas ascend more or less vertically toward the surface from their ca. 50–100 km deep source, i.e., through the lithospheric mantle and the overlying crust. This assumption is consistent with current models of magma dynamics and emplacement in arcs as well as in intraplate extensional/transtensional settings and in rifts (ANNEN et al., 2006; ZELLMER and ANNEN, 2008). In both cases, magmas are thought to ascend first as visco-elastic diapirs and then to fill subvertical fracture networks within the brittle upper crustal rocks. However, Castillo (2008) claimed that basaltic mag-mas produced beneath the axis of the Proto-Gulf of California migrated laterally toward the fossil trench within the lithospheric mantle (1) ~150 km below Santa Clara and Mesa San Carlos volcanic fields, and (2) ~100 km below San Ignacio volcanic field (Fig. 1). This model has been questioned by MAURY et al. (2009), who contended that corresponding magmas rose up through the Baja California astheno-spheric window as proposed by previous authors (BENOIT et al., 2002; PALLARES et al., 2007).

The second condition is that the K-Ar and/or Ar-Ar ages of volcanic rocks could be taken as indicative of that of the tectonic event responsible for their genesis, i.e., that melts are not stored within the crust long enough to allow tectonic changes to occur between partial melting and final crystallization. Based on a proposal by CAÑÓN-TAPIA and WALKER (2004), NEGRETE-ARANDA and CAÑÓN-TAPIA (2008) consider that Baja California melts could have been stored within either their source zones or the over-lying lithosphere during ~10^6 years. Thus, there could be a significant decoupling between the tectonic regim deduced from their composition and that occurring at the time of their emplacement. The same hypothesis has been considered for Sonora lavas by TILL et al. (2009) for subduction-related lavas post-dating by 4 Ma the end of the subduction of the Farallon plate at ca. 12.5 Ma. However, current thermal models predict that silicate melts can have segregation times (from their mantle or crustal source) in the range 10^3–10^6 years, but ascent times that are geologically almost instantaneous (ANNEN et al., 2006). The largest volcanoes on earth, i.e., the Hawaiian shield volcanoes, have an expected life

time of ~1 Ma (DEPAOLO and STOLPER, 1996), as examplified by the ~0.65 Ma time range obtained for the 3.1-km-deep drilling in Mauna Kea volcano (GARCIA et al., 2007). Of course, lower storage/residence time is likely to be expected for the much smaller Baja California and Sonora volcanoes.

The time span separating partial melting from crystallization at the surface can also be evaluated using short-lived U-series isotopes (e.g., HAWKES-WORTH et al., 2004). It is usually very short (a few 10^3 years, or even less) for mafic magmas from intraplate settings (SIGMARSSON et al., 2005) and arcs (TURNER et al., 2001; ZELLMER, 2008), e.g., for the Pinacate Quaternary basalts (Asmerom and Edwards, 1995). In arcs, it increases together with silica contents up to 10^4–10^5 years, with 2×10^5 years (0.2 Ma) as an usual upper limit for the more evolved rock types (HAWKESWORTH et al., 2004; ZELLMER et al., 2005; ZELLMER, 2008). This last time span is within the range of usual errors on K-Ar and Ar-Ar ages measured at ca. 10-12 Ma (± 0.05 to ± 0.40 Ma). Therefore, we will in the following discussion consider these ages as representative of the times of partial melting of the mantle or crustal sources of Baja California and Sonora lavas.

4.2. Insights from the Distribution of Lava Types in Northwestern Mexico

The spatial distribution of the above-described lava types is rather striking (Fig. 1). Indeed, they usually occur within several 100-km-long belts subparallel to the fossil trench and to the axis of the Gulf of California. Five of these belts can be recognized. They are, from west to east: (1) a belt in clear fore-arc position, including the alkali basalts and trachybasalts from San Quintín and San Carlos-Santa Catarina, the Santa Clara adakites and NEB and the Santa Margarita adakites; (2) the "magnesian andesite belt" extending from Jaraguay to La Purís-ima and the San Juanico-San Ignacio tholeiitic basaltic andesites; (3) the Comondu calc-alkaline arc; (4) a set of Plio-Quaternary calc-alkaline islands and volcanoes located along the western Gulf margin (Isla San Luis, Tres Virgenes, La Reforma, Cerro Mencenares, Isla Coronado); and (5) the central part of the Gulf, including the MORB-type basalts from

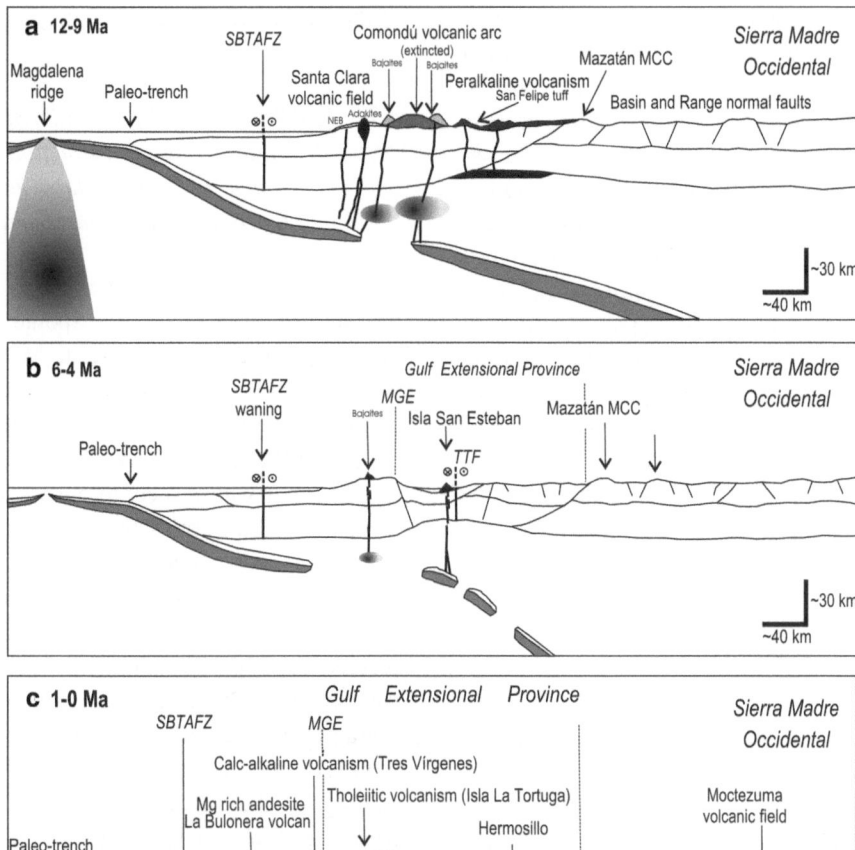

Figure 4
Three schematic W-E cross sections from the paleo-trench to Sierra Madre Occidental, showing the relationship between the evolution of volcanism and tectonics from 12 Ma to the present, at ca. 28°N latitude. **a** Between ca. 12 and 9 Ma, the activity of Magdalena ridge decreased and the right-handed motion between the Pacific and North America plates began along the Tosco-Abreojos fault zone. The localization of the Oligocene to Early Miocene metamorphic core complex (MCC) belt and schematic Basin and Range faulted structure are shown. Note that there is no Basin and Range structure in Baja California. **b** The plate motion is partitioned into the waning Tosco-Abreojos fault zone and the Gulf of California, here the Tiburón transform fault (TTF). The Gulf Extensional Province is located along the western Basin and Range Province. **c** At the present time, the dextral transform motion is principally controlled by faults within the Gulf, here the Ballena transform fault (BTF). Note the intracontinental Quaternary basaltic volcanism in the Moctezuma region. For **a**, **b** and **c**, see text for the source and origin of volcanism. *SBTAFZ* San Benito Tosco Abreojos Fault Zone, *MCC* metamorphic core complex, *MGE* Mean Gulf Escarpment, *TTF* Tiburón transform fault, *BTF* Ballena transform fault

Guaymas Basin and Tortuga Island and the calc-alkaline and adakitic lavas of Isla San Esteban. Such a belt array is much less obvious in Sonora, although the tholeiitic/transitional lavas and the peralkaline rhyolites and icelandites occur within a broad NW-SE trending band (Fig. 1).

The main implication of the belt array distribution described above is that the Early-Middle Miocene organization of the Pacific-Farallon subduction zone controlled the distribution of the post-subduction (Late Miocene to Quaternary) volcanism throughout Baja California. Especially adakites, NEB, magnesian andesites, tholeiitic andesites and alkali basalts/ trachybasalts were emplaced on front-arc position with respect to the axis of the Comondú arc (Fig. 1). This feature suggests that the slab tearing process responsible for the ascent of subslab magmas occurred in the front-arc position (PALLARES et al., 2007) rather than in the back-arc position, i.e., beneath the Proto-Gulf of California (CASTILLO, 2008). Another interesting feature is the back-arc position of Plio-Quaternary calc-alkaline and/or adakitic islands and volcanoes (Isla San Luis, Tres Virgenes, La Reforma, Cerro Mencenares, Isla Coronado), close to the axis of the Gulf, which may indicate that the thermal flux responsible for the partial melting of their sources was linked to the Gulf opening (DESONIE, 1992; BIGIOGGERO et al., 1987, 1995; CALMUS et al., 2008).

4.3. Spatial and Temporal Patterns of Volcanism in Northwestern Mexico

4.3.1 Early-Middle Miocene (Aquitanian to Langhian): active subduction and Basin and Range tectonics

Prior to ca. 13 Ma, calc-alkaline magmatism was active in all the studied areas (Fig. 3a), where it built up the last volcanic edifices of the Comondú arc and several sierras in Sonora. Other geochemical groups were not represented in Baja California with the exception of a single magnesian andesite flow, 14.6 Ma old, in La Purísima (BELLON et al., 2006). In western Sonora, tholeiitic to transitional basalts and basaltic andesites, very similar to the lithospheric mantle-derived Early Miocene basalts of the Basin and Range province in the southwestern USA (MILLER et al., 2000), were emplaced in several areas (Fig. 3a), e.g., in the substratum of the Pinacate and Moctezuma volcanic fields. Their geochemical signature is consistent with the involvement of lithospheric materials, possibly from a highly radiogenic (high $^{87}Sr/^{86}Sr$, low $^{143}Nd/^{144}Nd$) Precambrian source (VIDAL-SOLANO et al., 2008a). Thus, their occurrences stake the southward prolongation of the Basin and Range extensional province (PAZ-MORENO et al., 2003; VIDAL-SOLANO et al., 2008a). The lack of these tholeiitic to transitional lavas in Baja California is consistent with the fact that this area was not part of the Basin and Range province, as discussed above (Sect. 2).

4.3.2 Middle to Late Miocene (Serravallian-Messinian): End of Subduction and Basin and Range Extension, Slab-Tearing and Opening of an Asthenospheric Window

A major change is recorded at ca. 13 Ma throughout the studied area (Fig. 3b). Sporadic calc-alkaline activity persisted during the Late Miocene in several volcanic fields of Baja California and Sonora (until 5.8 Ma in Puertecitos, 7.3 Ma in Jaraguay and 8.3 Ma in Sierra El Aguaje), but it became volumetrically minor with respect to other lava types.

In Baja California, these include (1) lavas of presumed subslab mantle origin, such as the Mesa San Carlos and Mesa Santa Catarina trachybasalts (PALLARES et al., 2007; CASTILLO, 2008), the San Juanico-San Ignacio tholeiitic basaltic andesites (BENOIT et al., 2002; BELLON et al. 2006), and Santa Clara NEB (according to the interpretation of CASTILLO, 2008); (2) lavas derived from the partial melting of the subducted Pacific-Farallon oceanic crust, i.e., Santa Clara adakites according to the interpretation of AGUILLÓN-ROBLES et al. (2001) and followers (e.g., MAURY et al., 2009); (3) lavas of presumed supraslab mantle origin, such as the volumetrically major magnesian andesites (SAUNDERS et al., 1987; CALMUS et al., 2003; CASTILLO, 2008; PALLARES et al., 2008) and the Santa Clara NEB (AGUILLÓN-ROBLES et al., 2001; BENOIT et al., 2002); and finally (4) adakites derived from the mafic base of the Baja crust heated by subslab niobium-enriched melts, according to CASTILLO's (2008, 2009) model. Whatever the divergences regarding the sources of individual lava types, most authors (with the exception of NEGRETE-ARANDA and CAÑÓN-TAPIA, 2008) agree that the geochemical features of some of them suggest that they derive from the subslab (Pacific) asthenospheric mantle. Therefore, a tear-in-the-slab (evolving through time towards a slab window) is thought to have developed beneath the Peninsula at ca. 13 Ma (CALMUS et al., 2003; BELLON et al., 2006; MICHAUD et al., 2006; PALLARES et al., 2007, 2008; CASTILLO, 2008, 2009; MAURY et al., 2009). In addition to the ascent of subslab melts, it allowed the hot Pacific asthenosphere to flux into the window and to trigger the melting of parts of the metasomatized supraslab mantle (e.g., of the sources of the magnesian andesite suite).

In Sonora, tholeiitic to transitional basalts and basaltic andesites were emplaced between 13 and 7.2 Ma in several volcanic fields, e.g., Pinacate and Las Trincheras (Fig. 3b). Their geochemical features suggest their derivation from deep asthenospheric mantle, uplifting as the result of convective thinning and extension of the Basin and Range lithosphere, and therefore a "no-slab" regime beneath coastal Sonora (VIDAL-SOLANO et al., 2008a). The associated peralkaline rhyolites (12.6-12 Ma) and icelandites (10.9 Ma) are thought to derive from the open-system fractional crystallization of these basaltic magmas (VIDAL-SOLANO et al., 2008a, b) coupled with minor assimilation of Precambrian continental crust.

The locations of the asthenosphere-derived Middle to Late Miocene lavas from Baja California and Sonora suggest that an about 200-km-wide slab window parallel to the fossil trench developed between ca. 13-10 Ma from Mesa San Carlos, Mesa Santa Catarina and Santa Clara to the west towards Sierra de Mazatán and Sierra Lista Blanca to the east (Fig. 3b). The wider distribution of its magmatic markers in the northern part of the study area is consistent with its formation as a southern extension of the pre-existing Southern California window (WILSON et al., 2005; MICHAUD et al., 2006; PALLARES et al., 2007; VIDAL-SOLANO et al., 2008a, b; MCCRORY et al., 2009). However, available K-Ar and Ar-Ar data do not allow to track its propagation, as ages of 13–12 Ma have been measured on basalts and magnesian andesites from Sonora (VIDAL-SOLANO et al. 2008a; TILL et al., 2009) and southern Baja California (La Purísima; BELLON et al., 2006). The lack of Miocene adakites in Sonora and their occurrence in Santa Clara, Santa Rosalía and Jaraguay is consistent with their interpretation as slab melts. Indeed, oceanic metabasalts experience hydrous melting at relatively low pressures (1–2.5 GPa: DEFANT AND DRUMMOND, 1990; MARTIN, 1999; MARTIN et al., 2005) in hot subduction zones, where adakites are often found in front-arc position with respect to calc-alkaline lavas (DEFANT et al., 1992). Therefore, the Miocene Baja California

adakites were likely to derive from the upper (i.e., shallow) lip of the slab window (BENOIT et al., 2002; PALLARES et al., 2007), which underwent partial melting triggered by its thermal erosion by the ascending hot subslab asthenosphere (THORKELSON, 1996; THORKELSON and BREITSPRECHER, 2005). On the contrary, this partial melting was unlikely to occur in the much deeper lower (eastern) lip of the window, beneath Sonora. The occurrence of magnesian andesites and NEB in Baja California and their lack in Sonora are consistent with this interpretation, providing that these lavas derive from the supraslab mantle metasomatized by adakitic melts (BENOIT et al., 2002; CALMUS et al., 2003; PALLARES et al., 2008). The partial melting of this supraslab mantle has probably been triggered by the thermal input from the Pacific asthenosphere ascending through the slab window (PALLARES et al., 2007, 2008; CASTILLO, 2008).

4.3.3 Pliocene-Quaternary: Opening of the Gulf and Melting of Slab Slivers

Young volcanic activity in the northern and northeastern parts of the Pacific Mexican margin emplaced Quaternary alkali basalts in San Quintín, and tholeiitic, transitional and mildly alkali basalts together with trachybasalts in Pinacate and Moctezuma volcanic fields (Fig. 3c). The incompatible element patterns and isotopic (Sr, Nd, Pb) compositions of these lavas are typical of their OIB affinity, and they are thought to derive from plume-type asthenospheric mantle (ASMEROM and EDWARDS, 1995; LUHR et al., 1995; PAZ-MORENO et al., 2003; GOSS et al., 2008).

The numerous Plio-Quaternary magnesian andesite fields that trend NNW-SSE along Baja California from Jaraguay to La Purísima (Figs. 1 and 3c) resulted from the dehydration melting of amphibole-rich supraslab mantle metasomatized by either adakitic melts and/or by hydrous fluids (SAUNDERS et al., 1987; CALMUS et al., 2003; CASTILLO, 2008; PALLARES et al., 2008). With time, they became progressively depleted in Y and HREE, a feature consistent with the increase of the amount of residual garnet produced by the amphibole dehydration melting reaction (FRANCIS and LUDDEN, 1995; DALPÉ and BAKER, 2000; PALLARES et al., 2008). The origin of the melting is attributed to the high thermal flux linked to the uprise of Pacific asthenosphere in the "no-slab" regime that followed the detachment and sinking of the deep part of the Farallon plate (PALLARES *et al.*, 2007, 2008; CASTILLO, 2008).

The fact that the Quaternary Gulf tholeiites from Isla Tortuga (an. 16, Table 1) and Guaymas Basin differ from regular MORB by their selective enrichment in LILE and variable Sr isotopic ratios has been attributed to the presence of a "minor residual calc-alkaline component" in the mantle underlying the central and northern parts of the Gulf of California (SAUNDERS *et al.*, 1982a, b). In the same area, numerous Plio-Quaternary calc-alkaline and/or adakitic volcanoes (an. 3–4 and 7–8, Table 1) trend along the western margin of the Gulf (Puertecitos, Isla San Luis, Tres Virgenes, La Reforma, El Aguajito, Cerro Los Mencenares and Isla Coronado) and are even present in its central part (Isla San Esteban). Their occurrence suggests the presence below the Gulf area of slivers of subducted oceanic crust and of lithospheric mantle carrying a subduction-related geochemical imprint because of its interaction with either slab-derived hydrous fluids or adakitic melts (DESONIE, 1992; CALMUS *et al.*, 2008).

The Miocene tectonic and magmatic history of the area may account for the presence of slivers of oceanic crust and of subduction-modified mantle beneath the thinned continental crust of the central and northern Gulf of California. Prior to the opening of the Gulf, it was part of the western North American margin and was underlain by the subducting Farallon plate from ca. 25 to 13 Ma. The subcontinental lithospheric mantle may then have interacted with slab-derived hydrous fluids. During the opening of the slab window and the following sinking of the deep part of the Farallon plate, slivers of oceanic crust may have been introduced within this lithospheric mantle beneath Isla San Esteban, and possibly Cerro Mencenares and Isla Coronado (CALMUS *et al.*, 2008). Indeed, THORKELSON (1996) and THORKELSON and BREITSPRECHER (2005) have shown that the slab edges of an asthenospheric window are able to either melt or to leave restite fragments, which may become long-term residents of the continental lithospheric mantle. Then, the lithospheric thinning and rupture linked to the Pliocene opening of the Gulf

generated a high thermal regime, and asthenosphere-derived MORB-type tholeiites were emplaced in Isla Tortuga and in local spreading centers (Guaymas Basin, Lower Tiburón?). During their ascent, they interacted with the subduction-modified lithospheric mantle and were enriched in LILE, LREE and radiogenic Sr. The high thermal regime associated to asthenospheric uprise also induced partial melting of the Gulf heterogeneous lithospheric mantle, generating calc-alkaline basalts and basaltic andesites, or alternatively adakites from slivers of oceanic crust incorporated within this mantle (CALMUS *et al.*, 2008).

5. *Conclusions*

1. Prior to 13 Ma, the Neogene magmatic activity in northwestern Mexico, linked to the subduction of the Pacific-Farallon oceanic plate, emplaced the widespread and relatively homogeneous calc-alkaline Comondú and central coast of Sonora volcanic arc. In the extensional domain of western Sonora, tholeiitic to transitional basalts and basaltic andesites, very similar to the lithospheric mantle-derived Early Miocene basalts of the Basin and Range province in the southwestern USA, were emplaced in several areas (Fig. 3a).

2. The end of the Farallon subduction was marked by the emplacement of much more complex Middle to Late Miocene volcanic associations (Fig. 3b), between 13 and 7 Ma. Sporadic calc-alkaline activity persisted in Baja California (BC) and Sonora (SO), but became volumetrically minor with respect to other lava types. These include (a) lavas of presumed subslab mantle origin, such as alkali trachybasalts (BC), asthenosphere-derived tholeiitic/transitional basalts and basaltic andesites (BC, SO), and peralkaline rhyolites (comendites) and icelandites resulting from the open system fractional crystallization of these basaltic magmas (SO); (b) adakites derived from the partial melting of the subducted Pacific-Farallon oceanic crust (BC); and (c) magnesian andesites and niobium-enriched basalts (BC) derived from the melting of supraslab mantle metasomatized by adakitic melts.

3. We show that the spatial and temporal distribution of these lava types is consistent with the

development of a slab tear, evolving into a 200-km-wide slab window parallel to the fossil trench, within the young part of the subducted plate. Tholeiitic, transitional and alkali basalts of sub-slab origin ascended through this window (BC, SO), and adakites (BC) derived from the partial melting of the upper lip of the slab window triggered by its thermal erosion by the ascending hot subslab asthenosphere. The latter also triggered the melting of the hydrous fluid and/or slab melt-metasomatized supraslab mantle, generating calc-alkaline lavas (BC, SO), magnesian andesites (BC) and NEB (BC). The extension of the slab window below the continent is difficult to determine because it is controlled by several kinematic and geologic parameters. The first one is the geometry of the connection with the southern California asthenospheric window. The good fit between the location of the southern limit of the latter (WILSON *et al.*, 2005) and the northern limit of the Baja California slab window proposed by PALLARES *et al.* (2007) suggests that both slab windows are connected, and that the southern one is a younger extension of the northern and older one. In the case of the California slab window, it is possible to evaluate the velocity of opening based on the spreading ridge rate prior to ridge subduction, and on the spreading rate of remaining active ridge segments between Pacific and Farallon or between fragmented plates, such as the Guadalupe or Magdalena microplates. On the contrary, below northwestern Mexico, no ridge subduction occurred, and spreading along the ridge segments decreased between 12 and ca. 8 Ma west of southern Baja California peninsula. Then the opening rate of the slab window would depend of the sinking velocity of the eastern Farallon slab root, and of the possible retreat of Guadalupe and Magdalena to the west with respect to Baja California. The hypothesis of an initiating slab window at ca. 12.5 Ma below southernmost Baja California is proposed also by FLETCHER *et al.* (2007) in their tectonic evolution model depicted by three schematic cross sections at 16, 15-13, and 12.5 Ma at the latitude of the Magdalena fan. At 15-13 Ma these authors hypothesize a breakup of

the slab below the Comondú arc, where the dip of the slab increases.

The Middle and Late Miocene volcanism in Baja California is very different from the contemporaneous volcanism in Sonora. Two reasons can be reasonably proposed to explain such a difference. In Baja California, thermal conditions allowed the partial melting of the subducted Magdalena plate and generation of adakite and magnesian andesite. Below Sonora, on the one hand, the depth and temperature of the subducted slab were incompatible with the formation of such magmas. On the other hand, the presence of tholeiitic to transitional basalts and derived peralkaline rhyolites and icelandites in Sonora are related to the continental Basin and Range extension, which did not occur in Baja California, west of the Main Gulf Escarpment. Nevertheless, the age of this silicic magmatism and its distribution mainly along the western margin of the Gulf of California suggest that it was probably triggered by the initiation of breakup of the Gulf Extensional Province.

4. During the Plio-Quaternary (Fig. 3c), the "no-slab" regime following the sinking of the old part of the Farallon plate within the deep mantle allowed the emplacement of OIB-type tholeiitic/transitional (SO) and alkali basalts (BC, SO) of deep asthenospheric origin. The lithospheric thinning and rupture linked to the opening of the Gulf of California (GC) generated a high thermal regime associated to asthenospheric uprise and emplaced depleted MORB-type tholeiites (GC). This thermal regime also induced partial melting of the heterogeneous thinned lithospheric mantle of the Gulf area, generating calc-alkaline basalts and related lavas (BC, GC, SO), and locally adakites from slivers of oceanic crust incorporated within this mantle (BC, GC) (Fig. 4).

5. We believe that the geochemical study and dating of Neogene and Quaternary lavas may provide constraints on regional geodynamic reconstructions, keeping in mind that the composition of a magma is primarily dependent on that of its source(s), the degree of partial melting and the residual mineral assemblage. In a complex post-

subduction setting, lava compositions may therefore provide clues on the natures of their source(s), and, undirectly, on the various tectonic regimes that led to their partial melting.

Acknowledgments

Fieldwork and analytical expenses were funded by the French-Mexican ECOS project, by the Centre National de la Recherche Scientifique (CNRS) and the Université de Bretagne Occidentale (Unité Mixte de Recherche 6538 "Domaines océaniques"). The Instituto de Geología of the Universidad Autónoma de San Luis Potosí and the Estación Regional del Noroeste (ERNO), Instituto de Geología of the Universidad Nacional Autónoma de México also provided financial support for fieldwork. We thank Jesus Vidal-Solano for the communication of data on Sonora volcanism and for many discussions on the slab window models. The pertinent comments of Jesus Vidal-Solano and of an anonymous reviewer led to significant improvements of the final manuscript.

REFERENCES

AGUILLÓN-ROBLES, A. (2002), *Subduction de dorsale et évolution du magmatisme associé: exemple de la Basse Californie (Mexique) du Miocène au Quaternaire*, Thèse de Doctorat, Université de Bretagne Occidentale, Brest (France) 214 p. + annexes.

AGUILLÓN-ROBLES, A., CALMUS, T., BENOIT, M., BELLON, H., MAURY, R.C., COTTEN, J., BOURGOIS, J., and MICHAUD, F. (2001), *Late Miocene adakites and Nb-enriched basalts from Vizcaino Peninsula, Mexico: indicators of East Pacific Rise subduction below southern Baja California ?* Geology 29, 531–534.

ALONSO-PEREZ R., MÜNTENER, O., and ULMER, P. (2009), *Igneous garnet and amphibole fractionation in the roots of island arcs: experimental constraints on H_2O undersaturated andesitic liquids*, Contrib. Mineral. Petrol. 157, 541–558.

ANNEN, C., BLUNDY, J.D., and SPARKS, R.S.J. (2006), *The genesis of intermediate and silicic magmas in deep crustal hot zones*, J. Petrol. 47, 505–539.

ARCULUS, R.J. (1994), *Aspects of magma genesis in arcs*, Lithos 33, 189–208.

ARCULUS, R.J., LAPIERRE, H., and JAILLARD, E. (1999), *Geochemical window into subduction and accretion processes: Raspas metamorphic complex, Ecuador*, Geology 27, 547–550.

ASMEROM, Y., and EDWARDS, E.L. (1995), *U-series isotope evidence for the origin of continental basalts*, Earth Plan. Sci. Lett. 134, 1–7.

ATHERTON, M. P., and PETFORD, N. (1993), *Generation of sodium-rich magmas from newly underplated basaltic crust*, Nature 362, 144–146.

ATWATER, T., and STOCK, J. (1998), *Pacific-North American plate tectonics of the Neogene southwestern United States: an update*, Inter. Geol. Rev. 40, 375–402.

BATIZA, R. (1978), *Geology, petrology, and geochemistry of Isla Tortuga, a recently formed tholeiitic island in the Gulf of California*, Geol. Soc. Amer. Bull. 89, 1309–1324.

BATIZA R., FUTA K., and HEDGE C. E. (1979), *Trace element and strontium isotope characteristics of volcanic rocks from Isla Tortuga: a young seamount in the Gulf of California*, Earth Plan. Sci. Lett. 43, 269–278.

BELLON, H., AGUILLÓN-ROBLES, A., CALMUS, T., MAURY, R.C., BOURGOIS, J., and COTTEN, J. (2006), *La Purisima Volcanic Field, Baja California Sur, Mexico: Mid-Miocene to recent volcanism in relation with subduction and asthenospheric window opening*, J. Volcan. Geother. Res. 152, 253–272, doi:10.1016/j.j volgeores.2005.10.005.

BENOIT, M., AGUILLÓN-ROBLES, A., CALMUS, T., MAURY, R.C., BELLON, H., COTTEN, J., BOURGOIS, J., and MICHAUD, F. (2002), *Geochemical diversity of Late Miocene volcanism in southern Baja California, Mexico: implication of mantle and crustal sources during the opening of an asthenospheric window*, J. Geology 110, 627–648, doi:10.1086/342735.

BIGIOGGERO, B., CAPALDI, G., CHIESA, S., MONTRASIO, A., VEZZOLI, L., and ZANCHI, A. (1987), *Post-subduction magmatism in the Gulf of California: the Isla Coronados (Baja California Sur, Mexico)*, Inst. Lombardo (Rendiconti Scienze) B121, 117–132.

BIGIOGGERO, B., CHIESA, S., ZANCHI, A., MONTRASIO, A., and VEZZOLI, L. (1995), *The Cerro Mencenares volcanic center, Baja California Sur: source and tectonic control on postsubduction magmatism within the Gulf Rift*, Geol. Soc. Am. Bull. 107, 1108–1122.

BONINI, A.J., and BALDWIN, S.L. (1998), *Mesozoic metamorphic and middle to late Tertiary magmatic events on Magdalena and Santa Margarita Islands, Baja California Sur, Mexico: implications for the tectonic evolution of the Baja California continental borderland*, Geol. Soc. Am. Bull. 110, 1094–1104.

CALMUS, T., AGUILLÓN-ROBLES, A., MAURY, R.C., BELLON, H., BENOIT, M., COTTEN, J., BOURGOIS, J., and MICHAUD, F. (2003), *Spatial and temporal evolution of basalts and magnesian andesites ("bajaites") from Baja California, México: the role of slab melts*, Lithos 66, 77–105, doi:10.1016/S0024-4937(02)00214-1.

CALMUS, T., PALLARES, C., MAURY, R.C., BELLON, H., PÉREZ-SEGURA, E., AGUILLÓN-ROBLES, A., CARRENO, A.-L., BOURGOIS, J., COTTEN, J., and BENOIT, M. (2008), *Petrologic diversity of Plio-Quaternary post-subduction volcanism in Baja California: an example from Isla San Esteban (Gulf of California, México)*, Bull. Soc. Géolog. France 179, 465–481, doi:10.2113/gssgfbull.179.5.465.

CALMUS, T., POUPEAU, G., DEFAUX, J., and LABRIN, E. (1997), *Apatite fission track ages in Sonora, Mexico: a recording of Basin and Range events and opening of the Gulf of California*, GEOS 18-4, 293.

CAMERON, K.L., and CAMERON, M. (1985), *Rare earth element, $^{87}Sr/^{86}Sr$, and $^{143}Nd/^{144}Nd$ compositions of Cenozoic orogenic dacites from Baja California, northwestern Mexico, and adjacent west Texas: evidence for the predominance of a subcrustal component*, Contrib. Mineral. Petrol. 91, 1–11.

CAÑÓN-TAPIA, E., and WALKER, G.P.L. (2004), *Global aspects of volcanism: the perspectives of "plate tectonics" and "volcanic systems"*, Earth Sci. Rev. 66, 163–182.

CASTILLO, P.R. (2008), *Origin of the adakite-high Nb basalt association and its implications for post-subduction magmatism in Baja California, Mexico*, Geol. Soc. Am. Bull. 120, 451–462.

CASTILLO, P.R. (2009), *Origin of Nb-enriched basalts and adakites in Baja California, Mexico, revisited: reply*, Geol. Soc. Am. Bull. 121, 1470–1472.

CHIARADIA M., MÜNTENER O., BEATE B., and FONTIGNIE D. (2009), *Adakite-like volcanism od Ecuador: lower crust magmatic evolution and recycling*, Contrib. Miner. Petrol. 158, 563–588.

CONLY, A.G., BRENAN, J.M., BELLON, H., and SCOTT, S.D. (2005), *Arc to rift transitional volcanism in the Santa Rosalía Region, Baja California Sur, Mexico*, J. Volcan. Geotherm. Res. 142, 303–341.

Consejo de Recursos Minerales (CRM), (1992), *Monografía geológica-minera del Estado de Sonora*, Secretaría de Energía, Minas e Industria Paraestatal, 220 p., + anexos.

COTTEN, J., LE DEZ, A., BAU, M., CAROFF, M., MAURY, R.C., DULSKI, P., FOURCADE, S., BOHN, M., and BROUSSE R. (1995), *Origin of anomalous rare-earth element and yttrium enrichments in subaerially exposed basalts: evidence from French Polynesia*, Chemical Geology 119, 115–138.

DALPÉ, C., and BAKER, D. R. (2000), *Experimental investigation of large-ion-lithophile-element-, high-field-strength-element- and rare-earth-element-partitioning between calcic amphibole and basaltic melt: the effects of pressure and oxygen fugacity*, Contrib. Mineral. Petrol. 140, 233–250.

DEFANT, M.J., and DRUMMOND, M.S. (1990), *Derivation of some modern arc magmas by melting of young subducted lithosphere*, Nature 347, 662–665.

DEFANT, M.J., JACKSON, T.E., DRUMMOND, M.S., DE BOER, J.Z., BELLON, H., FEIGENSON, M.D., MAURY, R.C., and STEWARD, R.H. (1992), *The geochemistry of young volcanism throughout western Panama and southeastern Costa Rica: an overview*, J. Geol. Soc. London 149, 569–579.

DEFANT, M.J., and KEPEZHINSKAS, P. (2001), *Evidence suggests slab melting in arc magmas*, EOS (Am. Geophys. Union Transactions) 82, 67–69.

DEMANT, A., (1984), *The Reforma Caldera, Santa Rosalía area, Baja California. A volcanological, petrographical and mineralogical study*. In *Neotectonics and sea level variations in the Gulf of California area, Symposium*, Universidad Nacional Autónoma de México, Instituto de Geología, 75–96.

DESONIE, D.L., (1992), *Geological and geochemical reconnaissance of Isla San Esteban: post-subduction orogenic volcanism in the Gulf of California*, J. Volcan. Geotherm. Res. 52, 123–140.

DEPAOLO, D.J., and DALEY, E.E. (2000), *Neodymium isotopes in basalts of the southwest Basin and Range and lithospheric thinning during continental extension*, Chemical Geology 169, 157–185.

DEPAOLO, D.J., STOLPER, E.M. (1996), *Models of Hawaiian volcano growth and plume structure: implications of results from the Hawaii Scientific Drilling Project*, J. Geophys. Res. 101, 11643–11654.

DOKKA, R.K., MERRIAM, R.H. (1982), *Late Cenozoic extension of northeastern Baja California, Mexico*, Geol. Soc. Am. Bull. 93, 371–378.

DYMENT, J. (2003), *Anomalies magnétiques et datations des fonds océaniques: quarante ans après Vine et Matthews*, Rapport quadriennal 2000–2003 du Comité National Français de Géodésie et de Géophysique (CNFGG), 160–179.

FLETCHER, J.M., GROVE, M., KIMBROUGH, D., LOVERA, O., and GEHRELS, G.E. (2007), *Ridge–trench interactions and theNeogene tectonic evolution of the Magdalena Shelf and southern Gulf of California: insights from detrital zircon U–Pb ages from the Magdalena Fan and adjacent areas*, Geol. Soc. Amer. Bull. 119 (11–12), 1313–1336.

FORNARI, D.J., SAUNDERS, A.D., and PERFIT, MR. (1982), *Major-element chemistry of basaltic glasses recovered during deep-sea drilling project LEG-64*, Initial reports of the Deep Sea Drilling Project 64, 649–666.

FITTON, J.G., JAMES, D., and LEEMAN, W.P. (1991), *Basic magmatism associated with late Cenozoic extension in the western Unites States: compositional cariations in time and space*, J. Geophys. Res. 96 (13), 693–13,711.

FRANCIS, D., and LUDDEN, J. (1995) *The signature of amphibole in mafic alkaline lavas, a study in the Northern Canadian Cordillera*, J. Petrol. 36, 1171–1191.

GABB, W.M. (1882), *Notes on the Geology of Lower California*, Geol. Survey California, *II, Sec. H*, Cambridge Mass, John Wilson & Son, pp. 137–148.

GANS, P.B. (1997), *Large-magnitude Oligo-Miocene extension in southern Sonora: implications for the tectonic evolution of northwest Mexico*, Tectonics, 16 (3), 388–408.

GARCIA, M.O., HASKINS, E.H., STOLPER, E.M., and BAKER, M. (2007), *Stratigraphy of the Hawaii Scientific Drilling Project core (HSDP2): anatomy of a Hawaiian shield volcano*, Geochem. Geophys. Geosystems 8 (2), Q02G20, doi: 10.129/2006GC001379.

GARDUÑO-MONROY, V.H., VARGAS-LEDEZMA, H., and CAMPOS-ENRIQUEZ, J.O. (1993), *Preliminary geologic studies of Sierra El Aguajito (Baja California, Mexico): a resurgent-type caldera*, J. Volcan. Geotherm. Res 59 (1–2), 47–58.

GASTIL, G., KRUMMENACHEER, D., and MINCH, J. (1979), *The record of Cenozoic volcanism around the Gulf of California*, Geol. Soc. Am. Bull. 90, 839–857.

GASTIL, R.G., PHILLIPS, R.P., and ALLISON, E.C. (1975), *Reconnaissance geology of the state of Baja California*, Geol. Soc. Am. Memoir. 140–170.

GOSS, A.R., GUTMANN, J.T., VAREKAMP, J.C., and KAMENOV, G. (2008), *Pb isotopes and trace elements of the Pinacate volcanic field, northwestern Sonora, Mexico: a Basin and Range miniplume near the EPR spreading center*, Geol. Soc. Am., Abstracts with Programs 40, pp. 530.

GUTMANN, J.T. (2002), *Strombolian and effusive activity as precursors to phreatomagmatism: eruptive sequence at maars of the Pinacate volcanic field, Sonora, Mexico*, J. Volcan. Geotherm. Res. 113, 345–356.

HAUSBACK, B.P. (1984), *Cenozoic volcanism and tectonic evolution of Baja California Sur, Mexico*, In: FRIZZELL V.A. Jr. ed., *Geology of the Baja California Peninsula*. Pacific Section, Soc. Econ. Paleontol. Mineralog. 39, 219–236.

HAWKESWORTH, C.J., GEORGE R., TURNER, S., and ZELLMER, G. (2004), *Time scales of magmatic processes*, Earth Plan. Sci. Lett. 218, 1–16.

HENRY, C.D. (1989), *Late Cenozoic Basin and Range structure in western Mexico adjacent to the Gulf of California*, Geol. Soc. Am. Bull. 101, 1147–1156.

HENRY, C.D., and ARANDA GÓMEZ, J. J. (2000). *Plate interactions control middle-late Miocene, proto-Gulf and Basin and Range*

Reprinted from the journal

extension in the southern Basin and Range, Tectonophysics 318, 1–26.

HERZIG, C.T. (1990). Geochemistry of igneous rocks from the Cerro Prieto geothermal field, northern Baja California, Mexico, J. Volcan. Geotherm. Res. 42, 261–271.

HIDALGO, S., MONZIER, M., MARTIN, H., CHAZOT, G., EISSEN, J.-P., and COTTEN, J. (2007), Adakitic magmatism in the Ecuadorian volcanic front: petrogenesis of the Ilizina Volcanic Complex (Ecuador), J. Volcan. Geotherm. Res. 159, 366–392.

JÉGO, S., MAURY, R.C. POLVÉ, M., YUMUL Jr., G.P., BELLON, H., TAMAYO Jr., R.A., and COTTEN, J. (2005), Geochemistry of adakites from the Philippines: constraints on their origins, Resource Geol. 55, 163–187.

KEPEZHINSKAS, P.K., DEFANT, M.J., and DRUMMOND, M.S. (1996), Progressive enrichment of island arc mantle by melt-peridotite interaction inferred from Kamchatka xenoliths, Geochim. Cosmoch. Acta 60, 1217–1229.

LE BAS, M.J., LE MAITRE, R.W., STRECKEISEN, A., and ZANETTIN, B. (1986), A chemical classification of volcanic rocks based on the total alkali-silica diagram, J. Petrol. 27, 745–750.

LEE, J., MILLER, M.M., CRIPPEN, R., HACKER, B., and LEDESMA-VAZQUEZ, J. (1996), Middle Miocene extension in the Gulf Extensional Province, Baja California: evidence from the suthern Sierra Juarez, Geol. Soc. Am. Bull. 108, 505–525.

LONSDALE, P. (1991), Structural patterns of the Pacific floor offshore of peninsular California, In: DAUPHIN, J.P., SIMONEIT, B.A. eds., The Gulf and Peninsular Province of the Californias, Am. Ass. Petrol. Geol. Memoir. 47, 87–125.

LUHR, J.F., ARANDA-GÓMEZ, J.J., and HOUSH, T.B. (1995), San Quintin Volcanic field, Baja California Norte, México. Geology, petrology, and geochemistry, J. Geophys. Res. 100, 10,353–10,380.

MACPHERSON, C.G., DREHER, S.T., and THIRLWALL, M.F. (2006), Adakites without slab melting: high pressure differentiation of island arc magma, Mindano, the Philippines, Earth Plan. Sci. Lett. 243, 581–593.

MARTIN, H., (1999), Adakitic magmas: modern analogues of Archean granitoids, Lithos 46, 411–429.

MARTIN, H., SMITHIES, R.H., RAPP, R., MOYEN, J.-F., and CHAMPION, D. (2005), An overview of adakite, tonalite-trondhjemite-granodiorite (TTG), and sanukitoid: relationships and some implications for crustal evolution, Lithos 79, 1–24.

MARTÍN-BARAJAS, A., STOCK, J.M., LAYER, P., HAUSBACK, B., RENNE, P., and LOPEZ-MARTÍNEZ, M. (1995), Arc-rift transition volcanism in the Puertecitos Volcanic Province, northeastern Baja California, Mexico. Geol. Soc. Am. Bull. 107, 407–424.

MARTÍN, A., FLETCHER, J.M., LÓPEZ-MARTÍNEZ, M., and MENDOZA-BORUNDA, R. (2000), Waning Miocene subduction and arc volcanism in Baja California: the San Luis Gonzaga volcanic field, Tectonophysics 318, 27–51.

MAURY, R.C., CALMUS, T., PALLARES, C., BENOIT, M., GRÉGOIRE, M., AGUILLÓN-ROBLES, A., BELLON, H., and BOHN, M. (2009), Origin of the adakite-high Nb basalt association and its implications for post-subduction magmatism in Baja California, Mexico: discussion, Geol. Soc. Am. Bull. 122, 1465–1469, doi:10.1130/B30043.1.

MCCRORY, P.A., WILSON, D.S., and STANLEY, R.G. (2009), Continuing evolution of the Pacific-Juan de Fuca-North America slab window system–a trench-ridge-transform example from the Pacific rim, Tectonophysics 464, 30–42.

MEDINA, F., SUAREZ, F., and ESPINDOLA, J.M. (1989), Historic and Holocene volcanic centers in NW Mexico, A supplement to the IAVCEI catalogue, Bull. Volcan. 51, suppl. 1, 91–93.

MICHAUD, F., ROYER, J.-Y., BOURGOIS, J., DYMENT, J., CALMUS, T., BANDY, W., SOSSON, M., MORTERA-GUTIÉRREZ, C., SICHLER, B., REBOLLEDO-VIERA, M., and PONTOISE, B. (2006), Oceanic-ridge subduction vs. slab break-off: plate tectonic evolution along the Baja California Sur continental margin since 15 Ma, Geology 34, 13–16.

MIDDLEMOST, E.A.K. (1975), The basalt clan. Earth Sciences Review 11, 337–364.

MILLER, J.S., GLAZNER, A.F., FARMER, G.L., SUAYAH, I.B., and KEITH, L.A. (2000), A Sr, Nd, and Pb isotopic study of mantle domains and crustal structure in the Mojave Desert, California, Geol. Soc. Am. Bull. 112, 1264–1279.

MORA-KLEPEIS, G., and MCDOWELL, F. (2004), Late Miocene calc-alkaline volcanism in northwestern México: an expression of rift or subduction-related magmatism? J. South Am. Earth Sci. 17, 297–310.

MOYEN, J.-F. (2009), High Sr/Y and La/Yb ratios: the meaning of the "adakitic signature", Lithos 112, 556–574.

MÜNTENER, O., and ULMER, P. (2006), Experimentally derived high-pressure cumulates from hydrous arc magmas and consequences for the seismic velocity structure of lower arc crust, Geophys. Res. Lett. 33, L21308, doi:10.129/2006GL027629.

NEGRETE-ARANDA, R., and CAÑÓN-TAPIA, E. (2008), Post-subduction volcanism in the Baja California Peninsula, Mexico: the effects of tectonic reconfiguration in volcanic systems, Lithos 102, 392–414.

NEGRETE-ARANDA, R., CAÑÓN-TAPIA, E., BRANDLE, J.L., ORTEGA-RIVERA, M.A., LEE, J.K.W., SPELZ, R.M., and HINOJOSA-CORONA, A. (2010), Regional orientation of tectonic stress and the stress expressed by post-subduction high-magnesium volcanism in northern Baja California Peninsula, Mexico: tectonics and volcanism of San Borja volcanic field, J. Volcan. Geotherm. Res. 192, 97–115.

NIIDA, K., and GREEN, D.H. (1999), Stability and chemical composition of pargasitic amphibole in MORB pyrolite under upper mantle conditions, Contrib. Mineral. Petrol. 135, 18–40.

ORTEGA-GUTIÉRREZ, F., MITRE-SALAZAR, L.M., ROLDAN-QUINTANA, J., ARANDA-GÓMEZ, J.J., MORAN-ZENTENO, D., ALANIS-ALVAREZ, S., and NIETO-SAMANIEGO, A. (1992), Carta geológica de la República Mexicana, escala 1:2,000,000, quinta edición, Instituto de Geología UNAM y Sría. Energía e Industria Paraestatal, CRM, mapa + texto explicativo, 74 pp.

OSKIN, M. (2002), Tectonic evolution of the northern Gulf of California, Mexico, deduced from conjugated rifted margins of the Upper Delfin basin, Ph.D. Thesis, Pasadena, California Institute of Technology, 481 pp.

OSKIN, M., STOCK, J. (2003), Cenozoic volcanismand tectonics of the continental margins of the upper Delfin Basin, northeastern Baja California and western Sonora, In: JOHNSON, S.E. et al. eds., Tectonic Evolution of Northwestern Mexico and the Southwestern USA, Geol. Soc. Am. Spec. Paper 374, pp. 421–438.

PALLARES, C., MAURY, R.C., BELLON, H., ROYER, J.Y., CALMUS, T., AGUILLÓN-ROBLES, A., COTTEN, J., BENOIT, M., MICHAUD, F., and BOURGOIS, J. (2007), Slab-tearing following ridge-trench collision: evidence from Miocene volcanism in Baja California, México, J. Volcan. Geotherm. Res. 161, 95–117, doi:10.1016/j.jvolgeores.2006.11.002.

PALLARES, C., BELLON, H., BENOIT, M., MAURY, R.C., AGUILLÓN-ROBLES, A., CALMUS, T., and COTTEN, J. (2008), Slab-tearing following ridge-trench collision: evidence from Miocene volcanism in Baja California, Mexico, Lithos 105, 162–180.

PAZ-MORENO, F.A., and DEMANT, A. (1999), *The Recent Isla San Luis volcanic center: petrology of a rift-related volcanic suite in the northern Gulf of California, Mexico*, J. Volcan. Geotherm. Res. 93, 31–52.

PAZ-MORENO, F. A., DEMANT, A., COCHEMÉ J.-J., DOSTAL, J., and MONTIGNY, R. (2003), *The Quaternary Moctezuma volcanic field: a tholeiitic to alkali basaltic episode in the central Sonora Basin and Range Province, Mexico*, In: JOHNSON, S.E., PATERSON S.R., FLETCHER, J.M., GIRTY, G.H., KIMBROUGH, D.L., MARTÍN-BARAJAS, A., eds. *Tectonic evolution of northwestern México and southwestern USA*, Geol. Soc. Am. Sp. Paper 374, 439–455.

PECCERILLO, A., and TAYLOR, S.R. (1976), *Geochemistry of Eocene calc-alkaline volcanic rocks from the Kastamonu area, northern Turkey*, Contrib. Mineral. Petrol. 58, 63–81.

PERFIT, M.R., SAUNDERS, A.D., and FORNARI, D.J. (1982), *Phase chemistry, fractional crystallization, and magma mixing in basalts from the Gulf of California, deep-sea drilling project LEG-64*, Initial reports of the Deep Sea Drilling Project 64, 649–666.

PETFORD, N., and ATHERTON, M. P. (1996), *Na-rich partials melts from newly underplated basaltic crust: the Cordillera Blanca Batholith, Peru*, J. Petrol. 37, 1491–1521.

PROUTEAU, G., and SCAILLET, B. (2003), *Experimental constraints on the origin of the 1991 Pinatubo dacite*, J. Petrol. 44, 2203–2241.

PROUTEAU, G., SCAILLET, B., PICHAVANT, M., and MAURY, R.C. (2001), *Evidence for mantle metasomatism by hydrous silicic melts derived from subducted oceanic crust*, Nature 410, 197–200.

RAPP, R.P., LAPORTE, D., MARTIN, H., and SHIMIZU, N. (2006), *Reaction between slab-derived melts and peridotite in the mantle wedge: experimental constraints at 3.8 GPa*, Geochim. Cosmochim. Acta 70, (18), A517, doi:10.116/j.gca.2006.06.953.

RAPP, R.P., SHIMIZU, N., NORMAN, M.D., and APPLEGATE, G.S. (1999), *Reaction between slab-derived melts and peridotite in the mantle wedge: experimental constraints at 3.8 GPa*, Chemical Geology 160, 335–356.

ROGERS, G., and SAUNDERS, A.D. (1989), *Magnesian andesites from Mexico, Chile and the Aleutian Islands: implications for magmatism associated with ridge-trench collisions*, In: CRAWFORD, A.J. ed., Boninites and related rocks, London, Unwin Hyman, pp. 416–445.

ROGERS, G., SAUNDERS, A.D., TERRELL, D.J., VERMA, S.P., and MARRINER, G.F. (1985), *Geochemistry of Holocene volcanic rocks associated with ridge subduction in Baja California, Mexico*, Nature 315, 389–392.

ROLDÁN-QUINTANA, J., MORA-ALVAREZ, G., CALMUS, T., VALENCIA-MORENO, M., and LOZANO-SANTACRUZ, R. (2004), *El Graben de Empalme, Sonora, México: Magmatismo y tectónica extensional asociados a la ruptura inicial del Golfo de California*, Rev. Mex, Cienc. Geol. 21 (3), 320–334.

SAJONA, F.G., MAURY, R.C., BELLON, H., COTTEN, J., and DEFANT, M. (1996), *High field strength element enrichment of Pliocene-Pleistocene island arc basalts, Zamboanga Peninsula, western Mindanao (Philippines)*, J. Petrol. 37, 693–726.

SAMANIEGO, P., MARTIN, H., ROBIN, C., MONZIER, M., and COTTEN, J. (2005), *Temporal evolution of magmatism at Northern Volcanic Zone of the Andes: the geology and petrology of Cayambe Vocnic Complex (Ecuador)*, J. Petrol. 46, 2225–2252.

SAUNDERS, A.D., FORNARI, D. J., JORON, J.L., TARNEY, J., TREUIL, M. 1982a. *Geochemistry of basic igneous rocks, Gulf Of California, Deep Sea Drilling Project Leg. 641*. In: J. BLAKLEE, L.W. PLATT and L.N. STOUT, Eds, *Initial Reports of Deep Sea Drilling Project, Washington*, 64, part 2, 595–642.

SAUNDERS, A. D., FORNARI, D. J., and MORRISSON, M.A. (1982b), *The composition and emplacement of basaltic magmas produced during the development of continental-margin basins: the Gulf of California, Mexico*, J. Geol. Soc. London 139, 335–346.

SAUNDERS, A.D., ROGERS, G., MARRINER, G.F., TERRELL, D.J., and VERMA, S.P. (1987), *Geochemistry of Cenozoic volcanic rocks, Baja California, Mexico: implications for the petrogenesis of post-subduction magmas*, J. Volcan. Geotherm. Res. 32, 223–245.

SAWLAN, M.G. (1991), *Magmatic evolution of the Gulf of California rift*, In: DAUPHIN, J.P., SIMONEIT, B.A. eds., The Gulf and Peninsular Province of the Californias, Am. Ass. Petrol. Geol. Memoir 47, pp. 301–369.

SAWLAN, M.G., and SMITH, J.G. (1984), *Petrologic characteristics, age and tectonic setting of Neogene volcanic rocks in northern Baja California Sur, Mexico*, In: FRIZZEL, V.A, Jr., ed., Geology of the Baja California Peninsula: Pacific Section, Soc. Econ. Paleontol. Mineralogist, pp. 237–251.

SCHIANO, P., MONZIER, M., EISSEN, J.-P., MARTIN, H., and KOGA, K.T. (2010), *Simple mixing as the major control of the evolution of volcanic suites in the Ecuadorian Andes*, Contrib. Mineral. Petrol. 160, 297–312.

SCHMITT, A.K., STOCKLI, D.F., and HAUSBACK, B.P. (2006), *Eruption and magma crystallization ages of Las Tres Virgenes (Baja California) constrained by combined $^{230}Th/^{238}U$ and (U-Th)/He dating of zircon*, J. Volcan. Geotherm. Res. 158, 281–295.

SEN, C., and DUNN, T. (1994), *Dehydration melting of a basaltic composition amphibole at 1.5 and 2.0 Gpa: implication for the origin of adakites*, Contrib. Mineral. Petrol. 117, 394–409.

SIGMARSSON, O., CONDOMINES, M., and BACHÈLERY, P. (2005), *Magma residence time beneath the Piton de la Fournaise volcano, Reunion Island, from U-series disequilibria*, Earth Planet. Sci. Lett. 234, 223–234.

SONDER, L.J., JONES, C.H. (1999), *Western United States extension: How the West was widened*, Annual Rev. Earth Planet. Sci. 27, 417–462.

SPENCER, J. E., and NORMARK, W. R. (1989), *Neogene plate-tectonic evolution of the Baja California Sur continental margin and the southern Gulf of California, Mexico,*: In: WINTERER, E.L., HUSSONG, D.M., DECKER, R.W. eds., The eastern Pacific Ocean and Hawaii, Geol. Soc. Am., Geol. North Am. N, pp. 21–72.

STERN, R.J. (2002), *Subduction zones*, Rev. Geophys. 40, 1012, doi: 10.129/2001RG000108.

STOCK, J.M., and HODGES K.V. (1989), *Pre-Pliocene extension around the Gulf of California and the transfer of Baja California to the Pacific plate*, Tectonics 8, 99–115.

SUN, S.S., and McDONOUGH, W.F. (1989), *Chemical and isotopic systematics of oceanic basalts: implications for mantle composition and processes*, In: SAUNDERS, A.D., and NORRY, M.J. eds., Magmatism in the ocean basin, Geological Society of London Special Publication 42, pp. 313–345.

THORKELSON, D.J. (1996), *Subduction of diverging plates and the principles of slab window formation*, Tectonophysics 255, 47–63.

THORKELSON, D.J., and BREITSPRECHER, K. (2005), *Partial melting of slab window margins: genesis of adakitic and non-adakitic magmas*, Lithos 79, 25–41.

TILL, C.B., Gans, P.B., SPERA, F.J., MacMILLAN, I., and BLAIR, K.D. (2009), *Perils of petrotectonic modeling: a view from southern Sonora, Mexico*, J. Volcan. Geotherm. Res. 186, 160–168.

79

TURNER, S., EVANS, P., and HAWKESWORTH, C. (2001), *Ultrafast source-to-surface movement of melt at island arcs from* ^{226}Ra-^{230}Th *systematic*, Science 292, 1363–1366.

TURRIN, B.D., GUTMANN, J.T., and SWISHER III, C.C. (2008), *A 13 ± 3 ka age determination of a tholeiite, Pinacate volcanic field, Mexico, and improved methods for* $^{40}Ar/^{39}Ar$ *dating of young basaltic rocks*, J. Volcanol. Geotherm. Res. 177, 848–856.

UMHOEFER, P.J., DORSEY, R.J., WILLSEY, S., MAYER, L., and RENNE, P. (2001), *Stratigraphy and geochronology of the Comondu Group near Loreto, Baja California Sur, Mexico*, Sediment. Geol 144, 125–147.

VIDAL-SOLANO, J.R. (2005), *Le volcanisme hyperalcalin d'âge miocène moyen du nord-ouest du Mexique (Sonora). Minéralogie, géochimie, cadre géodynamique*, Thèse de Doctorat, Université Paul Cézanne, Aix-Marseille 3 (France), pp. 256.

VIDAL-SOLANO, J.R., PAZ-MORENO, F.A., IRIONDO, A., DEMANT, A., COCHEMÉ, J.-J. (2005), *Middle Miocene peralkaline ignimbrites in the Hermosillo region (Sonora, México). Geodynamic implications*, C. R. Geoscience 337, 1421–1430.

VIDAL-SOLANO, J.R., PAZ-MORENO, F.A., DEMANT, A., and LÓPEZ-MARTÍNEZ, M. (2007), *Ignimbritas hiperalcalinas del Mioceno medio en Sonora Central: revaluacion de la estratigrafía y significado del volcanismo terciaro*, Rev. Mex. Cien. Geol. 24, 47–67.

VIDAL-SOLANO, J.R., DEMANT, A., PAZ-MORENO, F.A., LAPIERRE, H., ORTEGA-RIVERA, M.A., and LEE, J.K.W. (2008a), *Insights into the tectonomagmatic evolution of NW Mexico: geochronology and geochemistry of the Miocene volcanic rocks from the Pinacate area, Sonora*, Geol. Soc. Am. Bull. 120, 691–708.

VIDAL-SOLANO, J.R., LAPIERRE, H., STOCK, J.M., DEMANT, A., PAZ-MORENO, F.A., BOSCH, D., BRUNET, P., and AMORTEGUI, A. (2008b), *Isotope geochemistry and petrogenesis of peralkaline Middle Miocene ignimbrites from central Sonora: relationship with continental break-up and the birth of the Gulf of California*, Bull. Soc. Géolog. France 179, 453–464.

WILSON, D.S., McCRORY, P.A., and STANLEY, R.G. (2005), *Implications of volcanism in coastal California for the Neogene deformation history of western North America*, Tectonics 24, TC3008, doi:10.129/2003TC001621.

YAXLEY, G.M., and GREEN, D.H. (1998), *Reactions between eclogite and peridotite: mantle refertilisation by subduction of oceanic crust*, Schweiz. Mineral. Petrogr. Mitt. 78, 243–255.

ZELLMER, G.F. (2008), *Some first-order ob servations on magma transfer from mantle wedge to upper crust at volcanic arcs*, Geol. Soc. London, Special Publications 304, 15–31.

ZELLMER, G.F., and ANNEN, C. (2008), *An introduction to magma dynamics*, Geol. Soc. London, Special Publications 304, 1–13.

ZELLMER, G.F., ANNEN, C., CHARLIER, B.L.A., GEORGE, R.M.M., TURNER, S.P., and HAWKESWORTH C.J. (2005), *Magma evolution and ascent at volcanic arcs: constraining petrogenetic processes through rates and chronologies*, J. Volcan. Geother. Res. 140, 171–191.

(Received February 14, 2010, revised July 8, 2010, accepted August 15, 2010, Published online November 9, 2010)

Pure Appl. Geophys. 168 (2011), 1331–1338
© 2010 The Author(s)
This article is published with open access at Springerlink.com
DOI 10.1007/s00024-010-0203-0

| Pure and Applied Geophysics

Double-difference Relocation of the Aftershocks of the Tecomán, Colima, Mexico Earthquake of 22 January 2003

VANESSA ANDREWS,[1] JOANN STOCK,[1] CARLOS ARIEL RAMÍREZ VÁZQUEZ,[2] and GABRIEL REYES-DÁVILA[2]

Abstract—On 22 January 2003, the $M_w = 7.6$ Tecomán earthquake struck offshore of the state of Colima, Mexico, near the diffuse triple junction between the Cocos, Rivera, and North American plates. Three-hundred and fifty aftershocks of the Tecomán earthquake with magnitudes between 2.6 and 5.8, each recorded by at least 7 stations, are relocated using the double difference method. Initial locations are determined using P and S readings from the Red Sismológica Telemétrica del Estado de Colima (RESCO) and a 1-D velocity model. Because only eight RESCO stations were operating immediately following the Tecomán earthquake, uncertainties in the initial locations and depths are fairly large, with average uncertainties of 8.0 km in depth and 1.4 km in the north–south and east–west directions. Events occurring between 24 January and 31 January were located using not only RESCO phase readings but also additional P and S readings from 11 temporary stations. Average uncertainties decrease to 0.8 km in depth, 0.3 km in the east–west direction, and 0.7 km in the north–south direction for events occurring while the temporary stations were deployed. While some preliminary studies of the early aftershocks suggested that they were dominated by shallow events above the plate interface, our results place the majority of aftershocks along the plate interface, for a slab dipping between approximately 20° and 30°. This is consistent with the slab positions inferred from geodetic studies. We do see some upper plate aftershocks that may correspond to forearc fault zones, and faults inland in the upper plate, particularly among events occurring more than 3 months after the mainshock.

Key words: Aftershocks, Colima, earthquakes, seismicity, relocation.

1. Introduction

Both the Cocos and Rivera Plates are subducting beneath the North American Plate, with a diffuse boundary between them, forming a triple junction near the state of Colima. The Cocos–North America subduction zone has a convergence rate of approximately 37 mm/year near the triple junction (DEMETS and WILSON, 1997). The rate of Rivera–North America subduction near the triple junction is considered to be less than this, but the exact value, as well as the degree of obliquity, is still debated (see discussion by KOSTOGLODOV and BANDY, 1995). Both of these trench segments are seismically active. In <80 years, this region has had five large earthquakes: two in 1932 ($M_w = 8.2$ and $M_w = 7.8$), one in 1973 ($M_w = 7.6$), one in 1995 ($M_w = 8.0$), and the Tecomán earthquake in 2003. The National Earthquake Information Center (NEIC) and Servicio Sismológico Nacional (SSN) reported a magnitude of 7.6 for this event.

Following the 1995 event, two sections of the subduction zone that had not ruptured since 1932 were identified: the Colima Gap, located offshore of the state of Colima, and the Vallarta Gap to the north (YAGI *et al.*, 2004; see Fig. 1). While the Vallarta Gap is clearly still open, it has not been agreed upon whether the 2003 Tecomán earthquake fully ruptured the Colima Gap.

The Tecomán earthquake caused 21 casualties, seven of which were in Colima city, and destroyed 1,065 buildings in Colima city (ZOBIN and PIZANO-SILVA, 2007). It caused significantly more damage in Colima city than in other cities closer to the epicenter, including Tecomán and Manzanillo, causing speculation about the possibility of faulting on multiple planes. Additionally, the presence of a precursory phase degraded the ability to constrain the depth and the fault plane dip, making it difficult to distinguish between an interplate and an intraplate event (GÓMEZ-GONZÁLEZ *et al.*, 2010).

---

[1] California Institute of Technology, 1200 E California Blvd, MC 252-21, Pasadena, CA 91125, USA. E-mail: vandrews@caltech.edu

[2] University of Colima, Av. Gonzalo de Sandoval 444, CP. 28045 Colima, Col, Mexico.

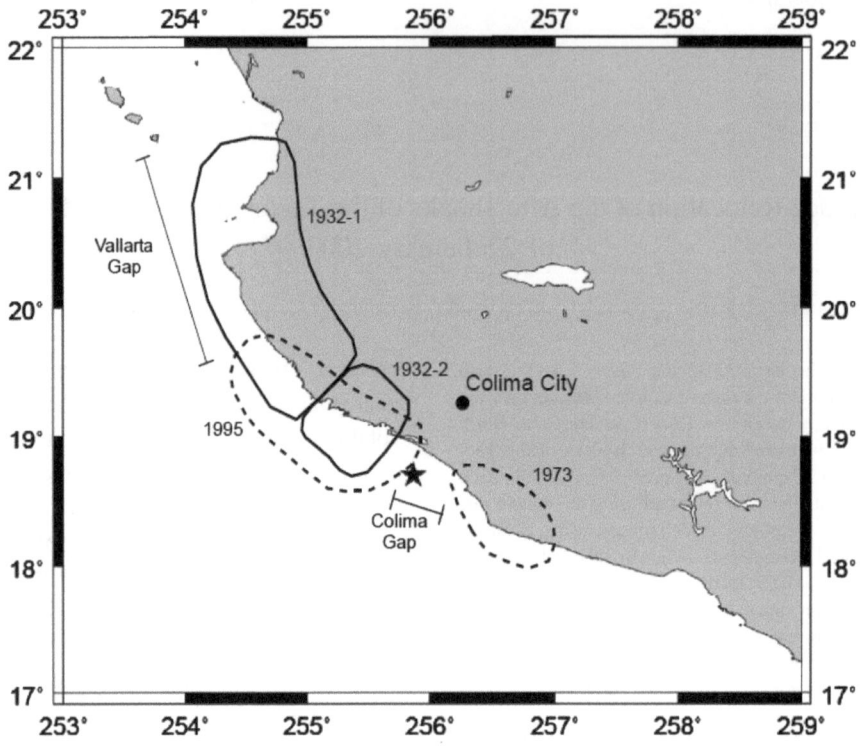

Figure 1
Approximate rupture areas for the 1932, 1973, and 1995 earthquakes, and epicenter for the 2003 earthquake (from YAGI *et al.*, 2004). The 1932 events are shown with a *solid line*, while the 1973 and 1995 events are shown with a *dashed line*. The Colima Gap lies between the rupture areas of the 1973 and 1995 events

The University of Colima maintains a network of short-period stations, currently consisting of 12 stations, called the Red Sismológica Telemétrica del Estado de Colima (RESCO), with the primary purpose of monitoring the Colima volcano (Fig. 2). Eleven stations were running during 2003. Amplification is variable from 60 to 120 dB. Analog data are continuously transmitted by radio to the University of Colima, where they are digitized and recorded. All phase-picking is performed manually.

Additionally, after the 2003 mainshock, temporary stations were deployed by various institutions, including the University of Colima, the Centro de Investigación Científica y de Educación Superior de Ensenada (CICESE) and the Universidad Nacional Autónoma de México (UNAM) (Fig. 2).

A preliminary study of the first 72 h of aftershocks of the Tecomán earthquake, using RESCO data together with data collected by another temporary deployment (University of Guadalajara, not used in this study) suggested that nearly all events

occurred in the upper plate (NÚÑEZ-CORNÚ *et al.*, 2004). Consequently, it was concluded that this earthquake was not an interplate event, and therefore it could not have closed the Colima Gap. However, due to a limited number of stations and unfavorable station geometry, errors in locations are large. NÚÑEZ-CORNÚ *et al.*, (2004) also assumed that the slab maintained a constant dip from the trench in constraining the interface location. SINGH *et al.*, (2003), using data from more regional stations, concluded that the mainshock rupture occurred in the Colima gap; however, their preliminary study was not able to provide details on the aftershock distribution or the geometry of the Wadati-Benioff zone.

In this paper, double difference relocation is used to improve the accuracy of aftershock locations for the months following the earthquake, and to better define the geometry of the Wadati-Benioff zone, using both RESCO data and phase picks from temporary stations. This also allows us to reexamine the importance of the upper plate faulting during the

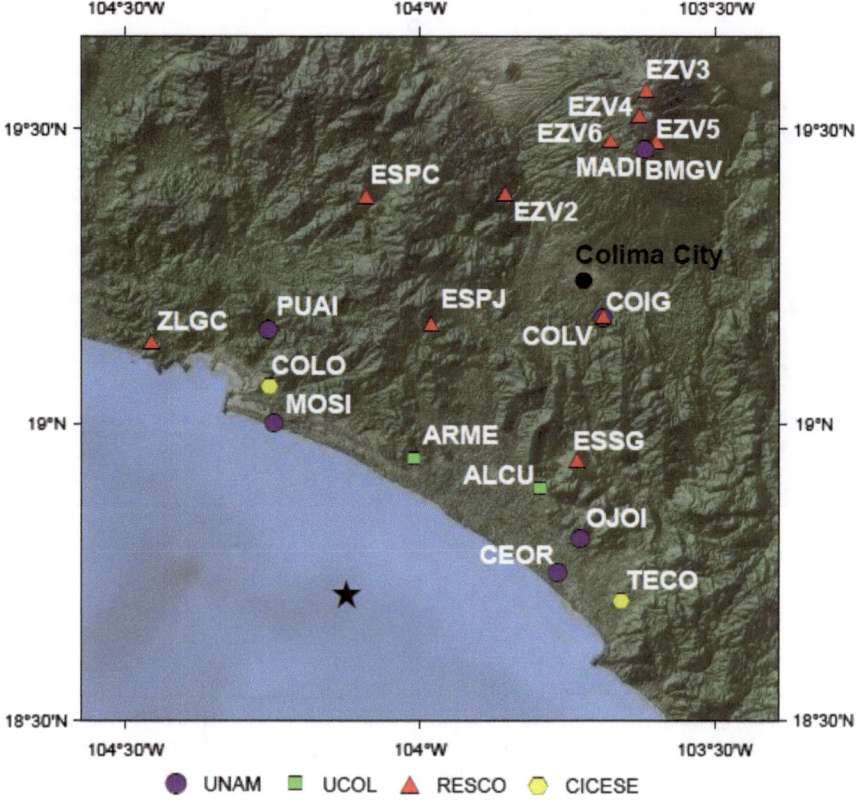

Figure 2

Location of stations used in event relocation. Stations EZV3, EZV4, EZV5, and EZV6 are concentrated near Colima Volcano. *Star* shows the mainshock epicenter. Topography is from Blue Marble Next Generation (visibleearth.nasa.gov) and ASTER GDEM. ASTER GDEM is a product of METI and NASA

aftershock sequence, and the degree of closure of the Colima Gap.

2. Data

Initial event locations were determined using the earthquake location program HYPO71 (Lee and Lahr, 1975), RESCO P and S picks, and a 1D crustal velocity model for the Jalisco block (Pacheco *et al.*, 2003; see Table 1). Of the 2990 events recorded by RESCO between January 22 and December 31, 2003, 617 events were selected for relocation, each of which was recorded by at least seven RESCO stations (Fig. 3a).

Additionally, 219 P-phase picks and 107 S-phase picks were added from 11 temporary stations deployed between 24 January and 31 January. There were 108 events in the time period selected for

Table 1

Velocity model used for determining initial locations (from Pacheco *et al., 2003)*

Layer thickness (km)	P-wave velocity (km/s)
0.4	4.60
8.1	5.69
10.2	6.27
17.3	6.71
Half-space	8.00

relocation, for an average of 3.0 additional phase readings per event. The aperture of the array was increased from 100 km with only the RESCO stations to a maximum of 130 km on January 25 and 26, the only days when all 11 of the temporary stations were deployed.

Due to mechanical failures caused by the strong coseismic shaking in Colima, not all of the RESCO stations were able to record the mainshock.

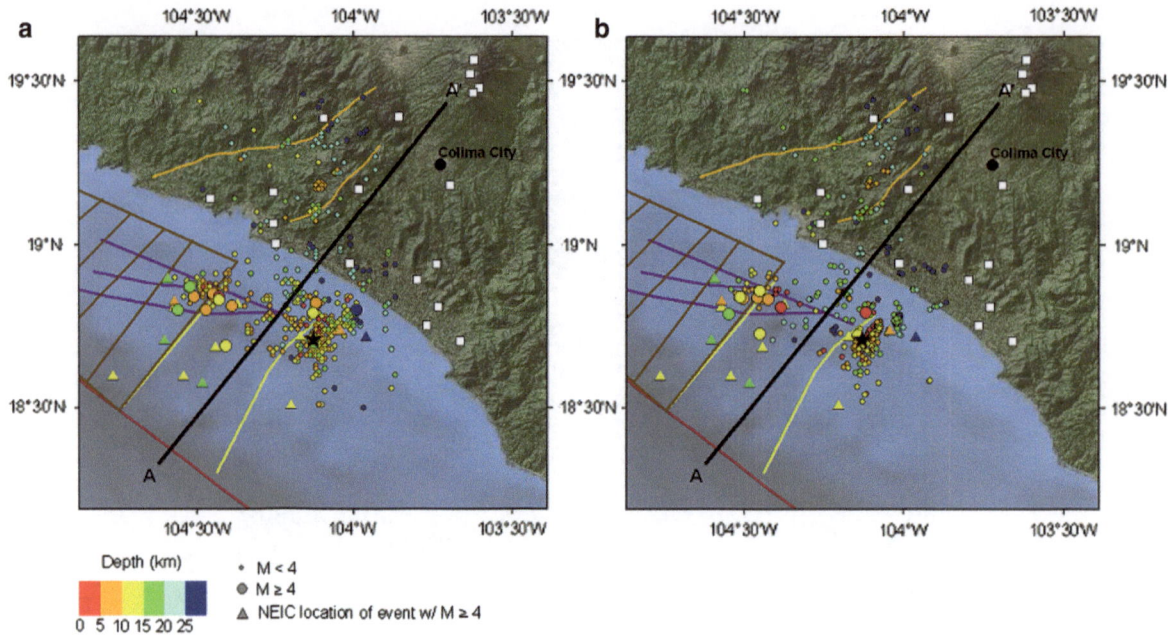

Figure 3

Original event locations (**a**) and relocations (**b**), denoted by *circles*, along with NEIC locations, denoted by *triangles*. *Larger circles* represent events with magnitudes ≥4.0, using magnitudes determined by the NEIC, while *triangles* represent the NEIC locations for those same events. *White squares* show the station locations. *Star* shows the mainshock epicenter. *A–A'* line shows the location of Fig. 4. Faults depicted are from BANDY *et al.*, (2005), and include the Tecomán graben (*yellow*), Middle America trench (*red*), Tamazula fault zone (*orange*), and a series of proposed strike-slip faults (*purple*) and proposed fore-arc sliver moving to the northwest (*brown*). Topography is from Blue Marble Next Generation (visibleearth.nasa.gov) and ASTER GDEM. ASTER GDEM is a product of METI and NASA

Relocation of the mainshock was not attempted, but rather, the hypocenter used by YAGI *et al.*, (2004) is accepted, at 18.71°N, 104.13°W, and 20 km depth. We consider this location to be superior to the NEIC location because of the inclusion of local stations.

Selected events were relocated using the double-difference method (WALDHAUSER and ELLSWORTH, 2000) and the progam hypoDD (WALDHAUSER, 2001). The double-difference method uses relative travel times between linked event pairs and iteratively adjusts the locations of the events to minimize the sum of travel-time residuals. While relative locations between events are significantly improved, the centroid of an event cluster is unchanged, so this method will not correct for systematic errors in the absolute location of events.

For fewer than about 100–200 events, depending on computational power, the matrix of travel time differences may be solved using singular value decomposition (SVD), while the matrices from larger sets of events must be solved using a conjugate

gradients method (LSQR). The advantage of SVD is that it allows for accurate error computation, whereas LSQR significantly underestimates errors, requiring independent error analysis (WALDHAUSER, 2001). Independent error analysis is beyond the scope of this project; rather, it is assumed that errors for large datasets are comparable to those for subsets of the data.

Focal mechanisms were attempted for offshore and nearshore events; however, the limited aperture of the array prevented us from obtaining well-constrained focal mechanisms for any of these events using first motion polarities and S–P ratios, and techniques using waveform fitting cannot be directly applied to short period data.

3. Results and Discussion

Of the 617 events selected for relocation, 350 were relocated by hypoDD, based on several stability criteria discussed by WALDHAUSER and ELLSWORTH

(2000) (Fig. 3b). The remaining events were discarded. All relocated events had an RMS misfit of <0.04 s, which we considered to be an acceptable level of misfit, so no additional events were discarded. Double difference relocation significantly reduces the uncertainties in the location of the events. For the 108 events between 24 January and 31 January, average uncertainties are reduced from 8.0 to 0.8 km in depth, from 1.4 to 0.3 km in the east–west direction, and from 1.4 to 0.7 km in the north–south direction.

For events not in this time frame, only RESCO data are used in the relocation, and uncertainties are computed using a randomly selected subset of 108 events. Average errors are reduced from 6.6 to 1.0 km in depth, from 1.6 to 0.5 km in the east–west direction, and from 1.6 to 1.0 km in the north–south direction. Errors for the entire dataset are probably significantly smaller, evidenced by lower RMS misfits for the entire dataset than for the subset, but, as discussed previously, accurate distance errors are not available for the entire dataset.

In a trench-perpendicular cross-section of the original event locations (Fig. 4c), the locations are too scattered to identify the seismic structure. In a cross-section of the relocated events (Fig. 4d), however, a number of events to the southeast of the line depicted in Fig. 3 appear to define a slab interface dipping between 20° and 30°. The position and dip of the slab are consistent with those inferred from geodetic studies (SCHMITT et al., 2007; HUTTON et al., 2001). This is one of the main important results from this study, because previous studies of the seismicity (e.g., PARDO and SUÁREZ, 1995; NÚÑEZ-CORNÚ et al., 2004) have differed in their interpretations of the dip of the shallow part of the Wadati-Benioff zone in this region, with dips varying from 12° to 48°. Other recent studies (YANG et al., 2009) have left the question open.

However, for events to the northwest of this line, which marks the approximate boundary between the Cocos and Rivera plates, the interface appears to dip much more shallowly, at least at depths shallower than 30 km. This is unexpected, as it is generally agreed that the Rivera Plate dips more steeply than the Cocos Plate at depths >40 km (PARDO and SUÁREZ, 1995; YANG et al., 2009). However, previous studies did not have much resolution on the shallower part of the plate interface. The geometry we observe is consistent with the Rivera plate having a more shallowly dipping seismogenic zone than the Cocos plate, with possibly a smaller radius of curvature before it bends down to have a steeper dip than the Cocos plate at depths below the seismogenic zone. This smaller radius of curvature may be related to the high, rugged topography of the Jalisco block, although more investigation, particularly geodynamic modeling, would be necessary to confirm this relationship.

In constructing a geodetic slip model for the 1995 Colima–Jalisco earthquake, which ruptured to the northwest of the Tecomán earthquake, HUTTON et al., (2001) found a best-fit dip of 9° down to a depth of 40 km, at which point the dip increased to 25°. This dip is consistent with that indicated by the seismicity, although the seismicity suggests that the increase in dip to 25° may start sooner, at approximately 20 km (Fig. 4d).

Our results do not support the hypothesis of NÚÑEZ-CORNÚ et al., (2010) that rupture on the plate interface did not begin until at least 72 h after the mainshock. Our relocations show many events near the slab interface starting on January 22, with no detectable difference in event distribution before and after the first 72 h.

While a large number of the events appear to occur on or near the slab interface, there are two additional clusters of events that are clearly located above the interface. The cluster of onshore events (Fig. 4b) is located near the Tamazula fault zone (GARDUÑO-MONROY et al., 1998) and most of the events in this cluster occurred more than 3 months after the mainshock, suggesting that these events may be unrelated to the mainshock. The cluster of offshore events trending subhorizontally at distances of 45–75 km from the trench appears to indicate additional crustal faulting in the North American plate. These events occurred near the NW side of the Manzanillo trough/El Gordo graben (Fig. 3) in a region of the forearc characterized by NE-striking normal faults and trench-parallel strike-slip faults (BANDY et al., 2005). More research is necessary to determine which of these upper plate faults slipped in these aftershocks.

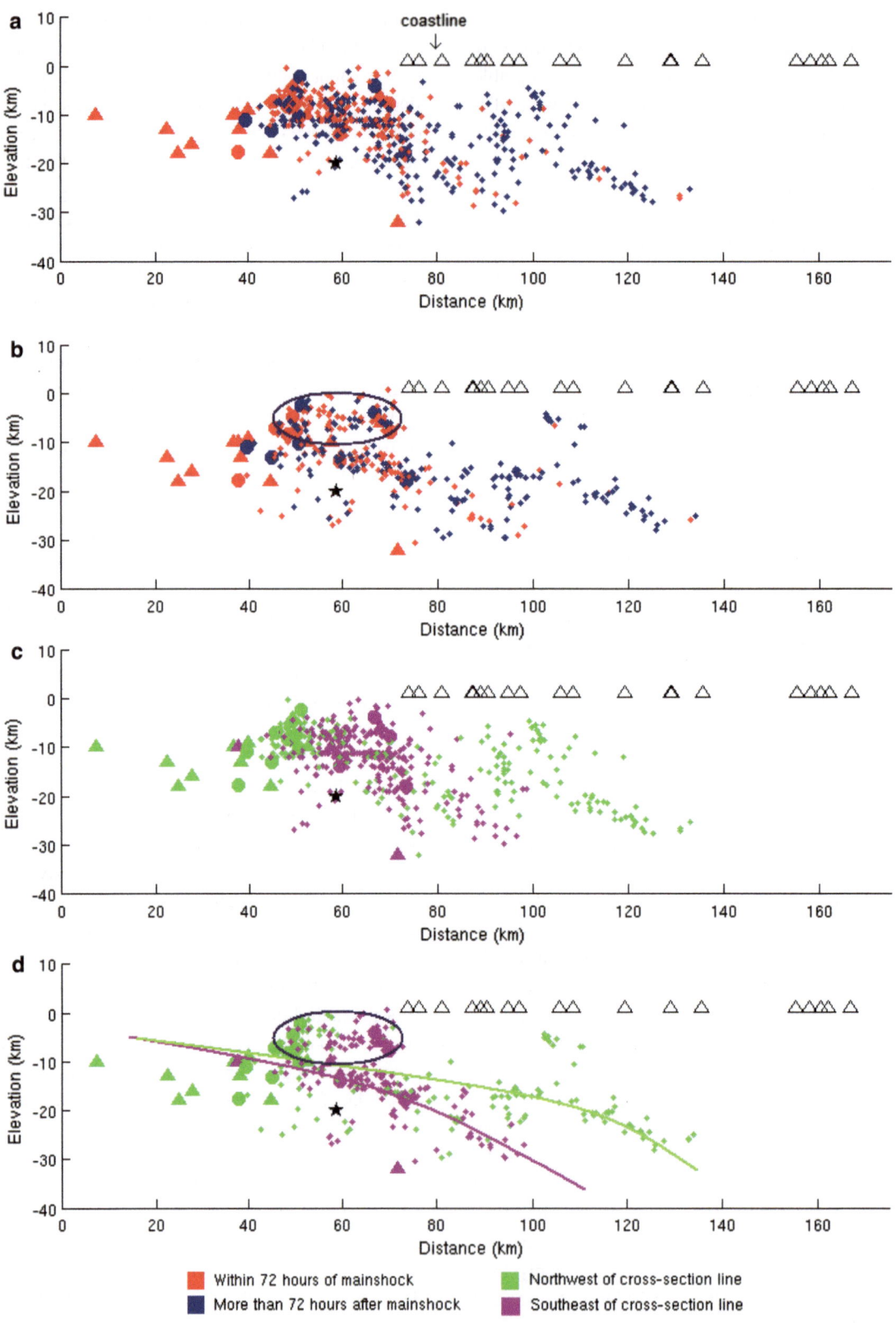

🟧 Within 72 hours of mainshock	🟩 Northwest of cross-section line
🟦 More than 72 hours after mainshock	🟪 Southeast of cross-section line

◀

Figure 4
Cross-sections from A–A′ (Fig. 3) of original event locations
(**a**, **b**) and relocations (**c**, **d**), denoted by *circles*, along with NEIC
locations, denoted by *triangles*. All events shown in Fig. 3 are
included, with a maximum distance of events from the projection
line of 78 km. *Larger circles* represent events with magnitudes
≥4.0, as determined by the NEIC, while *triangles* represent the
NEIC locations for those same events. In parts **a**, **b**, *red symbols*
represent events that occurred within 72 h of the mainshock, while
blue symbols represent events that occurred more than 72 h after
the mainshock. In parts **c**, **d**, *green symbols* represent events to the
northwest of the cross-section shown in Fig. 3, while magenta
symbols represent events to the southeast. *Star* shows the
mainshock epicenter. *Magenta* and *green curves* show an approximate fit to the subduction interface for the Cocos and Rivera
Plates, respectively. *Purple* ellipse outlines the shallow event
cluster discussed in the text. *Black triangles* show the projection of
the seismic stations. Distances are relative to the trench. The
coastline marked in part **a** is shown where it intersects the line
shown in Fig. 3; some stations appear to be offshore because the
coastline is not perpendicular to this line. While there are numerous
small events beneath the Colima Volcano, these events were
excluded from this study because they were not recorded by a
sufficient number of stations

Some of the events in this cluster have depths
<1 km, which is physically unreasonable, particu-
larly for offshore events. Most of these events have
only RESCO station readings, suggesting that the
locations might be adversely affected by the unfavorable geometry of the RESCO stations. However,
comparisons of subsets of the data relocated with and
without phase readings from temporary stations show
that while the addition of these stations moves out-
lying events closer to neighboring events, they are not
moved in a consistent direction. The most likely
explanation for these depth errors is that the velocity
model does not adequately match the shallow
velocity structure for shallow events.

The mainshock appears to be slightly below most
of the aftershocks. As mentioned previously, the
mainshock was not included in the relocation; thus
the accuracy of the relative location of the mainshock
to the aftershocks cannot be determined, and should
not be too highly interpreted.

We now consider an additional set of locations,
those determined by the NEIC using teleseismic data.
A comparison of RESCO locations to those determined by the NEIC shows that RESCO locations are
generally located inland of NEIC locations (Figs. 3, 4).
This is unsurprising, as a catalog search of events
in this region shows that the NEIC locations are

typically further offshore than would be predicted
from the location of the plate boundary, suggesting
inaccuracies in the NEIC velocity model. As the
centroid of the events is essentially unchanged by
relocation, relocating the events does not change this
trend.

The large number of events on or near the slab
interface indicates that the Tecomán earthquake did
break the Colima Gap. Thus, one might conclude that
the Vallarta Gap would be the most likely location for
the next earthquake on this system to occur, assuming
that the northwestward decrease in convergence rate
along the trench is not a significant factor. There is
one caveat to this conclusion, however. The 2003
aftershock distribution shows a sparsity of events in
the Tecomán graben (Fig. 3) which could indicate
that there may still be a small gap remaining between
the 2003 Tecomán earthquake sequence and the more
poorly known extent of the 1973 earthquake sequence
to the southeast.

4. Conclusions

The double-difference relocations of seismicity
for 11 months following the January 22, 2003
earthquake illustrates the following important characteristics of the Middle America trench subduction
system in the region offshore of the state of Colima,
Mexico. (1) The aftershocks define a Wadati-Benioff
zone with a dip of 20°–30°, indicating slip on the
main plate interface beneath the North America plate.
(2) Aftershocks in this location are present throughout the aftershock sequence, consistent with a model
in which the mainshock also ruptured the main plate
interface. (3) A difference in dip of the seismogenic
zone is detected on either side of the Tecomán graben, with the plate on the NW side dipping more
shallowly. This suggests that the Rivera plate dips
more shallowly than the Cocos Plate, at least for
depths <30 km. This may imply a difference in the
radius of curvature of the two plates as they descend
into the mantle. (4) A significant number of aftershocks occurred on the Rivera–North American plate
interface. (5) Clusters of shallow events that occurred
off of the main subduction interface, within this
time frame, are attributed to (a) active submarine

strike-slip or normal faults in the forearc of the Middle America Trench and (b) the onshore Tamazula fault zone. (6) The Colima Gap was at least partly closed, and may have been entirely closed, by the 2003 earthquake sequence.

Acknowledgments

Research supported by the US National Science Foundation under grant EAR-0510395 and by the California Institute of Technology's division of Geological and Planetary Sciences under contribution 10051.

REFERENCES

BANDY, W. L., MICHAUD, F., BOURGOIS, J., CALMUS, T., DYMENT, J., MORTERA-GUTIERREZ, C. A., ORTEGA-RAMÍREZ, J., PONTOISE, B., ROYER, J.-Y., SICHLER, B., SOSSON, M., REBOLLEDO-VIEYRA, M., BIGOT-CORMIER, F., DÍAZ-MOLINA, O., HURTADO-ARTUNDUAGA, A. D., PARDO-CASTRO, G., and TROUILLARD-PERROT C. (2005), *Subsidence and Strike-Slip Tectonism of the Upper Continental Slope off Manzanillo, Mexico*, Tectonophysics 398, 115–140.

DEMETS, C., and WILSON, D. S. (1997), *Relative Motions of the Pacific, Rivera, North American, and Cocos Plates Since 0.78 Ma*, J. Geophys. Res., 102 (B2), 2789–2806.

GARDUÑO-MONROY, V., H., SAUCEDO-GIRÓN, R., JIMÉNEZ, Z., GAVILANES-RUIZ, J. C., CORTES-CORTES, A., and URIBE-CIFUENTES, R. M. (1998), *La Falla Tamazula, Límite Suroriental del Bloque Jalisco, y sus Relaciones con el Complejo Volcánico de Colima, Mexico*, Rev. Mex. Ciencias Geológicas 15, 132–144.

GÓMEZ-GONZÁLEZ, J. M., MENDOZA, C., SLADEN, A., and GUZMÁN-SPEZIALE, M. (2010), *Kinematic Source Analysis of the 2003 Tecomán, México, Earthquake (Mw 7.6) using Teleseismic Body Waves*, Bol. Soc. Geol. Mex. 62, 249–262.

HUTTON, W., DEMETS, C., SÁNCHEZ, O., SUÁREZ, G., and STOCK, J. (2001), *Slip Kinematics and Dynamics During and After the 1995 October 9 $M_w = 8.0$ Colima-Jalisco Earthquake, Mexico, from GPS Geodetic Constraints*, Geophys. J. Int. 146, 637–658.

KOSTOGLODOV, V. and BANDY, W. (1995), *Seismotectonic Constraints on the Convergence Rate between the Rivera and North America Plates*, J. Geophys. Res. 100 (B9), 17977–17989.

LEE, W. H. K. and LAHR, J. C. (1975), *HYPO71 (Revised): A Computer Program for Determining Hypocenter, Magnitude, and First Motion Pattern of Local Earthquakes*, U.S. Geol. Surv. Open File Rept. 75–311.

NÚÑEZ-CORNÚ, F. J., REYES-DÁVILA, G. A., RUTZ LÓPEZ, M., TREJO GÓMEZ, E., CAMARENA-GARCÍA, M. A., and RAMÍREZ-VAZQUEZ, C. A. (2004), *The 2003 Armería, México Earthquake (M_w 7.4): Mainshock and Early Aftershocks*, Seism. Res. Lett. 75, 734–743.

NÚÑEZ-CORNÚ, F. J., RUTZ-LÓPEZ, M., MÁRQUEZ-RAMÍREZ, V., SUÁREZ-PLASCENCIA, C., and TREJO GÓMEZ, E. (2010), *Using an Enhanced Dataset for Reassessing the Source Region of the 2003 Armería, Mexico Earthquake*, Pure Appl. Geophys., doi: 10.1007/s00024-010-0178-x

PACHECO, J. F., BANDY, W., REYES-DÁVILA, G. A., NÚÑEZ-CORNÚ, F. J., RAMÍREZ-VÁZQUEZ, C. A., and BARRÓN, J. R. (2003), *The Colima, Mexico, Earthquake (M_w 5.3) of 7 March 2000: Seismic Activity Along the Southern Colima Rift*, Bull. Seism. Soc. Am. 93, 1458–1467.

PARDO, M., and SUÁREZ, G. (1995), *Shape of the Subducted Rivera and Cocos Plates in Southern Mexico: Seismic and Tectonic Implications*, J. Geophys. Res. 100, 12357–12373.

SCHMITT, S. V., DEMETS, C., STOCK, J., SÁNCHEZ, O., MÁRQUEZ-AZÚA, B., and REYES, G. (2007), *A Geodetic Study of the 2003 January 22 Tecomán, Colima, Mexico Earthquake*, Geophys. J. Int. 169, 389–406.

SINGH, S. K., PACHECO, J. F., ALCÁNTARA, L., REYES, G., ORDAZ, M., IGLESIAS, A., ALCOCER., S M; GUTIERREZ, C.,VALDÉS, C., KOSTOGLODOV, V., REYES, C., MIKUMO, T., QUAAS, R., and ANDERSON, J. G. (2003), *A Preliminary Report on the Tecomán, Mexico Earthquake of 22 January 2003 (M_w 7.4) and its Effects*, Seismol. Res. Lett. 74(3), 279–289.

WALDHAUSER, F. (2001), *hypoDD: A Program to Compute Double-Difference Hypocenter Locations (hypoDD version 1.0-03/2001)*, U.S. Geol. Surv. Open File Rept. 01-113.

WALDHAUSER, F., and ELLSWORTH, W. L. (2000), *A Double-Difference Earthquake Location Algorithm: Method and Application to the Northern Hayward Fault, California*, Bull. Seism. Soc. Am. 90, 1353–1368.

YAGI, Y., MIKUMO, T., PACHECHO, J., and REYES, G. (2004), *Source Rupture Process of the Tecomán, Colima, Mexico Earthquake of 22 January 2003, Determined by Joint Inversion of Teleseismic Body-Wave and Near-Source Data*, Bull. Seism. Soc. Am. 94, 1795–1807.

YANG, T., GRAND, S. P., WILSON, D., GUZMAN-SPEZIALE, M., GOMEZ-GONZALEZ, J. M., DOMINGUES-REYES, T., and NI, J. (2009), *Seismic Structure beneath the Rivera Subduction Zone from Finite-Frequency Seismic Tomography*, J. Geophys. Res. 114, B01302, doi:10.1029/2008JB005830.

ZOBIN, V. M. and PIZANO-SILVA, J.A. (2007), *Macroseismic Study of the M_w 7.5 21 January 2003 Colima, México, Across-Trench Earthquake*, Bull. Seism. Soc. Am. 97, 1221–1232.

(Received February 12, 2010, revised August 6, 2010, accepted August 9, 2010, Published online November 10, 2010)

Pure Appl. Geophys. 168 (2011), 1339–1353
© 2010 Springer Basel AG
DOI 10.1007/s00024-010-0202-1

Source Characteristics of the 22 January 2003 $M_w = 7.5$ Tecomán, Mexico, Earthquake: New Insights

Luis Quintanar,[1] Héctor E. Rodríguez-Lozoya,[2] Roberto Ortega,[3] Juan M. Gómez-González,[4]
Tonatiuh Domínguez,[5] Clara Javier,[6] Leonardo Alcántara,[7] and Cecilio J. Rebollar[8]

Abstract—Aftershock locations, source parameters and slip distribution in the coupling zone between the overriding North American and subducted Rivera and Cocos plates were calculated for the 22 January 2003 Tecomán earthquake. Aftershock locations lie north of the El Gordo Graben with a northwest-southeast trend along the coast and superimposed on the rupture areas of the 1932 ($M_w = 8.2$) and 1995 ($M_w = 8.0$) earthquakes. The Tecomán earthquake ruptured the northwest sector of the Colima gap, however, half of the gap remains unbroken. The aftershock area has a rectangular shape of 42 ± 2 by 56 ± 2 km with a shallow dip of roughly 12° of the Wadati-Benioff zone. Fault geometry calculated with the Nábélek (1984) inversion procedure is: (strike, dip, rake) = (277°, 27°, 78°). From the teleseimic body wave spectra and assuming a circular fault model, we estimated source duration of 20 ± 2 s, a stress drop of 5.4 ± 2.5 MPa and a seismic moment of $2.7 \pm .7 \times 10^{20}$ Nm. The spatial slip distribution on the fault plane was estimated using new additional near field strong motion data (54 km from the epicenter). We confirm their main conclusions, however we found four zones of seismic moment release clearly separated. One of them, not well defined before, is located toward the coast down dip. This observation is the result of adding new data in the inversion. We calculated a maximum slip of 3.2 m, a source duration of 30 s and a seismic moment of 1.88×10^{20} Nm.

Key words: Rupture process, focal mechanism inversion, waveform modeling.

C. J. Rebollar: deceased September 2006.

[1] Instituto de Geofísica, Universidad Nacional Autónoma de México, México D.F., Mexico. E-mail: luisq@ollin.geofisica.unam.mx
[2] Facultad de Arquitectura, Facultad de Ingeniería Civil, Universidad Autónoma de Sinaloa, Culiacán, Mexico.
[3] Centro de Investigación Científica y de Educación Superior de Ensenada, Unidad La Paz, La Paz, Mexico.
[4] Centro de Geociencias, Universidad Nacional Autónoma de México, Juriquilla, Querétaro, Mexico.
[5] Observatorio Vulcanológico de Colima, Universidad de Colima, Colima, Mexico.
[6] Comisión Federal de Electricidad, Gerencia de Estudios de Ingeniería Civil, Guadalajara, Mexico.
[7] Instituto de Ingeniería, Universidad Nacional Autónoma de México, México D.F., Mexico.
[8] Centro de Investigación Científica y de Educación Superior de Ensenada, Ensenada, Mexico.

1. Introduction

The subduction zone that encompasses the Mexican states of Jalisco and Colima, (Fig. 1a) has produced the largest earthquakes in Mexico. Offshore of the state of Colima, the Rivera, Cocos and North America plates interact (Fig. 1a). In this seismogenic area the 1932, magnitude 8.2, Jalisco earthquake occurred, the largest earthquake in Mexico in recent documented seismological history (SINGH *et al.*, 1985; SUÁREZ *et al.*, 1994). The 22 January 2003 ($M_w = 7.5$) Tecomán earthquake occurred between the rupture areas of the 9 October 1995 ($M_w = 8.0$) Colima–Jalisco earthquake and the 30 January 1973 ($M_w = 7.6$) Colima earthquake (see Fig. 1b) (REYES *et al.*, 1979; PACHECO *et al.*, 1997). The Tecomán earthquake was located by the University of Colima at 18.625° N, 104.125° W, at a depth of 10 km and with an origin time of 02 h 06 min and 35.0 s. This earthquake badly damaged the state of Colima, damaging and collapsing adobe houses, unreinforced masonry structures and construction in poorly consolidated fills, and producing landslides. Liquefaction was also observed in the Manzanillo port.

A preliminary study of the Tecomán earthquake was carried out by SINGH *et al.*, (2003). They reported a preliminary aftershock area, gross source characteristics, attenuation of strong motion data and an isoseismal map. NÚÑEZ-CORNU *et al.*, (2004) reported the aftershock locations of the first 72 h after the main shock and found that the seismic activity was mainly in the continental plate along two NW–SE trending zones. YAGI *et al.*, (2004) estimated the spatial and temporal slip over the fault plane of the Tecomán earthquake, inverting regional strong motion data as well as teleseismic body-

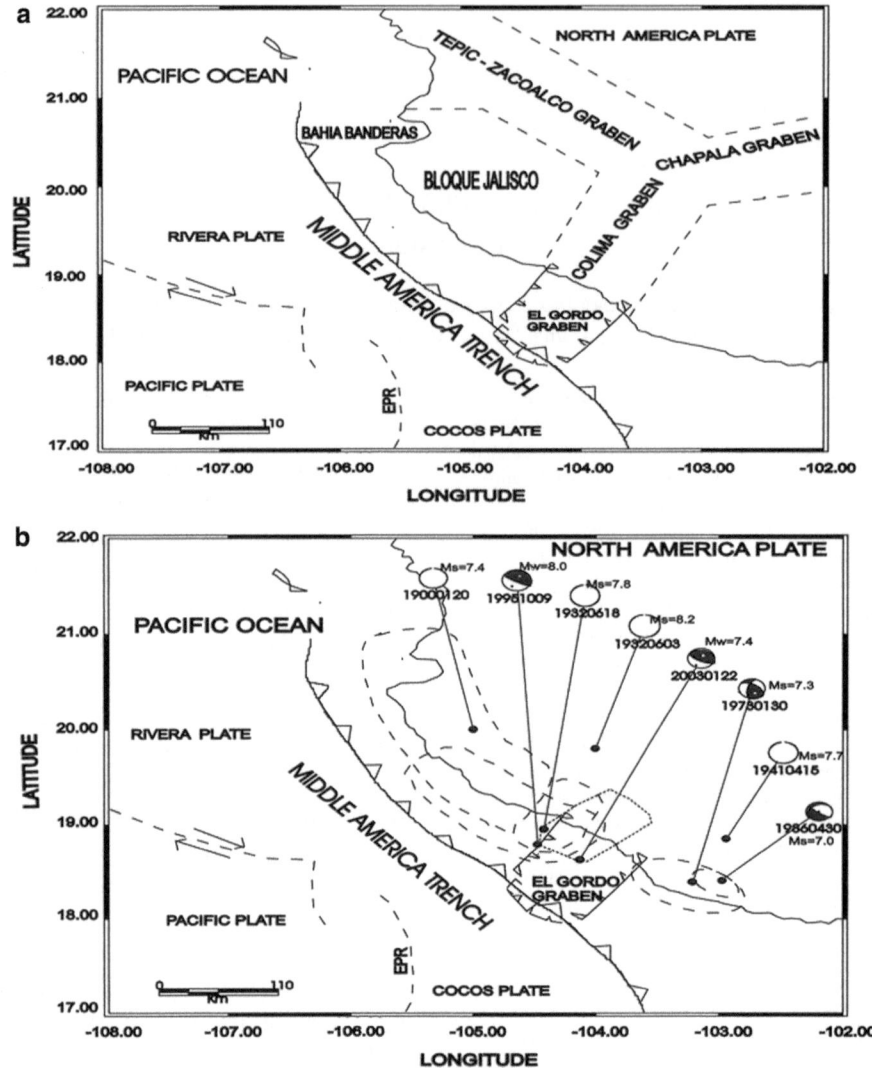

Figure 1

a Tectonic framework of the Colima–Jalisco region. **b** Location of earthquakes >7 that have occurred in the Colima–Jalisco region. *Light dashed lines* show the rupture area of the 1932 (M_w = 8.2) earthquake (SINGH *et al.*, 1985). *Heavy dashed line* is the rupture area of the 1995 (M_w = 8.0) earthquake (PACHECO *et al.*, 1997). Also shown are the aftershock areas of the 1973 (M_w = 7.3) and 1986 (1980) (M_w = 7.0) earthquakes. *Continuous line* is the aftershock area of the 2003 Tecomán earthquake. Available focal mechanisms are also shown. *EPR* east Pacific rise

waves. The closest near field strong motion station they used was located 138 km from the epicenter. In the inversion process they used a slightly modified fault plane geometry reported by the Harvard CMT solution. They found three stages in the rupture process: (1) the rupture nucleated near the hypocenter, (2) the first asperity, centering about 15 km southwest from the epicenter, was broken and (3)

the rupture propagated to the northeast and the second asperity was broken.

In the present study, we revise the aftershock area employing newly calculated hypocenters using arrival times recorded at the Colima seismic network (Red Sismica de Colima, RESCO) and nine portable seismic stations deployed by CICESE (Centro de Investigación Científica y de Educación Superior de

Ensenada). We found that the aftershocks lay mainly along the coupling zone of the plates and inside the subducted slab. We also calculated the stress drops and source duration of the main event with the spectral analysis of body P-waves recorded by broadband seismic stations located at teleseismic distances, following BEZZEGHOUD et al., (1989), assuming a circular fault model (BRUNE, 1970). In this study of the Tecomán event, we used near field strong motion waveforms recorded at the Manzanillo power plant located 54 km from the epicenter; data which have not been previously used. We combined this strong-motion data with the data used by YAGI et al., (2004) to obtain the slip on the fault plane and to analyze how much the additional near field data modified the previously reported slip distribution on the fault.

2. Tectonic and Historical Seismicity

West of the Mexican Volcanic Belt lies the junction of the North American, Pacific and Cocos plates. The Rivera and Cocos plates subduct at a rate of 5 cm/year (KOSTOGLODOV and BANDY, 1995) beneath North America, below the states of Colima and Jalisco. In the North American plate a tectonic block called the Jalisco Block (JB) is limited to the west by the Rivera plate, to the south and southeast by El Gordo-Colima graben and toward the east and northeast by the Chapala and Tepic-Zacoalco Grabens (Fig. 1a). Major seismic activity is due mainly to the collision of the Cocos–Rivera plates with the North American plate and occurs in the coupling zone. Depths of major underthrusting earthquakes range from 10 to 32 km. According to PARDO and SUÁREZ, (1995) seismic activity starts to decrease toward Bahia Banderas, located northwest of the JB, reaching depths on the order of 100 km (see Fig. 1a). The Wadati-Benioff zone increases its dip from ~25° in the southeast to 46° toward Bahia Banderas. Table 1 and Fig. 1b shows the location, rupture areas, and seismic moments of earthquakes of magnitude >7 that have occurred in this region.

3. Data Recording Systems

The Volcanic Observatory of the University of Colima maintains a short period and broadband telemetered network of seismic stations around the Colima volcano and along the Colima coast. This network, called Red Sísmica de Colima (RESCO), consists of 11 short period seismometers and 2 broadband CMG-40T Guralp seismometers. Data are continuously recorded and sent by radio links to the University of Colima where they are stored and analyzed. RESCO stations, which record at 100 samples per second, recorded the Tecomán event and all its aftershocks. A week after the occurrence of the main event, CICESE deployed a portable seismic network of nine short period GBV seismic recorders with a natural period of 4.5 Hz, as well as, a broadband Guralp CMG-40T seismometer that recorded continuously at 100 samples per second with a CMG-SAM2 acquisition module. All GBV seismic stations recorded in a triggered mode at 100 samples per second. Figure 2a shows the location of the seismic stations and Fig. 2b the location of the strong motion stations.

4. Aftershock locations and dip of the Wadati-Benioff zone

Aftershock locations were determined with the HYPO71 program using a one-dimensional crustal structure (shown in Table 2) routinely used by RESCO. Due to the occurrence of the aftershocks toward the ocean and the location of the seismic stations, azimuthal coverage is poor. We located aftershocks that included P and S wave arrival times at seismic stations located along the coast in order to better constrain focal depths. Figure 3 shows an example of the traces of an aftershock recorded in 19 seismic stations. Impulsive P wave arrivals as well as the onsets of S waves can be seen. We started locating aftershocks with initial fixed depths in the range from 10 to 50 km with steps of 5 km. From the HYPO71 outputs we chose those locations with average root mean square error of time residuals (rms) of 0.6 ± 0.3 s, average standard error of the epicenter

Table 1

Location and seismic moments of earthquakes of magnitude >7 that have occurred in the study region

Date	H:min.	Lat.	Lon.	Seismic moment (Nm)	Magnitude (Ms[c], M_w^a)	Depth (Km)
19000120	06:33	20.00	−105.00	–	7.4[c]	–
19320603	10:36	19.80	−104.00	9.10×10^{20}	8.2[c], 8.0[a]	16.0
19320618	10:12	18.95	−104.42	7.3×10^{20}	7.8[c]	16.0
19410415	19:09	18.85	−102.94	–	7.7	–
19730130	21:01:00	18.39	−103.21	3.0×10^{20}	7.3[bc]	32.0
19950910	15:35:53.0	19.05	−104.20	9.1×10^{20}	8.0	17.0
20030122	02:06:33.8	18.625	−104.125	1.60×10^{20}	7.4[a]	10.0

[a] RESCO

[b] Reyes et al. (1979)

[c] Pacheco and Sykes (1992)

(ERH) of 3.3 ± 1.8 km and average standard error of the focal depth (ERZ) of 3.1 ± 2.8. The errors are plus or minus one standard deviation. Ninety aftershocks fulfilled the above mentioned error criteria. Local magnitudes of the located events range from 2 to 3.

Figure 4 shows the location of the aftershocks in a plan view, from which it can be seen that the activity lies north of El Gordo Graben and that the aftershock area encompasses parts of the rupture areas of the 1932 and 1995 events. The area between the limits of the rupture areas of the 1995 and the 1973 earthquakes is the Colima seismic gap. The northwest area of this gap ruptured with the Tecomán earthquake. Toward the southeast, roughly half of the gap remains unbroken.

Figure 5 shows vertical cross-sections of aftershocks projected perpendicular and parallel to the middle America trench (MAT) as well as the projection of the P-axes of the Tecomán event and two of the largest aftershocks of magnitudes 5.7 and 5.3. Also shown is the projection of the 1995 $M_w = 8.0$ Colima–Jalisco event. The hypocenter (where the rupture started) of the Tecomán event lies within the coupling zone between the subducted Cocos–Rivera plates and the overriding North American plate. The boundary between the subducted slab and the continental crust was taken from cross section 'K' of Pardo and Suárez, (1995). The source mechanisms of the main event and two of the largest aftershocks were all thrust events. The projections of the P-axes of the main event and the largest aftershocks show that these compressional axes are oblique to the

interface of the overriding and subducted slab. One of the aftershocks (focal depth of 5 km) was an inter-plate event, while the second large aftershock (focal depth of 14 km) was an intraplate event. Fault plane solutions of the largest aftershocks were taken from Singh *et al.*, (2003). Using the obtained aftershocks' depth, we inferred that the Wadati-Benioff zone in the coupling area has a dip of 12° up to a distance of 85 km from the trench (see Fig. 5a); beyond that point the Wadati-Benioff increases to a dip of 50° according to Pardo and Suárez, (1995). Most of the aftershock activity of the 2003 Tecomán sequence occurred within the subducted plate and along the overriding North American plate. The thick line on Fig. 5a depicts our interpretation of the coupling zone. On the other hand, Fig. 5b shows a cross-section parallel to the MAT, from this figure we roughly calculated a rupture area of 44 by 58 km. Aftershock depths range from 5 to 38 km.

5. Focal Mechanism Determination

To obtain the fault geometry we carried out a waveform inversion with the point source approximation using P and SH recorded at teleseismic distances following Nábělek (1984). We used data from stations at epicentral distances between $30° < \Delta_P < 91°$ and $35° < \Delta_{SH} < 83°$ for P and SH waves, respectively. Various corrections were included for a direct comparison between synthetic and observed seismograms. Such corrections take care of the geometrical spreading, the instrumental response,

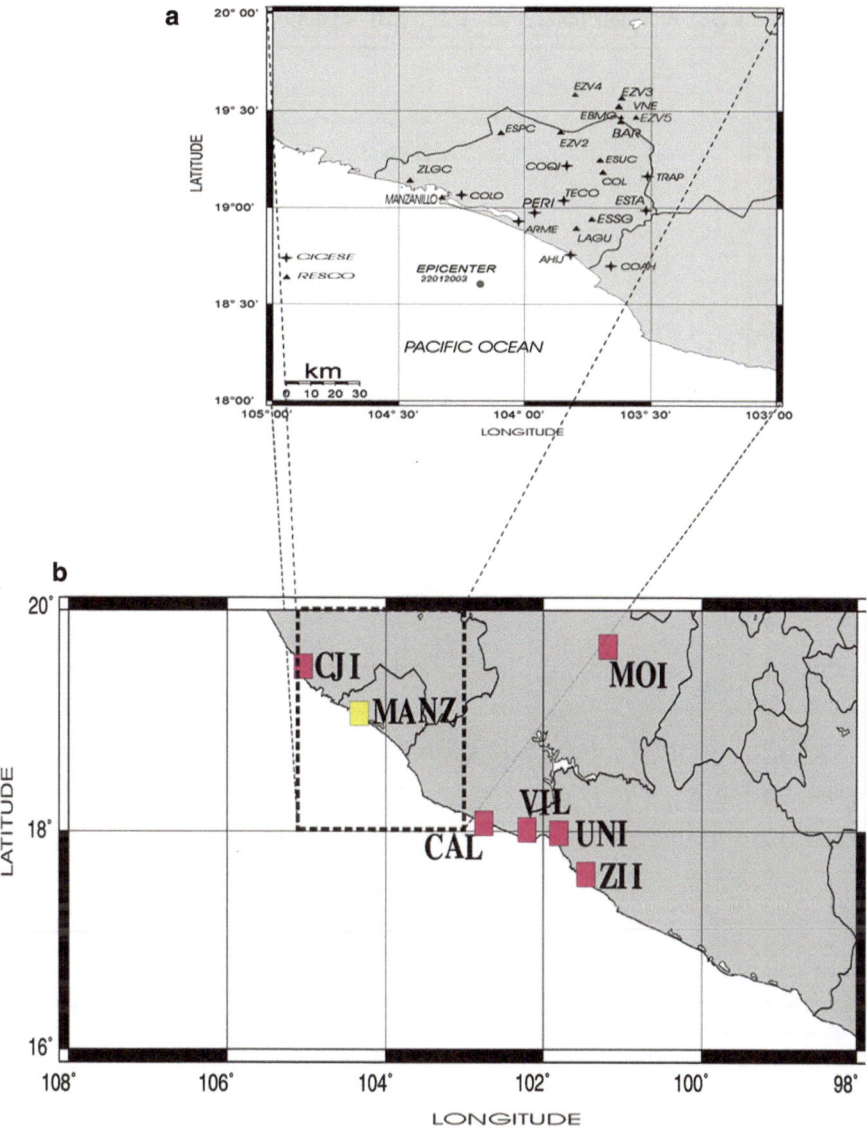

Figure 2

a Location of RESCO (*full triangles*) seismic stations and CICESE (*crosses*) portable stations. **b** Map showing the strong motion instruments from UNAM (*full squares*). *Empty square* shows the location of the Manzanillo strong-motion recorder located in the Manzanillo power plant

Table 2

Velocity model used for aftershocks location

P-wave velocity (km/s)	Layer depth (km)
1.7	0.0
2.7	2.0
3.6	3.0
4.6	4.0
5.7	6.0
6.0	12.0
7.4	18.0
7.8	35.0

the amplitude and phase distortions which are due to a layered crust. The anelastic attenuation of the Earth (FUTTERMAN, 1962) is responsible for these distortions. To correct for it, we used a frequency independent Q average seismic quality factor along the ray, which is assumed a constant value of $t^* = T/Q$, where T is the ray travel time. Here we used $t^* = 1.0$ and 4.0 s for P and SH waves, respectively.

The effects of the structure below the receiver were neglected and we consider the response to be

COL   Plot start time: 2003 1 27   3:51 43.000

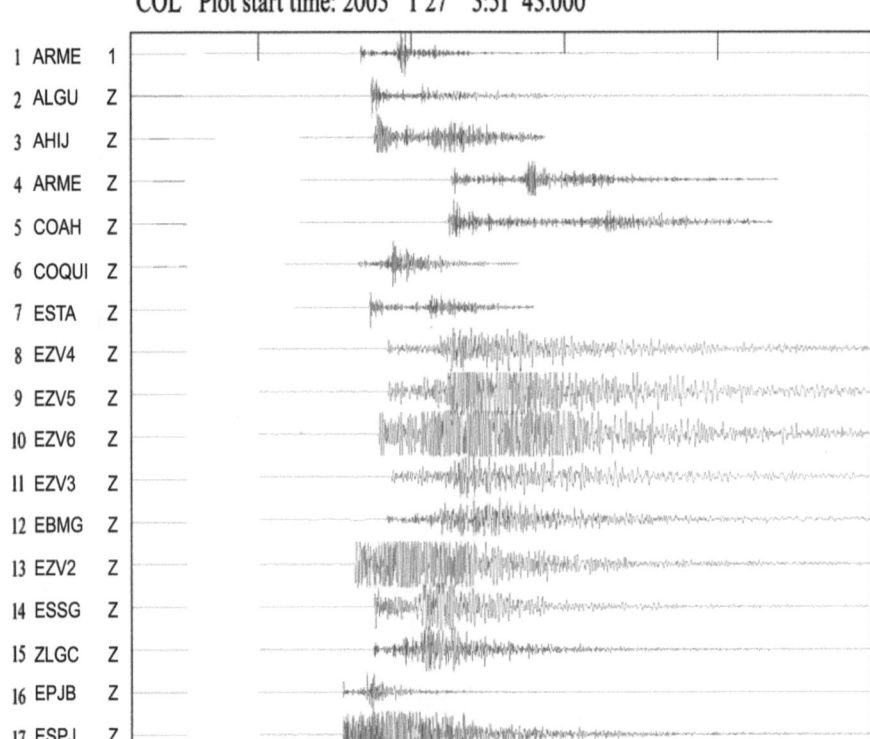

Figure 3

Example of an aftershock recorded by the 19 seismic stations (only the vertical component is shown). Impulsive *P*-wave arrivals and clear onsets of some S waves can be seen; traces are normalized to one. Names of the seismic stations are indicated

controlled mostly by the source radiation pattern at the near source structure. Green's functions around the source were computed and a stratified velocity model was assumed (Table 3) based on the crustal structure of REYES *et al.*, (1979). A simple half-space model at the receiver was used ($V_p = 6.0$ km/s, $V_s = 3.46$ km/s, $\rho = 2.8$ g/cm^3 and a Poisson ratio of 0.25). Seismograms were band-pass filtered using a third-order Butterworth filter from 0.01 to 0.4 Hz and corrected for instrument responses. Sixty seconds of vertical P and transverse SH wave displacements were obtained through integration of the velocity records. Good station coverage made it possible to obtain the fault geometry. The event was modeled using 7 overlapping triangles with 2.5 s of rise time.

We obtained the following mechanism: (strike, dip, rake) = (277°, 27°, 78°), and a scalar seismic moment of 1.3×10^{20} Nm and a apparent source time function of 20 s. By searching for the best centroid depth, the residual error minimized at 25 km, which is similar to that reported by YAGI *et al.*, (2004). Waveform fitting was good even for SH waves, which are usually difficult to fit. Figure 6 shows a comparison between observed and synthetic waveforms and the apparent source time function. P-wave fitting around the fault plane is reasonably good. Southward stations, in the dilatation quadrant close to the sub-vertical nodal plane (PTCN and RPN), constrained the strike of the south dipping plane. This is important because most of the stations

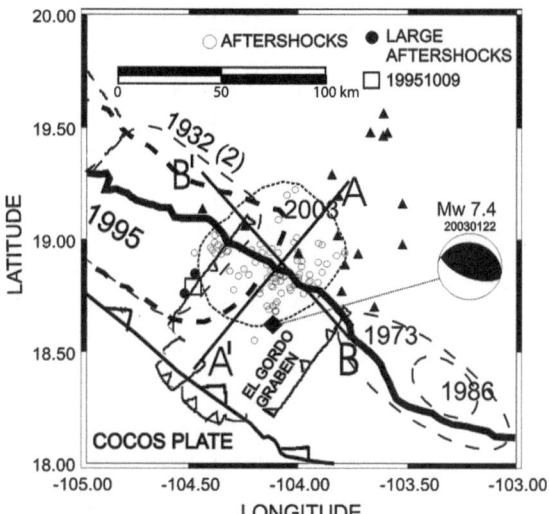

Figure 4

Map showing our best aftershock locations (*empty circles*). *Large full square* shows the epicenter location of the Tecomán earthquake. Its fault plane solution is also shown. *Large empty square* shows the epicenter location of the Colima–Jalisco earthquake of 1995 ($M_w = 8.0$). Profiles AA′ and BB′ are cross sections onto which aftershocks were projected shown in Fig. 5. *Triangles* show the locations of seismic stations

were distributed into the compression quadrant. A similar good quality fitting was found for SH-waves; stations PTCN, DSB, HRV, and MTE fixed one of the nodal planes. The NÁBĚLEK, (1984) point source waveform fitting was good to constrain the fault geometry, however, it clearly underestimated the seismic moment and the apparent source time function (also shown in the middle). Finite two-dimension source inversion (described later) showed that some patches of slip are detected by near field stations which raised the scalar moment, and that displacement could not be detected by the point source approximation at teleseismic distances, especially those displacements near the edge of the fault plane.

6. Static Stress Drop Estimate

We calculated the P-wave spectra of the Tecomán earthquake recorded at broadband stations at teleseismic distances. We used IRIS and GEOSCOPE seismic stations in order to estimate source parameters. P-wave windows of 70 s, which included 10 s of noise prior to the P-wave arrival and 60 s of the P

wave trains, were used in the spectral analysis. No filter was applied to the P wave trains. After calculating the fast Fourier transform, we estimated the seismic moment rate at each seismic station by:

$$\frac{\partial}{\partial t}M(f) = \frac{4\pi\rho\alpha^3 a}{g(\Delta, h)C(i)}\left[\frac{Q(f)W(f)}{I(f)S_{rad}(f)}\right] \quad (1)$$

where ρ is the density at the source, α is the P wave velocity at the source, a is the radius of the earth, $C(i)$ is the surface amplification at the ith seismic station, $Q(f)$ is the seismic attenuation, $g(\Delta, h)$ is the geometrical dispersion, $W(f)$ is the spectral displacement, $I(f)$ is the instrument response and $S_{rad}(f)$ is the radiation pattern as a function of frequency of the P, pP and sP waves.

Data was corrected for instrument response, attenuation, and radiation pattern. Following BEZZEGHOUD et al., (1989), we substituted the effective radiation pattern $S_{rad}(f)$ with the approximation:

$$S_{rad}(0) = R^P + R^{pP} + R^{sP} \quad (2)$$

where R^P, R^{pP}, R^{sP} are the P, pP, and sP radiation pattern, respectively. The fault geometry used to calculate the radiation pattern was strike 277°, dip 27° and rake 78°. According to the hypocenter (25 km), the pP and sP wave trains arrived 7 and 10 s after the direct P wave. The attenuation operator t^* was assumed to be 1.0. The static seismic moment was calculated from the following expression:

$$M_0 = \frac{\partial}{\partial t}M(0) = \frac{4\pi\rho\alpha^3 a}{g(\Delta, h)C(i)}\left[\frac{Q(0)W(0)}{I(0)S_{rad}(0)}\right] \quad (3)$$

Figure 7 shows the plot of the logarithmic mean of corrected moment rate spectra determined from the broadband displacement records of 15 P-waveforms recorded at teleseismic stations. An average seismic moment of $2.7 \pm 0.7 \times 10^{20}$ Nm ($2.7 \pm 0.7 \times 10^{27}$ dynes cm) was determined. From the spectra we obtained a corner frequency of $f = 0.05 \pm 0.005$ Hz and a slope at higher frequencies of -2.4. The rupture time needed for the rupture to propagate along the fault is approximately:

$$T_R = \frac{L}{v_R} \quad (4)$$

Where T_R is the rupture time, which is the inverse on the corner frequency, L is the length of the fault

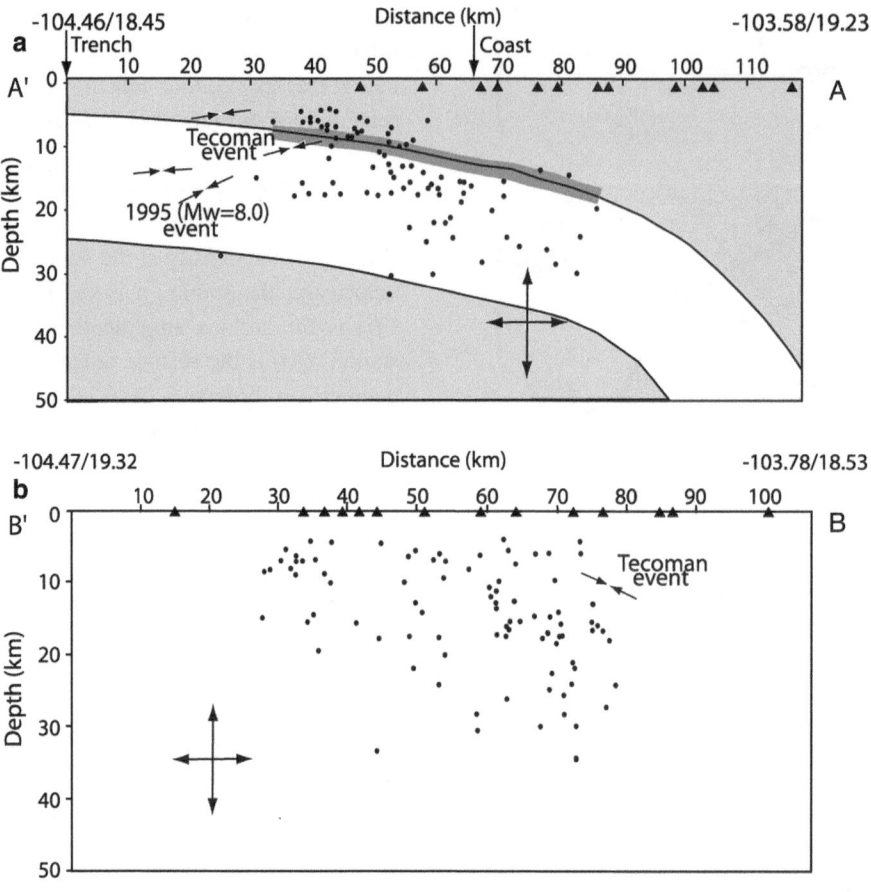

Figure 5

a Cross section of seismicity along profile AA'. **b** cross section of seismicity along profile BB'. *Arrows* show the projection of the P axes for the Tecomán earthquakes and two of its largest aftershocks. Subducting plate geometry was taken from PARDO and SUÁREZ, (1995). See Fig. 4 for profile locations

Table 3

Velocity model used for focal mechanism determination with teleseismic records

Thickness (km)	V_p (km/s)	V_s (km/s)	ρ (g/cm^3)
6.0	5.80	3.35	2.68
19.0	6.40	3.69	2.78
10.0	7.00	4.04	2.85
–	8.00	4.62	3.00

and v_R is the rupture velocity. Therefore, the rupture time is 20 ± 2 s and the rupture length is 66 ± 5 km, assuming a rupture velocity of $0.9V_s$ (3.3 km/s). This calculated rupture length lies within the range of the aftershock length calculated with the aftershock activity (56 ± 2 km). The stress drop on a circular fault (BRUNE, 1970) with radius R is given by

$$\Delta \sigma = \frac{7}{16} \frac{M_0}{R^3} \qquad (5)$$

From the aftershock area we calculated a fault length of 56 ± 2 km so the fault radius R is 28 ± 2 km. Therefore, the stress drop of the Tecomán earthquake is 5.4 ± 2.5 MPa. This value contrasts with the higher value (~ 10 MPa) obtained by YAGI *et al.*, (2004) using regional and local records.

7. Slip Distribution Inversion

BERESNEV, (2003) argued that incorrect assumptions of crustal structure, rupture velocity, fault geometry and array geometry will generate geologic artifacts in the inversion of seismic data for slip

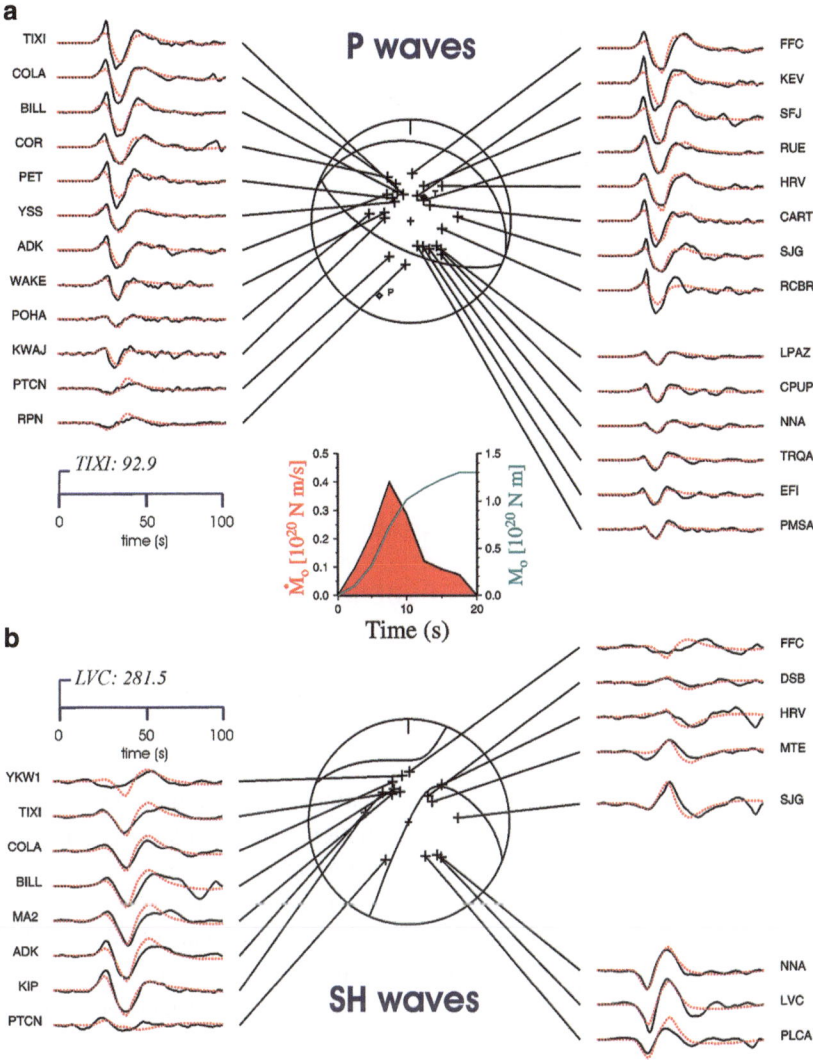

Figure 6

Point source inversion from body wave broadband data. **a** Observed P wave seismograms (*black lines*) and synthetics (*red lines*). The source time function was about 20 s and the corresponding seismic moment is 1.3×10^{20} Nm. **b** Observed SH wave seismograms (*solid lines*) and synthetics (*dashed lines*). The solution corresponds to an underthrust event with (strike, dip, rake) = (277°, 27°, 78°). The axes indicate the amplitude of the waves in microns

distribution of finite faults. On the other hand, IDE *et al.*, (2005) suggested that the quality and quantity of the data is critical to constrain the source-imaging problem. So, in line with those opinions, we decided to do the source imaging of the Tecomán event with better-constrained fault geometry and the inclusion of near field strong motion data not used previously in the study of the Tecomán event. Figure 8 shows the acceleration records of the free-field strong motion station located at the Manzanillo power plant (MNZ).

We also used data recorded at IRIS and GEOSCOPE broadband stations, strong motion records from the Servicio Sismologico Nacional (SSN) and Instituto de Ingenieria, UNAM (Fig. 2). To obtain the slip over the rupture area of the Tecomán earthquake, we used the method described by YAGI *et al.*, (1999, 2004) following the formulation of HARTZELL and HEATON, (1983), YOSHIDA (1989, 1992). The rupture process is simulated as a spatio-temporal slip distribution over the fault plane. The fault plane is divided

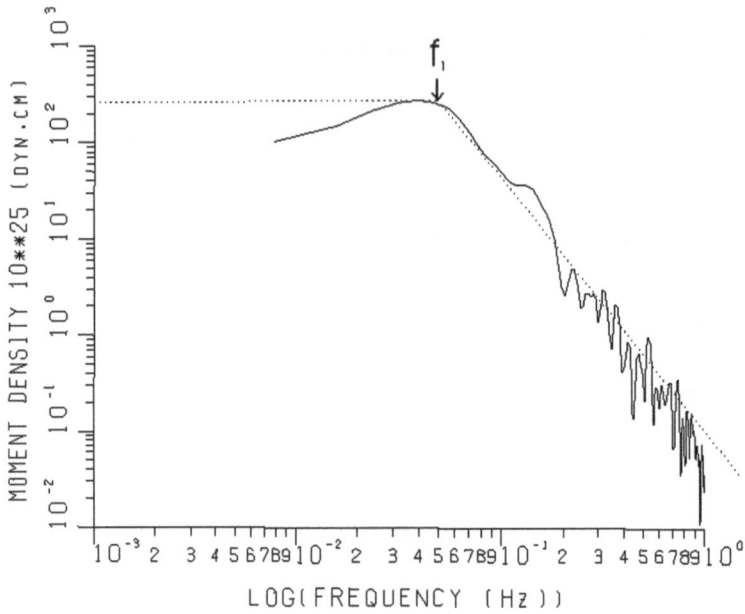

Figure 7

Average displacement spectra of teleseismic P waveforms of the Tecomán earthquake. f is the corner frequency and dashed lines indicate the trend at lower and higher frequencies. The slope at higher frequencies is -2.4

into $M \times N$ sub-faults with length xx and width yy. The slip-rate function at each sub-fault is described with B splines that are a series of L triangle functions with rise time τ. Slip vectors at each sub-fault are taken as vectors with components given by λ_0, the assumed slip angle. The synthetic waveform at the jth station is given by

$$W_j^{obs}(t_i) = \sum_m \sum_n \sum_l \sum_k X_{mnlk} g_{mnkj}$$
$$\times (t_i - (l-1)\tau - T_{mn}) + e,$$

where X_{mnlk} is the lth component of slip at mnth subfault at kth time step. g_{mnkj} is the Green's function from a point source at the mn-subfault, T_{mn} is the start time of the slip of the basis functions at each sub-fault and e is assumed to be a Gaussian error with variance σ. Green's functions for teleseismic body waves are calculated using the KIKUCHI and KANAMORI, (1991) method. Green's functions for near field ground motion are calculated by the discrete wave number method developed by KOHKETSU, (1985). Table 4 shows the one-dimensional elastic models used in the calculation of the Green's functions. This model is the same used by YAGI et al. (2004) for teleseismic

stations and has been modified lightly for strong motion stations to take into account the inclusion of the MNZ accelerograms.

In the inversion process we used the fault geometry previously calculated with the NÁBĚLEK, (1984) method given by (strike, dip, slip) = (277°, 27°, 78°). We used a larger fault area in order to follow the same procedure of YAGI et al., (2004), even though the observed aftershock area is smaller. We considered a rupture area with a width of 70 km, a length of 100 km and divided the fault plane into 14 × 20 subfaults with length of 5 km and width of 5 km. Twenty five kilometers depth of the initial front rupture was assumed. The slip-rate function on each sub-fault was expanded into a series of twelve triangles with a rise time of 2.0 s and the slip angle kept unchanged during the rupture. We used a rupture wave velocity of 3.3 km/s.

It is instructive to investigate how well the MNZ accelerogram fits with our estimated fault geometry (strike, dip, slip) = (277°, 27°, 78°) and YAGI et al. (2004) fault geometry (strike, dip, slip) = (300°, 20°, 93°). Figure 9 shows the comparison of the observed and synthetic accelerograms of MNZ with both fault

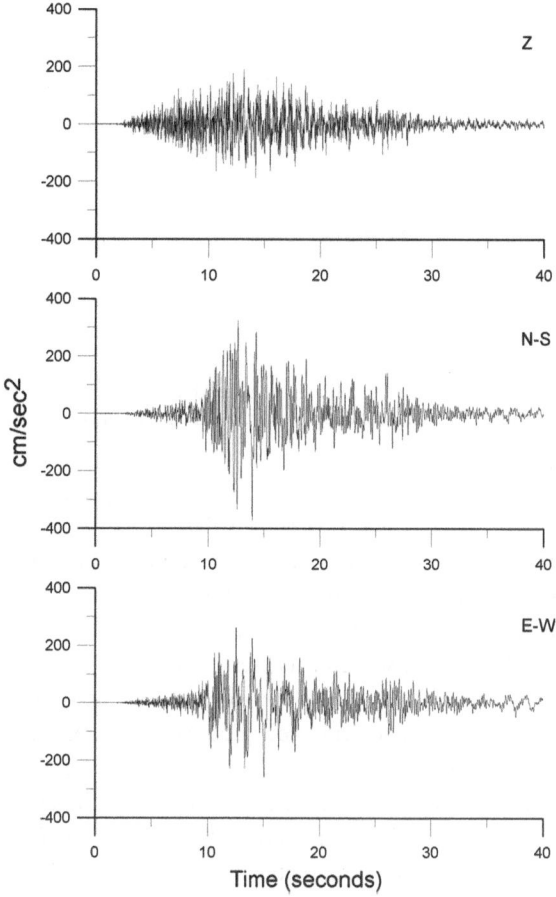

geometries. Both geometries reproduce the main amplitudes, but our fault geometry fit better the later parts of the accelerograms of the vertical and north–south components; however both geometries failed to adequately fit the east–west component, even though our fault geometry better fit the initial onset of the first arrivals.

We performed the inversion with the near field strong motion data and the far field P-waveforms. Comparisons of the observed and synthetic waveforms, shown in Fig. 10, indicate that the fits are good. The seismic moment was 1.88×10^{20} Nm equivalent to a moment magnitude of 7.5. The source time duration lasted nearly 30 s, but the majority of the energy was released during the first 20 s, in agreement with the source time duration calculated with the teleseimic spectra and NÁBĚLEK (1984) point source inversion. Figure 11 show a horizontal projection of the slip distribution on the fault plane, which consisted mainly of the same seismic moment release found by YAGI et al., (2004), however, it shows two additional zones of seismic moment release up-dip and down-dip which are barely shown in YAGI et al., (2004) inversion. The maximum fault slip was 3.2 m.

Figure 8

Acelerograms recorded in the Manzanillo (MNZ) power plant located 54 km from the epicenter

Table 4

Velocity model used for rupture process inversion with local, regional and teleseismic records

α (km/s)	β (km/s)	ρ (10^3 kg/m³)	Qα	Qβ	Thickness (km)
For teleseismic body wave					
1.5	0.0	1.0			1.0
5.54	3.20	2.50			3.0
5.69	3.29	2.70			5.0
6.27	3.62	2.80			9.70
6.71	3.87	2.90			17.30
8.10	4.68	3.30			–
For strong ground motion					
5.69	3.37	2.70	500	250	9.0
6.27	3.54	2.80	600	300	9.7
6.71	3.82	2.90	800	400	17.3
8.00	4.53	3.25	1,200	600	–

8. Summary

We used data recorded by a permanent network of short period and broadband seismic stations (RESCO) and nine portable seismic stations in order determine the aftershock area of the Tecomán earthquake. Figure 4 shows our aftershock locations. A depth of 10 km for the hypocenter of the Tecomán earthquake was calculated with P-wave arrivals of local stations; we think that the rupture could have started at this depth. The depths of the main event and the aftershocks defined a shallow dip of the Wadati-Benioff zone of roughly 12° (Fig. 5). Beyond the aftershock area, the Wadati-Benioff dip increases to 50° according to PARDO and SUÁREZ, (1995), however we did not located any aftershock activity at the bend of the subducted slab. We inferred an approximate aftershock area with a rectangular shape of 42 ± 2 by 56 ± 2 km. Aftershocks were located toward the north of the location of the epicenter of the Tecomán

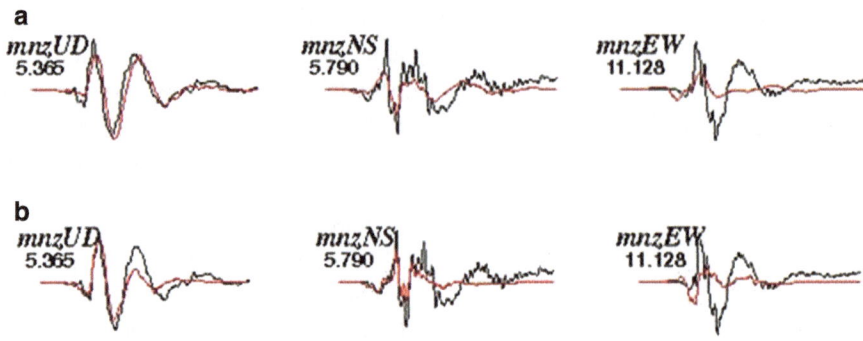

Figure 9
Comparison of modeling for the strong-motion record of MNZ station located 54 km from the epicenter. **a** Modeling with the focal mechanism solution obtained in this work using Nábělek technique. **b** Results obtained using YAGI *et al.*, (2004) geometry

event, and they enclose the southeast aftershock area of the 9 October 1995 ($M_w = 8.0$) Colima–Jalisco earthquake. The Tecomán earthquake ruptured approximately half of the Colima gap. Most aftershocks occurred inside the subducted slab, and our aftershock locations (Fig. 5) have features similar to those reported by NÚÑEZ-CORNU *et al.*, (2004) and YAGI *et al.*, (2004). For example, they report an elongated trend of aftershocks from southeast to northwest and three clusters of activity; however, their locations have a trend toward the southwest. Our aftershock locations are spread in an extended area close to the seashore with some of the features reported by NÚÑEZ-CORNU *et al.*, (2004) and YAGI *et al.*, (2004).

Source parameters of the Tecomán earthquake were calculated from the spectra of P-waveforms recorded at teleseismic distances. A corner frequency was observed at $f = 0.05 \pm 005$ Hz, a stress drop of 5.4 ± 2.5 MPa (54 ± 25 bars) and a seismic moment of $2.7 \pm 0.70 \times 10^{20}$ Nm ($2.7 \pm 0.7 \times 10^{27}$ dynes cm) were calculated assuming that BRUNE's (1970) circular fault model holds. We note the lower stress drop value calculated with teleseismic data compared with that obtained using local and regional records (YAGI *et al.*, 2004, SINGH *et al.*, 2003). This result would imply that Tecomán earthquake was deficient in high-frequency radiation. This, however, appears not to be the case from the reported severe damage to buildings in the localities situated at distances where along-trench rupturing subduction earthquakes of the same magnitude usually do not produce any

significant damage (ZOBIN and PIZANO-SILVA, 2007). We think, therefore, that a low rupture velocity is more likely the cause of the low stress drop value obtained from spectral analysis.

Besides the waveforms used by YAGI *et al.*, (2004) we included, in the inversion of near field body waves, the waveforms recorded at the Manzanillo power plant (MNZ), a strong motion station located 54.0 km from the epicenter. Those body waveform were not used by YAGI *et al.*, (2004). In the inversion process we assumed a rectangular fault area of 70 by 100 km and a depth of the initial break of 25 km. Figure 11 shows a horizontal projection of the fault rupture from our inversion. The slip distribution shows the same asperities reported by YAGI *et al.*, (2004), however, two additional zones of seismic moment release, not so energetic in their inversion, appeared clearly defined in our analysis as a significant zone of seismic moment release up-dip and down-dip. The total source duration was 30 s and a seismic moment of 1.88×10^{20} Nm ($M_w = 7.5$). A significant seismic moment release was observed at the end of the rupture process; this seismic moment released was located down-dip below the Manzanillo seismic station and up-dip. The source time duration of 20 s calculated with the spectra of teleseismic body-waves and with the inversion of P and SH waves using the NÁBĚLEK, (1984) method agree with the main source energy release. On the other hand, source time duration obtained with the inversion of teleseismic and near-source strong motion stations is 30 s, however the maximum seismic moment release

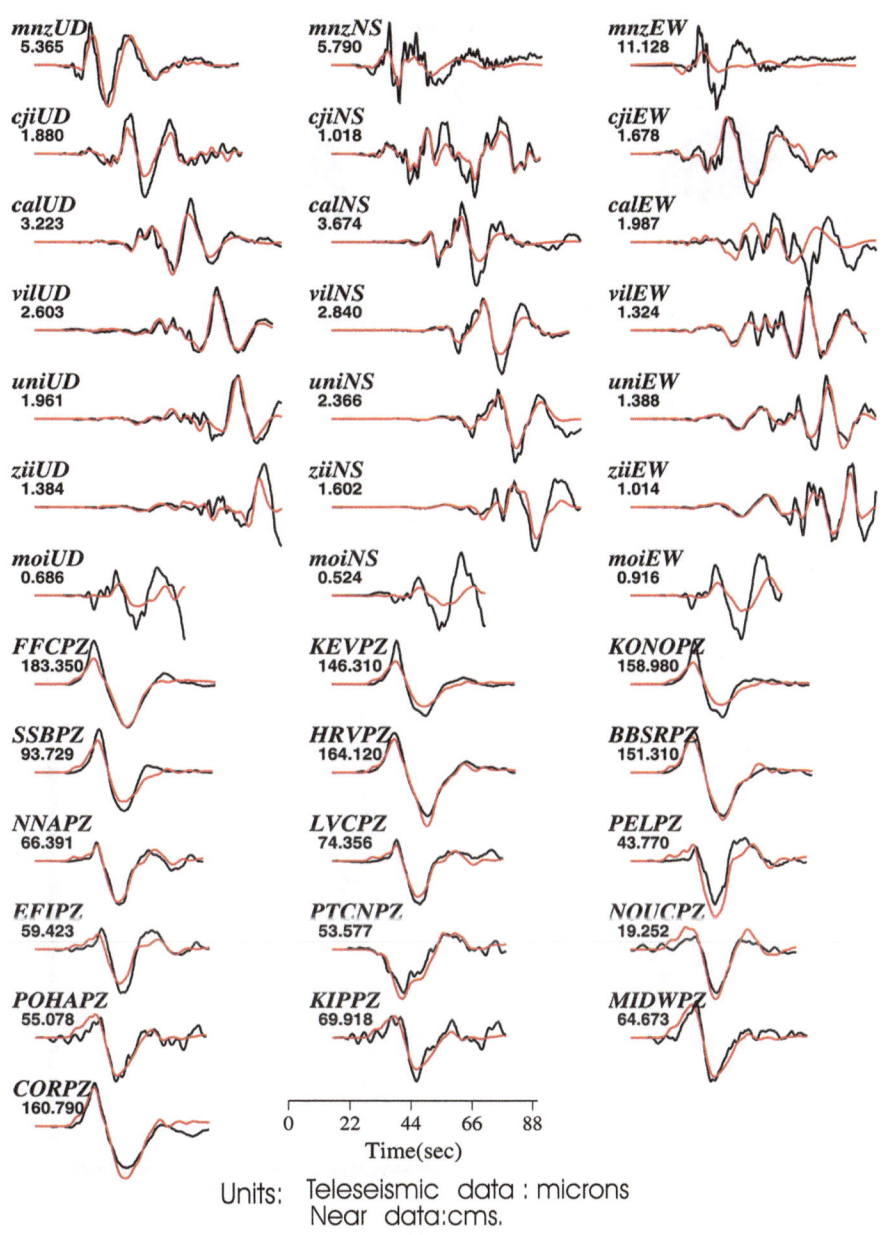

Units: Teleseismic data : microns
Near data:cms.

Figure 10

Comparison of the observed waveforms (*black line*) with the calculated waveforms (*red line*) obtained from the inversion. Seven upper records correspond to strong motion data of Mexican stations. Six lower seismograms are teleseismic records. *Numbers* below the station name are the maximum amplitude in centimeters and microns of strong-motion and teleseismic records, respectively

lasted about 20 s (see Fig. 11b). We think that the large source duration obtained with the two dimensional method of inversion is due to the effect of the contribution of energy recorded on the strong-motion (MNZ) stations which was not observed at teleseismic stations.

It is worth mentioning that the uncertainties suggested by BERESNEV, (2003) about the slip inversions using body waves did not have much effect in our study. We did not find great disagreement in the inversion process for the spatial slip distribution. On the other hand, we found an improvement in the

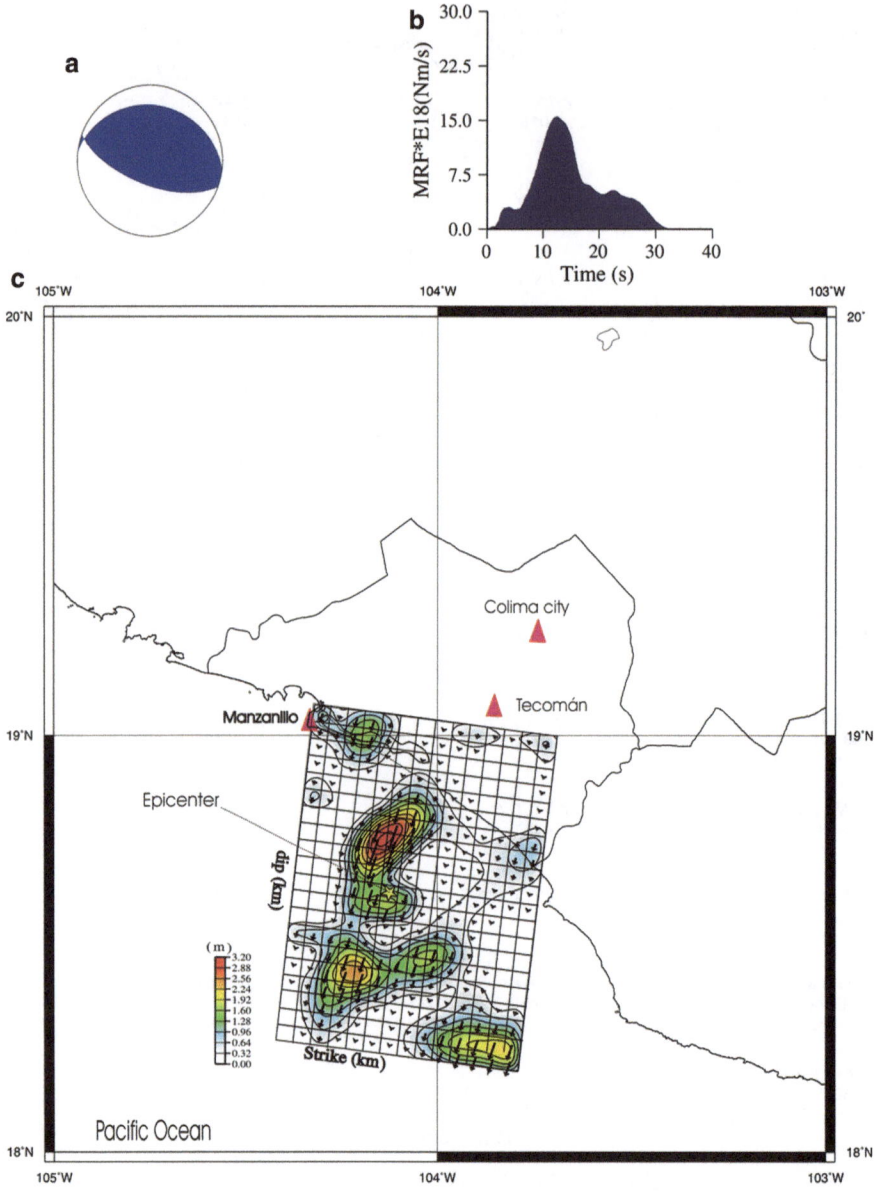

Figure 11

a Fault plane solution used for the slip distribution inversion on the fault plane. **b** Source time function obtained in this inversion. **c** Horizontal projection of the fault showing the slip distribution. *Triangle* shows the location of the Manzanillo power plant, where it is located the MNZ strong motion recorder, Colima and Tecomán cities

spatial slip distribution by adding the new near source body waveforms into the inversion, therefore, the quality and quantity of waveforms in the near and far field will greatly improve our knowledge of the source process as suggested by IDE *et al.*, (2005).

Acknowledgments

We would like to thank Arturo Pérez-Vertti for the fieldwork and Antonio Mendoza for helping to handle the digital data. Thanks to the following networks for providing seismic data through the Incorporated Research Institutions for Seismology (IRIS) Data Management Center: FDSN, GEOFON, Geoscope, IDA, GTSN, UI, USGS. We acknowledge El Consejo Nacional de Ciencia y Tecnologia (CONACYT) for sponsoring part of this project (Proyecto 37038-T). One of us (HERL) was supported by a scholarship from PROMEP.

REFERENCES

BERESNEV, I. A. (2003), *Uncertainties in Finite-Fault Slip Inversion: To What Extent to Believe?*, Bull. Seismol. Soc. Am. 93, 2445–2458.

BEZZEGHOUD, M., DESCHAMPS, A. and MADARIAGA, R., Broad–band P-wave Signals and Spectra from Digital stations. In *Digital Seismology and Fine Modeling of the Lithosphere* (eds. Cassinis, R., Nolet, G., and Panza, G.F.) (Plenum Press, New York, 351–374, 1989).

BRUNE, J.N. (1970), *Tectonic stress and the spectra of seismic shear waves from earthquakes*, J. Geophys Res. 75, 4997–5009.

FUTTERMAN, W.I. (1962). *Dispersive body waves*, J. Geophys. Res. 67, 5279–5291.

SUÁREZ. G., GARCIA–ACOSTA, V. and GAULON, R. (1994), *Active crustal deformation in the Block, Mexico: evidence for great historical earthquake in the 16th century*, Tectonophysics 234, 117–127.

HARTZELL, S. H. and HEATON, T.H. (1983), *Inversion of strong ground motion and teleseismic waveform data for the fault rupture history of the 1979 Imperial Valley, California earthquake*, Bull. Seismol. Soc. Am. 73, 1553–1583.

IDE, S., BEROZA, G.C. and McGUIRE, J.J., *Imaging earthquake source complexity, Geophysical Monograph 157, Seismic earth: Array analysis of broadband seismograms* (eds. Levender, A. and Nolet, G. (American Geophysical Union, Washington, D. C., pp, 2005).

KIKUCHI, M., and KANAMORI, H. (1991), *Inversion of Complex Body Waves-III*, Bull. Seism. Soc. Am. 81, 2335–2350.

KOHKETSU, K. (1985), *The extended reflectivity method for synthetic near-field seismograms*, J. Phys. Earth 33, 121–131.

KOSTOGLODOV, V., and BANDY, W. (1995), *Seismotectonic constraints on the rate between the Rivera and North American plates*, J. Geophys. Res. 100, 977–990.

NÁBĚLEK, J. (1984), *Determination of earthquake source parameters from inversion of body waves*, PhD. Thesis (Mass. Inst. of Technol., Cambridge), 346 pp.

NÚÑEZ-CORNU, F. J., REYES-DÁVILA, G.A., RUTZ LÓPEZ, M., TREJO GOMEZ, R., CAMARENA-GARCÍA, M.A. and RAMÍREZ-VAZQUEZ, C.A. (2004), *The 2003 Armeria, Mexico Earthquake (M_w = 7.4): Mainshock and early aftershocks*, Seismological Research Letters 75, pp. 506–605.

PACHECO, J. F., and SYKES, L. (1992), *Seismic moment catalog of large shallow earthquakes, 1900 to 1989*, Bull. Seismol. Soc. Am. 82, 1306–1349.

PACHECO, J., SINGH, S. K., DOMINGUEZ, J., HURTADO, A., QUINTANAR, L., JIMÉNEZ, Z., YAMAMOTO, J., GUTIÉRREZ, C., SANTOYO, M., BANDY, W., GUZMAN, M., KOSTOGLODOV, V., REYES, G. and RAMÍREZ, C. (1997), *The October 9, 1995 Colima–Jalisco, Mexico earthquake (Mw 8): an aftershock study and comparison of this earthquake with those of 1932*, Geophys. Res. Lett. 24, 2223–2226.

PARDO, M., and SUÁREZ, G. (1995), *Shape of the subducted Rivera and Cocos plates in Southern Mexico: Seismic and tectonic implications*, J. Geophys. Res. 100, 12357–12373.

REYES, A., BRUNE, J. N. and LOMNITZ, C. (1979), *Source mechanism and aftershock study of the Colima, Mexico earthquake of January 30, 1973*, Bull. Seismol. Soc. Am. 69, 1819–1840.

SINGH, S.K., PONCE, L. and NISHENKO, S. P. (1985), *The great Jalisco, Mexico, earthquakes of 1932: subduction of the Rivera plate*, Bull. Seism. Soc. Am. 75, 1301–1313.

SINGH, S. K., PACHECO, J. F., ALCÁNTARA, L., REYES, G., ORDAZ, M., IGLESIAS, A., ALCOCER, S. M., GUTIERREZ, C., VALDÉS, C., KOSTOGLODOV, V., REYES, C., MIKUMO, T., QUAAS, R., ANDERSON, J.G. (2003), *Contributions. A Preliminary Report on the Tecomán, Mexico Earthquake of 22 January 2003 (Mw 7.4) and Its Effects*. Seism. Res. Lett. 74, 279–289.

YAGI, Y., KIKUCHI, M., YOSHIDA, S. (1999), *Comparison of the coseismic rupture with the aftershock distribution in the Hyuganada earthquakes of 1996*, Geophys. Res. Lett. 26, 3161–3164.

YAGI, Y., MIKUMO, T., PACHECO, J., and REYES, G. (2004), *Source rupture process of Tecoman, Colima, Mexico earthquake of January 22, 2003, determined by joint inversion of teleseismic body-wave and near-source data*, Bull. Seismol. Soc. Am. 94, 1795–1807.

YOSHIDA, S. (1989), *Waveform inversion using ABIC for the rupture process of the 1983 Hindu-Kush earthquake*, Phys. Earth Planet. Inter. 56, 389–405.

YOSHIDA, S. (1992), *Waveform inversion for rupture process using a non-flat seafloor model: application to 1986 Andreanof Islands and 1985 Chile earthquakes*, Tectonophysics 211, 45–59.

ZOBIN, V.M. and PIZANO-SILVA, J.A. (2007), *Macroseismic Study of the M_w 7.5 21 January 2003 Colima, México, Across-Trench Earthquake*, Bull. Seismol. Soc. Am. 97, 1221–1832.

(Received February 11, 2010, revised September 22, 2010, accepted September 23, 2010, Published online November 16, 2010)

Pure Appl. Geophys. 168 (2011), 1355–1361
© 2010 Springer Basel AG
DOI 10.1007/s00024-010-0172-3

Influence of Rivera-Cocos Plate Boundary Geodynamics on Earthquake Intensity Patterns: the 9 October 1995 (Mw 8.0) and 21 (22) January 2003 (Mw 7.5) Earthquakes

VYACHESLAV M. ZOBIN[1]

Abstract—This study examines two large thrust subduction earthquakes occurring within the Rivera-Cocos plate boundary which struck the western coast of México on 9 October 1995, Mw 8.0, and 21 (22 GMT) January 2003, Mw 7.5. The Modified Mercalli (MM) earthquake intensities observed during these earthquakes were surprising for some towns located in the Mexican coastal zone. During the smaller Mw 7.5 2003 earthquake, MM intensity VII was observed for towns of Colima, Villa de Alvarez and Ixtlahuacán, while during the larger Mw 8.0 1995 earthquake, their MM intensities were only IV–V, V and V–VI, respectively. We construct the macroseismic patterns for these two earthquakes and discuss the possible reasons for the significant difference in the outline of the MM VII isoseismals, such as the tectonic setting of epicentral zones and the directivity of rupture processes along and across the coastal line.

Key words: Geodynamics, Rivera-Cocos plate boundary, earthquake intensity pattern.

1. Introduction

This study examines two large thrust subduction earthquakes (9 October 1995, Mw 8.0 and 21 (22 GMT) January 2003, Mw 7.5) which struck the western coast of México (Table 1). The distance between their epicenters was 43 km; their distance from the coastal line was 30 and 15 km, respectively (Fig. 1). The Modified Mercalli (MM) earthquake intensities observed during these earthquakes were surprising for some towns located in the Mexican coastal zone (ZOBIN and VENTURA-RAMÍREZ, 1998; ZOBIN and PIZANO-SILVA, 2007). During the smaller Mw 7.5 2003 earthquake, MM intensity VII was observed for towns of Colima, Villa de Alvarez and Ixtlahuacán, while during the larger Mw 8.0 1995

earthquake, their MM intensities were only IV–V, V and V–VI, respectively. The reason for this difference may lie in the difference in tectonic structures where the earthquake source rupturing occurs.

In this paper, I discuss the tectonic setting of the epicentral areas where these two earthquakes occurred, the rupturing characteristics during these earthquakes and their intensity patterns along the Mexican coast.

2. General Characteristics of the Rivera-Cocos Plate Boundary and Rupturing in the Sources of Two Large Earthquakes

2.1. Plate Tectonic Elements of the Region

Figure 1 shows the 1963–1995 distribution of earthquake epicenters near the northern part of the western coast of México where the two large 1995 and 2003 earthquakes occurred. The main tectonic element of this region, the oceanic Rivera plate, subducts beneath the North American plate and produces large destructive earthquakes. The earthquake epicenters outline the limits of the Rivera plate along the Tamayo and Rivera Fracture zones, East Pacific Rise and Middle American trench and its boundary with Cocos plate. The Rivera-Cocos plate boundary zone is a complicated across-trench tectonic structure (Fig. 2) consisting of a few grabens and horsts (BANDY *et al.*, 2005). It run across the Middle American trench (El Gordo graben) and has its continental continuation within the Colima rift. For simplicity, we call this boundary zone El Gordo Graben (EGG, Fig. 1). Our two large earthquakes had their epicenters within the EGG (Figs. 1, 2).

[1] Observatorio Vulcanológico, Universidad de Colima, Colima 28045, México. E-mail: vzobin@ucol.mx

Table 1

Characteristics of two large earthquakes of 1995 and 2003

Date (yyyy/mm/dd)	Latitude (N)	Longitude (W)	Depth (km)	Mw	Ms	Seismic moment (N-m)	Focal mechanism
1995/10/09	18.81	−104.54	20	8.0	7.3	1.15×10^{21}	Thrust strike = 302 dip = 9 slip = 92
2003/01/21	18.79	−104.13	10	7.5	7.6	2.05×10^{20}	Thrust strike = 108 dip = 79 slip = 86

Earthquake locations were determined by the Colima regional seismic network; the magnitudes, seismic moments and focal mechanisms were taken from the Harvard CMT catalog (http://www.globalcmt.org, visit 08 January, 2010)

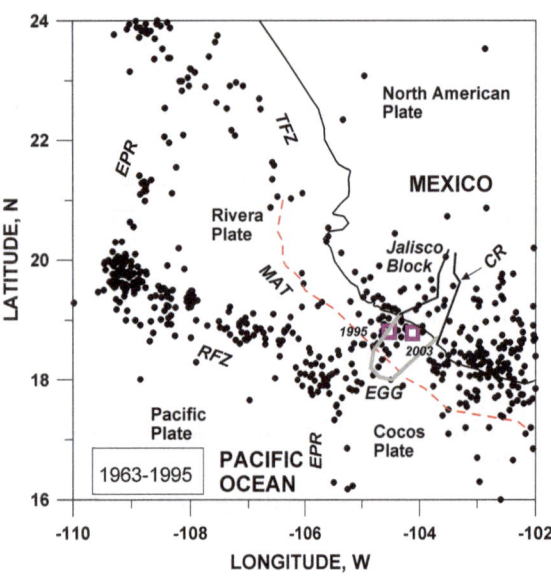

Figure 1

The principal lithospheric plates near the western coast of México and the 1963–1995 earthquake epicenters located by the US Geological Survey. The open squares show the epicenters of two large earthquakes of 09 October, 1995 (Mw 8.0) and 21 January, 2003 (Mw 7.5) located by the Colima regional seismic network RESCO. *TFZ* Tamayo Fracture zone; *RFZ* Rivera Fracture zone; *MAT* Middle American trench (*dashed line*); *EPR* East Pacific rise; *CR* Colima rift. *EGG* El Gordo graben (outlines shown by a *heavy grey line*)

Plate motion studies indicate the direction of convergence between the Rivera and North American plates as northwestward along the Mexican subduction zone (BANDY *et al.*, 2005). The EGG zone is characterized by NE-SW oriented structures. Figure 2 shows that the 1995 earthquake epicenter was located within the Manzanillo horst representing the NW border of the EGG zone. The 2003 earthquake epicenter is situated within the Tecoman trough in the central part of the EGG zone.

2.2. Aftershock Areas and Rupturing in the Sources of Two Large Earthquakes

Figure 3a shows the areas of the first-month aftershock epicenters of the 1995 and 2003 earthquakes according to the National Seismological Service of México. It is seen that the 1995 aftershock activity developed northwestward along the trench structures that corresponds to the direction of convergence between the Rivera and North American plates. The 2003 aftershock clustering was observed mainly within the EGG zone between the trench and coast.

The source rupture process of these two large earthquakes has been reconstructed by the inversion of near-source strong motions and teleseismic body-wave data (MENDOZA and HARTZELL, 1999; YAGI *et al.*, 2004). The coseismic slip isolines and the main broken asperities, indicating the rupture segments with the largest slip values during the rupturing, are shown in Fig. 3b. According to (MENDOZA and HARTZELL, 1999), the rupturing of the 1995 source was directed northwestward from the epicenter along the trench. Two main broken asperities were observed along the trench line at a distance of about 70 and 130 km from the epicenter. The rupturing of the 2003 source (YAGI *et al.*, 2004) was directed along the NE–SW structures of the EGG zone. Two main broken asperities were observed along the line connecting them with the epicenter position at a distance of about 15 km to the southwest and 25 km to the northeast from the epicenter.

These observations show that the 1995 earthquake source, originating within the NW border of the EGG zone, developed according to the direction of convergence between the Rivera and North American plates, while the 2003 earthquake source, originating within the central part of the EGG zone, developed

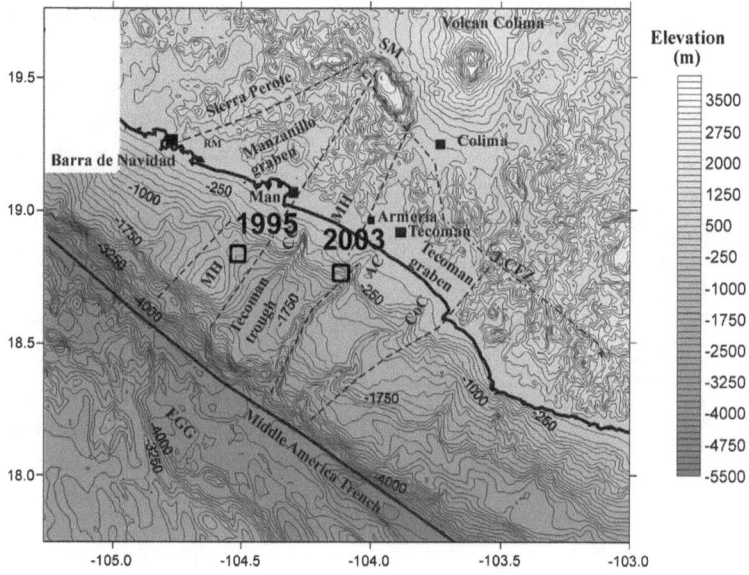

Figure 2

The main morphotectonic features of the Rivera-Cocos boundary zone (taken from BANDY *et al.*, 2005, with some additions). *AC* Armeria submarine canyon; *CC* Cuyutlan submarine canyon; *CoC* Coahuayana submarine canyon; *LCFZ* La Cumbre fault zone; *MH* Manzanillo horst; *SM* Sierra Manantlan; *EGG* El Gordo graben; *Man* Manzanillo; *RM* Río Marabasco. *Filled squares* mark locations of the main towns of the area; *open squares* show the epicenters of the two large earthquakes of 1995 and 2003 located by the Colima regional seismic network RESCO

along the Rivera-Cocos plate boundary across the subduction zone structures.

3. The MM Intensity Patterns for Two Large Earthquakes and Intensity Attenuation

3.1. Methodology of Earthquake Intensity Estimation

Earthquake intensity estimations were based on two conditions. The first was a personal inspection of a house and an interview with the persons who were in this house during the earthquake. The selection of a house for inspection was based on the characteristic type of construction for a locality, not on its level of damage. The second condition required a correction of intensity for the type and age of the building to avoid the effects of the increase in intensity for old adobe buildings or poor-quality, unreinforced masonry buildings as well as a decrease in intensity for modern reinforced buildings. All final estimations of intensity formed a homogeneous sample of residential houses.

The original version of the MM scale describes generally the damage to buildings for the different grades of intensity, but without a reference to a particular type of construction; therefore, we use the classification of buildings (types A, B and C) described in (ZOBIN and VENTURA-RAMÍREZ, 1998):

1. Type A: good quality. The buildings are designed with some lateral resistance to ground shaking. Columns are installed along the walls at a distance of 2 m and in the corners of the construction.
2. Type B: intermediate quality. The buildings are not designed to resist ground shaking. The walls have resistant elements only in the corners of the construction.
3. Type C: poor quality. Old buildings made out of adobe and cinder blocks. No resistant elements.

A simple use of the MM description of damage was not enough to estimate comparable values of intensity in small villages where all construction was of C type and in neighboring sites with construction of better quality. The MM scale gives the "damage slight in poorly built buildings" only beginning from MM VI. In the condition of old Mexican buildings made of adobe, this damage may be observed beginning from MM IV–V, estimated from the

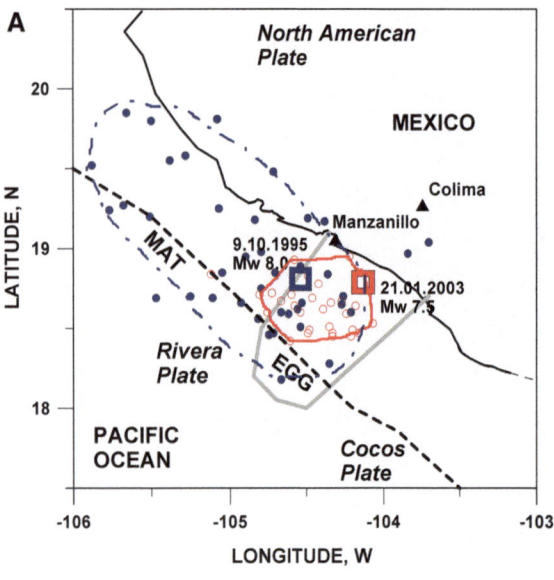

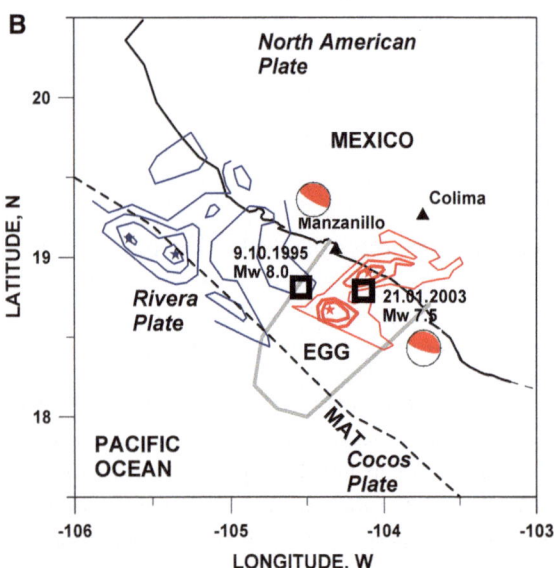

Figure 3

Aftershock areas (**a**) and rupturing (**b**) of two large earthquakes of 9 October 1995 (Mw 8.0) and 21 January 2003 (Mw 7.5). *Open squares* show the epicenters of two large earthquakes of 1995 and 2003 located by the Colima regional seismic network RESCO. The first-month aftershock epicenters in A (located by the National Seismological Service of México) are shown as *filled* (1995) and *open* (2003) points. In **b**, the isolines of co-seismic slip within the *rupture plane* are shown (for the 1995 event, according to Mendoza and Hartzell, 1999; for the 2003 event, according to YAGI *et al.*, 2004). The main asperities, broken during the earthquake rupturing, are shown as *stars*. The Harvard CMT focal mechanisms are shown near the epicenters. The towns of Colima and Manzanillo are shown by *triangles*. Other symbols are the same as in Fig. 1

characteristics of human behavior and response of objects. Corrections of intensity for the type and age of the building were introduced to avoid account for the effect of poor quality rural construction.

Initial estimates of earthquake intensity were referenced to Type B masonry. For this study, the corrections proposed earlier by ZOBIN and VENTURA-RAMÍREZ (1998) during the macroseismic study of the 1995 earthquake were introduced for other types of masonry: +0.5 MM intensity for Type A masonry and −0.5 for Type C masonry. If the construction was more than 20 years old, its category of masonry was decreased by one grade (from A to B, from B to C). For each site, the mean intensity was calculated for all constructions after the correction for type of masonry.

3.2. Characteristics of the Macroseismic Patterns of Two Large Earthquakes

The macroseismic patterns of the 1995 and 2003 earthquakes were based on personal interviews made in 56 and 83 localities, respectively. Figure 4 shows the spatial distribution of the mean MM intensities of the two large earthquakes. The isoseismal maps represent the effects of the earthquakes along a 500-km coastal line of the Pacific Ocean, with an area within the adjoined territories of about 120,000 km^2.

For the 1995 earthquake, three zones of equal intensity were distinguished, from intensity MM VI–VII to IV (Fig. 4a). The event is characterized by asymmetry of zone of maximum intensity, elongated in NW direction, according to the position of the epicenter. The zone of maximum intensity runs parallel to the maximum co-seismic slip isolines. The position of the asperities far from the epicenter led to the formation of the MM VII intensity sub-zones, situated opposite to the epicenter and asperities positions within the MM VI zone. Generally, the northwestward orientation of isoseismals along the trench structures can be clearly seen.

For the 2003 earthquake, four zones of equal intensity were distinguished, from intensity MM VII to IV (Fig. 4b). The zone of maximum intensity

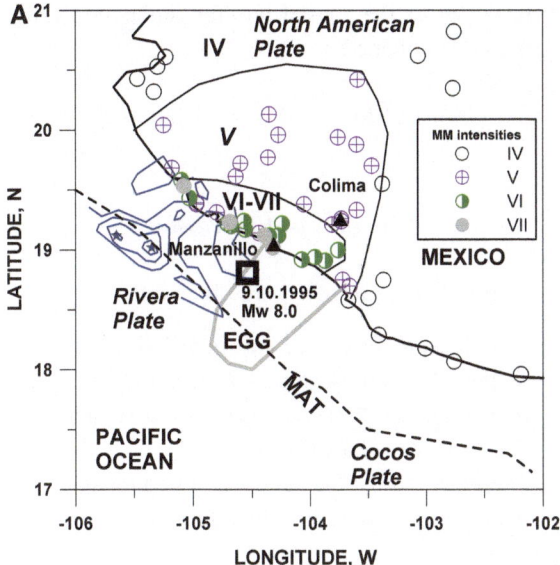

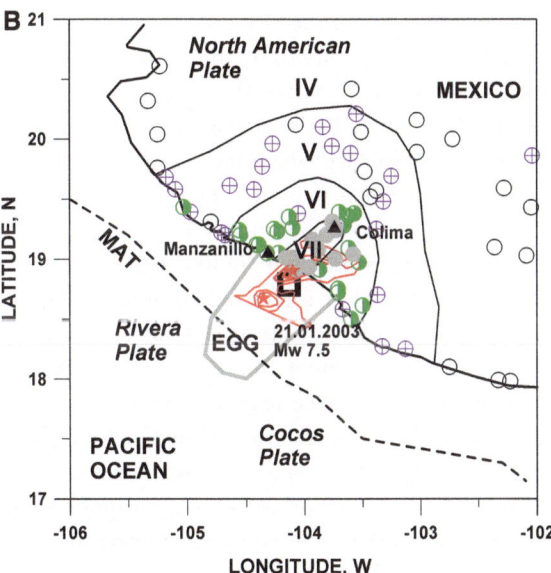

Figure 4

The earthquake intensity patterns observed during the large earthquakes of **a** 9 October 1995 (Mw 8.0) and **b** 21 January 2003 (Mw 7.5). The isolines of co-seismic slip within the *rupture plane* are shown. IV–VII are the Modified Mercalli (MM) intensities. Other symbols are as in Fig. 3

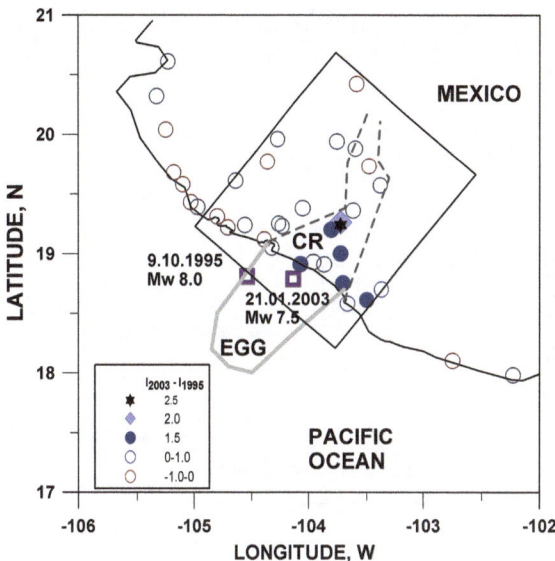

Figure 5

The intensity difference $I_{2003}-I_{1995}$ observed for the 1995 and 2003 earthquakes in the same localities. The data selected for the calculation of the attenuation of MM intensity across the coastal line are limited within the *square*. The outlines of Colima rift (*dashed lines*) and EGG (*heavy line*) are shown

aftershocks, and the position of two asperities broken during the mainshock rupturing as well as the area of maximum co-seismic slip show that all these seismotectonic features have the same NE trend that was noted for the isoseismals of the earthquake. This suggests that the NE trend in the isoseismals is the result of the NE trend of the earthquake source rupturing.

3.3. Earthquake MM Intensity Attenuation with Distance

The two large earthquakes were close in magnitude, Ms, 7.3 and 7.6 (Table 1), which allows us to study the attenuation of MM intensity. The specific features of intensity patterns caused by the difference in rupture directivity do not allow calculating the MM intensity attenuation simply by taking all intensity observations. Figure 5 shows the differences between the MM intensity estimations $I_{2003}-I_{1995}$ for 39 localities. It is seen that within the Colima rift, these differences were between 0 and 2.5. We consider that $I_{2003}-I_{1995} \geq |1|$ may be the result of differences in rupture directivity and local effects.

elongates in the NE direction. The isoseismals of intensities MM V–VI and IV–V are also characterized by this trend. The distribution of intensities on both sides of the epicenter is symmetrical.

Comparison of isoseismals with the position of the El Gordo Graben where the earthquake epicenter was located, the distribution of the first-month

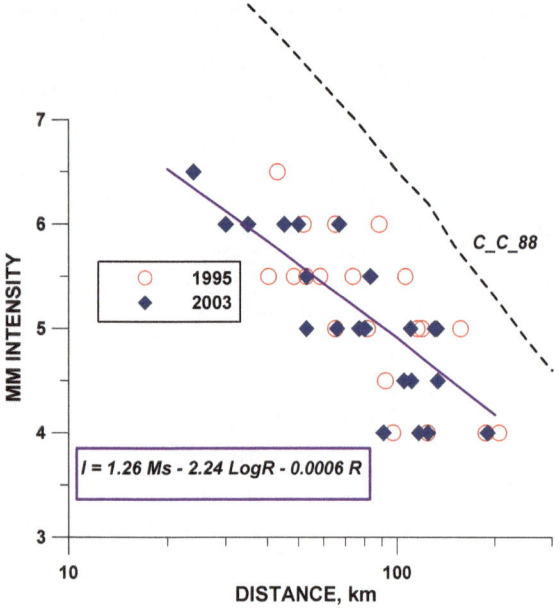

distance across the coastal line for subduction earthquakes was obtained:

$$I = 1.26\,\text{Ms} - 2.24\log R - 0.0006R \qquad (2)$$

4. Results and Discussion

The following results were obtained:

1. A noticeable difference was revealed in the shape
   of the MM VII isoseismals for the 1995 along-
   trench and the 2003 across-trench earthquakes of
   the Mexican subduction zone.
2. The equation for the subduction earthquakes MM
   intensity attenuation with distance across the
   coastal line was calculated.

The difference in the shape of the MM VII isoseismals may be caused by the different rupture
directivities for two earthquakes conditioned by the
specific features in the geodynamics of the Rivera-
Cocos plate boundary where the transverse structure
of El Gordo graben intersects the general Mexican
subduction zone. As was noted by SCHMITT et al.
(2007), given their proximity in time and space, these
two earthquakes offer a rare opportunity to contrast
earthquakes in nearly identical tectonic settings.

The obtained curve of earthquake intensity
attenuation with distance may be used together with
the earthquake acceleration attenuation curves or
instead of them. This is important for México, where
the coastal accelerograph network is insufficient for
registration of large earthquakes. During the two
large Colima earthquakes, only two accelerograph
records (both in Manzanillo) were obtained within the
first 100 km from the epicenters. The curve (2)
obtained in this study differs from the curve proposed
about 20 years ago by CHÁVEZ and CASTRO (1988) for
the intensity attenuation of subduction earthquakes
across the trench. It was based on the macroseismic
studies of 18 magnitude Ms 6.4 to 8.0 earthquakes
with thrust focal mechanism and depths between 15
and 20 km that occurred during the period from 1845
to 1979. Their curve is shown in Fig. 6 together with
our data. Comparison with our results reveals an
over-estimation of the earthquake intensity using this
curve. This difference may be caused, from one side,

Figure 6
Attenuation of MM intensity with distance during the large
earthquakes of 9 October 1995 (Mw 8.0) (*circles*) and 21 January
2003 (Mw 7.5) (*diamonds*). The *heavy line* shows the calculated
attenuation *curve* (2). The dashed line (C_C_88) shows the
attenuation for the Mexican subduction zone earthquake with Ms
7.5 according to Chávez and Castro (1988)

Therefore, to have a homogeneous set of data, we
took as our basis for calculating of the attenuation
law only those observations obtained for the same
localities for both earthquakes with differences in the
estimated intensity between 1 and −1.

We calculated the earthquake intensity attenuation across the coastal line, or trench, that goes
parallel to the coastal line, using 44 selected intensity
estimations (22 localities). Figure 6 shows the attenuation of intensities with a distance measured from
the hypocenter. We have calculated the intensity-
distance equation for the points of Fig. 6 in the
following form, standard for macroseismic equations
(MUSSON and CECIC, 2002):

$$I = a\text{Ms} - b\log R - cR \qquad (1)$$

where $R = (A^2 + h^2)^{1/2}$ is the hypocentral distance
(in km), A is the epicentral distance (in km), h is the
focal depth (in km), coefficients b and c are the
coefficients of geometric spreading of seismic waves
and absorption, respectively. As a result, the equation
characterizing the MM intensity attenuation with

by the absence of the Cocos-Rivera plate boundary earthquakes in the CHÁVEZ and CASTRO (1988) data set and (more importantly) by the use of over-estimated intensities based on the newspaper descriptions of the earthquake effects, which usually describe the largest destructive effects.

Our results may be of importance for the problem of seismic zonation. According to the map of the seismic regionalization of México (Manual de Diseño por Sismo 1993), a broad (about 100 km) zone D of high seismic intensity runs along the western Pacific coast where the Mexican subduction zone earthquakes occur. This map was constructed using the standard curves of the attenuation of earthquake acceleration with distance based on the data obtained from along-trench earthquakes, mainly from the Guerrero zone (ORDAZ et al., 1987). Therefore, the expected intensities for the region of Colima city may be under-estimated for the case of across-trench earthquakes. The same situation may arise also for the coastal region of Southern México, where the Orozco Fracture zone, O'Gorman fracture zone and Tehuantepec ridge intersect the trench structure separating the Michoacan, Guerrero and Oaxaca segments of the Mexican subduction zone (PARDO and SUÁREZ, 1995).

Acknowledgments

I thank Raul Castro and the anonymous reviewer for valuable comments.

REFERENCES

BANDY, W.L., MICHAUD, F., JACQUES BOURGOIS, J. et al. (2005), *Subsidence and strike-slip tectonism of the upper continental slope off Manzanillo, Mexico*, Tectonophysics *398*, 115–140.

CHÁVEZ, M. and CASTRO, R. (1988), *Attenuation of Modified Mercalli intensity with distance in Mexico*, Bull Seismol Soc Am *78*, 1875–1884.

Manual de Diseño por Sismo (1993). Comisión Federal de Electricidad, México, D.F., 162 p.

MENDOZA C. and HARTZELL S. (1999), *Fault-Slip Distribution of the 1995 Colima-Jalisco, Mexico, Earthquake*, Bull Seismol Soc Am *89*, 1338–1344.

MUSSON, R.M.W. and I. CECIC, Macroseismology, In *International Handbook of Earthquake and Engineering Seismology, Part A*, (eds. W.H.K. Lee, H. Kanamori, P.G. Jenning, and C. Kisslinger), (Academic Press, Amsterdam,pp 807–822, 2002).

ORDAZ, M. et al. (1987) *Espectros de respuesta para el estado de Guerrero* (Informe Interno Instituto de Ingeniería, UNAM, México).

PARDO, M. and SUÁREZ, G. (1995). *Shape of the subducted Rivera and Cocos plates in southern México: Seismic and tectonic implications*, J Geophys Res *100*, 12357–12373.

SCHMITT, S.V., DeMets, C., STOCK, J., SÁNCHEZ, O., MÁRQUEZ-AZÚA, B. and REYES, G. (2007). *A geodetic study of the 2003 January 22 Tecomán, Colima, México earthquake*, Geophys J Int *169*, 389–406.

YAGI, Y., T. MIKUMO, J. PACHECO, and G. REYES (2004), *Source rupture process of the Tecoman, Colima, México earthquake of 22 January 2003, determined by joint inversion of teleseismic body-wave and near-source data*, Bull Seismol Soc Am *94*, 1795–1807.

ZOBIN, V.M. and VENTURA-RAMÍREZ, J.F. (1998). *The macroseismic field generated by the Mw 8.0 Jalisco, México, earthquake of 9 October 1995*, Bull Seismol Soc Am *88*, 703–711.

ZOBIN, V.M. and J.A. PIZANO-SILVA (2007). *Macroseismic study of the Mw 7.5 21 January 2003, Colima, México, across-trench earthquake*, Bull Seismol Soc Am *97*, 1221–1232.

(Received January 29, 2010, revised April 16, 2010, accepted May 25, 2010, Published online June 25, 2010)

Reprinted from the journal

Pure Appl. Geophys. 168 (2011), 1363–1372
© 2010 Springer Basel AG
DOI 10.1007/s00024-010-0193-y

Active Deformation along the Southern End of the Tosco-Abreojos Fault System: New Insights from Multibeam Swath Bathymetry

François Michaud,[1] Thierry Calmus,[2] Gueorgui Ratzov,[1] Jean-Yves Royer,[3] Marc Sosson,[1]
Florence Bigot-Cormier,[1] William Bandy,[4] and Carlos Mortera Gutiérrez[4]

Abstract—The relative motion of the Pacific plate with respect to the North America plate is partitioned between transcurrent faults located along the western margin of Baja California and transform faults and spreading ridges in the Gulf of California. However, the amount of right lateral offset along the Baja California western margin is still debated. We revisited multibeam swath bathymetry data along the southern end of the Tosco-Abreojos fault system. In this area the depths are less than 1,000 m and allow a finer gridding at 60 m cell spacing. This improved resolution unveils several transcurrent right lateral faults offsetting the seafloor and canyons, which can be used as markers to quantify local offsets. The seafloor of the southern end of the Tosco-Abreojos fault system (south of 24°N) displays NW–SE elongated bathymetric highs and lows, suggesting a transtensional tectonic regime associated with the formation of pull-apart basins. In such an active tectonic context, submarine canyon networks are unstable. Using the deformation rate inferred from kinematic predictions and pull-apart geometry, we suggest a minimum age for the reorganization of the canyon network.

Key words: Tosco-Abreojos fault, seafloor morphology, submarine canyon, plate boundary, Baja California peninsula, Mexico.

1. Introduction

Motion partitioning between the North America plate, Baja California Peninsula and the Pacific plate since Middle-late Miocene is still subject to discussion (Fig. 1). Mammerickx and Klitgord (1982)

consider that after subduction ceased along the northwestern margin of Mexico, several microplates detached from the Farallon plate and subsequently accreted to the Pacific plate. This interpretation has two consequences: (1) the demise of spreading along the Farallon/Pacific ridge segments, and (2) the formation of a new plate boundary between the Pacific and North America plates. This new plate boundary formed along the mid-slope region of the continental margin of Baja California and accommodated the northwest-directed transform motion of the Pacific plate with respect to the North America plate (Stock and Hodges, 1989). Along southern Baja California, the transition between the seafloor spreading/subduction regime and the transform regime was progressive. Magnetic profiles across the ridge segment at 26°20′N, 115°W evidence a progressive slowing of the spreading rate that started at 14 Ma and completely ceased at 8–7 Ma (Michaud et al., 2006). The strike-slip motion along the Tosco-Abreojos transform fault reached its maximum rate after the spreading stopped (8–7 Ma; Michaud et al., 2006). This model does not exclude some possible strike-slip motion prior to 8 Ma (since 14–12 Ma; Spencer and Normark, 1979), before spreading completely stopped.

After the complete cessation of the subduction beneath the North American plate, the Pacific margin off southern Baja California Sur acted as a transform boundary between the North American and Pacific plates (Spencer and Normark, 1979; Normark et al., 1987; Spencer and Normark, 1989). The total amount of right lateral offset is still being debated (Yeats and Haq, 1981; Atwater and Stock, 1998; Moore and Curray, 1982; Marsaglia, 2004; Michaud et al., 2006; Fletcher et al., 2007).

---
[1] Géoazur, UPMC, UNS, CNRS, La Darse, B.P. 48, 06230 Villefranche sur mer, France. E-mail: micho@geoazur.obs-vlfr.fr

[2] Instituto de Geología, Universidad Nacional Autónoma de México, Apartado Postal 1039, 83000 Hermosillo, Sonora, Mexico.

[3] Domaines Océaniques UMR 6538, IUEM, CNRS, Université de Brest, Place Copernic, 29280 Plouzané, France.

[4] Instituto de Geofísica, Universidad Nacional Autónoma de México, C.P. 04510 Mexico, D.F., Mexico.

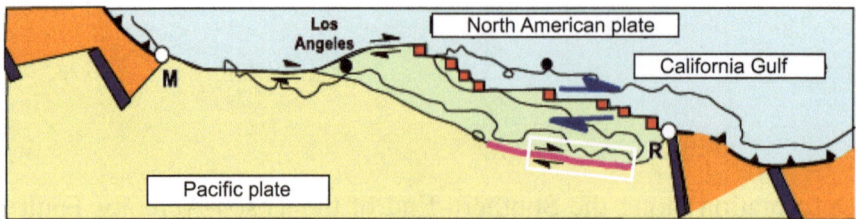

Figure 1

Geodynamic framework of Baja California (*green*). Fragments of Farallon plate that are not yet completely transferred onto the Pacific plate (*brown*). The motion between the Pacific (*yellow*) and the North American plates (*blue*) is partitioned between the Gulf of California (spreading center in *red*) and the western margin of the Peninsula. *Red line* Tosco-Abreojos fault system, *white rectangle* study area and location of Fig. 2. *M* Mendocino triple junction, *R* Rivera triple junction

Since the Early Pliocene, most of the transcurrent motion between the Pacific and North America plates has been accommodated in the Gulf of California. However, along the western margin of the Baja California peninsula (Fig. 1), some active faults, such as the Tosco-Abreojos fault system in the southern part of the Baja California margin, still accommodate a small part of the motion (DeMets, 1995; DeMets and Dixon, 1999; Dixon et al., 2000; Fletcher et al., 2000; Fletcher and Munguia, 2000; Michaud et al., 2004; Michaud et al., 2005, 2007). The relative displacement between the Pacific plate and Baja California Sur is consistent with a right lateral motion along the margin and with the focal mechanisms of earthquakes occurring in this region (Fig. 2; DeMets and Dixon, 1999; Dixon et al., 2000; Munguia et al., 2006); however, its rate is still debated. Michaud et al. (2004) show that the Pacific plate moves relative to Baja California Sur at a rate of 3–7 mm/year along a direction of motion of N130° (i.e. 130° clockwise from geographic north). Recently, Platt-ner et al. (2007) measured geodetic velocities ranging from 6.4 mm/year in the northern part of Baja California (32°N) to 3.7 mm/year in its southern part (24°N).

Several faults were identified along Baja California after the work of Spencer and Normark (1979). In this paper, we will refer to them as the "Tosco-Abreojos fault system" or TAFS. Multibeam bathymetry is a powerful tool for mapping active transcurrent fault systems (Fournier et al., 2009). Multibeam bathymetry data acquired after this early work (Fig. 2) allowed for a more detailed characterization of the TAFS. Specifically, the TAFS consists of a series of scarps that are aligned with elongated

and asymmetric basins located in the middle of the continental slope, and where recent sediments are deformed (Michaud et al., 2004). The extension of the TAFS to the south (south of 25°N) is more problematic. Using multibeam bathymetry data, Fletcher, et al. (2007) showed that the southern end of the fault system is divided into several faults within the upper slope and on the continental shelf (Fig. 2). However, none of these studies presents direct evidence of lateral movement of the sea floor that could allow for a direct measurement of the displacement along the faults.

The aim of this paper is to take advantage of the higher resolution capabilities of the multibeam bathymetric data in the shallower areas (continental shelf and upper slope), to identify laterally displaced morphological structures and to quantify local displacements.

2. Methodology

We compiled multibeam data collected in the area of the Tosco-Abreojos fault system (Fig. 2) during the FAMEX (R/V L'Atalante), MONA (R/V Marion Dufresne), OXMZ01MV and MOCEO5MV (R/V Melville) cruises, in addition to other data from NGDC-NOAA. This compilation was partially published by Michaud et al. (2007) at a 200 m grid spacing for scientific targets located at a water depth of several thousand meters. However, during a cruise the number of beams is generally constant and the swath width depends on the depth. Thus, for the same acquisition system and geometry, the seafloor is more densely sampled in shallow water than in deep water.

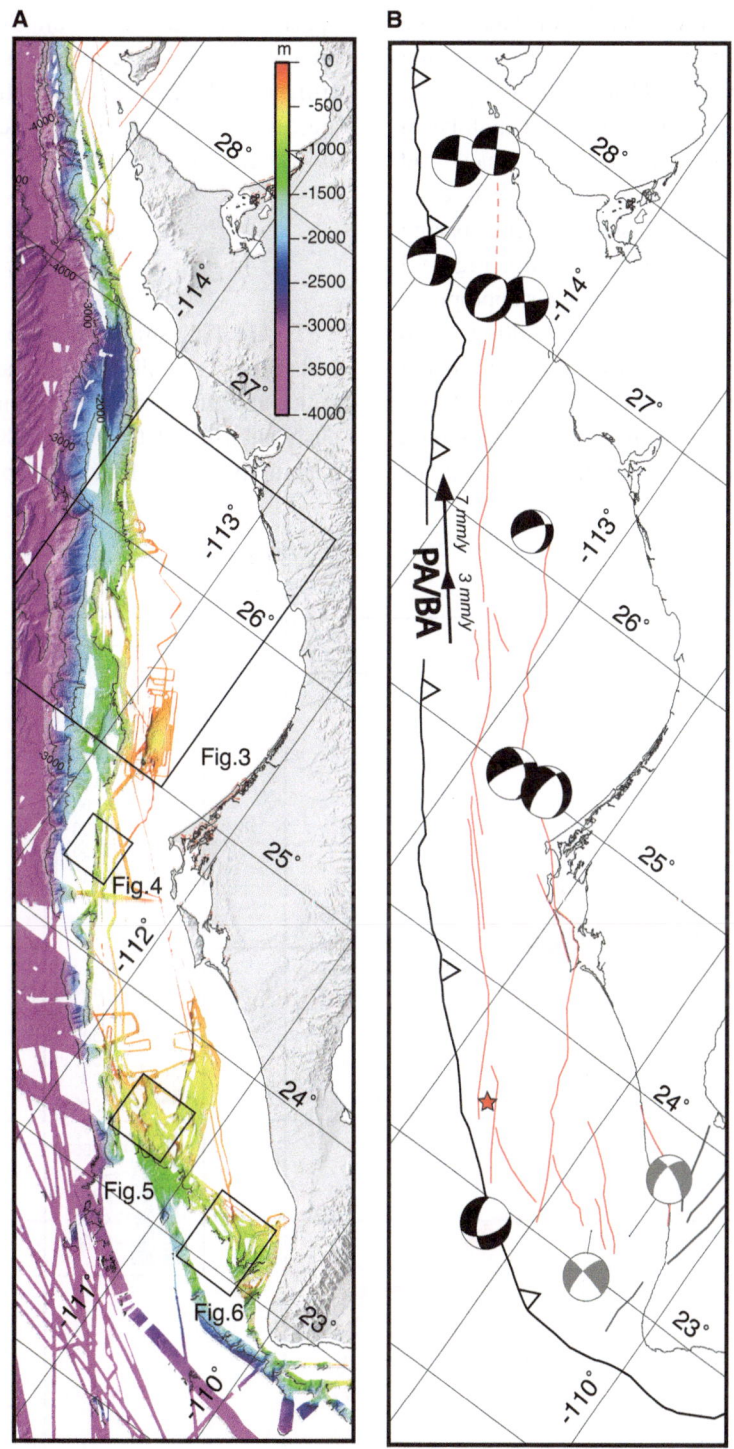

Figure 2

a Compilation of multibeam bathymetry data along the Tosco-Abreojos fault system (*red*, according to FLETCHER *et al*., 2007). The data come principally from the FAMEX (*R/V L'Atalante*), MONA (*R/V Marion Dufresne*) and OXMZ01MV and MOCEO5MV cruises (*R/V Melville*). We added some data available at the NGDC-NOAA. **b** *red lines* Tosco-Abreojos fault system from FLETCHER *et al*. (2007). Focal mechanisms are from the Harvard catalogue (*black*) and from MOLNAR (1973) (in *grey*). The Pacific/Baja motion vector is from MICHAUD *et al*. (2007). The *red star* location of the core where the sedimentation rate has been established by BLANCHET *et al*. (2007)

F. Michaud et al.   Pure Appl. Geophys.

Consequently, the shallow water depth data can be gridded at smaller grid-cell spacings (Fig. 3). Using the CARAIBES software (CARtography Adapted to Imagery and BathymEtry of Sonars and multibeam echosounders) of Ifremer, we constructed a new grid with a 60 m cell spacing (Fig. 4). This was done without the need to interpolate (i.e. the spacing between data points was less than 60 m). Figure 3 shows clearly that this grid-cell spacing is appropriate only for water depth shallower than 1,200 m. Such grid-cell spacing allows us to observe in more detail the deformed seafloor features along the TAFS (Fig. 4), especially in the southern end of this fault system where it cuts the continental shelf and the

upper slope (FLETCHER *et al.*, 2007). The higher resolution of the data also makes it easier to identify artifacts (to be expected given the mix of data sets we are using); unfortunately, such artifacts are not always easy to eliminate.

3. Results

Figure 4 illustrates that the new high resolution bathymetric grid allows us to more easily identify the faults and offset structures compared to the previously published data. Here, a 200 m bathymetric high is abruptly cut and laterally displaced. Unfortunately,

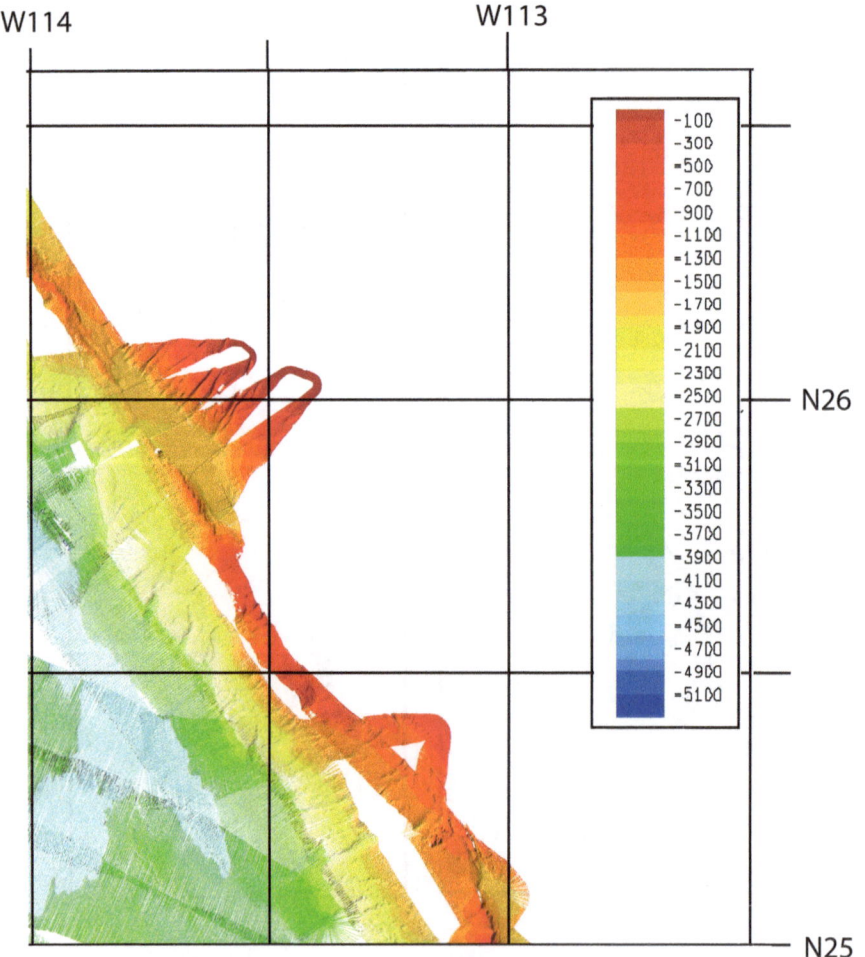

Figure 3
Results of a regridding, 60 m grid-spacing, of the FAMEX data: presented without interpolation (location on Fig. 2). The areas with water depths shallower than 1,500 m (*orange* and *red*) support a 60 mgrid-cell size (continuous image) while at water depths below 1,500 m (*green* and *blue*) many grid cells are empty (discontinuous image)

Reprinted from the journal   116

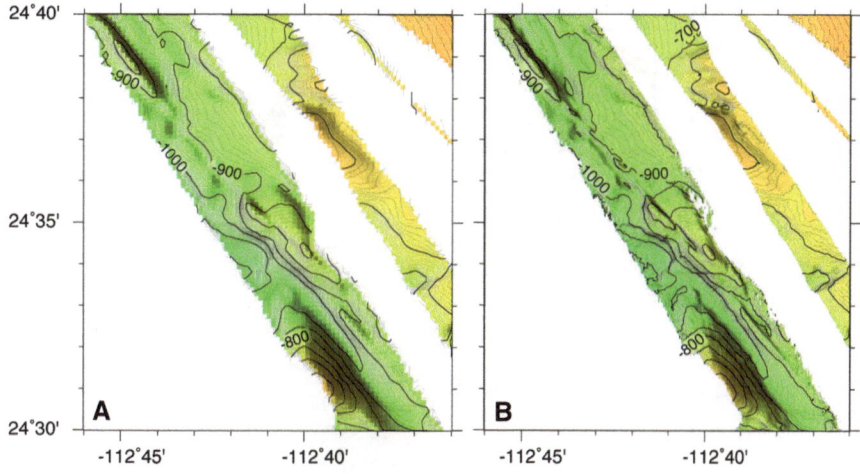

Figure 4
Comparison between multibeam bathymetry. **a** 150 m × 150 m grid, and **b** 60 m × 60 m grid (contour interval 10 m, location Fig. 2). The 60 m × 60 m grid clearly shows 200 m-high relief that is abruptly cut by right-lateral faulting. However, due to incomplete swath bathymetry coverage, it is difficult to identify the two parts of the reliefs displaced by the fault; thus it is difficult to estimate the displacement across the fault

to either side of the fault the morphology of the slope is different (contour lines are closer on one side) and we cannot estimate the offset.

Southward (at 23°30′N, Fig. 5), two parallel series of ∼N170° and ∼N125° trending scarps bound a basin. The southwestern scarps have a throw of up to ∼300 m, whereas the northeastern ones are much smoother and have a throw lower than ∼150 m. The two series of scarps are offset and bound a 7.2 km wide basin. Four elongated bathymetric depressions (grey areas on Fig. 5b) can be interpreted as smaller sub-basins aligned parallel to the general trend. The southeastern limit of the basin shows a series of antiform crests aligned in a ∼N100° direction. In the same area, a canyon is clearly cut: the canyon axis shows a 2 km right lateral offset. At the intersection with the fault, the canyon axis exhibits a 70 m topographic rise, likely due to an upward vertical component of the SW block. This suggests that the submarine erosion in the canyon is not sufficient to compensate for the seafloor deformation induced by tectonic activity. Moreover, to the west (23°20′N–111°15′W), southwest verging erosional scours on the seafloor attest to mass wasting, thus evidencing a seafloor uplift of the area.

Between latitudes 23°07′N and 23°17′N, an elongated 10 km long basin, similar to that observed in Fig. 5, is bounded by a 300 m-high steep and

N120° trending scarp (label a in Fig. 6). It is considered to be an active fault scarp by FLETCHER et al. (2007). At 110°29.5′W 23°13.5′N, the trending of this scarp changes to a N160° direction. A parallel scarp faces the relief eastwards. The SW gently inclined surface of this basin is cut by a preserved winding canyon with a N060° general trend (Fig. 6). This canyon, disconnected from both the land and lower slope, is preserved at the top of this relief. Downstream, this abandoned canyon can be followed for over ∼20 km: it cannot be followed upstream. The flanks of this abandoned canyon segment (∼13° slope) are less steep than the active ∼N180°-trending land-connected canyons (∼22° slope) as shown on Fig. 6.

4. Discussion

4.1. Pull-apart basin identification

The northernmost basin in Fig. 5 exhibits N125° trending offset scarps attesting to local transtension. The observed elongated sub-basins support the presence of a transtensional regime: these sub-basins have a geometry comparable to one of those observed by WU et al. (2009) in sandbox models of transtensional pull-apart basins, which outlines a negative flower

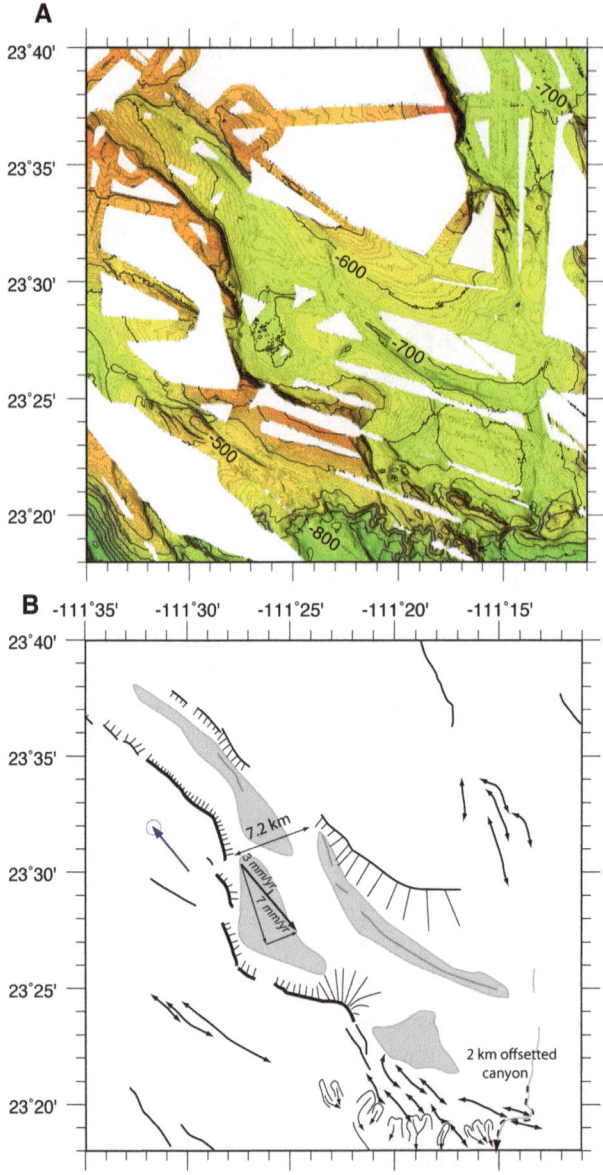

Figure 5

a Multibeam bathymetric data (60 m × 60 m grid, contour interval 10 m; location in Fig. 2) clearly shows a pull-apart basin. Southward, a canyon is cut and displaced 2 km in a right-lateral sense. **b** Structural sketch map with the present-day Pacific/Baja motion vector (*blue*) derived from PLATTNER *et al.* (2007). The vectors along, and perpendicular to, the fault correspond to the upper and lower bounds of the strike slip and normal components, respectively (after MICHAUD *et al.*, 2004)

structure. In this case, the N170° trending fault will act with a normal component and cause the basin trough. Meanwhile, the series of antiform crests aligned along a ∼N100° direction evidence a local transpression. The transpression is also supported by mass wasting which, in turn, is suggestive of seafloor uplift. Indeed, the area west of the displaced canyon

displays a set of NW–SE crests which can be interpreted as Riedel shears associated with the main fault; however, partial bathymetric data coverage on the other parallel faults do not show such evidence. Within the uncertainties, the directions of the transpressive faults are consistent with the PA/BA present-day predicted direction of motion from PLATTNER

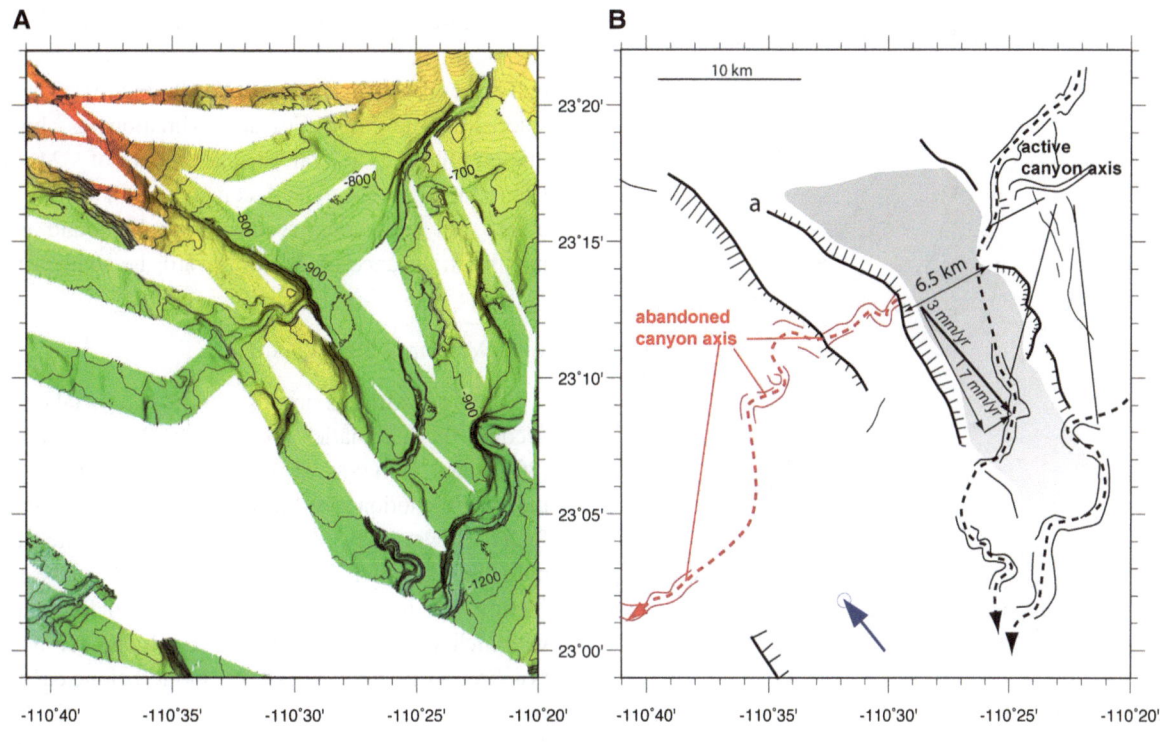

Figure 6

a Multibeam bathymetric data (size 60 m × 60 m grid, contour interval 10 m; location Fig. 2). Clearly shown are canyon segments abandoned due to fault activity. **b** Structural sketch map with the present-day Pacific/Baja motion vector (*blue*) derived from PLATTNER *et al*. (2007). The vectors along, and perpendicular to, the fault correspond to the *upper* and *lower* bounds of the strike slip and normal components, respectively (after MICHAUD *et al*., 2004). The abandoned canyon network is shown in *red*; the new canyon network in *black*. *a* represents the scarp bounding the elongated relief with the abandoned canyon

et al. (2007; blue arrow in Fig. 7), although this azimuth may locally be revised by several degrees (red arrow in Fig. 7) to account for both transtensional and transpressional areas; the present-day azimuth may also not correspond to that present at the beginning of the pull-apart basin opening.

4.2. Age and Kinematics of the Pull-Apart Basins

The motion of the Pacific plate relative to southern Baja California is on the order of 3–7 mm/year along a N140° direction (MICHAUD *et al*., 2004; PLATTNER *et al*., 2007). Figures 5 and 6 display the Pacific/Baja motion vectors derived from the PLATTNER *et al*. (2007) GPS solution, which at this latitude predicts a motion of 4.6 mm/year along a N320° direction (combining their PA1-IG00 to IG00-BAC1 rotation vectors); this value is half-way between the upper and lower bound of MICHAUD *et al*. (2004). We note that the general orientation of the strike-slip

faults agrees with these predictions (Figs. 2, 4). Assuming that the kinematic model is correct, that the rate of motion remained constant and that the whole motion was accommodated along the pull-apart basin shown on Fig. 5 (7.2 km basin width, along N070°, perpendicular to the faults which bound the basin), we find a minimum age of between 2.7 and 6.4 Ma using 7 and 3 mm/year, respectively. Similarly, from Fig. 6 (6.5 km along N065°), we deduce a minimum age of between 3.4 and 7.8 Ma. These results are consistent with the onset of Pacific/Baja transcurrent motion at about 12–8 Ma (LONSDALE, 1989, 1991; MICHAUD *et al*., 2006). Nevertheless, as shown by FLETCHER *et al*. (2007), there are other dead or active parallel faults in this area that may have accommodated part of the movement.

If we compare the 2 km offset of the canyon (Fig. 5) with the kinematics (3–7 mm/year), we conclude that the fault has been active for

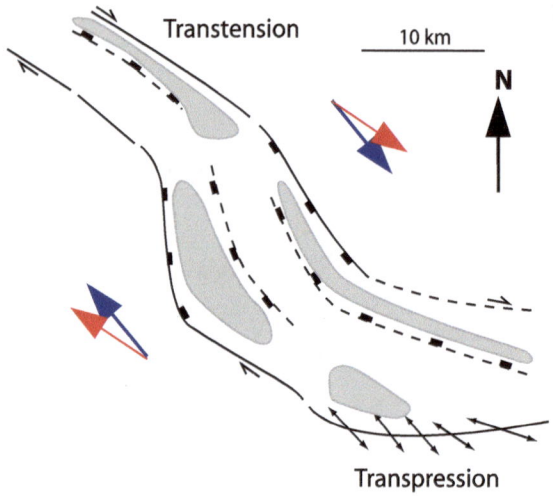

Figure 7
Interpretation of Fig. 5. The *blue arrow* corresponds to the present-day PA/BA kinematic predictions of PLATTNER *et al.* (2007). The *red arrow* is the motion vector that would better fit the observed fault scarps bounding the pull-apart basins

290,000–690,000, years which fits well with age of extension the pull-apart basin.

4.3. Canyon Morphology and Relative Age

Submarines canyons are common features which incise continental margins. The Pleistocene age is proposed for the San Lucas canyon located on the southern tip of Baja California (NORMARK and CURRAY, 1968). In Fig. 6, it is possible to distinguish two main groups of canyons: abandoned canyons and active canyons whose last activity was during the Holocene. The abandoned canyon segments located on the bathymetric high are very well preserved (up to 160 m deep incision) and seem to be recent, whereas the abandoned canyon segment located on the downstream basin is much smoother, and less than 50 m deep. Westward of the study area, the hemi-pelagic sedimentation rate measured in a small basin is 0.32 m/kyr (BLANCHET *et al.*, 2007). Seismic profiles realized northward across the margin (SPENCER and NORMARK, 1979; NORMARK *et al.*, 1987; MICHAUD *et al.*, 2007) show that basins have high rates of sedimentary filling while bathymetric highs (such as the block cut by the abandoned canyon) have slow sedimentation rates. This implies that the age of abandonment could be much older than Pleistocene,

as suggested by sediment-starved structural highs (NORMARK and CURRAY, 1968). If we consider the pull-apart opening age, 3.4 and 7.8 Ma (Fig. 6), and consider the same kinematics approximations as above, this implies a lateral displacement of ∼24 km. Consequently, if the land connection of the abandoned canyon is preserved it is to be found at a maximum distance of 24 km toward the southeast.

4.4. Deformation Chronology

The age estimates proposed in this study take into account the kinematics proposed by MICHAUD *et al.* (2004) and PLATTNER *et al.* (2007), but assume that the whole motion was accommodated along single faults. However, FLETCHER *et al.* (2007) pointed out numerous parallel faults that share the strike-slip motion. It is unknown if the faults acted individually, in which case our estimates are reliable, or acted synchronously, implying slower motion along each of them. The maximum age is, however, limited to 14–12 Ma to 8 Ma corresponding to the transition from a subduction to a strike-slip regime in the area (SPENCER and NORMARK, 1979; MICHAUD *et al.*, 2006). Our dating (2.7–7.8 Ma) is in agreement with the general geodynamical framework of the area regardless of whether the faults acted separately or together. With the caveat given above about the TAFS geometry, the consistency of our upper bound age (7.8 Ma) with that of the demise of spreading of the Farallon/Pacific Ridge off Baja California (7–8 Ma) would suggest that the TAFS has been active since then, with an average rate in the order of 3 mm/year. However, if we consider some possible strike-slip motion prior to 7.8 Ma (i.e 14–12 Ma, SPENCER and NORMARK, 1979) and assuming the same caveat as above, then the predicted average motion rate would be lower.

5. Conclusion

The reprocessed and reanalyzed multibeam bathymetry data along the shelf and upper slope of the Baja California margin evidence active right-lateral faults. The area close to the southern end of the TAFS (south of 24°N) displays NW–SE elongated bathymetric highs and lows, indicating a transtensional regime with the formation of sedimentary pull-apart basins adjacent to the faults. Using the published kinematic predictions, we obtain ages which are compatible with the geodynamical framework. The canyon morphology represents a good marker to quantify the deformation, but full bathymetric coverage is required to search for markers on other faults in order to better constrain the kinematics. Moreover, in the absence of absolute dating or relative dating of stratigraphic units, the age of the canyon systems is somewhat speculative. Sedimentary cores and seismic profiles are therefore a prerequisite to constrain the slip rates, to estimate the total amount of right-lateral offset, and to establish the timing of the fault activity along the southern end of the TAFS.

Acknowledgments

The authors wish to thank the two anonymous reviewers for their comments. We thank the crews of the R/V Marion Dufresne (MONA cruise, chief scientist Luc Beaufort), R/V Atalante (FAMEX cruise, chief scientist F. Michaud), and R/V Melville (OXMZ01MV cruise chief scientist Alexander Van-Green and MOCE05MV cruise chief scientist Dennis Clark). The two R/V Melville cruises are from Scripps Institution of Oceanography and data are available on NOAA web site. Thanks to Alexandre Dano for data processing.

REFERENCES

ATWATER, T., and STOCK, J. (1998), *Pacific-North America plate tectonics of the neogene southwestern United States: an update*, Int Geol Rev *40*, 375–402

BLANCHET, C.L., THOUVENY, N., VIDAL, L., LEDUC, G., TACHIKAWA, K., BARD, E., and BEAUFORT, L. (2007), *Terrigenous input response to glacial/interglacial climatic variations over South Baja California: a rock magnetic approach*, Quat Sci Rev *26*, 3118–3133

DEMETS, C. (1995), *A reappraisal of seafloor spreading lineations in the Gulf of California: Implications for the transfer of Baja California to the Pacific plate and estimates of Pacific-North America motion*, Geophys Res Lett *22*, 3545–3548

DEMETS, C., and DIXON, T.H. (1999), *New kinematic models for Pacific-North America motion from 3 Ma to present, I: evidence for steady motion and biases in the NUVEL-1A model*, Geophys Res Lett *26*, 1921–1924

DIXON, T., FARINA, F., DEMETS, C., SUAREZ-VIDAL, F., FLETCHER, J., MARQUEZ-AZUA, B., MILLER, M., SANCHEZ, O., and UMHOEFER, P. (2000), *New kinematic models for Pacific-North America motion from 3 Ma to Present II: Evidence for a "Baja California shear zone"*, Geophys Res Lett *27*, 3961–3964

FLETCHER, J.M., and MUNGUIA, L. (2000), *Active continental rifting in southern Baja California, Mexico: implications for plate motion partitioning and the transition to seafloor spreading in the Gulf of California*, Tectonics *19*(6), 1107–1123

FLETCHER, J. M., EAKINS, B. A., SEDLOCK, R. L., MENDOZA-BORUNDA, R., WALTER, R. C., EDWARDS, R. L., AND DIXON, T. H. (2000), *Quaternary and Neogene slip history of the Baja-Pacific plate margin: Bahia Magdalena and the southwestern borderland of Baja California*, Eos (Transactions, American Geophysical Union) *81*, F1232.

FLETCHER, J.M., GROVE, M., KIMBROUGH, D., LOVERA, O., and GEHRELS, G. (2007), *Ridge-trench interactions and the Neogene tectonic evolution of the Magdalena shelf and southern Gulf of California: Insights from detrital zircon U-Pb ages from the Magdalena fan and adjacent areas*, Geol Soc Am Bull *119*, 1313–1336. doi:10.1130/B26067.1

FOURNIER, M., CHAMOT-ROOKE, N. R., RODRIGUEZ, M., PETIT, C., HUCHON, P., BESLIER, M.-O., HAZARD, B., The Arabia-India plate boundary unveiled, American Geophysical Union, Fall Meeting 2009. Eos Trans. AGU, v. *90*(52), Fall Meet. Suppl., Abstract T51C-1533

LONSDALE, P. (1989), Geology and tectonic history of the Gulf of California In: *The Eastern Pacific Ocean and Hawaii: The Geology of North America N* (eds. Winterer, Hussong, and Decker) (Geological Society of America, Boulder, Colorado, 1989) pp. 499–521

LONSDALE, P. (1991), Structural patterns of the Pacific floor offshore of peninsular California, In: *The Gulf and Peninsular Province of the Californias, American Association of Petroleum Geologists Memoir 47* (eds. Dauphin, and Simoneit) (AAPG, Tulsa, Oklahoma, 1991) pp. 87–125

MAMMERICKX, J., and KLITGORD, K. D. (1982), *East Pacific Rise evolution from 25 my B.P. to present*, J Geophys Res *87*, 6751–6758

MARSAGLIA, K. M. (2004), *Sandstone detrital modes support Magdalena Fan displacement from the mouth of the Gulf of California*, Geology *32*, 45–48

MICHAUD, F., SOSSON, M., ROYER, J.-Y., CHABERT, A., BOURGOIS, J., CALMUS, T., MORTERA, C., BIGOT-CORMIER, F., BANDY, W., DYMENT, J., PONTOISE, B., and SICHLER, B. (2004), *Motion partitioning between the Pacific plate, Baja California and the North America plate: The Tosco-Abreojos fault revisited*, Geophys Res Lett *31*, L08604

MICHAUD, F., CALMUS, T., SOSSON, M., ROYER, J.-Y., BOURGOIS, J., CHABERT, A., BIGOT-CORMIER, F., BANDY, W., MORTERA, C., AND DYMENT, J. (2005), La zona de falla Tosco-Abreojos: un sistema lateral derecho activo entre la placa Pacifico y la peninsula de Baja California, Boletin de la Sociedad Geologica Mexicana, Volumen Conmerorativo del Centenario: Grandes Fronteras Tectonicas de Mexico, T. LVII, 1, p. 53–63

MICHAUD, F., ROYER, J.-Y., BOURGOIS, J., DYMENT, J., CALMUS, T., BANDY, W., SOSSON, M., MORTERA-GUTIÉRREZ, C., SICHLER, B., REBOLLEDO-VIERA, M., and PONTOISE, B. (2006), *Oceanic-ridge subduction vs. slab break off: Plate tectonic evolution along the Baja California Sur continental margin since 15 Ma*, Geology *34*(1), 13–16. doi:10.1130/G22050.1

MICHAUD, F., CALMUS, T., ROYER, J.-Y., SOSSON, M., CHABERT, A., BIGOT-CORMIER, F., BANDY, W., MORTERA, C., and DYMENT, J. (2007), Right lateral active fault between southern Baja California and the Pacific plate: the Tosco-Abreojos fault, In *Celebrating the Centenary of the Geological Society of Mexico: Geological Society of America, special paper 442* (eds. Alaniz-Alvarez, and Nieto-Samaniego) (Geological Society of America, Boulder, Colorado, 2007) pp. 287–300

MOLNAR, P. (1973), *Fault plane solutions of earthquakes and direction of motion in the Gulf of California and in the Rivera fracture zone*, Geol. Soc. Am. Bull. *84*, 1651–1658.

MUNGUIA, L., GONZALES, M., MAYER, S., and AGUIRRE, A. (2006), Seismicity and state of stress in the La Paz-Los Cabos Region, Baja California Sur, Mexico, Bulletin of Seismological Society of America *96*(2), 624–636

NORMARK, W., and CURRAY, J. (1968), *Geology and structure of the Tip of Baja California, Mexico*, Geol Soc Am Bull *79*, 1589–1600

NORMARK, W., SPENCER, J., and INGLE, J. (1987), Geology and Neogene history of the Pacific margin of Baja California Sur, In *Geology and Resource Potential of the Continental Margin of Western North America, Earth Science Series, 6* (eds. Scholl, Grantz, Vedder) (Circum Pacific Counsel Energy Min. Resources, Houston, Tx, 1987) pp. 449–472

PLATTNER, C., MALVERVISI, R., DIXON, T.H., LAFEMINA, P., SELLA, G.F., FLETCHER, J., and SUAREZ-VIDAL, F. (2007), *New constraints on the relative motion between the Pacifoc Plate and Baja California microplate (Mexico) from GPS measurements*, Geophys J Int *170*, 1373–1380

SPENCER, J.E., AND NORMARK, W. (1979), *Tosco-Abreojos fault zone: a Neogene transform plate boundary within the Pacific margin of southern Baja California, Mexico*, Geology *7*, 554–557

SPENCER J.E., and NORMARK, W. (1989), Neogene plate-tetonic evolution of the Baja California Sur continental margin and the southern Gulf of California, Mexico, In *The Eastern Pacific Ocean and Hawai: The Geology of North America, N* (eds. Winterer, Hussong, and Decker) (Geological Society of America, Boulder, Colorado, 1989) pp. 489–497

STOCK, J. M., AND HODGES, K. V. (1989), *Pre-Pliocene extension around the Gulf of California and the transfer of Baja California to the Pacific plate*, Tectonics *8*, 99–115

WU, J. E., MCCLAY, K., WHITEHOUSE, P., and DOOLAY, T. (2009), *4D analogue modeling of transtensional pull-apart basins*, Marine Petroleum Geol *26*, 1608–1623

YEATS, R. S., and HAQ, B. U. (1981), Deep-Sea drilling off the Californias: Implications of Leg 63, Initial Reports of the Deep Sea Drilling Project *63*, 949–961

(Received February 10, 2010, revised July 29, 2010, accepted July 29, 2010, Published online September 11, 2010)

Pure Appl. Geophys. 168 (2011), 1373–1389
© 2010 Springer Basel AG
DOI 10.1007/s00024-010-0206-x

▌Pure and Applied Geophysics

Imaging the Seismic Crustal Structure of the Western Mexican Margin between 19°N and 21°N

RAFAEL BARTOLOMÉ,[1] JUANJO DAÑOBEITIA,[1] FRANÇOIS MICHAUD,[2] DIEGO CÓRDOBA,[3] and LUIS A. DELGADO-ARGOTE[4]

Abstract—Three thousand kilometres of multichannel (MCS) and wide-angle seismic profiles, gravity and magnetic, multibeam bathymetry and backscatter data were recorded in the offshore area of the west coast of Mexico and the Gulf of California during the spring 1996 (CORTES survey). The seismic images obtained off Puerto Vallarta, Mexico, in the Jalisco subduction zone extend from the oceanic domain up to the continental shelf, and significantly improve the knowledge of the internal crustal structure of the subduction zone between the Rivera and North American (NA) Plates. Analyzing the crustal images, we differentiate: (1) An oceanic domain with an important variation in sediment thickness ranging from 2.5 to 1 km southwards; (2) an accretionary prism comprised of highly deformed sediments, extending for a maximum width of 15 km; (3) a deformed forearc basin domain which is 25 km wide in the northern section, and is not seen towards the south where the continental slope connects directly with the accretionary prism and trench, thus suggesting a different deformational process; and (4) a continental domain consisting of a continental slope and a mid slope terrace, with a bottom simulating reflector (BSR) identified in the first second of the MCS profiles. The existence of a developed accretionary prism suggests a subduction–accretion type tectonic regime. Detailed analysis of the seismic reflection data in the oceanic domain reveals high amplitude reflections at around 6 s [two way travel time (twtt)] that clearly define the subduction plane. At 2 s (twtt) depth we identify a strong reflection which we interpret as the Moho discontinuity. We have measured a mean dip angle of 7° ± 1° at the subduction zone where the Rivera Plate begins to subduct, with the dip angle gently increasing towards the south. The oceanic crust has a mean crustal thickness of 6.0–6.5 km. We also find evidence indicating that the Rivera Plate possibly subducts at very low angles beneath the Tres Marias Islands.

Key words: Rivera Plate, Middle America Trench, crustal structure, seismic imaging, subduction plate, BSR.

1. Introduction

The tectonics, seismicity and magmatism in the western part of central Mexico is largely controlled by the subduction of the Cocos and Rivera oceanic plates beneath the North American (NA) Plate, the Middle America Trench (MAT) being the morphologic expression of the subduction contact (Fig. 1). The Rivera Plate is a key structural element to understand the complex geodynamic interactions that occur at the western coast of Mexico. ATWATER (1970) was the first to suggest the existence of this microplate. Since then, several authors have shown that the Rivera Plate is kinematically distinct from the NA and Cocos Plates (EISSLER and MCNALLY, 1984; BANDY and YAN, 1989; DEMETS and STEIN, 1990), although the precise location of the Rivera-Cocos boundary is still controversial since no clear bathymetric features can be clearly associated with the plate boundary (EISSLER and MCNALLY, 1984; BOURGOIS and MICHAUD, 1991). Seafloor accretion occurs along the western boundary Rivera plate, at the Pacific-Rivera Rise (PRR), whereas to the east, the lithosphere of the Rivera plate is consumed at the trench where the plate has been dated as late Miocene (~9 Ma) based on seafloor magnetic anomalies (KLITGORD and MAMMERICKX, 1982). The age of the Cocos Plate varies along the MAT, with jumps in ages occurring across several fracture zones. Thus, the younger and shallower Cocos crust near the Rivera Plate is dated as 10 Ma whereas at 90°W it is dated at 25 Ma old (COUCH and WOODCOCK, 1981).

[1] Centre Mediterrani d'Investigacions Marines i Ambientals (CMIMA), Unidad de Tecnología Marina-CSIC, Passeig Marítim de la Barceloneta 37-49, 08003, Barcelona, Spain. E-mail: rafael@utm.csic.es
[2] Géosciences Azur, UMR 6526, Université Pierre et Marie-Curie, La Darse, BP 48, 06235 Villefranche-sur-Mer Cedex, France.
[3] Universidad Complutense de Madrid, 28040 Ciudad Universitaria, Madrid, Spain.
[4] CICESE, Carr. Tijuana-Ensenada km 107, 22800 Ensenada, Baja California, Mexico.

Reprinted from the journal

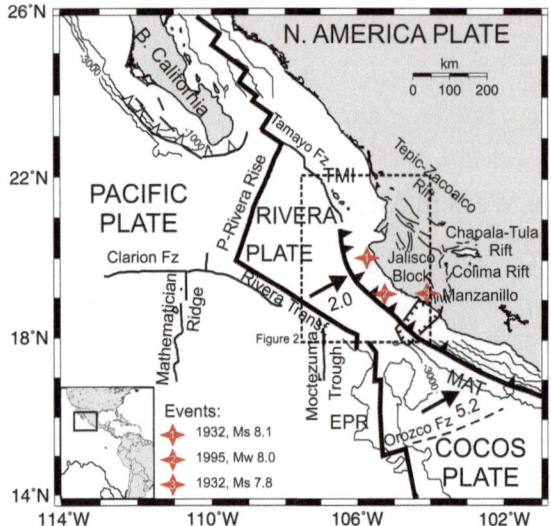

Figure 1

General tectonic setting of western Mexico. Four lithospheric plates act in the area: Pacific, Rivera, Cocos and North American. *Arrows* indicate relative convergence rates (cm/year) between the oceanic and continental plates. *Red stars* show rupture zones for the 1932 and 1995 events (PACHECO *et al.*, 1997). *Black square* highlights the area of Fig. 2. *TMI* Tres Marias Islands, *MAT* Middle America Trench, *EPR* East Pacific Rise

There are still points to be clarified regarding the Rivera Plate subduction patterns. Among them, the convergence direction and rate (KOSTOGLODOV and BANDY, 1995; DEMETS and WILSON, 1997; BANDY *et al.*, 1998; DEMETS and TRAYLEN, 2000), the recent strike-slip deformation (BANDY *et al.*, 2005), the seismic activity (NIXON, 1982), and a plausible evolution model for the junction area of the East Pacific Rise (EPR), the Rivera Transform and the MAT (MAMMERICKX, 1984; BOURGOIS *et al.*, 1988; DEMETS and WILSON, 1997; BANDY *et al.*, 1998; MICHAUD *et al.*, 2001; BANDY *et al.*, 2005). In terms of seismic data, only a few single channel seismic profiles are available in the literature (BOURGOIS *et al.*, 1988; BOURGOIS and MICHAUD, 1991; KHUTORSKOY *et al.*, 1994; MICHAUD *et al.*, 1996) and one threefold profile (BANDY *et al.*, 2005). These data show active shallow strike-slip tectonics and a subsidence in the upper (first second twtt) continental slope area at 18.5°N, although the subducting Rivera Plate crustal structure remains unidentified. Therefore, the studies of the internal structure of the overriding plate have relied primarily on potential field data (BANDY *et al.*, 1993;

BANDY *et al.*, 1999). The subduction of the Rivera Plate remains a matter of debate due to the poorly controlled shape of the subducted plate (EISSLER and MCNALLY, 1984; PARDO and SUÁREZ, 1993; BANDY *et al.*, 1999), and the low background seismicity compared to the rest of the MAT to the south. Regarding this low seismicity, there are few historical earthquakes at the Rivera-NA plate boundary, which raises the possibility that the Rivera Plate subducts aseismically (NIXON, 1982). However, some of the largest destructive earthquakes reported in recent history occurred offshore of the Jalisco region (19°N), such as the 1932 (Ms 8.1 and Ms 7.8) Jalisco earthquakes (SINGH *et al.*, 1985) and the 9 October 1995 (M_w 8.0, 17 km depth epicentre and 5 m of tsunami) Colima earthquake in the southeast flank of the region (COURBOULEX *et al.*, 1997). The main event of the 1995 earthquake was a subduction related thrust event that activated normal faults along the northwest margin of Manzanillo (19°N and 105°W, Fig. 1), clearly showing present day plate convergence. However, it is not clear if the M_w 7.4 event of 22 January 2003 occurred along the current thrust interface between the Rivera and NA Plates (SINGH *et al.*, 2003).

The angle of subduction of the Rivera Plate has been computed by accurately relocating hypocenters (e.g., PARDO and SUÁREZ, 1993, 1995). The inferred Wadatti-Benioff zone indicates a steeper dip for the Rivera Plate than the adjacent Cocos Plate. This has led to the proposal of a step in the slab between the Cocos and Rivera Plates at the present time (FERRARI *et al.*, 2001) due to the slab rollback mechanism, acting on the Rivera Plate underlying the NA Plate, during a period of very low convergence rate (19 mm/year on average) between 8.5 and 4.6 Ma (DEMETS and TRAYLEN, 2000). The seismicity causes a significant seismic hazard to the coastal regions of Mexico, as well as areas considerably inland, including Mexico City (CURRIE *et al.*, 2002). A strong understanding of the Mexico subduction region (geometry, seismic zone of subduction, faults and rupture width, etc.) and detailed information about the crustal structure along the Rivera Plate are required to carry out seismic hazard studies. Other constraints are comprehension of the limits of the lateral and vertical extent of the Rivera Plate as well as plate tectonic models, including the opening of the Gulf of California.

For these reasons, in 1996, the Spanish *R/V Hespérides* and the Mexican *R/Vs Altair and Humboldt* surveyed the northwestern Mexican margin between 16°N and 30°N in a geophysical experiment named the Crustal Offshore Research Transect by Extensive Seismic Profiling, or CORTES-P96 (DAÑOBEITIA *et al.*, 1997). Swath bathymetry, backscattering and multichannel seismic reflection data profiles (Fig. 2) were acquired within the Rivera subduction zone extending from 21°N to 19°N. In this paper, we present five post-stack time migrated MCS sections (Fig. 3) that show for the first time in this area, seismic images from the Earth's surface down to the Mohorovicic discontinuity (Moho) that enable us to seismically characterize the crustal structure of the overriding and subducting plates offshore Puerto Vallarta, describe the spatial distribution of the incoming plates (sediment, crust and mantle lithosphere) and identify some of the processes occurring in this area.

2. Geological Setting

The western margin of Mexico between 26°N and 23°N is structurally composed of a transform margin made up of a set of spreading centers linked by a fault system (Fig. 1). The right lateral fault system is the southward prolongation of the San Andreas Fault across the Gulf of California. The 70 km Tamayo transform fault connects the southernmost spreading segment in Gulf of California (the Alarcon Rise) with the PRR at 23°N. This spreading center, located between 23°N and 19°N, is the western plate boundary between the Rivera and Pacific Plates and its spreading velocities have fluctuated between intermediate and fast rates over the past 10 Ma (average full spreading rate of around 9.7 cm/year, DeMets and Traylen, 2000). The Rivera Transform is the southern plate boundary between the Rivera and Pacific Plates and the MAT the eastern plate boundary between the Rivera and NA Plates.

The geodynamic history since 25 Ma starts with the collision of the Pacific-Farallon seafloor-spreading center against the convergent western margin of

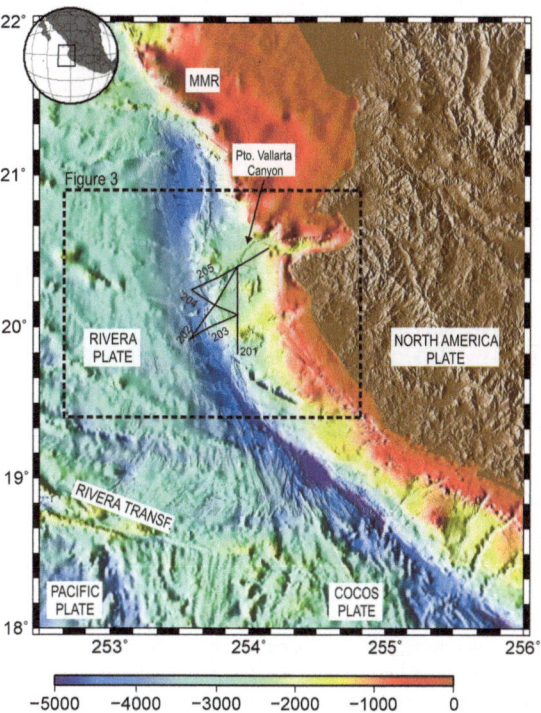

Figure 2

Location of ship tracks of the CORTES geophysical profiles, numbered from 201 to 205, acquired offshore of Puerto Vallarta and analyzed in the present study. Additional bathymetry data is from surveys of the R/V Jean Charcot and R/V Marion Dufresne (Mona survey), the NGDC database and Gebco (2008) (30 min arc cell grid resolution). *Shaded Bathymetry* has been displayed illuminated from the west. *MMR* Maria Magdalena Rise. *Black square* shows the study area

NA. This collision marked the beginning of a major change in the tectonic evolution and volcanism of western NA (Atwater, 1970, 1989). Some authors propose a final reorganization in the Middle Miocene (6.5–3.5 Ma) triggered by the jump of the EPR to its present location, transferred by the development of the Rivera Transform (Mammerickx, 1984; Mammerickx and Klitgord, 1982; Lonsdale, 1991). Alternatively, Lonsdale (1995) proposes that the northward propagation of the EPR to its present location took place without any clear early connection with the PRR, suggesting a broad diffuse dextral shear zone connecting the two spreading centers before the full development of the Rivera Transform. Even if the connection between the EPR and the Rivera Transform is imprecise, multibeam and backscatter data collected during the CORTES-P96 experiment pointed out that the EPR reached the

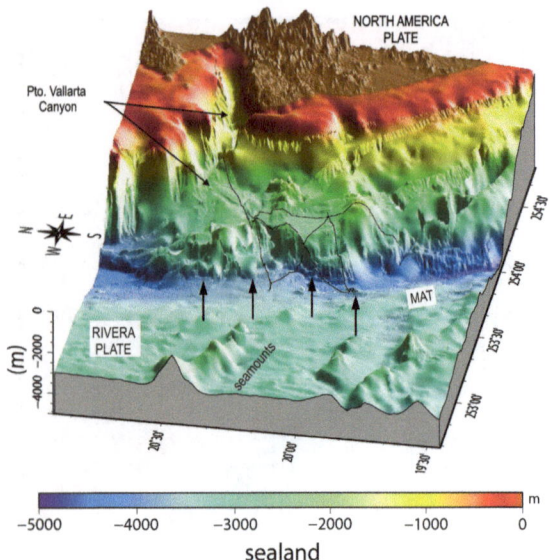

Figure 3

3D shaded relief bathymetry of the study area, looking eastward. *MCS lines* have been projected on the seafloor to aid in the interpretation of seismic data. *Black arrows* mark the deformation front, emphasizing the increasing distance towards the trench with latitude. The collision with MAT of seamount chains oriented perpendicular to the MAT, observed along the margin, push the frontal sediment prism towards the coast as is observed between the second and third black arrow. Note the sediment infill of the trench by the Puerto Vallarta Canyon. Bathymetric data is described in Fig. 2

trench (MICHAUD *et al.*, 2001). This suggests that the EPR extended northward of the Rivera Transform, consistent with the findings of BANDY (1992) based on marine magnetic data, and subducted beneath the NA plate before 2 Ma without the need of any connection with the Rivera Transform. The convergence rate reported at the MAT shows significant variations during the last 10 Ma, with convergence stopping between 2.5 and 1 Ma (DEMETS and TRAYLEN, 2000). The current reported value of Rivera-NA relative convergence rate is a matter of debate. It ranges from 2.0 cm/year, based on the accurately located hypocenters of local and teleseismic earthquakes (PARDO and SUÁREZ, 1995), to 5 cm/year based on seismotectonic relationships (KOSTOGLODOV and BANDY, 1995) which relate seismic characteristics of subduction zones (maximum magnitudes, maximum seismic depths, etc.) to plate tectonic parameters (convergence rates, age of the oceanic lithosphere, etc.). The most recent estimations derived from

reconstruction of magnetic anomalies for the last 0.78 Ma, give Rivera Plate convergence rates of 3.3–4.3 cm/year along the NA Plate in the area of the Jalisco Block (BANDY and PARDO, 1994; DEMETS and WILSON, 1997; BANDY *et al.*, 1998; DEMETS and TRAYLEN, 2000).

The mechanism (slab-pull) that powers the subduction and moves the plate is well known and is caused by slightly more dense mantle lithosphere of the Rivera Plate than that underlying North American Plate, with one peculiarity. As density excess increases as the lithosphere ages and thickens, and the lithosphere becomes negatively buoyant (namely, subducting) when it is 10–30 million years old (DAVIES, 1992), the <10 Ma buoyant crust of the Rivera Plate resists subduction. This resistance results in a strong coupling between plates having higher magnitude earthquakes and shallower dips compared to those plates subducting old lithospheres. The strain regime behind the magmatic arc for the subducting of the young Rivera lithosphere is expected to be strongly compressional (folding and thrusting), and the maximum magnitude earthquakes must occur under these conditions (young and fast lithosphere) beneath the continental crust arcs (STERN, 2002).

The direction of convergence between the Rivera and NA Plates at the MAT becomes progressively more oblique (in a counter-clockwise sense relative to the trench-normal direction) northward along the end of the MAT (19°N–21°N), commonly termed the Jalisco subduction zone (JSZ) (e.g. BANDY, 1992; LONSDALE, 1995; KOSTOGLODOV and BANDY, 1995; DeMets and WILSON, 1997; DeMETS and TRAYLEN, 2000). This area comprises the northern part of the MAT between the Tres Marias Islands and the southern tip of the Colima Rift (Fig. 1). The depth of the slab under western Mexico is poorly constrained, but local earthquake data indicate a Benioff zone for the Rivera Plate bending from around 10° at 20 km depth to 50° at 40 km depth, whereas the Cocos Plate is sub-horizontal at the same depth (PARDO and SUÁREZ, 1993). This change of dip angle is unrelated to the thermal structure of the slabs and the present rate of subduction due to the similar age and comparable subduction rates of both plates. A rollback of the slab when the relative convergence between the

Rivera and NA Plates decreased to a very low rate at the end of Miocene was proposed as a plausible explanation (FERRARI et al., 2001).

The Jalisco Block (JB) is bounded landwards by the Tepic-Zacoalco Rift on the NE and by the Colima Rift on the SE (Fig. 1). Rifting and volcanism have been proposed to be associated with the northwestward detachment of the JB from the NA Plate (LUHR et al., 1985) at the very low rate of <5 mm/year (DEMETS and STEIN, 1990; BANDY and PARDO, 1994), starting in the Pliocene–Quaternary. Recent strike-slip deformation within the forearc region of the JSZ has also been reported by shallow (<1 s twtt penetration) seismic data collected offshore the Jalisco Block (BANDY et al., 2005). That study also provided additional evidence for recent subsidence within the area offshore of Manzanillo in agreement with previous works (MERCIER DE LEPINAY et al., 1997; RAMÍREZ-HERRERA and URRUTIA-FUCUGAUCHI, 1999).

The uplift-subsidence history proposed for the entire margin along the JSZ from the Tres Marias Islands to the Manzanillo area starts with uplift and emergence before the late Miocene (6.5 Ma) and continuing subsidence during the late Miocene–Lower Pliocene until the Pliocene–Quaternary boundary (~2 Ma). Regional subsidence of an active margin is generally related to tectonic erosion. After the Pliocene–Quaternary boundary, the Tres Marias began to be uplifted (McCLOY et al., 1988), whereas the Manzanillo area started to subside, certainly due to the fragmentation associated with the development of the JB during the Pliocene–Quaternary, at a subsidence rate of 0.35 mm/year (MERCIER DE LEPINAY et al., 1997). For the Holocene period, in the active tectonic margin environment of the Jalisco area, a new model is proposed where the coseismic subsidence (centimetre-scale) produced by large offshore earthquakes is rapidly recovered during the postseismic and interseismic periods, yielding a general coastal uplift (RAMIREZ-HERRERA et al., 2004). In other words, the slow long-term uplift is considered the most important factor to the total net vertical motion (uplift) of the Jalisco area. Based on the first reported organisms dated with radiocarbon along the Jalisco coast, an average rate of about 3 mm/year for tectonic uplift since 1,300 years BP has been computed (RAMIREZ-HERRERA et al., 2004).

3. MCS Data

The data set analysed in this study is composed of five MCS profiles (profiles 201–205, Fig. 2), totalling 280 km and acquired during the CORTES-P96 cruise, is located in the subducting domain of the Rivera Plate off Puerto Vallarta, at the contact with the NA Plate. The five seismic profiles used in this work have three different orientations with respect to the direction of the trench axis: profiles 203 and 205 are perpendicular, profiles 202 and 204 are oblique and profile 201 is parallel to the MAT (Fig. 2). The profiles were acquired using a 2.4 km long streamer with 96 channels at 2 ms sampling interval, towed 165 m behind the ship at 9 m depth. The source used by the Spanish R/V Hesperides was a tuned airgun array of 50 l (3,000 in^3). The shooting rate was 30 s. Quality control and preliminary stacking were done onboard, while final processing of MCS data through 9 s (twtt) and a 16-fold coverage stack and migration was done at the Department of Geophysics of the Institute of Earth Sciences *Jaume Almera* and in the Unidad de Tecnologia Marina-CSIC (Barcelona, Spain).

The seismic processing flow includes band-pass filtering (5–9–58–62.5 Hz), editing of noisy traces, divergence correction, and internal mute to reduce the water bottom multiple effects. After CMP sorting, a minimum phase predictive deconvolution was performed (50–350 ms). Stacking velocity analysis every 200 CDP maximum interval and NMO correction were performed. The profiles were finally Kirchhoff post-stack time migrated. The final MCS sections were band-pass filtered, normalized and trace mixed.

In the following sections, the components of the subduction zone are described, and their interactions are discussed.

4. Analysis of MCS Data and Discussion

4.1. Shallow Crustal Structure

The northernmost profile 205 (Fig. 4) provides the first ever reported seismic image of the active margin in the area offshore of Puerto Vallarta. The upper and western part of profile 205 was already published in a study of the heat flow through the margin (MINSHULL et al., 2005). From offshore to onshore (NE orientation) we have identified the following morphological features: deformation front (CMP 1,500), trench (CMP 2,500), accretionary prism (CMP 2,500–3,600), a well-developed forearc basin (CMP 3,600–5,600), backstop limit at (CDP 4800), and continental slope and mid slope terrace (CMP 5,600–6,400). We also recognized the presence of erosion as mass wasting affecting the slope is also identified (CMP 5,700–6,300).

The joint interpretation of profiles 205 (Fig. 4), 204 (Fig. 5), 202 (Fig. 6) and 203 (Fig. 7) in the oceanic domain indicate that the trench-fill sediments progressively thicken landward. The maximum thickness of the trench-fill decreases towards the south being around 2 s twtt in profile 205 (CMP 2,400, Fig. 4), 1.52 s twtt in profile 204 (CMP 2,200, Fig. 5), 0.85 s twtt in profile 202 (CMP 14,000, Fig. 6), and 0.6 s twtt in profile 203 (CMP 1,200, Fig. 7), respectively. Assuming an average P-wave velocity of 2.5 km/s for the sedimentary cover as suggested by the velocity analysis of the MCS data, we obtain a maximum thickness of 2.5 km for section 205, 1.9 km for 204, 1.1 km for 202 and 0.75 km for 203. The different thickness and deformation of the shallower reflectors between the northern and southern profiles may be explained by increasing infill of terrigenous sediments associated with drainage from the head of the Puerto Vallarta Canyon (Fig. 3), pointed to in profile 205 (CMP 3,700 and 4,100; Fig. 4) and profile 201 (CMP 4,700, Fig. 8). The trench-fill sediments show a significant deformation seaward of the trench in the northern sections 205 (CMP 1,900–2,400, Fig. 4) and 204 (CMP 2,100–2,400, Fig. 5), indicating a recent deformation at the northern end of the MAT. The observed sediment thickness in the trench off Puerto Vallarta, despite the decreasing tendency observed southwards due to the infill of the Puerto Vallarta Canyon, is similar to that observed in the Cascadia and southwest Japan subduction zones, which are of comparable age and covered with 1.5–3.5 km of sediment (WANG et al., 1995). Nevertheless, opposite of what has been observed in the trench off Puerto Vallarta, a lack of a thick cover of sediments (~200 m) has been noticed to the south and

especially in the area of the boundary between the Rivera and Cocos Plates. Of potential importance is the sediment thickness on top of the plate that slows the rate of cooling of the oceanic lithosphere and thus changes its thermal structure. A poor sediment cover for the Rivera Plate has been assumed during the computation of the thermal modeling in the Mexico subduction zone published by CURRIE et al. (2002). Therefore, in light of these new crustal images, thermal models may need to be revised, at least in the northern part of the Rivera Plate, to be consistent with the new seismic observations. In addition, the crust of the Rivera Plate may be warm enough to melt (<10 Ma and slowly subducted), possibly aided by greater friction across the subduction interfaces, producing adakites such as the ones investigated by FERRARI et al. (2001). This also has to be considered during the modelling. One should bear in mind that sediments in most subducted zones melt but subducted crust only melts when it is young and hot enough (JOHNSON and PLANK, 1999).

The southernmost profiles 202 (CMP 15,000, Fig. 6) and 203 (CMP 400, Fig. 7) clearly illustrate a set of normal faults reaching the surface. The structure of the subducting plate consists of grabens and horsts that are products of the extensional stress due to bending of the plate, which might facilitate sediment subduction and the percolation of fluids. These faults, arriving up to the seafloor, might indicate recent tectonic activity related to the dip augmentation of the Rivera Plate since 8.5 Ma, when bending occurred at the time the Rivera Plate slowed the convergence (7.2–4.8 Ma) as recorded by the trenchward migration of the volcanic front (FERRARI et al., 2001). Fluids are likely to percolate and circulate along these faults deep into the overriding plate, probably favouring the formation of gas hydrates and significantly cooling the oceanic plate as shown by marine heat flow observations (ZIAGOS et al., 1985; PROL-LEDESMA et al., 1989; KHUTORSKOY et al., 1994). The increase of the amount of fluids (including asthenospheric rising) started in the Pliocene and released from the subducting slab of the Rivera Plate due to the increase of the convergence is confirmed by the subsequent volcanic rate increment (FERRARI et al., 2001). The distance from the trench axis to the deformation front differs from one section to another as well. From north to south, the

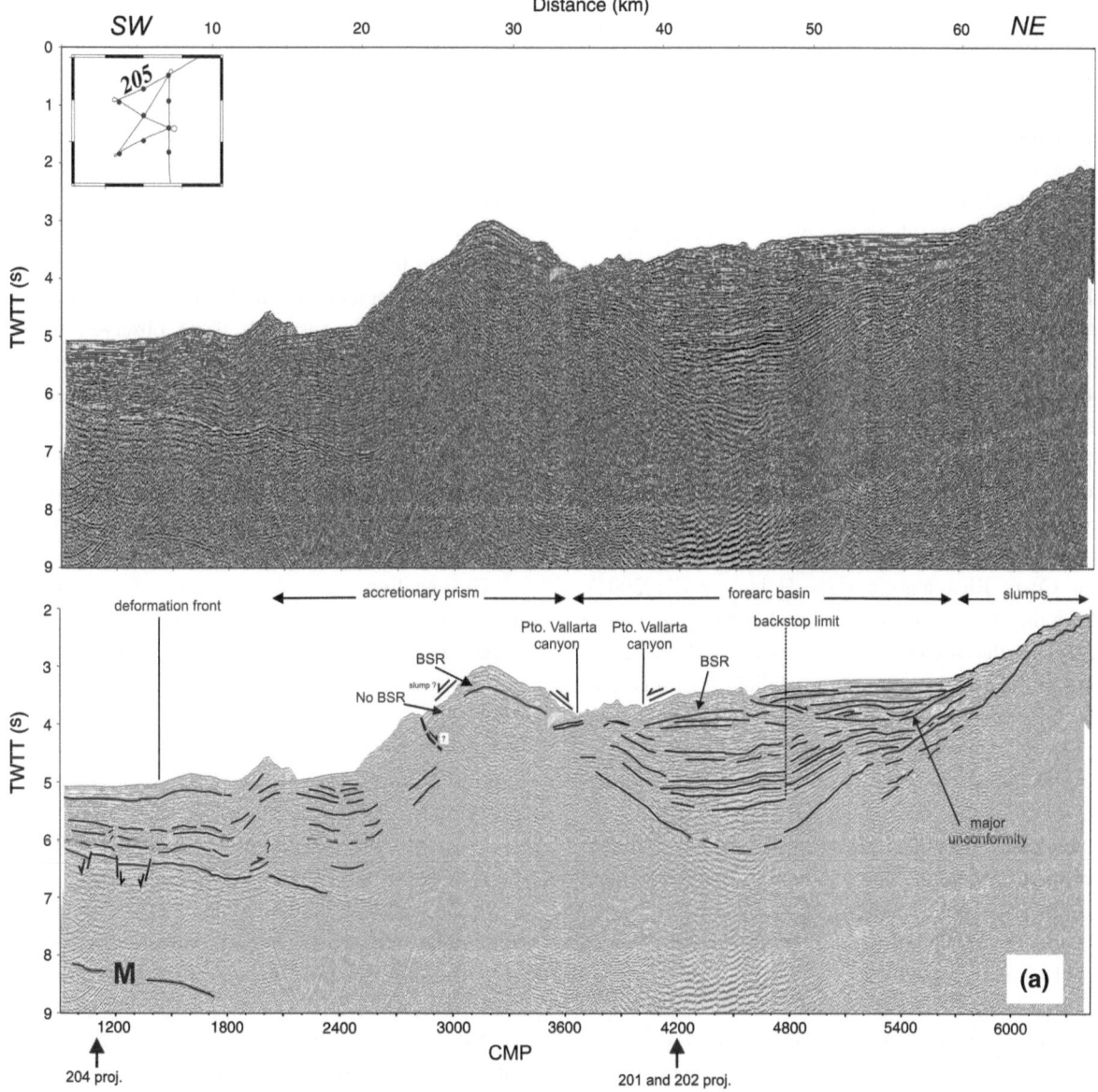

Figure 4

Top panel shows the un-interpreted time migration section of profile 205 (see the *text* for processing sequence). *Left-inner box* (as well as Fig. 2) shows the location of the profile; *bottom panel* displays the interpreted profile, with noticeable features such as the Moho discontinuity detected around 8.0–8.5 s twtt (CMP 1,000–1,800), the interruption of the BSR due to the Puerto Vallarta Canyon (CMP 3,500–4,000) or a high slump (CMP 3,000) at the accretionary prism. *M* Moho discontinuity

deformation front is 12 km westward from the trench axis in section 205 (CMP 1,500, Fig. 4), 5 km in profiles 204 (CDP 1,400, Fig. 5) and 202 (14,100, Fig. 6), and the distance is undetectable in the southern profile 203 (Fig. 7), as we can summarize in Fig. 3 by the black arrows. The basement is also faulted beneath the sedimentary layers at 1.1–1.3 s

twtt (1.3–1.6 km depth) in the northern sections (CMP 1,200, Fig. 4; CMP 3,000, Fig. 5).

Nevertheless, the existence of an accretionary prism observed in seismic sections indicates a subduction–accretion tectonic regime along this segment of the Rivera margin, in contrast with what has been observed previously (MERCIER DE LEPINAY *et al.*,

1997) off Manzanillo. Whereas subduction–accretion characterizes the Rivera margin in the study area and off Acapulco (100°W) according to the results from DSDP Leg 66 (MOORE et al., 1982), subduction–erosion controls the subduction of the Rivera–Cocos Plates off Manzanillo (104°W), during at least the last 8 Ma (FERRARI et al., 2001). This difference in tectonic regimes, of areas only 500 km apart, has been related with a variation in the geometry of the slab along the trench: the downgoing slab is steeper beneath the JB than off Acapulco (PARDO and SUÁREZ, 1993). However, we observe chains of seamounts stretching hundreds of km seaward of, and perpendicular to, the MAT, offshore (Fig. 3). When subducted, these seamounts destabilize the overlying seafloor and leave a morphological trace across the continental slope that marks their path beneath the surface, as has been described along the MAT along the Costa Rica margin (VON HUENE et al., 2004). A chain of seamounts is pointed directly at profile 202 and the southern part of profile 201 where a morphological elevation of the seafloor is observed in the bathymetric and seismic data (Figs. 3, 6, 8). We speculate that a seamount collision and subduction is responsible for the lack of an extensive accretionary prism observed in line 204 (Fig. 5, CMP 2,100). Subducted seamounts push back the frontal sediment prism after entering the subduction zone leaving a gap the size of the seamount in the continental slope. Seamount subduction is a key point for the study of local tsunami generation.

Landwards, the continental domain is characterized by a forearc basin exhibiting two primarily sequences, an undeformed upper sequence (CMP 4,700–5,700, Fig. 4) separated from a deformed lower sequence by a major unconformity (CMP 3,600–5,700, Fig. 4). The upper undeformed sequence ranges in thickness from few meters until 0.5 s twtt and is mainly composed of sub-horizontal strata. The lower sequence ranges from 1 to 2 s twtt thick and appears to be part of a large and thick basin. The stratigraphic units off Manzanillo sampled during submersible dives were a granitic and volcanic, well-rounded conglomerate overlying pre-Eocene (Late Cretaceous–Palaeocene) plutonic rocks that correlate with the onshore coastal plutonic belt of the Jalisco Block (MERCIER DE LEPINAY et al., 1997). A similar

unconformity exists in the well studied stratigraphy of the Tres Marias Islands, 250 km northward of the Manzanillo area, where McCLOY et al. (1988) described massive fine- to medium-grained sandstone, deposited in a non-marine or shallow-marine environment, on top of pre-Eocene granites. The same planktonic assemblages and lithology (siltstone, mudstone and sandstone layers), indicating upper Miocene to lower Pliocene ages, have been described above the conglomerate in both areas. Correlating sedimentary sequences with those observed in the seismic sections suggests that the lower seismic sequence is part of the upper Miocene to lower Pliocene marine basin possibly associated with the opening of the Gulf of California. This could explain why this basin is not present southward as illustrated along the profiles 204, 202 and 203. The above stratigraphic sequence shows the plutonic rocks extending to the trench axis.

The shallow crustal structure of section 201 (Fig. 8) reveals a continental mid slope terrace underlain by sediments with a maximum thickness of 0.5 s twtt (about 375 m), which shows a change of reflectivity that we interpret to be the top of the continental basement. The location of the southern limit of the forearc basin at CMP 5,200 in the profile 201 could indicate a change in the geodynamic setting such as the southern end of the influence of the opening of the Gulf of California.

In the shallow continental domain, strong bottom simulating reflections (BSR) have been clearly imaged along the five MCS profiles (Figs. 4, 5, 6, 7, 8). The presence of a BSR is frequently related to the presence of gas hydrates in the marine sediments, although gas hydrates have also been encountered in regions without BSR (YUAN and EDWARDS, 2000). There are at least two types of origins for the occurrence of BSR. One is related to the presence of gas hydrates causing a negative acoustic impedance contrast between sediments containing gas hydrate and free gas underneath the gas hydrate stability zone (PECHER et al., 1996). Therefore, these BSRs have reversed polarity. The other origin has been related with the strong positive acoustic impedance contrast between silicate rich sediments of the different diagenetic stages: opal A, opal CT and quartz (KASTNER et al., 1977). Therefore, diagenesis-related BSRs have the same polarity as the seafloor

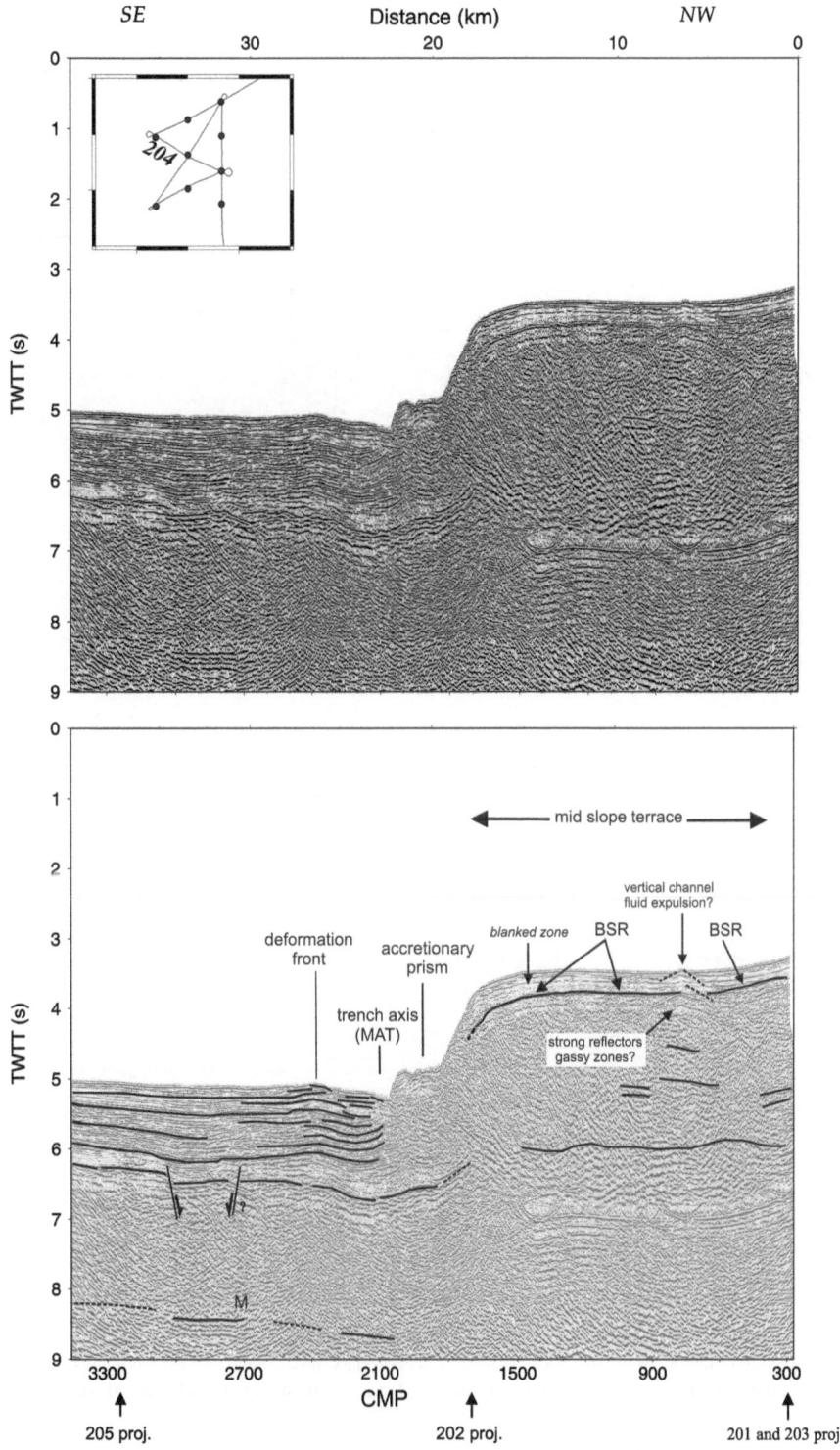

Figure 5

Top panel shows the un-interpreted time migration section of profile 204 (see the *text* for processing sequence). *Left-inner box* as well as Fig. 2, shows the location of the profile; *bottom panel* displays the interpreted profile, with noticeable features such as high-reflective zones beneath 4 s twtt (CMP 700) that could represent gas-charged zones likely expulsed at vertical channels (see *text* for more details) and the Moho discontinuity around 8.0–8.5 s twtt (CMP 2,000–3,400). *M* Moho discontinuity

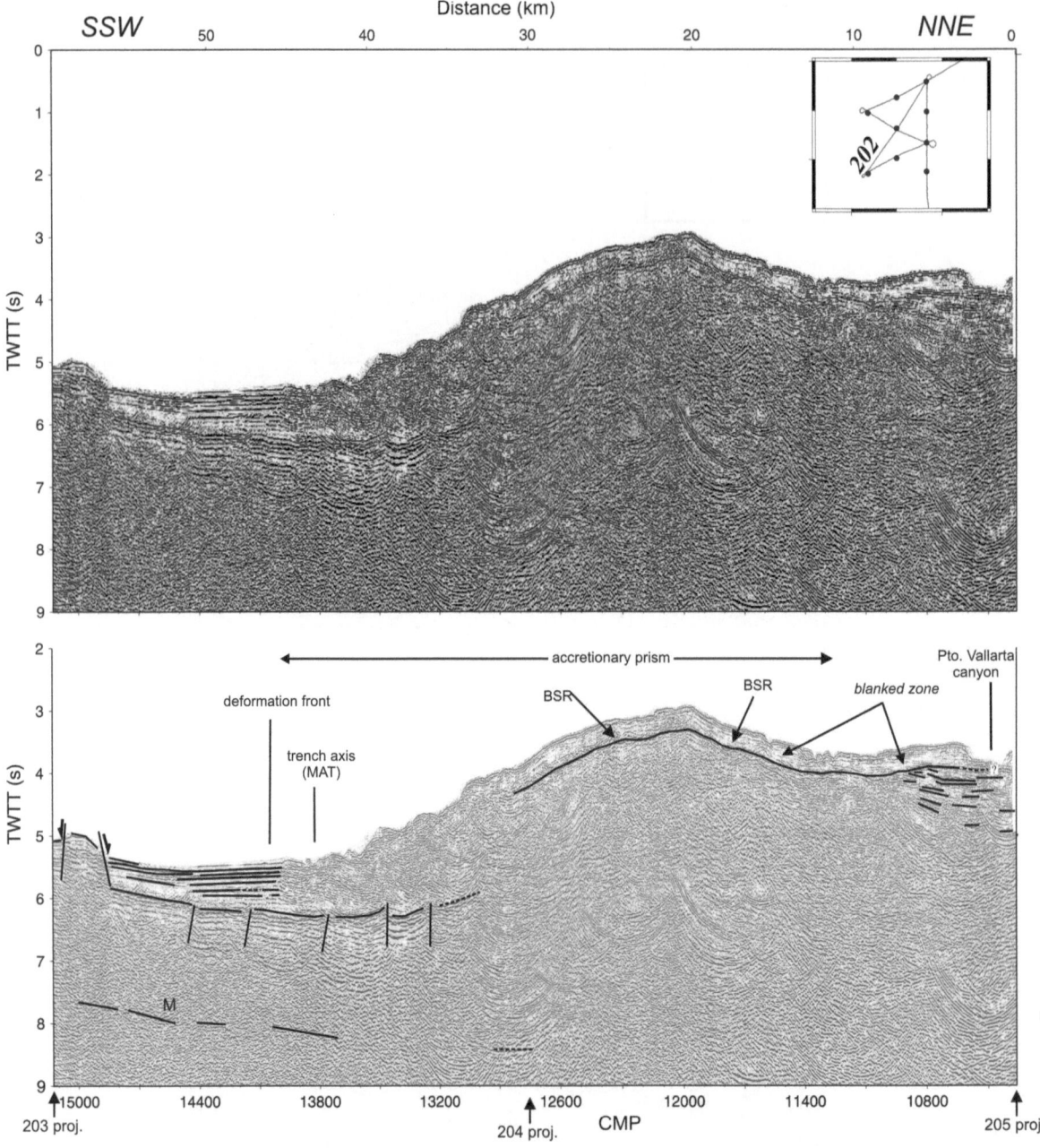

Figure 6
Top panel shows the un-interpreted time migration section of profile 202 (see the *text* for processing sequence). *Left-inner box* as well as Fig. 2, shows the location of the profile; *bottom panel* displays the interpreted profile, with noticeable features such as a highly deformed sequence interpreted as the deformation front, the BSR along the accretionary prism here crossed obliquely, the faulted oceanic crust subducting the prism and the Moho discontinuity around 7.5–8.5 s twtt (CMP 15,000–13,800). *M* Moho discontinuity

reflection. The BSR identified in seismic data off Puerto Vallarta have a reversed polarity with respect to the seafloor (MINSHULL *et al.*, 2005), consequently related with the presence of natural gas hydrates. The

formation of natural gas hydrates is well-known and widely described as a global phenomenon that has been recognized all along the Eastern Pacific margin [e.g. Costa Rica (PECHER *et al.*, 1998); northern

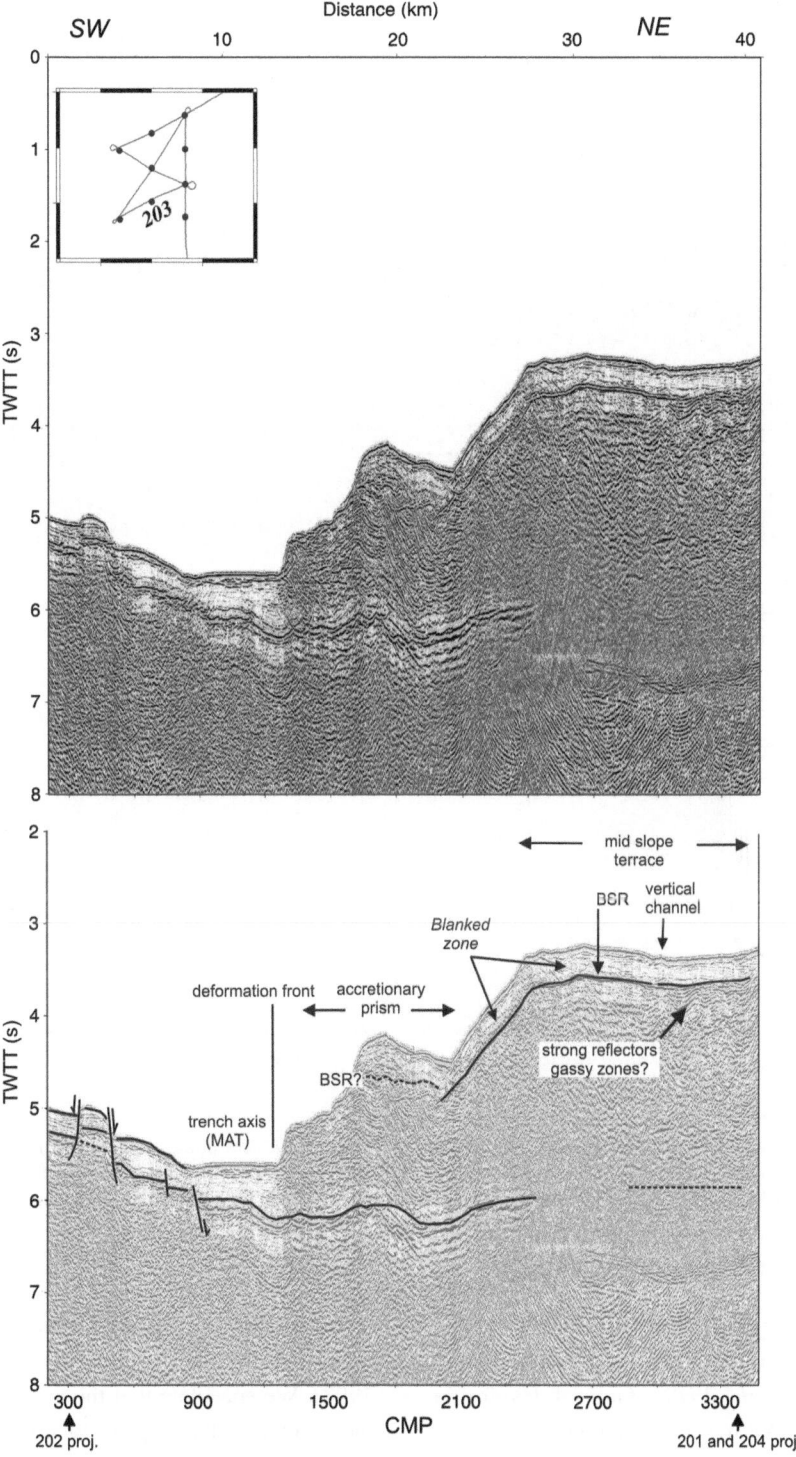

Figure 7
Top panel shows the un-interpreted time migration section of profile 203 (see the *text* for processing sequence). *Left-inner box* as well as Fig. 2, shows the location of the profile; *bottom panel* displays the interpreted profile, with noticeable features such as the high reflective zones beneath 4 s twtt (CMP 3,000) that could represent gas-charged zones and a surface expression of near-vertical channels. In this southernmost profile, the deformation front coincides with the trench axis

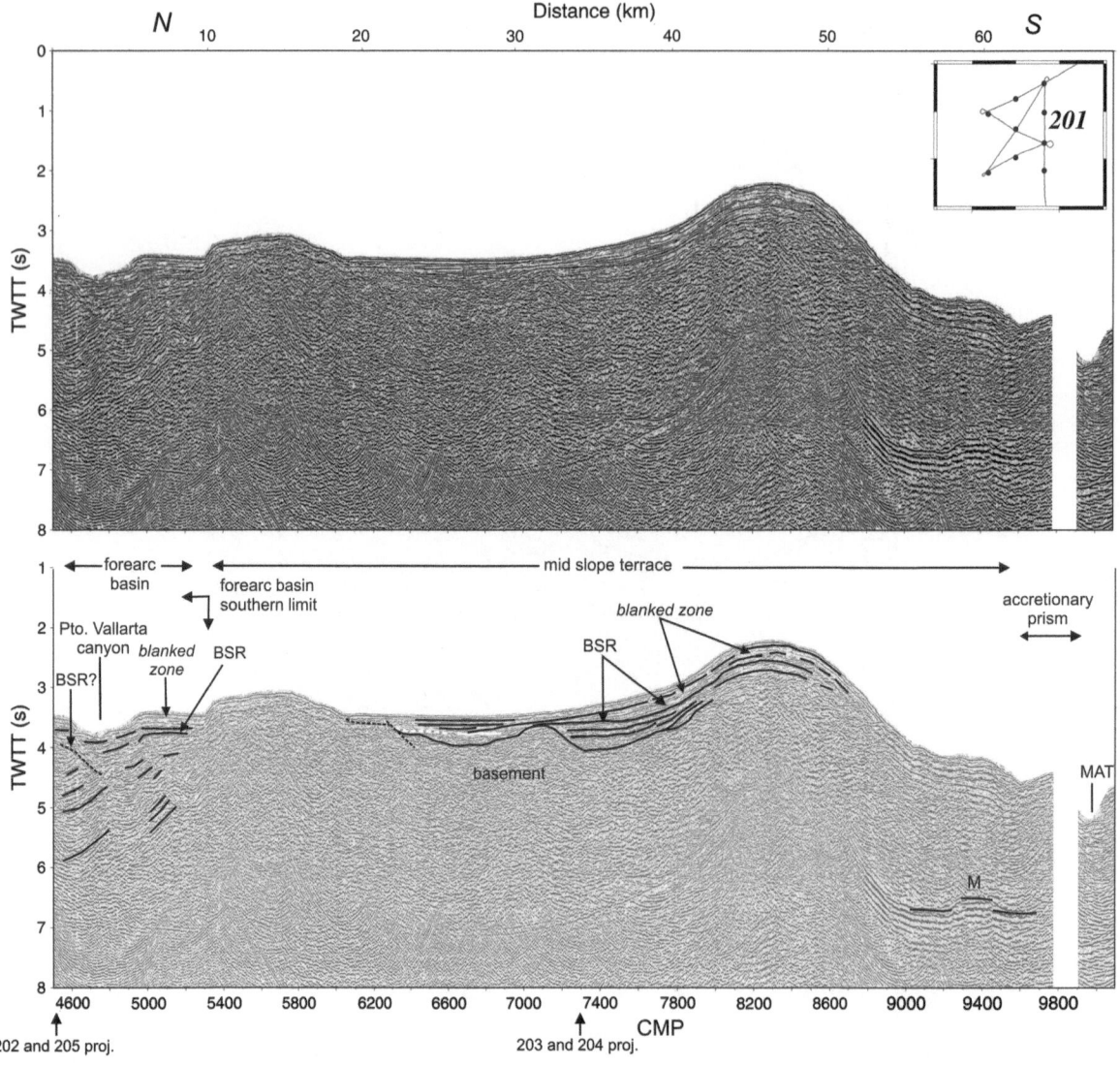

Figure 8

Top panel shows the un-interpreted time migration section of profile 201 (see the *text* for processing sequence). *Left-inner box* as well as Fig. 2, shows the location of the profile; *bottom panel* displays the interpreted profile, with noticeable features such as a low reflectivity of the blanked zone above the BSR (CMP 7,600–8,500) and the Moho discontinuity around 7 s twtt on the southernmost part of the line (CMP 9,000–9,700). *M* Moho discontinuity

California (BROOKS *et al.*, 1991); Peru (PECHER *et al.*, 1996); Pacific ocean off Mexico, Guatemala and Panama (SHIPLEY *et al.*, 1979); Gulf of California (LONSDALE, 1985)]. We have detected the widespread distribution of the BSR not only in the accretionary prism offshore of Puerto Vallarta, but also in the forearc basin, the continental slope and the mid slope terrace. A BSR is, however, absent in areas affected by slumping, as can be observed in profile 205 (CDP 3,500–4,000, Fig. 3). A similar situation has been described in the Peruvian margin (PECHER *et al.*, 1998). We speculate that the gas may escape through the fractures that appear to cross the base of gas hydrate stability zone near CDP 800 in profile 204 (Fig. 5). Vertical fluid expulsion channels and sea-floor venting sites have also been recognized in northern Cascadia (HYNDMAN *et al.*, 2001) and the Sea of Okhotsk (LÜDMANN and WONG, 2003) linked with

BSR. When these structures reach the seafloor, they are associated either with pockmarks or with dome-like structures (profile 204, CMP 700). The pale reflectivity above the BSR, known as a *blanked zone* (e.g., KORENAGA *et al.*, 1997), is clearly recognized in all the profiles at 0.2 s twtt below the BSR and it has been associated with hydrate cementation or porosity filling suppressing the impedance contrasts between the layers (Figs. 5, 6, 7, 8).

4.2. Deep Crustal Structure

Seismic imaging is the best method to investigate the structure of subduction zones and becomes highly dependent on the material properties (seismic velocities) and the depth of the earthquakes in the specific case of seismic tomographic imaging. Producing images of a young lithosphere with shallow earthquakes becomes challenging, but in the meantime, we have imaged the deep crustal structure using active multichannel seismics. The analysis and interpretation of the MCS profiles allows us to identify, for the first time in the area of the subduction Rivera Plate, the major structures of the margin at deep crustal levels.

Seismic images show crustal thickness of 3 s twtt for the subducting Rivera Plate (roughly ~6.5 km thick) and measured dip angles of subduction, from

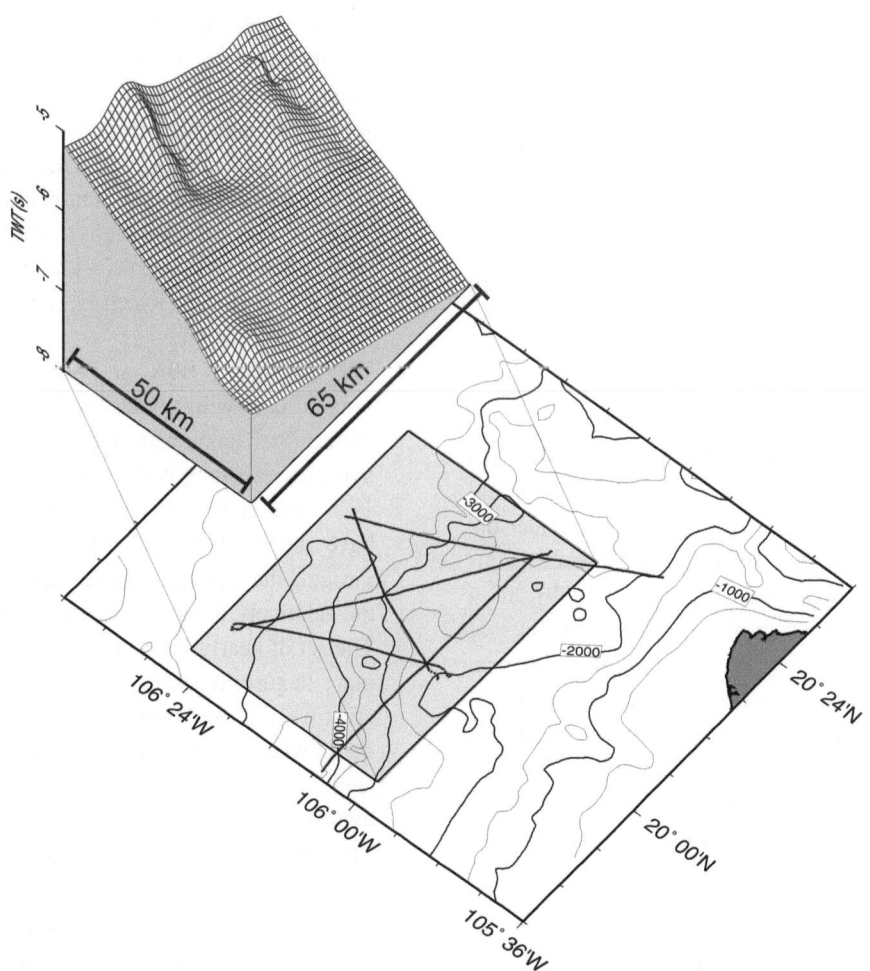

Figure 9

Block diagram showing the top of the subducted slab of the Rivera Plate, with a dip angle of 6° in the northern part of the block and 8° in the south under the Jalisco Block. This dip distribution suggests that the Rivera Plate is also being subducted to the north. *Blue-areas* are well controlled by the seismic lines whereas *black-areas* are inferred from the interpolation. Note the presence of subducted ridges

north to south, of 6° in profile 205 (Fig. 4), 6.5° in profile 204 (Fig. 5), and 8° in profile 203 (Fig. 7). This decreasing tendency towards the north is related to the termination of the subduction north of profile 205 in the oceanic domain. These values are consistent with the 10° dipping Wadatti-Benioff zone at 20 km depth estimated from local seismicity data by PARDO and SUÁREZ (1993).

To illustrate the geometry of the top of the subducted slab off Puerto Vallarta, we compute a 3D block diagram (Fig. 9) using the depth (in seconds twtt) obtained from the MCS sections and using a nearest neighbour average algorithm. The area well controlled by existing data is displayed in blue in the block diagram, whereas the interpolated areas are shown in black. The 3D block points out the eastward subduction of the Rivera Plate beneath the continent with an average dip angle of 7° ± 1° as previously discussed. Surprisingly, it seems also to subduct, or bend, towards the north. This result agrees with an oblique convergence between the Rivera and NA Plates at 20.95°N 106.25°W (KOSTOGLODOV and BANDY, 1995). New additional seismic data and geologic sampling north of Puerto Vallarta will facilitate answering specific questions such as the age of Maria Magdalena Rise (MMR, Fig. 2), a bathymetric feature of about 500 m relief located at the northern ending of the Jalisco subduction zone in the area of Tres Marias Islands. The uplift of the Tres Marias escarpment may be helped by the gently northward convergence of the Rivera Plate under the North American Plate, where upper Miocene hemipelagic muds crops out (LONSDALE, 1989) as a consequence of a history of successive processes of subduction and opening of the Gulf of California. Also, the nature of the underlying crust (oceanic or continental), which is still under discussion (LONSDALE, 1995) and may be related to a previous location of the EPR before a westward jump during the opening of the Gulf of California, could be resolved acquiring new geophysical and geological data.

5. Conclusions

The new seismic data presented herein provide new insight into the crustal structure at the contact between the Rivera and North American Plates, offshore of Puerto Vallarta. We summarized our conclusions as follows:

1. The Rivera Plate subducts beneath the NA Plate with, from north to south in the trench axis, a dip angle ranging from 6° to 8°. Although data are not conclusive, the subduction seems to continue northward and the Tres Marias Islands could be uplifted as a consequence of this subduction.

2. Tectonic observations based on MCS data indicate that there are significant structural differences from north to south of 20.2°N. In the trench-fill wedge, the sedimentary infill is 2.5 km thick near the trench axis, and there is evidence of normal faulting at the top of the oceanic crust. The topography of the slope is quite rough due to slumping and a well developed, 25 km-wide, forearc basin is constituted of two primary sequences. These two sequences have been split by a major unconformity; the lower unit is of Miocene age. South of 20.2°N the forearc basin is not seen; instead, the upper continental crust is composed of one stratigraphic sequence with a faulted thin sedimentary cover (0.75 km) below the seafloor, west of the trench, which is another indication of recent deformation.

3. We identify a BSR at 0.25–0.4 s twtt in the seismic data with reverse polarity with respect to the seafloor, demonstrating the occurrence of gas hydrates offshore northwestern Mexico between 19°N and 21°N. Vertical expulsion morphologies have been observed proving fluid circulation through the detected active fault systems despite the relatively thick sedimentary coverage.

4. The set of nearly vertical normal faults, formed by plate flexure, reaching the surface in the oceanic crust west of the Middle America Trench seems to help, when subducted, the circulation of fluids (fluid percolation) and the formation of gas hydrates in the continental domain.

5. The presence of a well developed accretionary prism suggests a subduction–accretion type regime instead of a subduction erosion regime as is present along the convergent margin of the Rivera Plate off Manzanillo. This difference may be related with a variation in the geometry of the downgoing slab along the trench.

Acknowledgments

Funding is from the Spanish National Research Project (ANT94-0182-C02-01/02). Additional funding comes from INSU-France, ORSTOM-IRD, France bilateral cooperation (HF1997-077), and CSIC/CONACYT 0894PT, the Secretaría de Marina of Mexico, CICESE project 644107 and by CIRIT (Project 1995SGR00438). We acknowledge the captain and crew of the R/V *Hespérides*, *Altair* and *Humboldt*, and all the scientific and technical staff that participated in the CORTES-P96 seismic experiments for their support during the experiment, and the Mexican authorities for facilitating work within their exclusive economic zone. The figures were done using the Generic Mapping Tools-GMT, Wessel, P. & Smith, W., (1995). R. Bartolome acknowledges the financial support of the Spanish Ministry of Science and Innovation (Ramon y Cajal program).

REFERENCES

ATWATER, T. (1970), *Implications of plate tectonics for the Cenozoic tectonic evolution of western North America*, Geol. Soc. Am. Bull. 81, 3513–3536.

ATWATER, T. (1989), *Plate Tectonic history of the Northeast Pacific and Western North America*, in The Eastern Pacific Ocean and Hawaii—The Geology of North America N, Winterer, R. W.E., Hussong, L., Decker D.M., Eds: Geological Society of America, Boulder, pp. 21–72.

BANDY, W.L., and YAN, C.Y. (1989), *Present-day Rivera-Pacific and Rivera-Cocos relative plate motions (abstract)*, Eos trans. Am. Geophys. Union, 70, 1342.

BANDY, W.L. (1992), *Geological and geophysical investigation of the Rivera-Cocos plate boundary: implications for plate fragmentation*, Ph.D. Dissertation, Texas A&M University, College Station, 195 p.

BANDY, W., MORTERA-GUTIERREZ, C.A., and URRUTIA-FUCUGAUCHI, J. (1993), *Gravity field of the southern Colima Graben, Mexico*, Geofis. Intern. 32, 561–567.

BANDY, W., and PARDO, M. (1994), *Statistical examination of the existence and relative motion of the Jalisco and Southern Mexico blocks*, Tectonics 13, 755–768.

BANDY, W.L., KOSTOGLODOV, V., and MORTERA-GUTIERREZ, C.A. (1998), *Southwest migration of the instantaneous Rivera-Pacific Euler pole since 0.78 Ma*, Geof. Int. 37, 153–169.

BANDY, W., KOSTOGLODOV, V., HURTADO-DIAZ, A., and MENA, M. (1999), *Structure of the southern Jalisco subduction zone, Mexico, as inferred from gravity and seismicity*, Geofísica Internacional 38 (3), 127–136.

BANDY, W.L., MICHAUD, F., BOURGOIS, J., CALMUS, T., DYMENT, J., MORTERA-GUTIÉRREZ, C.A., ORTEGA-RAMÍREZ, J., PONTOISE, B., ROYER, J.-Y., SICHLER, B., SOSSON, M., REBOLLEDO-VIEYRA, M., BIGOT-CORMIER, F., DÍAZ-MOLINA, O., HURTADO-ARTUNDUAGA, A.D., PARDO-CASTRO, G., and TROUILLARD-PERROT, C. (2005), *Subsidence and strike-slip tectonism of the upper continental slope off Manzanillo, Mexico*, Tectonophysics 398, 115–140.

BOURGOIS, J., RENARD, V., AUBOUIN, J., BANDY, W., BARRIER, E., CALMUS, T., CARFANTAN, J. C., GUERRERO GARCIA, J., MAMMERICKX, J., MERCIER DE LEPINAY, B., MICHAUD, F., and SOSSON, M. (1988), *Fragmentation en cours du continent Nord Américain: les frontières sous marines du bloc de Jalisco (Mexique)*, C. R. Acad. Sc. Paris 307 II, 1121–1130.

BOURGOIS, J., and MICHAUD, F. (1991), *Active fragmentation of the North American plate at the Mexican Triple Junction Area off Manzanillo (Mexico)*, Geomarine Letters 11, 59–65.

BROOKS, J.M., FIELD, M.E., and KENNICUTT II, M.C. (1991), *Observations of gas hydrates in marine sediments, offshore northern California*, Marine Geology 96, 103–109.

COUCH, R., and WOODCOCK, S. (1981), *Gravity and structure of the continental margins of southwestern Mexico and northwestern Guatemala*, J. Geophys. Res. 86, 1829–1840.

COURBOULEX, F., SINGH, S.K., PACHECO, J.F., and AMMON, C.J. (1997), *The 1995 Colima-Jalisco, Mexico, earthquake (Mw=8): a study of the rupture process*, Geophys. Res. Lett. 24, 1019–1022.

CURRIE, C.A., HYNDMAN, R.D., WANG, K., KOSTOGLODOV, V. (2002), *Thermal models of the Mexico subduction zone: Implications for the megathrust seismogenic zone*, Journal of Geophy. Res. 107, No. B12.

DAÑOBEITIA, J.J., CÓRDOBA, D., DELGADO-ARGOTE, L.A., MICHAUD, F., BARTOLOMÉ, R., FARRÁN, M., and CORTES-P96 Working Group (1997), *Expedition gathers new data on crust beneath Mexican West Coast*, EOS Trans. Am. Geophys. Union 78 (49), 565–572.

DAVIES, G. F. (1992), *On the emergence of plate tectonics*, Geology 20,963–966.

DEMETS, C., and STEIN, S. (1990), *Present-Day kinematics of the Rivera Plate and Implications for Tectonics in Southwestern Mexico*, J. Geophys. Res. 95 (B13), 21931–21948.

DEMETS, C., and WILSON, D. (1997), *Relative motions of the Pacific, Rivera, North American and Cocos plates since 0.78 Ma*, J. Geophys. Res. 102, 2789–2806.

DEMETS, C., and TRAYLEN, S. (2000), *Motion of the Rivera plate since 10 Ma relative to the Pacific and North American plates and the mantle*, in The Influence of plate interaction on post-Laramide magmatism and tectonics in Mexico, FERRARI, L., STOCK, J., URRUTIA-FUCUGAUCHI J., Eds: Tectonophysics 318 (1–4), 119–159.

EISSLER, H. K., and McNALLY, K. C. (1984), *Seismicity and tectonics of the Rivera Plate and implications for the 1932 Jalisco, Mexico, earthquake*, J. Geophys. Res. 89, 4520–4530.

FERRARI, L., PETRONE, CH., and FRANCALANCI, L. (2001), *Generation of oceanic-island basalt type volcanism in the western Trans-Mexican volcanic belt by slab rollback, asthenosphere infiltration and variable flux meeting*, Geology 29, 507–510.

GEBCO, 2008 Bathymetric dataset, http://www.gebco.net/.

HYNDMAN, R.D., SPENCE, G.D., CHAPMAN, R., RIEDEL, M., and EDWARDS, R.N. (2001), *Geophysical studies of marine gas hydrate in northern Cascadia, in: Natural Gas Hydrates: Occurrence, and Detection Geophysical Monograph 124*, American Geophysical Union.

JOHNSON, M.C., and PLANK, T. (1999), *Dehydration and melting experiments constrain the fate of subducted sediments*, Geochem. Geophys. Geosyst., vol. 1, p.1999GC000014.

Kastner, M., Keene, J.B., and Gieskes, J.M. (1977), *Diagenesis of siliceous oozes-I. Chemical controls on the rate of opal-A to opal-CT transformation an experimental study*, Geochimica et Cosmochimica Acta 41, 1041–1059.

Khutorskoy, M.D., Delgado-Argote, L.A., Fernández, R., Kononov, V.I., and Polyak, B.G. (1994), *Tectonics of the offshore Manzanillo and Tecpan basins, Mexican Pacific, from heat flow, bathymetric and seismic data*, Geofís. Intern. 33, 161–185.

Klitgord, K., and Mammerickx, J. (1982), *Northern East Pacific rise: magnetic anomaly and bathymetric framework*, J. Geophys. Res. 100, 24367–24392.

Korenaga, J., Holbrook, W.S., Singh, S.C., and Minshull, T.A. (1997), *Natural gas hydrates on the southeast U.S. margin: Constraints from full waveform and travel time inversions of wide-angle seismic data*, J. Geophys. Res. 102, 15345–15365.

Kostoglodov, V., and Bandy, W. (1995), *Seismotectonic constraints on the convergence rate between the Rivera and North America plates*, J. Geophys. Res. 100, 17977–17989.

Lonsdale, P. (1985) *A transform Continental Margin rich in hydrocarbons, Gulf of California*, The American Association of Petroleum Geologists Bulletin 69 (7), 1160–1180.

Lonsdale, P. (1989), *Geology and tectonic history of the Gulf of California*, in The Eastern Pacific Ocean and Hawaii, Geological Society of America Winterer, E.L., Hussong, D.M., Decker, R.W., Eds: The Geology of North America, pp. 499–521.

Lonsdale, P., (1991) *Structural patterns of the Pacific floor offshore of peninsular California*, in The Gulf and Peninsular province of the Californias, Dauphin, J.P., Simoneit, B.R.T., Eds: Am. Assoc. Pet. Geol. Mem. 47, pp. 87–125.

Lonsdale, P. (1995), *Segmentation and disruption of the East Pacific Rise in the mouth of the Gulf of California*, Mar. Geophys. Res. 17, 323–359.

Lüdmann, T., and Wong, H.K. (2003), *Characteristics of gas hydrate occurrences associated with mud diapirism and gas scape structures in the northwestern Sea of Okhotsk*, Marine Geology 201, 269–286.

Luhr, J.F., Nelson, S.A., Allan, J.F., and Carmichael, I.S.E. (1985), *Active rifting in SW Mexico: manifestations of an incipient eastward spreading-ridge jump*, Geology 13, 54–57.

Mammerickx, J., and Klitgord, K. (1982), *Northern East Pacific Rise: evolution from 25 m.y.B.P. to the Present*, J. Geophys. Res. 87, 6751–6759.

Mammerickx, J. (1984), *The Morphology of propagating spreading centers: New and old*, J. Geophys. Res. 89, 1817–1828.

McCloy, C., Ingle, J.C., and Barron, J.A. (1988), *Neogene stratigraphy, foraminifera, diatoms, and depositional history of María Madre Island Mexico: Evidence of Early Neogene marine conditions in the southern Gulf of California*, Marine Micropaleontology 13, 193–212.

Mercier de Lepinay, B., Michaud, F., and the Nautimate team (1997), *Large Neogene Subsidence along the Middle America trench off Mexico (18°-19°N): Evidence from Submersible Observations*, Geology 25 (5), 387–390.

Michaud, F., Mercier de Lepinay, B., Bourgois, J., and Calmus, T. (1996), *Evidence for active extensional tectonic features within the Acapulco trench fill at the Rivera-North America plate boundary*, C.R.Acad. Sci., Paris t. 321 série IIa, 521–528.

Michaud, F., Dañobeitia, J.J., Bartolome, R., Carbonell, R., Delgado-Argote, L., Cordoba, D., and Monfret, T. (2001), *Did the East Pacific rise subduct beneath the North America plate (western Mexico)?*, Geo-Marine Letters 20, 168–173.

Minshull, T., Bartolomé, R., Byrne, S., and Dañobeitia, J. J. (2005), *Low heat flow from young oceanic lithosphere at the Middle America Trench off Mexico*, Earth and Planetary Science Letters, 239, 33–41, doi:10.1016/j.epsl.2005.05.045.

Moore, J. C., Watkins, J. S., and Shipley, T. H (1982) *Summary of accretionary processes, Leg 66 DSDP: Offscraping, underplating, and deformation of the slope apron: Initial Reports of the Deep Sea Drilling Project, v. 66*, U.S. Government Printing Office Washington D.C., pp. 825–836.

Nixon, G.T. (1982), *The relatioship between Quaternay volcanism in central Mexico and the seismicity and structure of subducted oceanic lithosphere*, Geol. Soc. Am. Bull. 93, 514–523.

Pacheco, F.J., Singh, S.K., Dominguez, J., Hurtado, A., Quintanar, L., Jimenez, Z., Yamamoto, J., Gutierrez, C., Santoyo, M., Bandy, W., Guzman, M., and Kostoglodov, V. (1997), *The October 9, 1995 Colima-Jalisco, Mexico earthquake (Mw 8): An aftershock study and a comparison of this earthquake with those of 1932*, Geophys. Res. Lett. 24 (17), 2223–2226.

Pardo, M., and Suárez, G. (1993), *Steep subduction geometry of the Rivera plate beneath the Jalisco Block in western Mexico*, Geophys. Res. Lett. 20, 2391–2394.

Pardo, M., and Suárez, G. (1995), *Shape of the subducted Rivera and Cocos plates in southern Mexico: Seismic and tectonic implications.* J. Geophys. Res. 100, 12,357–12,373.

Pecher, I. A., Minshull, T. A., Singh, S.C., and von Huene, R. (1996), *Velocity structure of a bottom simulating reflector offshore Peru: Results from full waveform inversion*, Earth Planet. Sci. Lett. 139, 459–469.

Pecher, I. A., Ranero, C. R., von Huene, R., Minshull, T. A., and Singh, S. C. (1998), *The nature and distribution of bottom simulating reflectors at the Costa Rican convergent margin*, Earth Planet. Sci. Lett. 139, 459–469.

Prol-Ledesma, R. M., Sugrobov, V. M., Flores, E. L., Juarez, G., Smirnov, Y. B., Gorshkov, A. P., Bondarenko, V. G., Rashidovm, V. A., Nedopekin, L. N., and Gavrilov, V. A. (1989), *Heat flow variations along the Middle America Trench*, Mar. Geophys. Res. 11, 69–76.

Ramírez-Herrera, M.T., and Urrutia-Fucugauchi, J. (1999), *Morphotectonic zones along the coast of the Pacific continental margin, southern Mexico*, Geomorphology 28, 237–250.

Ramirez-Herrera, M. T., Kostoglodov, V., and Urrutia-Fucugauchi, J. (2004), *Holocene emerged notches and tectonic uplift along the Jalisco coast, Southwest Mexico*, Geomorphology 58, 291–304.

Shipley, T. H., Houston, M. H., and Buffer, R.T. (1979), *Seismic evidence for widespread possible gas hydrate horizons on continental slopes and rises*, Am. Assoc. Pet. Geol. Bull. 63, 2204–2213.

Singh, S. K., Ponce, L., and Nishenko, S. (1985), *The great jalisco, Mexico earthquake of 1932: subduction of the Rivera plate*, Bull. Seismic. Soc. Amer. 75, 1301–1313.

Singh, S.K., Pacheco, J.F., Alcántara, L., Reyes, G., Ordaz, M., Iglesias, A., Alcocer, S.M., Gutierrez, C., Valdés, C., Kostoglodov, V., Reyes, C., Mikumo, T., Quaas, R., and Anderson, J. G. (2003), *A Preliminary Report on the Tecomán, Mexico Earthquake of 22 January 2003 (Mw7.4) and its effects*, Seismological Research Letters 74, 279–289.

Stern. R.J. (2002), *Subduction zones*, Reviews of Geophysics, 40, 4.

VON HUENE, R., RANERO, C. R., and WATTS, P. (2004), *Tsunamigenic slope failure along the Middle America Trench in two tectonic settings*, Marine Geology 203, 303–317.

WANG, K., HYNDMAN, R.D., YAMANO, M. (1995), *Thermal regime of the southwest Japan subduction zone: Effects of age history of the subducting plate*, Tecnonophysics 248, 53–69.

YUAN, J., and EDWARDS, R. N. (2000), *The assessment of marine gas hydrates through electrical remote sounding: Hydrate without a BSR?*, Geophysical Research Letters 27 (16), 2397–2400.

ZIAGOS, J. P., BLACKWELL, D. D., and MOOSER, F. (1985), *Heat flow in southern Mexico and the thermal effects of subduction*, J. Geophy. Res. 90, 5410–5420.

(Received February 15, 2010, revised August 13, 2010, accepted August 16, 2010, Published online November 16, 2010)

Pure Appl. Geophys. 168 (2011), 1391–1413
© 2010 Springer Basel AG
DOI 10.1007/s00024-010-0208-8

The Mid-Rivera-Transform Discordance: Morphology and Tectonic Development

WILLIAM L. BANDY,[1] FRANÇOIS MICHAUD,[2] CARLOS A. MORTERA GUTIÉRREZ,[1] JÉRÔME DYMENT,[3] JACQUES BOURGOIS,[4] JEAN-YVES ROYER,[5] THIERRY CALMUS,[6] MARC SOSSON,[2] and JOSE ORTEGA-RAMIREZ[7]

Abstract—To better define the morphotectonic elements and tectonic development of the Mid-Rivera-Transform Discordance, multibeam bathymetric, seafloor backscatter, multichannel seismic reflection and total field marine magnetic data were collected along the entire Rivera Transform west of 107°W during the BART and FAMEX campaigns of the N.O. L'Atalante conducted in 2002. These data show that, although the transform tectonized zone of the Rivera Transform west of 107°30'W is a single continuous morphologic basin, this basin consists of two distinct morphotectonic domains: an eastern domain which morphologically is a deep rhombochasm within which organized seafloor spreading has occurred, and a western 'leaky transform' domain. These new data, in conjunction with the results of previous studies, support the idea that the Rivera-Pacific Euler pole is migrating southward towards the eastern half of the Rivera Transform, and further indicate a recent (<0.14 Ma), and most likely ongoing, clockwise reorganization of the principle transform displacement zones of the Rivera Transform west of 108°W. We propose that the Mid-Rivera-Transform Discordance owes its origin to this eastward progressing, clockwise reorganization of the transform segments that is occurring in response to recent changes in Rivera-Pacific relative plate motion.

Key words: Rivera Transform, Plate motions, Morphology, Multi-beam bathymetry.

[1] Instituto de Geofísica, Universidad Nacional Autónoma de México, 04510 México D.F, México. E-mail: bandy@geofisica.unam.mx

[2] Géosciences Azur, La Darse, BP48, 06235 Villefranche-sur-Mer, France.

[3] Institut de Physique du Globe de Paris, 4 Place Jussieu, 75252 Paris, France.

[4] Institut des Sciences de la Terre Paris (ISTEP), UPMC-CNRS, Université Pierre et Marie Curie, 4 Place Jussieu, 75252 Paris, France.

[5] CNRS, Domaines Océaniques, Institut Universitaire Européen de la Mer, 29280 Plouzané, France.

[6] ERNO, Instituto de Geología, Universidad Nacional Autónoma de México, 83000 Hermosillo, México.

[7] Geophysics Laboratory, Instituto Nacional de Antropología e Historia, México D.F, México.

1. Introduction

The presence of an intra-transform spreading center along the Rivera Transform at 108.25°W (Fig. 1) was first proposed based on local seismicity studies conducted in the 1970s (REID, 1976; PROTHERO et al., 1976; PROTHERO and REID, 1982). Prior to the N.O. L'Atalante BART/FAMEX campaigns of 2002 (e.g. BANDY et al., 2005, 2008), the existing multi-beam and conventional echo sounder data were consistent with this proposal (e.g. LONSDALE, 1995), although the intra-transform spreading center and the surrounding area had yet to be fully surveyed with high resolution multi-beam bathymetric systems.

The presence of intra-transform spreading centers is not uncommon (e.g. HOLCOMBE et al., 1973; KASTENS et al., 1979; SEARLE, 1983; HOLCOMBE and SHARMAN, 1983; EMBLEY and WILSON, 1992; POCKALNY et al., 1997; TEN BRINK et al., 1999). However, this intra-transform spreading center is of particular interest in that a marked discordance occurs; both the orientations of the transform and earthquake slip vectors across the spreading center change by ~15°. This discordance (the 'Mid-Rivera-Transform Discordance') has hindered our understanding not only of the plate configuration in this area but also of the relative motion between the Rivera and Pacific plates, and indirectly, between the Rivera and North American plates. Specifically, (1) the gross morphologic trend of the Rivera Transform west of this spreading center is 126°, whereas to the east it is 108° (MICHAUD and BOURGOIS, 1995; MICHAUD et al., 1997) and (2) the slip vector orientations of earthquakes occurring immediately west of this area lie between 115° and 130° whereas to the east they lie between 105° to 112° (Fig. 2). The discordance is also indicated by (1) the different character and distribution of

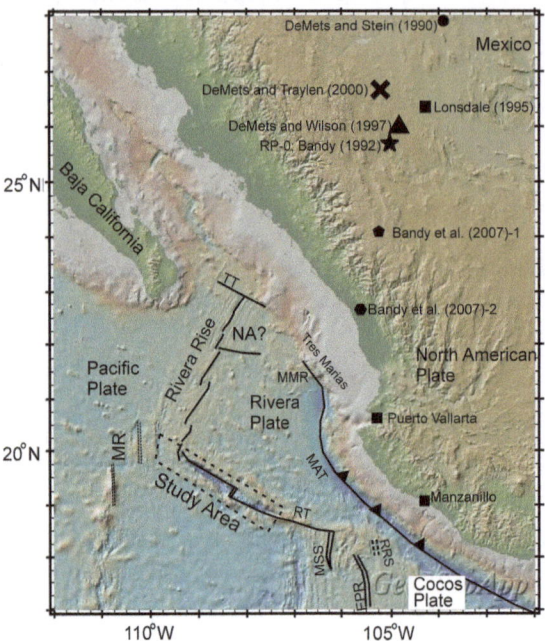

Figure 1

Study area location. Shown are the locations of the Rivera-Pacific Euler poles determined by DeMets and Stein (1990), DeMets and Wilson (1997), DeMets and Traylen (2000), Lonsdale (1995), Bandy (1992). Also shown are two poles of Bandy et al. (2007) which incorporated the FAMEX and BART data into their determination. EPR, East Pacific Rise; MAT, Middle America Trench; MR, Mathematician Ridge; MSS, Moctezuma Spreading Segment; NA, North American Plate; RRS, relict ridge segment; RT, Rivera Transform; TT, Tamayo Transform. Base map from GeoMapApp (http://www.geomapapp.com)

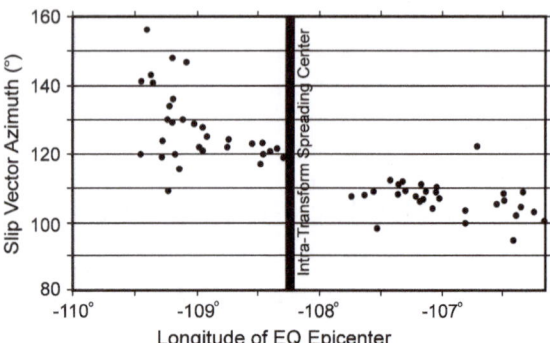

Figure 2

Slip Vector Azimuth (*black dots*) versus longitude of the epicenter location of large earthquakes occurring along the Rivera Transform. Slip vectors calculated from the CMT solutions, obtained from the Harvard CMT database, include events occurring prior to December 2009. Epicenter locations are from the USGS PDE catalogue. The PDE locations are preferred over the locations reported in the CMT database due to the reduced scatter in the slip vector versus longitude plot. Note that the slip vectors located east of the spreading center lie, in general, between 100° and 112°; whereas to the west, they generally lie between 118° and 130° until 109.2°W, at which point the range in slip vectors increases dramatically (115°–155°). Also note the lack of slip vectors in the region extending ~50 km to the east of the intra-transform spreading center

earthquakes (Reid, 1976) to either side of this spreading center (Fig. 3): specifically, teleseismically recorded events are more numerous to the west whereas larger magnitude events occur more frequently to the east, and a gap in the occurrence of events is observed just east of the spreading center and (2) none of the existing Rivera-Pacific Euler poles (Fig. 1) adequately predicts the orientation of the Rivera Transform both to the east and to the west of the spreading center (e.g. Michaud et al., 1997; Bandy et al., 2008). As a result, several proposals exist for the plate configuration at the northern terminus of the Middle America Trench (Larson, 1972; Reid, 1976; Bandy, 1992; Lonsdale, 1995; Michaud and Bourgois, 1995; Michaud et al., 1997).

The purpose of this study is to map, in more detail, the morphotectonic elements and shallow subsurface geology of the entire Rivera Transform

west of 107°W in order to clarify its structure and tectonic development, and thus clarify the origin of the discordance zone and the plate configuration in this area. The data used are the previously unpublished multi-beam bathymetric and seafloor backscatter strength data, multi-channel seismic reflection data and total field magnetic data collected during the 2002 BART/FAMEX campaigns aboard the N/O L'Atalante. West of the discordance, the results are consistent with a recent, gradual clockwise reorientation of the relative motion between the Rivera and Pacific plates (Lonsdale, 1995). In addition, they indicate that no recent major plate motion reorientation has occurred east of the discordance zone, except near the Moctezuma spreading segment (MSS) where a small counter-clockwise reorientation may have occurred (Bandy et al., 2008). These results are, in general, consistent with previous studies which illustrate that the Rivera-Pacific Euler pole has been migrating towards the eastern transform segment (Bandy and Yan, 1989; DeMets and Stein, 1990; Bandy, 1992; Bandy et al., 1998b; Bandy and Hilde; 2000; Bandy et al., 2000; DeMets and Traylen, 2000). We propose that the discordance

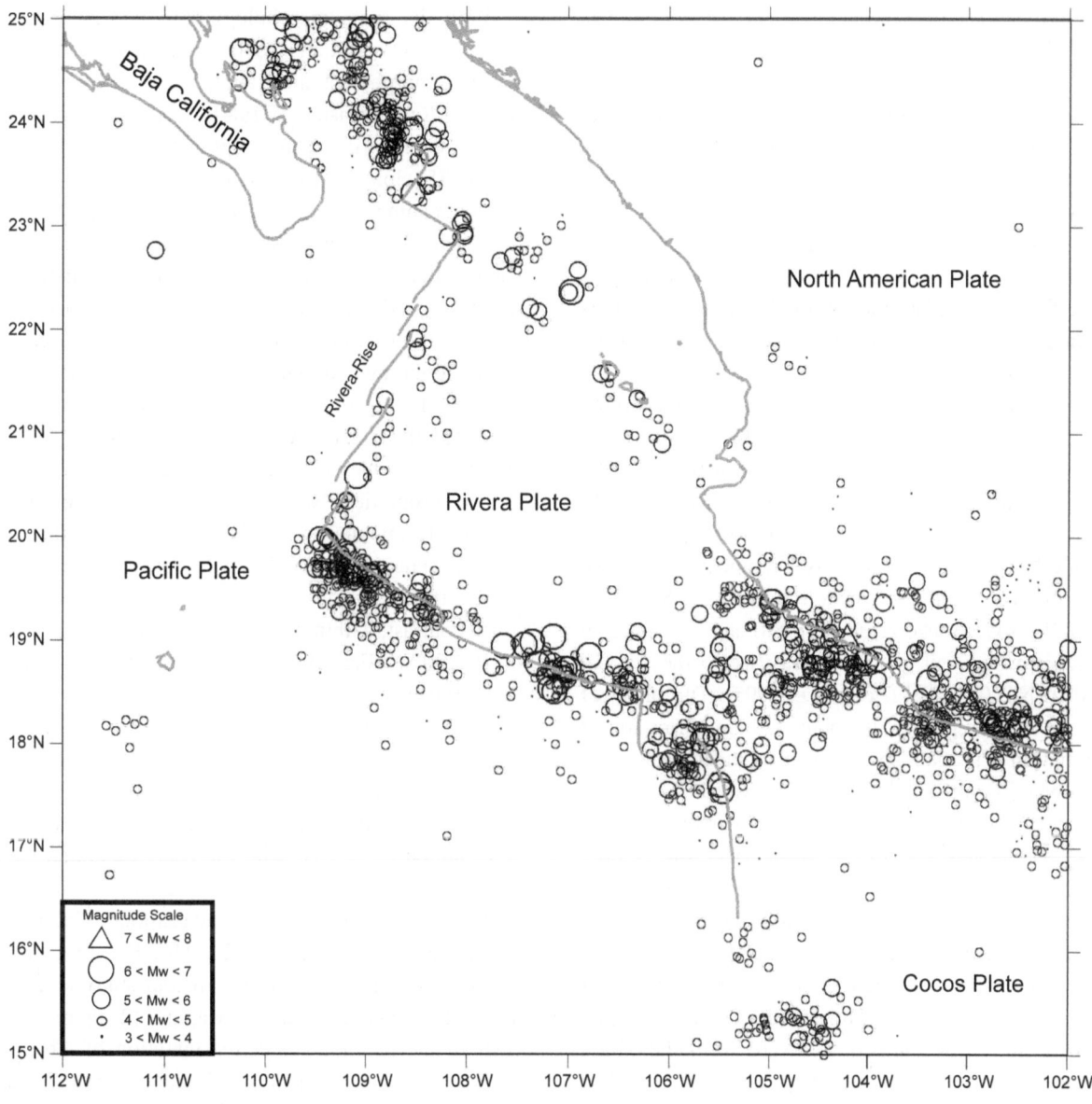

Figure 3

Earthquake epicenter locations. Note that the number of events along the Rivera Transform is greater west of the ITSC, whereas larger magnitude events (Mw >6.0) occur more frequently to the east. Again, note the scarcity of events just east of the intra-transform spreading center. The earthquake locations were obtained from the NEIC catalogue and cover the time period from 1 January 1976 until 1 December 2009

may be the result of a slow adjustment to very recent (<0.15 m.y.b.p.) changes in Rivera-Pacific plate motion. Such discordances by slow reorganization of individual transform segments to changing plate motion have been proposed elsewhere; for example, along the Blanco Transform where its western segment appears to be presently undergoing reorganization (BRAUNMILLER and NÁBĚLEK, 2008).

2. Tectonic Setting

The existence and geometry of the Rivera Plate has been debated since ATWATER (1970) first proposed its existence. Plate motion studies (e.g. REID, 1976; DEMETS and STEIN, 1990; BANDY 1992; LONSDALE, 1995) support the existence of the Rivera Plate. However, several studies (e.g. LARSON, 1972; MICHAUD and BOURGOIS, 1995; MICHAUD *et al.*, 1997) propose that the Rivera Plate is not an independent plate, but is instead part of the North American Plate. Their reasoning for this is twofold. First, as previously mapped, the bathymetric trend of the western part of the Rivera Transform (which forms the southern boundary of the Rivera Plate) is better fit by Pacific-North America motion, whereas the eastern part of the transform is not well-fit by this motion. Second, none of the Rivera-Pacific relative plate motion models predict motions that adequately fit the bathymetric orientation of the entire Rivera Transform.

The Rivera Transform (Fig. 4) is a recently formed (1.5–3.5 Ma) (e.g., MAMMERICKX and KLITGORD, 1982; LONSDALE, 1991; BANDY, 1992; LONSDALE, 1995; BANDY and HILDE, 2000; DeMETS and TRAYLEN, 2000), left stepping, fast slipping (∼70 mm/year), transform fault that presently connects two segments of the East Pacific Rise (EPR); the MSS to the east at 18°30.8′N, 106°16.5′W (BOURGOIS *et al.*, 1988) and the Southern Elenerth Segment of the Rivera Rise to the west at 20°03′N, 109°25′W (LONSDALE, 1995). The offset of these ridge segments is about 370 km. Prior to the initiation of seafloor spreading along the MSS at about 0.7 Ma (BANDY *et al.*, 2008), the Rivera Transform connected with a now inactive ridge segment (RRS on Fig. 1) located east of the MSS, near the Middle America Trench (BOURGOIS *et al.*, 1988; BANDY, 1992; BANDY *et al.*, 2008). Thus, seafloor spreading underwent a westward relocation at about 0.7 Ma, similar to the relocation that occurred at about 3.5 Ma near the Middle America Trench off Puerto Vallarta, where spreading jumped westward from the Maria Magdalena Rise to the presently active Rivera Rise (MAMMERICKX, 1980; LONSDALE, 1991, 1995). These relocations of seafloor spreading centers may somehow be related to the approach of the spreading center to the Middle America Trench. However, in recent 2-D numerical models (BURKETT and BILLEN, 2009) such an oceanward relocation of a spreading center as it approached a subduction zone was not observed. Regardless of its cause, this relocation affected the geometry of the Rivera Transform by reducing the ridge offset distance, which has been demonstrated to be a very important parameter controlling the morphology of fracture zones (MAUDUIT and DAUTEUIL, 1996; LIGI *et al.*, 2002).

From its gross morphology (Fig. 4), the Rivera Transform was initially proposed to consist of two distinct morphotectonic domains (LARSON, 1972; REID, 1976; MAMMERICKX, 1980). In the first domain, located west of 107°30′W, the transform is marked by a NW–SE elongate basin, containing an intra-transform spreading center. According to REID (1976), the basin was the result of oblique motion between the Pacific and Rivera plates; i.e., a 'leaky transform' (MENARD and ATWATER, 1969). The extensional component of leaky transforms has been proposed to be due to thermal contraction (TURCOTTE, 1974), plate motions slightly oblique to the trend of the fracture zone (MENARD and ATWATER, 1969; REID, 1976; GARFUNKEL, 1986) or due to mechanical and rheological properties, such as ridge offset distance, lithosphere thickness and/or the thermal structure of the lithosphere (MAUDUIT and DAUTEUIL, 1996; LIGI *et al.*, 2002).

In contrast, within the second domain, located east of 107°30′W (Fig. 4), the basin disappears and the transform was originally mapped as consisting of a series of discontinuous ridges. BANDY *et al.* (2008) remapped the entire transform tectonized zone (the transform nomenclature used is that of MACDONALD *et al.* (1986) who define this zone as "the wide zone of deformation associated with the time-integrated tectonics of the transform fault") in the second domain using multi-beam bathymetric and seafloor reflectivity data collected during the 2002 BART/FAMEX campaigns. Several notable results of that study are: (1) the MSS formed at about 0.7 Ma, and (2) there is no clear evidence for a major reorientation of Rivera-Pacific relative motion in this domain since 0.7 Ma, with the possible exception of the area immediately adjacent to the MSS where the principal transform displacement zone (PTDZ: the active trace of the transform fault) is observed to climb up the northern wall of the transform tectonized zone as it approaches the MSS. Normally, the PTDZ is expected to be located at the base of the transform wall (WILSON and DeMETS, 1998), thus, this observation may indicate a small amount of recent, counter-clockwise reorientation of Rivera-Pacific relative motion near the MSS.

The part of the EPR located between the Rivera and Tamayo transforms has also been referred to as the Pacific-Rivera Spreading Center (LONSDALE, 1991)

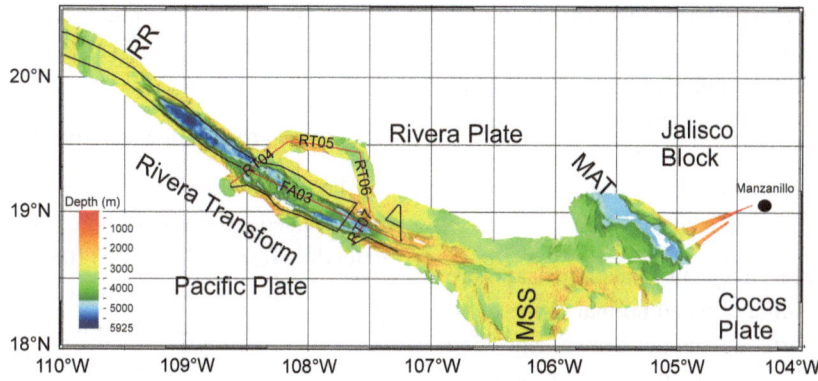

Figure 4

Ship tracks (*black and red lines*) of the BART/FAMEX campaign superimposed on a 3D bathymetric relief map of the Rivera Transform. Ship tracks are not shown east of 107.5°W as the data have been previously presented (BANDY *et al.*, 2008). Labeled, red, ship tracks mark those along which multichannel seismic reflection data were collected. The bathymetric relief map was constructed solely from multi-beam data collected during the FAMEX and BART campaigns of the NO L'Atalante. Grid spacing is 200 × 200 m. See *inset* for color scale. Note the discordance in the trend of the transform at about 108.5°W. MSS, Moctezuma Spreading Segment; MAT, Middle America Trench; RR, Rivera Rise

or Rivera Rise (DAUPHIN and NESS, 1991). At about 2 Ma, the axis of the Rivera Rise began to reorient clockwise and abruptly broke into seven short ridge segments (Fig. 1) (e.g. KLITGORD and MAMMERICKX, 1982; MAMMERICKX and KLITGORD, 1982; LONSDALE, 1995). Although some debate exists (WILSON and DEMETS, 1998; BANDY *et al.*, 1998a), LONSDALE (1995) showed that this rotation and segmentation is continuing at present as evidenced (1) by the difference between the strike of the axes of the Rivera Rise segments and the strike of the Jaramillo magnetic lineation to either side of the rise, (2) by the recent segmentation of the Elenerth segment into northern and southern segments and (3) by a recent plate reorganization during which the northwest corner of the Rivera plate appears to have been attached to the North American plate. The kinematics of this reorganization has been described by a southwest migration of the Rivera-Pacific Euler pole since at least Chron 2A (BANDY and YAN, 1989; DEMETS and STEIN, 1990; BANDY, 1992; BANDY *et al.*, 1998b; DEMETS and TRAYLEN, 2000).

3. Data and Methods

The data used in this study consist of previously unpublished multi-beam bathymetric, seafloor backscatter, multi-channel seismic reflection and total field magnetic data collected in April and May of 2002 during the FAMEX and BART campaigns of the N/O L'Atalante. The ship tracks along which the previously unpublished data were collected are illustrated in Fig. 4.

The multi-beam bathymetric and seafloor backscatter data were collected with a dual SIMRAD EM-12 multi-beam system and the raw data were processed using the CARAIBES software while onboard by IFREMER technicians to produce a 200 × 200 m grid of bathymetric values. The backscatter data were processed to produce a 50 × 50 m grid of seafloor backscatter amplitudes.

The 2-D seismic reflection profiles (see Fig. 4 for profile locations) collected during the campaigns are three-fold data, acquired employing 300 in^3 GI guns tuned in harmonic mode and a hydrophone streamer with six hydrophone groups (48 hydrophones per group) spaced 50 m apart. The spacing between stacked traces is 25 m (note that on the seismic sections, the spacing between CMP locations is 50 m). The data were sampled at 4 ms and recorded in SEG-Y format.

The data were processed using the following processing sequence:

1. Geometry assignment
2. Spherical divergence correction

3. 10–70 Hz bandpass filter with a high and low rolloff rate of 18 dB/octave
4. NMO correction (assuming a water velocity of 1,500 m/s and a sediment velocity of 2,000 m/s)
5. Stack
6. Gazdag phase-shift migration using a constant velocity of 1,500 m/s (GAZDAG, 1978).

Total field, marine magnetic lineation data were projected along a profile oriented normal to the trend of the intra-transform spreading center and then modeled using the 2-D magnetic modeling program coded by J.L. LaBrecque, which is based on the fast Fourier transform method of SCHOUTEN and McCAMY (1972). The time scale used is that of CANDE and KENT (1995). In all models the magnetized layer is assumed to be 500 m thick, having a time invariant strength of magnetization equal to 10 A/m. The depth in meters below the sea surface to the top of the magnetized layer, D, is assumed to follow the relationship $D = 2,500 + 350\sqrt{t}$, where t is the age of the crust in millions of years. A transition zone 1.0 km wide is assumed between each reversal. Present day inclinations and declinations are obtained from MERRILL and McELHINNY (1983). The remnant inclination is assumed to be identical to the present day inclination, and the remnant declination is assumed to be zero.

4. Morphology

The new data (Fig. 4) indicate that although the area west of 107°30′W is a continuous basin, it is not a single morphotectonic domain as previously thought; instead, it consists of two very distinct domains. Morphologically, the two domains are readily delineated by the marked differences in their depth and strike. The first domain, henceforth referred to as the 'Intra-Transform Spreading Basin' (ITSB), is situated east of ∼108°40′W. This basin has an overall strike of ∼115° and contains the intra-transform spreading center, analogous to the Cascadia Depression within the Blanco Transform (DeCHARON, 1989; EMBLEY and WILSON, 1992). The second morphotectonic domain, henceforth referred to as the 'Leaky Transform Basin' or LTB, is located west of ∼108°40′W and overall is deeper than the ITSB. The strike of the escarpment forming the southern wall of LTB is 122°.

4.1. Intra-Transform Spreading Basin

The ITSB (Fig. 5) is a NW–SE elongated, rhomboid-shaped depression approximately 115 km long extending from 107°18′W to 108°40′W, and it exhibits a fairly constant width of between 18 to 20 km. With the exception of the NW corner, the ITSB is bounded by steep, large offset escarpments (or 'walls') which mark the outer limits of the transform-tectonized zone (TTZ). The gross geometry of the ITSB is similar to large extensional basins or rhombochasms that are formed in major strike-slip systems within continental lithosphere (e.g., MANN et al., 1983; SYLVESTER, 1988; ROBERTS and YIELDING, 1994), albeit the ITSB has formed entirely within oceanic lithosphere.

The southwest limb of the rhombochasm of the ITSB extends from 107°18′W to 108°31′W (Fig. 5). Between 107°40′W and 107°55′W, the southern wall along this limb strikes ∼115° and is very steep, straight and continuous. At 107°55′W the southern wall changes orientation to 120° and maintains this orientation until 108°31′W. Between 107°57′W and 108°31′W a narrow elongate bathymetric trough (referred to as 'Principal Transform Displacement Zone 1 East' or PTDZ-1E) is observed at the base of the southern wall. The low amplitude (Fig. 6) of the backscatter strength of the seafloor in this trough indicates that it is probably sediment filled. Between 108°W and 108°20′W a series of wavy, discontinuous faults are observed (Fig. 6) just north of PTDZ-1E. This zone of faulting (collectively referred to as PTDZ-2E) exhibits an overall strike of 125°, with the strikes of individual faults ranging from 117° to 137°; approximately 10° clockwise of PTDZ-1E. PTDZ-2E merges with PTDZ-1E at 108°W.

The NW limb of the rhombochasm extends from 108°29′W to 108°43′W and strikes at an azimuth of 145°. Here, the southern wall of the ITSB consists of a series of down-to-the-basin normal faults and becomes progressively less steep to the NW. A broad, sediment filled trough (the 'western' trough) lies at the base of the wall (Figs. 5, 6). Seismic reflection data (Figs. 7, 8a) indicate that this trough is filled by at least 0.5 s (two-way-travel-time, or TWTT) of

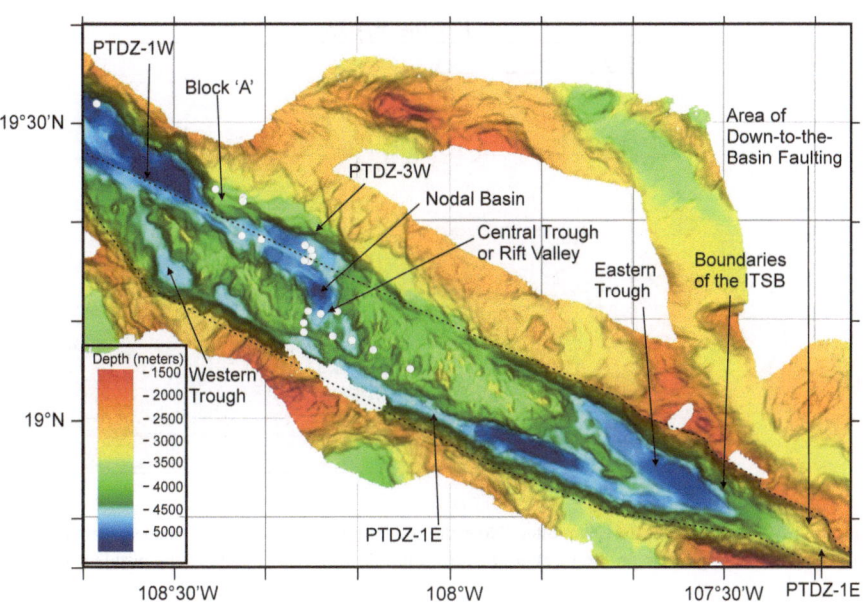

Figure 5
Bathymetric relief map of the ITSB. White filled circles are the epicenters of the earthquakes recorded in the OBS study of REID (1976), PROTHERO *et al.* (1976) and PROTHERO and REID (1982). *Dashed outline* marks the boundary of the ITSB. ITSC, Intra-transform spreading center

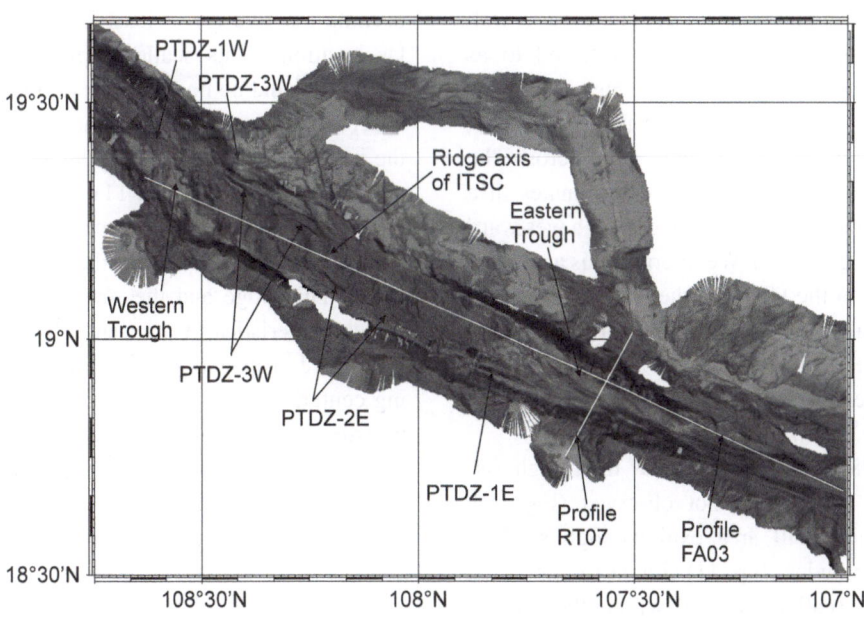

Figure 6
Seafloor backscatter mosaic image of the ITSB. *Light and dark areas* represent low and high backscatter amplitudes, respectively. *White lines* mark the location of seismic reflection profiles, FA03 and RT07, presented in this study. ITSC, Intra-Transform Spreading Center

sediments, which corresponds to 500 m of sediments assuming that the velocity of sound in the sediments is 2 km/s.

The NE limb of the rhombochasm (Fig. 5) lies between 108°43′W and 107°45′W. Between 108°43′W and 108°05′W the NE limb of the rhombochasm does

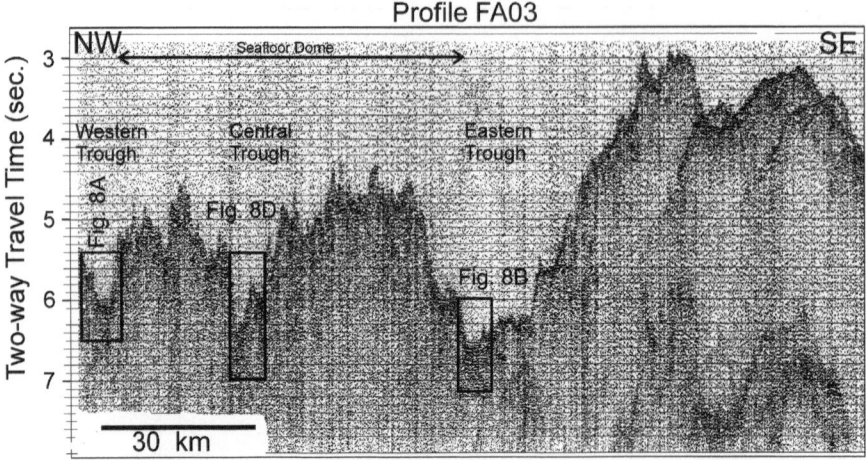

Figure 7
Seismic reflection profile FA03. Every 10th trace is plotted. Note the rough, unsedimented, dome-shaped seafloor between the eastern and western troughs. See Fig. 6 for profile location. *Boxes* on profile mark those parts of the profile shown in Fig. 8

not coincide with either the northern or the southern wall. Instead, it is marked by a narrow bathymetric trough oriented at an azimuth of 115°. Seafloor reflectivity data indicate that this trough is a zone of highly sheared seafloor (Figs. 6, 9). Thus, it is a well developed PTDZ and is henceforth referred to as PTDZ-1W. Westward, this trough (Fig. 4) merges with the base of the southern wall at 19°28′N, 108°40′W. Eastward, the trough cuts across the depression until 108°12′W where it terminates in a deep trough which is flanked to the east by a westward curving ridge (Figs. 5, 10). We interpret this area to be the intersection of the PTDZ of the Rivera Transform with the rift valley of the intra-transform spreading center, and the deep basin to be the nodal basin commonly observed at ridge-transform intersections (e.g. MACDONALD *et al.*, 1986).

Between 108°12′W and 108°05′W the trough associated with PTDZ-1W is not observed (Figs. 10, 11); instead this small area exhibits high seafloor backscatter amplitudes (Fig. 11) similar to that of the majority of the seafloor located within the rhombochasm (Fig. 11). A trough, which we associate with the eastern extension of PTDZ-1W, appears east of 108°05′W, albeit at much shallower depths. This trough also exhibits a sheared seafloor like that observed within PTDZ-1W. Given both the high backscatter strength and the separation of PTDZ-1W with its eastern extension, we interpret that the area

between 108°12′W and 108°07′W (labeled 'New Crust' on Fig. 11) is an additional area of seafloor spreading within the transform tectonized zone, the spreading segment being initiated after PTDZ-1W became inactive (or at least inactive at this particular location). This seafloor spreading lies north of PTDZ-1W and no nodal basins or transform faults are observed. Thus, it is unclear how, or if, it links with the intra-transform spreading center to the west. One particular ridge (Figs. 10, 11) of the spreading fabric (abyssal hill) in this area appears to be continuous with an abyssal hill within the rhombochasm to the south. This ridge and the surrounding seafloor have the characteristics of a recent, short lived period of northward propagation of the intra-transform spreading center. Since no obvious connection to the intra-transform spreading center to the west is observed, this spreading segment appears to be presently inactive. A narrow elongate bathymetric trough is observed at the base of the northern wall east of 108°05′W.

The SE limb of the rhombochasm (Fig. 5) lies between 107°45′W and 107°18′W, the point at which the northern wall begins to become progressively less steep. Here, the top of the northern wall has a zig-zag pattern with an overall strike of 120°; the base of the wall strikes 130°. Between 107°18′W and 107°35′W, the northern wall is comprised of a series of down-to-the-basin normal faults typical of pull-apart

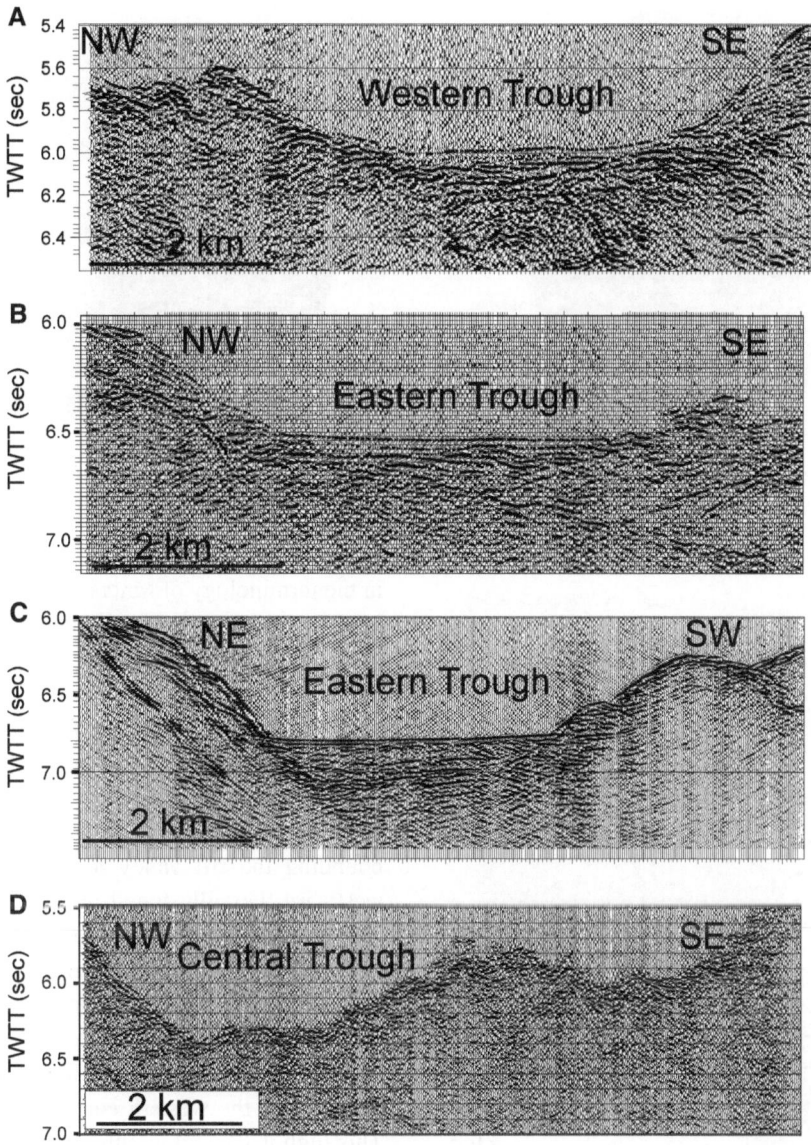

Figure 8

Selected areas of seismic reflection profiles. **a** Portion of profile FA03 illustrating the sediment infilling the basin located at the base of the NW limb of the rhombochasm. **b** Portion of profile FA03 illustrating the sediment infilling the basin located at the base of the SE limb of the rhombochasm. **c** Southern part of profile RT07 illustrating the sediment infilling the basin located at the base of the SE limb of the rhombochasm. **d** Portion of profile FA03 illustrating the Intra-transform spreading center and the lack of sedimentary infill within the central trough. See Figs. 6 and 7 for profile locations

systems. The easternmost of these faults strikes nearly north–south. The orientation of the faults progressively rotates counterclockwise, obtaining a strike of N40°W for the westernmost fault. Like the NW limb of the rhombochasm, a broad sediment filled trough (the 'eastern' trough) lies at the base of the wall along this limb. Seismic reflection data (Fig. 8b, c) indicate that this basin is filled by at least 0.5 s (TWTT) of sediments, which corresponds to ~500 m of sediments. The similar amount of sediments contained within the eastern and western troughs suggests that they formed at roughly the same time. This, as well as

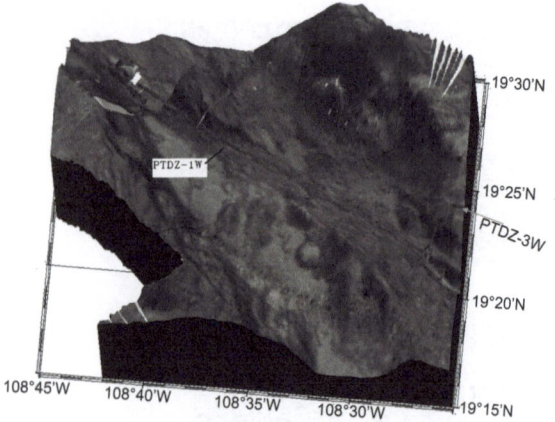

Figure 9
Seafloor backscatter image draped on 3-D bathymetry illustrating the sheared seafloor within PTDZ-1W along the NE limb of the ITSB

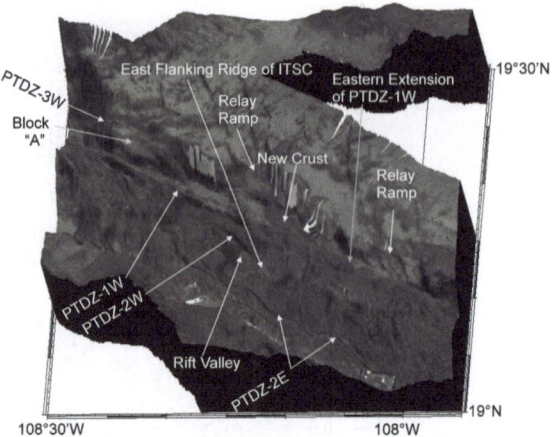

Figure 11
Seafloor backscatter image draped on 3-D bathymetry in the area of the ITSC

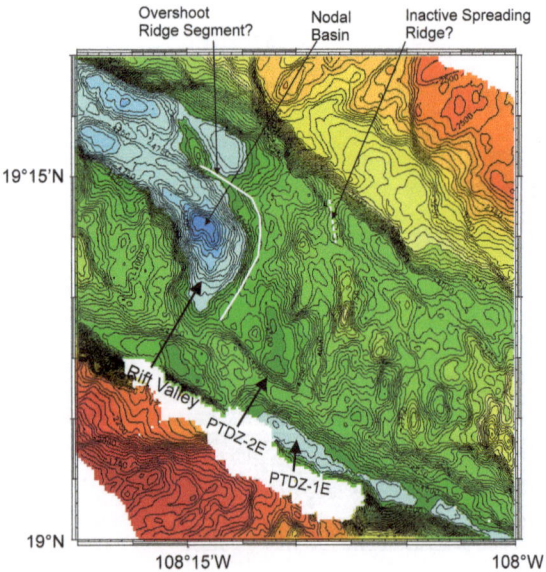

Figure 10
Bathymetric contour map of the ITSC. Contour interval, 50 m

their similar orientations that are oblique to the orientation of the PTDZs, is consistent with their formation by a pull-apart mechanism.

The seafloor within the rhombochasm is domed-shaped and exhibits high backscatter strength (Figs. 5, 6). No significant sediments are observed over the dome in the seismic reflection data (Fig. 7). This dome is broken between 108°10'W and 108°15'W by a deep, sediment free (Fig. 8d) trough which we interpret to be the rift valley, or the neovolcanic zone

in the terminology of MACDONALD *et al.* (1986), of the intra-transform spreading center. This fits fairly well with the offset in the seismicity (Fig. 5) recorded during the OBS deployment of REID (1976). The adjacent seafloor spreading fabric is also consistent with the interpretation of the existence of an intra-transform spreading center, as the fabric is aligned roughly parallel to this ridge. At 19°13'N, the ridge bounding the rift valley to the east bends sharply westward (best illustrated in Figs. 10, 11) and overshoots the nodal basin as is commonly observed at the junctions of spreading centers and transforms. Such an overshoot ridge is observed at the intersection of the Rivera Transform and the Rivera Rise (MICHAUD *et al.*, 1990). A bathymetric high, striking N27°W is observed at the western end of the overshoot ridge. This high is flanked to either side by small, sedimentary filled depressions (i.e. depressions with low seafloor backscatter amplitude). Although this ridge appears to be connected to the overshoot ridge, and thus to the active spreading center, its character on the seafloor backscatter data is not that expected for an active spreading center. We suspect that it is a piece of old seafloor formed prior to the development of the transform valley.

A NW–SE lineament cuts across a small block (labeled 'Block A' in Figs. 5, 11) and extends southeastward along the northern wall, north of PTDZ-1W. This lineament is aligned with the faults forming the northern wall to the east and the block is presently down-dropped towards the basin. Seismicity was recorded in this area during the OBS deployment of REID (1976), which indicates that this lineament is presently active. This lineament, PTDZ-3W, extends eastward only to the area of the high reflectivity seafloor noted previously at 108°05′W.

The seafloor north of the intra-transform spreading center, out of the transform tectonized zone, is disrupted by a series of 130°–155° striking faults that intersect the northern wall west of 107°56′W (Fig. 11). This faulting produces discontinuities in the northern wall between 107°56′W and 108°04′W and between 108°11′W and 108°15′W. At these locations the northern wall is replaced by relay ramps or transfer zones (ROBERTS and YIELDING, 1994; DAVISON, 1994), which plunge into the depression, thus indicating that the faulting is more recent than the northern wall. Between these two ramps the northern wall is again marked by a prominent escarpment, however, the escarpment strikes ∼ 131°, roughly parallel to the NW–SE trending faults. It seems likely that this disruption reflects an attempt of the spreading center to propagate further northward. No seismicity was recorded in this area during the OBS deployment of REID (1976) which, at face value, indicates that this deformation is presently inactive.

4.2. Leaky Transform Basin (LTB)

The LTB (Figs. 12, 13) is a NW–SE elongated, ellipsoidal structure, roughly 100 km long, 18 km wide, extending from 109°13′W to 108°25′W (or to 108°05′W if one chooses to include in this domain all areas of the depression located north of PTDZ-1W). Like the ITSB, the LTB is bounded by steep, large offset (as much as 3 km) escarpments. Overall, the depth of the LTB is greater than the ITSB, reaching depths of ∼6 km. The southern wall of the LTB strikes 127° between 108°45′W and 109°10′W. West of 109°10′W the southern wall strikes 140°. The LTB's northern wall has an overall strike of 127° between 108°30′W and 108°50′W, parallel to its southern wall in this area. At 108°50′W the northern wall is broken by a more gently dipping ramp leading down into the LTB. On the west side of the ramp the northern wall strikes 115°.

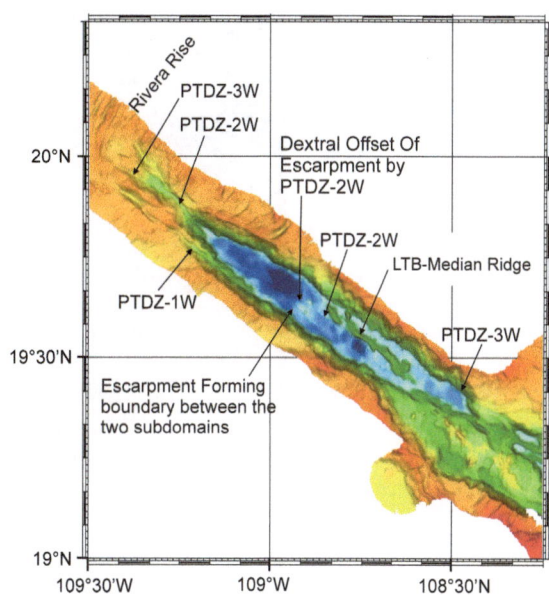

Figure 12
Bathymetric relief map of the LTB. Values range from 1,800 m (*red*) to 5,900 m (*dark blue*)

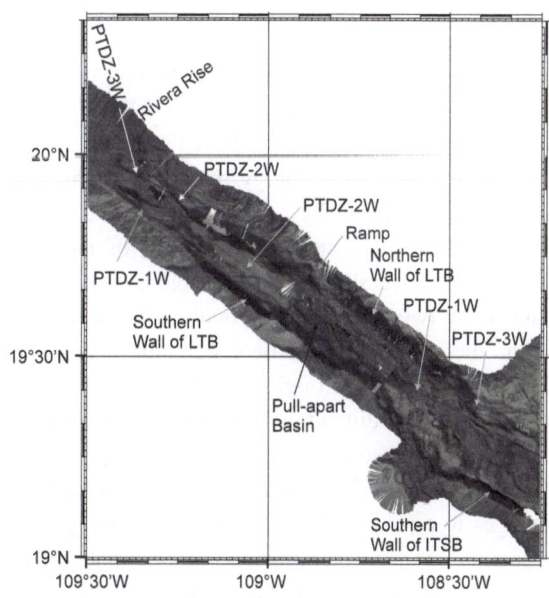

Figure 13
Seafloor backscatter image mosaic of the LTB. *Light and dark areas* represent low and high backscatter amplitudes, respectively

The LTB can be further divided into a western and eastern sub-domain based on the differences in their depths and seafloor reflectivity. The two

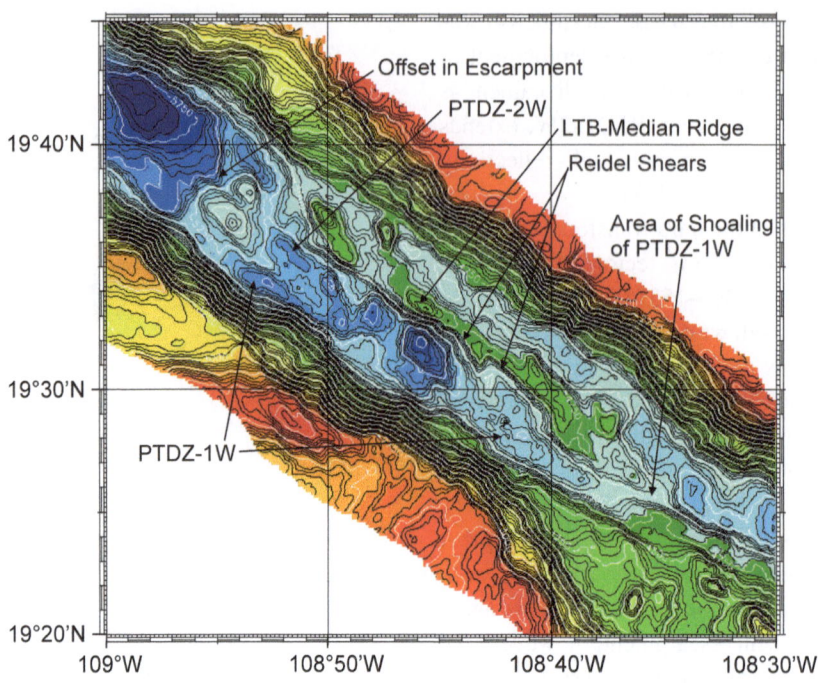

Figure 14
Bathymetric contour map of the eastern sub-domain of the LTB. Contour interval, 50 m

sub-domains are separated by a prominent, down-to-the-NW escarpment located at 108°57′W (Fig. 12). The relief on this escarpment is ∼1 km.

The seafloor morphology in the eastern sub-domain is dominated by a narrow ridge (the "LTB-Median Ridge") located in the center of the LTB (Fig. 12): the overall trend of this ridge is ∼122°. It is flanked by two troughs, the trough to the southwest being the deeper of the two. Seafloor backscatter (Fig. 13) within both troughs is high, suggesting either that they lack a sediment infill or that the sediments are highly disrupted so as to produce a rough seafloor. At its western end the character of the LTB-Median Ridge changes to a pair of short, NNW–SSE striking ridge segments. A close examination of the bathymetry (Fig. 14) and seafloor backscatter image (Fig. 15) reveals that the LTB-Median Ridge consists of a series of en-echelon segments, striking ∼130° which are collectively referred to as PTDZ-2W. These segments have the appearance of Riedel shears associated with dextral strike-slip motion. The westernmost of these lineaments does not coincide with the LTB-Median Ridge. Instead, the lineament lies to the south and, at 19°38.5′N, 108°55.5′W, it right-laterally offsets

by ∼2.5 km the prominent escarpment separating the two sub-domains.

PTDZ-1W runs along the base of the southern wall of the LTB until 108°57′W (the boundary between the two morphotectonic sub-domains). This is best observed in the bathymetry as PTDZ-1W and the southern wall are blurred on the backscatter image (Fig. 15). Westward from 108°57′W, PTDZ-1W climbs up and out of the LTB, terminating south of the Rivera Rise at 19°55′W, 109°26′W (Figs. 12, 13). Between 108°40′W and 108°57′W, PTDZ-1W strikes at an azimuth of 122° and the seafloor fabric (Fig. 15) between PTDZ-1W and PTDZ-2W is characteristic of a pull-apart basin indicating that, during at least one time period, PTDZ-1W and PTDZ-2W were both active in this area. As pointed out by one of the reviewers of this manuscript, an interesting disruption of the bathymetric trough associated with PTDZ-1W occurs at 19°25′N, 108°36′W in the area where PTDZ-1W and PTDZ-2W converge (Fig. 14). Here, the trough associated with PTDZ-1W shoals due to the prolongation of the LTB-Median Ridge into the trough, and the trough exhibits a left-step. This left step in the trough might be taken to indicate a

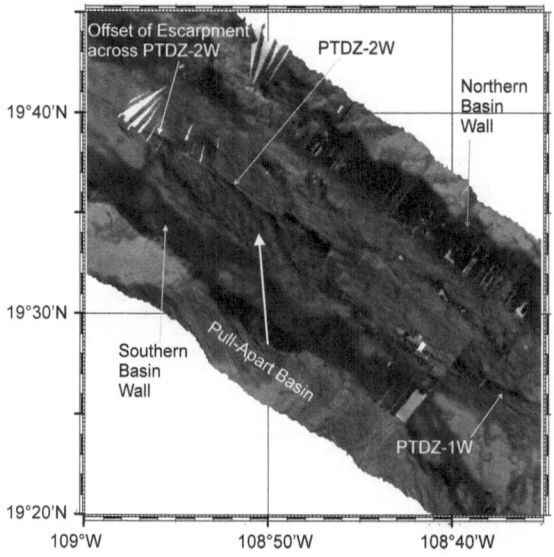

Figure 15
Seafloor backscatter image draped on 3-D bathymetry in the eastern sub-domain of the LTB illustrating the pull-apart zone located between PTDZ-1W and PTDZ-2W. Unfortunately, the southern basin wall and PTDZ-1W are blurred because the angle between ship and the steep southern wall was too small to properly record the backscatter strength. Also note the polarity reversal in parts of the image, for example, at 19°27′N, 108°40′W

left-stepping offset in PTDZ-1W; however, this is clearly not the case. A close examination of the reflectivity data (Figs. 9, 13) indicates that the zone of

shearing associated with PTDZ-1W cuts across this zone of shoaling and is straight, exhibiting no noticeable offsets. This observation is important in that it clearly indicates that the trough and sheared seafloor of PTDZ-1W formed first and were later uplifted during the formation of the LTB-Median Ridge associated with PTDZ-2W.

The seafloor in the western sub-domain lies deeper than that of the eastern-sub-domain. The seafloor within this sub-domain is, for the most part, smooth and featureless and is marked by low backscatter amplitudes (Fig. 16). However, a prominent NW–SE striking high amplitude lineament is observed on the seafloor backscatter image. This lineament extends westward from the point at which the escarpment separating the two sub-domains is offset right-laterally (and is thus the western continuation of PTDZ-2W) across the basin to 19°50′N, 109°12.6′W, where it merges with the narrow trough that extends up and out of the LTB. This trough strikes at an azimuth of 139° and extends to the Rivera Rise (Fig. 12).

4.3. Rivera Transform-Rivera Rise Junction

The morphology (Fig. 17) of the intersection between the Rivera Transform and Rivera Rise is similar in many respects to other ridge-transform

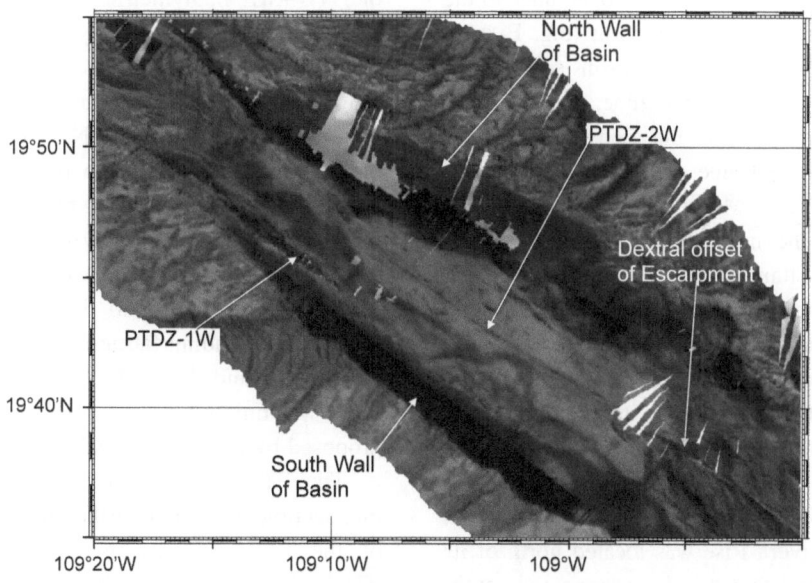

Figure 16
Seafloor backscatter image draped on 3-D bathymetry in the western sub-domain of the LTB illustrating the trace of PTDZ-2W across the sediment basin

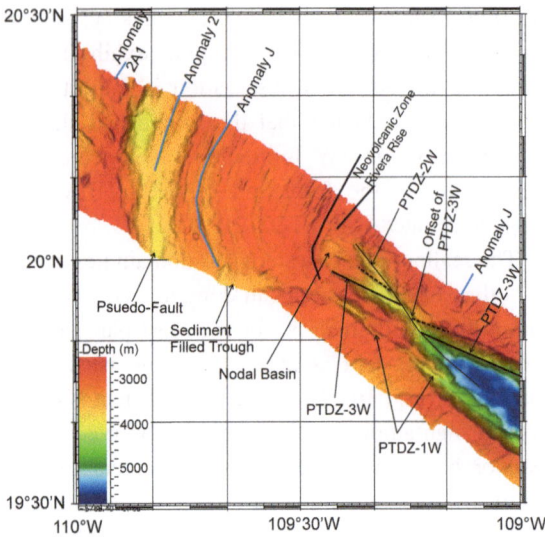

Figure 17

Bathymetric contour map at the intersection of the Rivera Transform and the Rivera Rise. Note the offset in PTDZ-3W across PTDZ-2W indicating that PTDZ-2W is the younger of the two structures. *Solid line* for PTDZ-3W assumes that this PTDZ lies at the base of the northern wall of the LTB east of the offset, and that west of the offset it lies within the trough which intersects the nodal basin to the west. We favor this interpretation. However, another possibility (*dashed lines*) is that PTDZ-3W is located at the top of the northern wall of the LTB east of the offset, and that west of the offset it corresponds to a narrow ridge which also intersects the nodal basin to the west. Both interpretations indicate that PTDZ-3W is offset by 5–7 km by PTDZ-2W

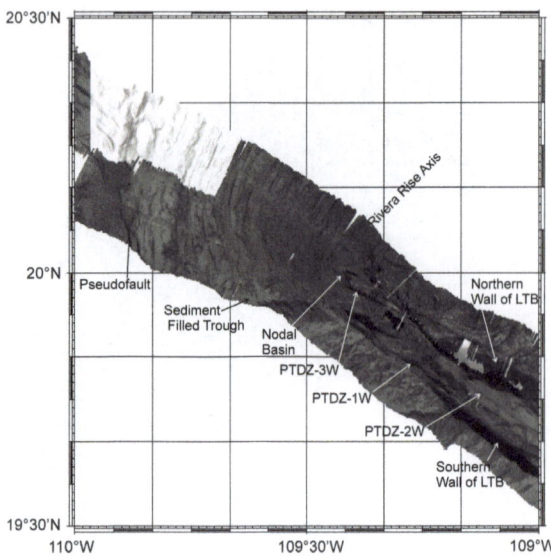

Figure 18

Seafloor backscatter image draped on 3-D bathymetry at the intersection of the Rivera Transform and the Rivera Rise

intersections, and is quite similar in appearance to the western end of the Blanco Transform (e.g. EMBLEY and WILSON, 1992). A deep nodal basin is present at the terminus of the neovolcanic zone, and the ridge forming the west wall of the neovolcanic zone extends along the west side of the nodal basin. As previously noted (e.g. MICHAUD *et al.*, 1990), this ridge overshoots the nodal basin, similarly to that observed in the Eltanin Fracture Zone (LONSDALE, 1986), and curves eastward and terminates against an east–west oriented, sediment filled trough (Fig. 18). The seafloor spreading fabric formed at the Rivera Rise terminates at this trough, so it appears to mark the boundary separating lithosphere of two different ages.

LONSDALE (1995) proposed that the southern terminus of the Rivera Rise was located north of its present location and that the rise propagated southward in response to the formation of the Rivera Transform and the abandonment of spreading along

the northern Mathematician ridge segment (Fig. 1). The new bathymetry, along with the magnetic lineations located west of the Rivera Rise, are consistent with his proposal. The pseudofault formed by this propagation is observed west of the Rivera Rise at 109°50'W (Fig. 17). Magnetic Anomaly $2A_1$ lies NW of the pseudofault and terminates against an area of NW–SE bathymetric lineations at 20°20'N, 109°55'W. This most likely marks the location of the southern terminus of the Rivera Rise at 2.15 Ma. Magnetic anomaly 2 is observed to terminate against this pseudofault at 20°10'N, 109°50'W indicating the Rivera Rise was in the process of propagating southward at 1.8 Ma. Thus, the southward propagation began sometime between Anomaly $2A_1$ and Anomaly 2 or at ~2 Ma. Anomaly J extends to the sediment filled trough to the south, as does, most likely, the spreading fabric roughly midway between Anomaly 2 and Anomaly J. Thus, this propagation event was finished by ~1.4 Ma, consistent with that proposed by LONSDALE (1995).

The PTDZs noted east of the ridge transform intersection continue into the area of the intersection of the Rivera Transform with the Rivera Rise. PTDZ-1W terminates ~10 km south of the nodal basin, at the eastern end of the sediment filled trough. Interestingly, these two features are not aligned and therefore the sediment filled trough does not appear to be the fracture zone associated with PTDZ-1W. Two troughs which terminate at the nodal basin are interpreted to correspond to PTDZ-3W. The trough associated with PTDZ-2W terminates north of the nodal basin. PTDZ-2W right-laterally offsets PTDZ-3W by between 5 to 7 km indicating that PTDZ-2W is younger.

5. Discussion

5.1. The Principal Transform Displacement Zones: Relative Ages

Presently, the location resolution of the seismicity occurring along the Rivera Transform is inadequate to determine conclusively which of the PTDZs are presently active. BANDY et al. (2008) noted, for the eastern part of the Rivera Transform near 107°W, a 20–30 km northward bias in the location of teleseismically recorded earthquakes in that area consistent with the bias noted by SINGH and LERMO (1985) for large events occurring along the Middle America Trench. Given the small separation between the PTDZs in the study area (Fig. 19), the resolution of event locations needs to be less than 2 or 3 km to be useful in determining the active PTDZ(s). Thus,

installing a local OBS network along the transform is desirable.

The slip vectors of large events (Fig. 2) might be considered useful in determining which of the PTDZ(s) are currently active, although large uncertainties of ±10° to ±15° are commonly assigned to these slip vectors (e.g. GORDON, 1995). Between 107°30'W and 108°W (the area where only PTDZ-1E is observed) only four slip vectors have been determined, three of which show a slip vector azimuth of ~109°, 4° counterclockwise from that of PTDZ-1E (115°). Since PTDZ-1E is the only observed PTDZ in this area, this difference might indicate a counterclockwise bias in the slip vector azimuths. However, a counterclockwise bias is contrary to that expected in right-slipping transforms where slip vectors tend to be a few degrees clockwise of the transform azimuth (GORDON, 1995). Another possible explanation for this difference is that, given that event locations derived from teleseismic records have an uncertainty of 30–40 km, these events might instead occur east of the ITSB. The slip vectors are consistent with the strike of the transform in that area (BANDY et al., 2008). No slip vectors have been determined in the area between 108°W and the intra-transform spreading center, i.e. in the area where PTDZ-2E is observed. Thus, we cannot use slip vectors to determine if PTDZ-2E is presently active. Between the intra-transform spreading center

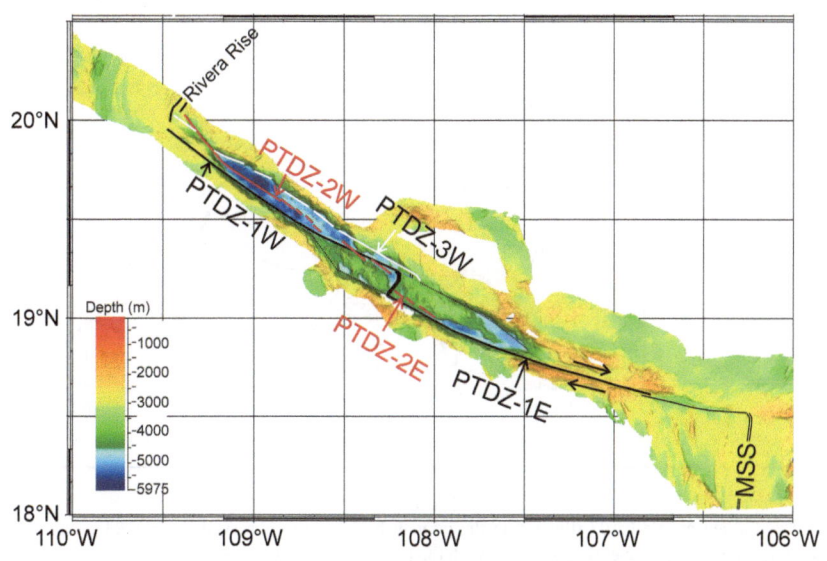

Figure 19
Summary of the PTDZs of the Rivera Transform

and 109°W, slip vectors range from 118°–130°. The trend of PTDZ-2W (130°) and PTDZ-3W (120°) lie within this range, whereas the range in slip vectors is slightly clockwise of the trend of PTDZ-1W (115°). However, given the large uncertainty commonly assigned to slip vectors, one cannot rule out the possibility that PTDZ-1W is presently active. Between 109°W and the Rivera Rise there is a large scatter in the slip vectors (110° and 155°). Since the trends of all three PTDZs lie within this range, the large scatter of the slip vectors near the Rivera Rise might indicate activity on multiple PTDZs. However, it might also be caused by a complex lithospheric structure near the Rivera Rise. In summary, in the study area, slip vectors by themselves cannot be used to distinguish which of the PTDZs are presently active.

The relative ages of the PTDZs can be gleaned from their morphologies and structures. Figure 19 summarizes the major morphotectonic elements of the Rivera Transform. The PTDZ of the Rivera Transform (PTDZ-1E) enters from the east at 107°30′W, south of the sediment filled trough, and runs along the base of the southern wall until 108°W where it splits into two trends: the first (PTDZ-1E) continues along the base of the southern wall; whereas the second (PTDZ-2E) trends in a more northerly direction. PTDZ-1E is clearly the eastern strike slip fault of the

pull-apart system that was responsible for the early formation of the ITSB. We can obtain an approximate age of initiation of this system from the width of the ITSB (∼100 km) if we assume that the seafloor within the ITSB was formed at the intra-transform spreading center [which is not entirely clear from the magnetic anomaly lineations (Fig. 20)] and if we assume that the rate of Rivera-Pacific relative motion in this area has averaged ∼70 km/m.y. If these assumptions are correct, then PTDZ-1E began to form at 1.4 Ma. This age seems reasonable in that it roughly coincides with the age of formation of the Rivera Transform proposed by BANDY (1992) and LONSDALE (1995) and with the age at which the southward propagation of the Elenerth segment of the Rivera Rise ceased as noted in the present study. Thus, PTDZ-1E is the oldest of the Rivera Transform PTDZs. PTDZ-2E appears to be quite young: it is comprised of a series of sinuous, en-echelon faults (possibly the Riedel shears) characteristic of the early stages in the development of a major strike slip system (SYLVESTER, 1988).

Of the three PTDZs located west of the intra-transform spreading center, PTDZ-1W is clearly the western strike slip fault of the pull-apart system that was responsible for the early formation of the ITSB, and thus is contemporaneous with PTDZ-1E which,

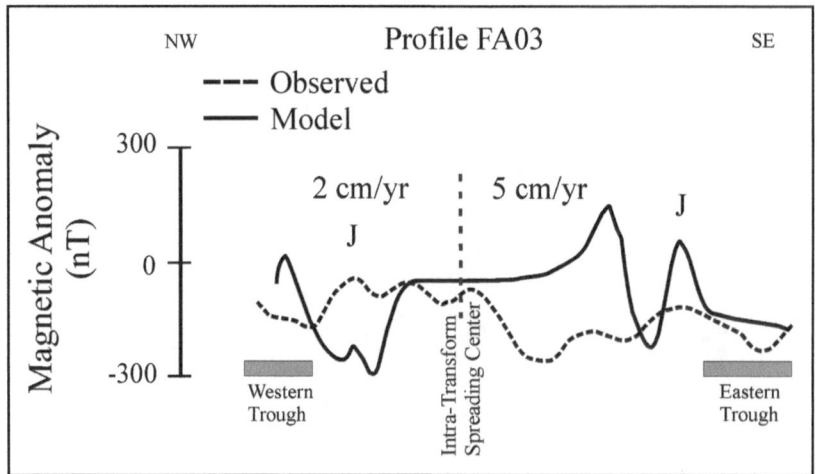

Figure 20
Magnetic anomalies recorded along profile FA03 illustrating the poor fit between the observed and modeled anomalies. The reason for the poor fit is unclear as the morphologic data is clearly consistent with the proposal of recently formed seafloor within the rhombochasm. Perhaps a deep towed magnetic survey would resolve this ambiguity. Also note that the model implies asymmetric spreading at the Intra-Transform spreading center

we estimate, formed at about 1.4 Ma. That PTDZ-1W is the oldest of the three western PTDZs is consistent with the extensive shearing observed within this zone. It is unclear if it is active at present, however the small pull-apart structures present in the area where it overlaps with PTDZ-2W indicates that it was, at least for some period of time, active contemporaneously with PTDZ-2W.

Although the amount of offset is hard to determine, the observation that, near the Rivera Rise, PTDZ-3W is offset in a dextral sense across PTDZ-2W, clearly indicates that PTDZ-2W is younger than PTDZ-3W. Also, as with PTDZ-2E, the en-echelon morphology of PTDZ-2W is characteristic of the early stages in the development of a major strike slip fault. The age of initiation of PTDZ-2W can be better gleaned from its offset ($\sim$2–3 km) of the escarpment that separates the western and eastern sub-domain of the western morphotectonic domain. If we again assume that the rate of Rivera-Pacific relative motion in this area has averaged $\sim$70 km/m.y., then PTDZ-2W was formed very recently, i.e. within the past 50,000 years.

The time of initiation of PTDZ-3W was clearly intermediate between that of PTDZ-1W and PTDZ-2W, however it is very hard to glean its age of initiation from the available data. However, it is clear that it was initiated contemporaneously with the propagation of the intra-transform spreading center northward beyond PTDZ-1W. The width of the seafloor generated at this spreading center north of PTDZ-1W is $\sim$10 km, thus if the northern spreading segment is still active, then PTDZ-3W was initiated at $\sim$0.14 Ma. If the northern part of the spreading center is inactive, then 0.14 Ma represents a minimum age for the initiation of PTDZ-3W. The seismicity study of REID (1976) indicates that PTDZ-3W is still seismically active in the area near the intra-transform spreading center. However, this activity may be related to the down-dropping (normal-faulting) of the basin margins and not to strike-slip motion.

5.2. Development of the Rivera Transform and the Origin of the Mid-Rivera-Transform Discordance

In this section we present a chronology for the tectonic development of the major morphotectonic elements that comprise the Rivera Transform. This chronology rests firmly on the observations and results of MAMMERICKX and KLITGORD (1982), LONSDALE (1995) and others that the Rivera Rise has been undergoing a clockwise reorientation since 2 Ma and that this reorientation has continued into Chron 1 time as inferred by BANDY (1992) and evidenced by LONSDALE (1995). These observations imply that the Rivera Transform near the Rivera Rise must also be reorienting clockwise, the amount of clockwise rotation increasing with time. We also use as a foundation the robust result of the various kinematic studies that the Rivera-Pacific Euler pole has been migrating southward (BANDY and YAN, 1989; DEMETS and STEIN, 1990; BANDY, 1992; BANDY et al., 1998b, 2000, 2007; DEMETS and TRAYLEN, 2000) towards the Rivera Transform since Chron 2A time. In our chronology we propose that there has been a counterclockwise reorientation of Rivera-Pacific relative motion near the MSS (BANDY et al., 1998a) as is evidenced by the unequivocal observation that the PTDZ of the Rivera Transform climbs up the northern wall of the transform tectonized zone as it approaches the MSS (see Fig. 7 of BANDY et al. (2008)). If correct, this, along with the results of the present study, indicates that the Rivera-Pacific Euler pole has been migrating towards the eastern part of the Rivera Transform, somewhere between 107° and 108°W where no plate motion reorientation is evidenced (BANDY et al., 2008).

Since a clear understanding of the spatial distribution of the sense (i.e. clockwise or counterclockwise) and the amount of reorientation associated with a southward migration of the Euler pole is critical to our chronology, as well as to our proposal as to the origin of the Mid-Rivera-Transform Discordance, we feel it is worthwhile to illustrate this with an example. Figure 21 illustrates the distribution of the sense and amounts of plate motion reorientation associated with a southward migration of the Rivera-Pacific Euler pole towards the eastern part of the Rivera Transform. In this figure we have superimposed the Rivera-Pacific relative motion flow-lines (i.e. the small circles about the poles) for the Rivera-Pacific Euler poles RP-2A and RP-0 of BANDY (1992) (which is a SW shift of the pole towards the Rivera Transform at 107°W) on the bathymetric map constructed from the FAMEX/BART bathymetry. As illustrated, a large

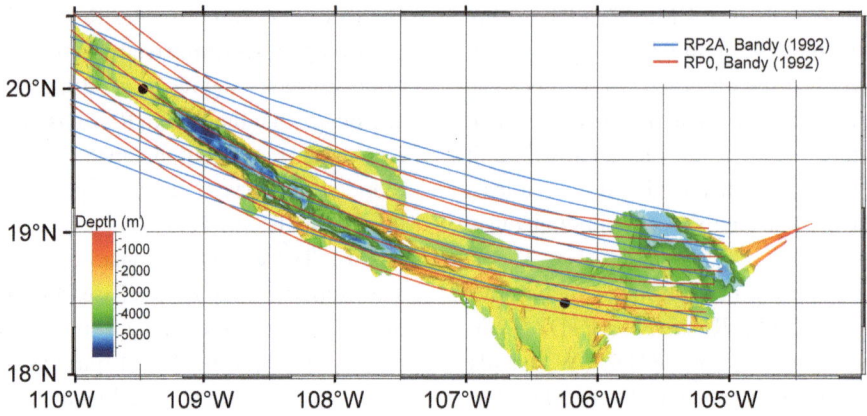

Figure 21

Rivera-Pacific relative motion flow-lines for Euler poles RP2A and RP0 of BANDY (1992) overlain on the bathymetry. These flow-lines illustrate, in general, the reorientation of motion that would result from a SW migration of the Rivera-Pacific Euler pole towards the eastern part of the Rivera Transform. Note that near the Rivera Rise, one would expect a large amount of clockwise reorientation of the relative motion and that the amount of clockwise reorientation would progressively decrease eastward along the transform until ∼107°15′W, where no reorientation is predicted. East of 107°15′W, one would expect a small amount of counterclockwise reorientation of relative motion, the amount of counterclockwise reorientation progressively increasing eastward of this point. Note that the flow lines of RP0 indicate that the trend of the transform would be better fit if the pole were located closer to the transform

amount of clockwise reorientation of Rivera-Pacific relative motion occurs near the Rivera Rise, whereas a small amount of counterclockwise reorientation of this motion occurs at the MSS. Further, no noticeable reorientation of motion occurs along the transform near 107°W, which indicates that the Euler pole is moving along a great circle towards this point. Interestingly, this example is roughly what we observe in reality, however, the exact migration path of the Rivera-Pacific Euler pole still needs to be accurately determined. The determination of the migration path of the Rivera-Pacific Euler pole is necessarily left for future work in that a complete coverage of multi-beam bathymetry along and adjacent to the Rivera Rise is critical for such a determination: presently 100% multi-beam bathymetric coverage is lacking in that area.

It is interesting to note that a comparison of the flow lines of pole RP-0 with the overall trend of the Rivera Transform suggests that the transform would be better fit by an Euler pole located closer to the transform (i.e. the flow lines need to be more accurate). This is indeed what was found by BANDY et al. (2007), who redefined the present day Rivera-Pacific Euler pole (Fig. 1) by incorporating the newly collected bathymetric data. The flow lines (Fig. 22) of their pole 2007-1 best fit the overall curvature of

the transform. However, the predicted directions do not fit the orientation of PTDZ-2W, which we propose is the youngest of the PTDZs. The predictions of their Euler pole 2007-2 (Fig. 22), located even closer to the transform, best fit the orientation of PTDZ-2W. Also, as shown in BANDY et al. (2008), the plate motion directions predicted by pole 2007-2 better fit the transform orientations to the east of the ITSB. We interpret these results as indicating that the southward migration of the Rivera-Pacific Euler pole is continuing at present.

Figure 23 illustrates the proposed chronology. Shortly after its formation at 1.4 Ma, the transform was comprised of two overlapping PTDZs: PTDZ-1W to the west and PTDZ-1E to the east. A pull-apart basin had formed in the zone of overlap of the PTDZs, after which the intra-transform spreading center began to form within the rhombochasm.

With the clockwise rotation of Rivera-Pacific motion, the stresses progressively increased with time until they were sufficient to produce the first reorientation of the morphotectonic elements of the transform west of the intra-transform spreading center. The time of initiation of this reorganization is tentatively placed at 0.14 Ma. Figure 23 (middle diagram) illustrates the morphotectonic elements after the first reorganization. During this reorganization,

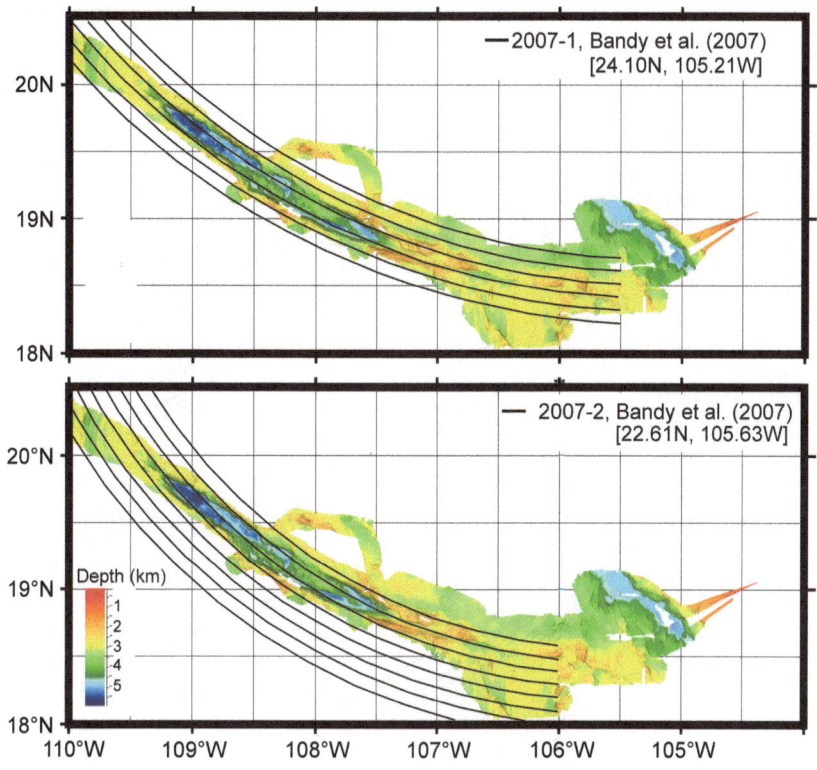

Figure 22
Rivera-Pacific relative motion flow-lines for Euler poles 2007-1 and 2007-2 (BANDY et al., 2007) overlain on the bathymetry. Note that the flow-lines of pole 2007-1 better fit the gross trend of the transform (taking into account the ITSC) than do the flow-lines of either RP0 or 2007-2. However, the flow lines of pole 2007-2 better fit PTDZ-2W near the Rivera Rise. Also, the flow lines of pole 2007-2 better fit the trend of the Rivera Transform east of 107°30'W (BANDY et al., 2008)

the intra-transform spreading center propagated northward. This resulted in the relocation of the PTDZ from PTDZ-1W to PTDZ-3W, whose orientation is presently 12° clockwise of the strike of PTDZ-1W. As PTDZ-3W is the fault forming the northern wall of the LTB, it is likely that LTB was also initiated during this reorganization.

East of the intra-transform spreading center, the change in stress apparently was not sufficient at this time to produce a reorganization of the PTDZ in this area, and the PTDZ remained along PTDZ-1E. This is consistent with a SW migration of the Euler pole as the amount of reorientation of the direction of plate motion is expected to decrease eastward along the transform. Thus, in this chronology, the Mid-Rivera-Transform Discordance was initiated by the eastward progressing reorganization of the transform segments in response to changing plate motions as the Rivera-Pacific

Euler pole migrated towards the eastern transform segment.

The clockwise change in Rivera-Pacific plate motion did not stop with the first reorganization. Therefore the stresses again built up to where a second clockwise reorganization of the transform occurred. During this reorganization (Fig. 23, bottom) the PTDZ west of the intra-transform spreading center rotated an additional 3° clockwise with the formation of PTDZ-2W. PTDZ-2W intersects the intra-transform spreading center in the deep, westward curving basin located just south of PTDZ-1W. Thus, it appears that the northern part of the intra-transform spreading center is inactive. If so, the PTDZ has relocated from PTDZ-3W to PTDZ-2W. The pull-apart structures located in the overlap area between PTDZ-1W and PTDZ-2W indicate that at least part of PTDZ-1W was reactivated during this reorganization.

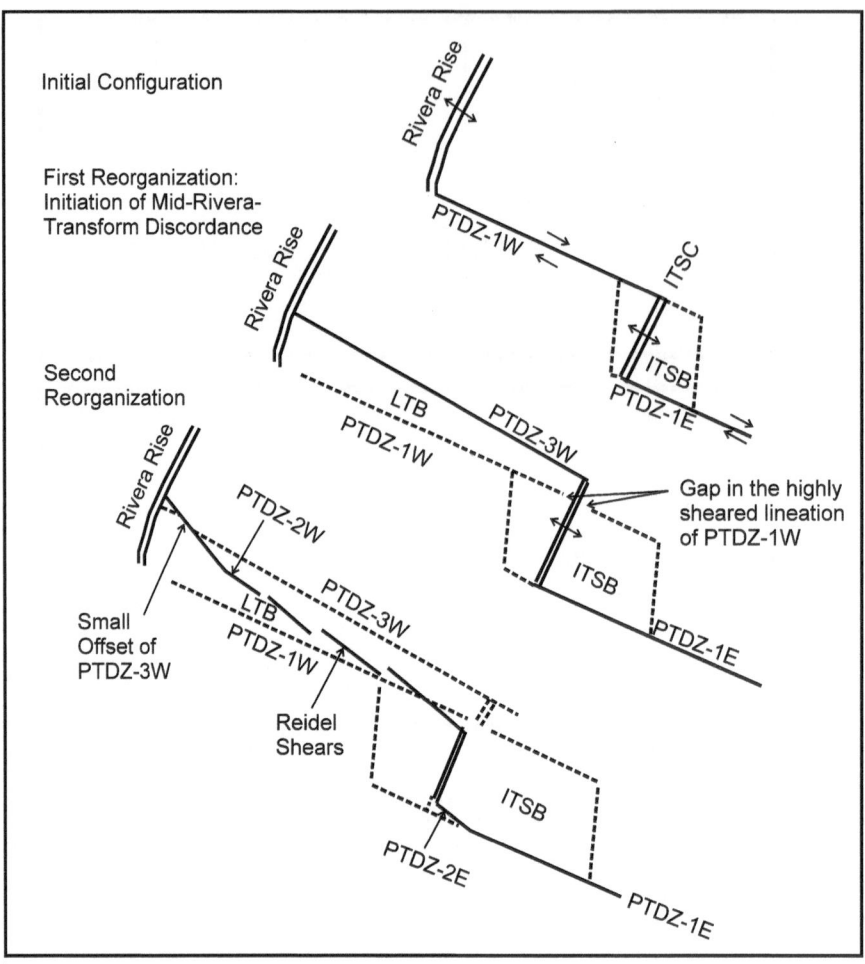

Figure 23
Schematic representation of the development chronology of the PTDZs of the Rivera Transform that is consistent with the clockwise rotation of the Rivera Rise and western part of the Rivera Transform. Configuration of the PTDZs at ~1.4 Ma (shortly after the initial formation of the Rivera Transform) *(top)*. Configuration of the PTDZs at ~0.15 Ma (shortly after the first reorganization) *(middle)*. Configuration of the PTDZs at ~0.05 Ma (shortly after the second reorganization) *(bottom)*

East of the intra-transform spreading center, the change in stress was now sufficient to form PTDZ-2E (which is roughly parallel to PTDZ-2W adjacent to the ITSC). Presently PTDZ-2E does not extend east of 108°W, and the PTDZ remains along PTDZ-1E, east of 108°W. Apparently, the change in stress is still not sufficient at present to produce a plate boundary reorganization east of 108°W, and thus the Mid-Rivera-Transform Discordance still exists. Thus, the morphology of the transform is consistent with an eastward progressing reorganization of the Rivera-Pacific plate boundary produced in association with a migration of the Rivera-Pacific Euler pole towards the eastern segment of the Rivera Transform.

It is interesting to note that with the formation of PTDZ-2W and PTDZ-2E, the length of the intra-transform spreading center decreased by almost half. With more time, the intra-transform spreading center may be completely deactivated, the eastern part of the Rivera Transform may completely readjust to the changing plate motions, and the mid-transform discordance may be eliminated. It is also interesting to note that the now abandoned ridge segments off southern Baja California also underwent a clockwise rotation prior to their cessation and subsequent attachment to the Pacific plate (e.g., MICHAUD *et al.*, 2006); perhaps the Rivera Plate is undergoing an analogous situation.

6. Conclusions

The main conclusions of this study include:

1. The transform tectonized zone of the Rivera Transform west of 107°30'W is a single continuous morphologic basin. However, this basin consists of two distinct morphotectonic domains: an eastern domain which morphologically is a deep rhombochasm within which organized seafloor spreading has occurred, and a western 'leaky transform' domain.

2. The initial formation of the Rivera Transform was not instantaneous. Instead, it developed between 2 and 1.4 Ma. From 1.4 Ma until ∼0.14 Ma the transform consisted of a pair of overlapping fault segments connected by an intra-transform spreading center located within a deep rhombochasm formed in the overlap/pull-apart zone.

3. Since ∼0.14 Ma, at least two periods of clockwise reorientation of the PTDZs of the western segment of the Rivera Transform have occurred, the degree of reorientation diminishing eastward along the transform.

4. During the first reorganization, the intra-transform spreading center propagated northward but appears to have since retreated back to roughly its previous position.

5. The morphology of the transform indicates an eastward progressing reorganization of the Rivera-Pacific plate boundary. This is kinematically consistent with a migration of the Rivera-Pacific Euler pole towards the eastern segment of the Rivera Transform.

6. We propose that the Mid-Rivera-Transform Discordance was initiated by the eastward progressing, clockwise, reorganization of the transform segments. Presently, the effects of this reorganization are observed west of 108°W, and near the MSS. Kinematically, this reorganization can be described as the response to changing plate motions as the Rivera-Pacific Euler pole migrated towards the eastern transform segment.

7. Local seismic networks need to be deployed to better constrain the presently active PTDZs of the Rivera Transform.

Acknowledgments

We thank the captain and crew of N/O L'Atalante and the staff of the ship operations section of IFREMER for their valuable assistance. We also thank the two anonymous reviewers for their comments which have improved the manuscript. Financial support was provided by *Centre National de la Recherche Scientifique* (CNRS), and by CONACyT grants 36681-T, #50235, R34906-T and 25709T, and UNAM DGAPA grants # IN104707, IN114602, IX117504, IN104199, IN110897, IN108110 and IX111304.

REFERENCES

ATWATER, T. (1970), *Implications of plate tectonics for Cenozoic tectonic evolution for western North America*, Geol. Soc. Am. Bull. 81, 3513–3536.

BANDY, W.L. (1992), *Geological and geophysical investigation of the Rivera-Cocos plate boundary: Implications for plate fragmentation*, Ph.D. Dissertation, Texas A&M University, College Station, 195 pp.

BANDY, W.L., and HILDE, T.W.C. (2000), Morphology and recent history of the ridge propagator system located at 18°N, 106°W, In *Cenozoic Tectonics and Volcanism of Mexico, Geological Society of America Special Paper 334* (eds. Delgado-Granados, H., Aguirre-Díaz and Stock, J.M), (Geological Society of America, Boulder, Colorado 2000) pp. 29–40.

BANDY, W.L., and YAN, C.-Y. (1989), *Present-day Rivera-Pacific and Rivera-Cocos relative plate motions (abs.)*, EOS Transactions of the American Geophysical Union 70, 1342.

BANDY, W.L., MICHAUD, F., DYMENT, J., MORTERA-GUTIERREZ, C.A., BOURGOIS, J., CALMUS, T., SOSSON, M., ORTEGA-RAMÍREZ, J., ROYER, J.-Y., PONTOISE, B., and SICHLER, B. (2008), *Multibeam bathymetry and sidescan imaging of the Rivera Transform-Moctezuma spreading segment junction, northern East Pacific Rise: New constraints on Rivera-Pacific relative plate motion*, Tectonophysics 454, 70–85, doi:10.1016/j.tecto.2008.04.013.

BANDY, W.L., MICHAUD, F., DYMENT, J., MORTERA GUTIERREZ, C., BOURGOIS, B., CALMUS, T., SOSSON, M., ORTEGA RAMIREZ, J. ROYER, J.-C., PONTOISE, B., and SICHLER B. (2007), *New constraints on Rivera-Pacific relative motion from multibeam bathymetric data along the MSS and Rivera Transform (abstract)*, 2007 Joint Assembly of the AGU, Acapulco, Mexico, Meeting Abstracts, S31A-12.

BANDY, W.L., MICHAUD, F., BOURGOIS, J., CALMUS, T., DYMENT, J., MORTERA-GUTIERREZ, C.A., ORTEGA-RAMÍREZ, J., PONTOISE, B., ROYER, J.-Y., SICHLER, B., SOSSON, M., REBOLLEDO-VIEYRA, M., BIGOT-COMIER, F., DIAZ-MOLINA, O., HURTADO-ARTUNDUAGA, A.D., PARDO-CASTRO, G., and TROUILLARD-PERROT, C. (2005), *Subsidence and strike-slip tectonism of the upper continental slope off Manzanillo, Mexico*, Tectonophysics 398, 115–140.

BANDY, W.L., HILDE, T.W.C., and YAN, C.Y. (2000), The Rivera-Cocos plate boundary: Implications for Rivera-Cocos relative motion and plate fragmentation, In Cenozoic Tectonics and Volcanism of Mexico, Geological Society of America Special Paper 334 (eds. DELGADO-GRANADOS, H., AGUIRRE-DÍAZ and STOCK, J.M), (Geological Society of America, Boulder, Colorado 2000) pp. 1–28.

BANDY, W.L., KOSTOGLODOV, V.V., MORTERA-GUTIERREZ, C.A., and URRUTIA-FUCUGAUCHI, J. (1998a), Comment on "Relative motions of the Pacific, Rivera, North American, and Cocos plates since 0.78 Ma" by Charles DeMets and Douglas S. Wilson, J. Geophys. Res. 103, 24,245–24,250.

BANDY, W.L., KOSTOGLODOV, V., and MORTERA-GUTIÉRREZ, C.A. (1998b), Southwest migration of the instantaneous Rivera-Pacific Euler pole since 0.78 Ma, Geofisica Internacional 37, 153–169.

BOURGOIS, J., RENARD, V., AUBOUIN, J., BANDY, W., BARRIER, E., CALMUS, T., CARFANTAN, J.-C., GUERRERO, J., MAMMERICKX, J., MERCIER DE LEPINAY, B., MICHAUD, F., and SOSSON, M. (1988), The East Pacific Rise-Rivera Fracture Zone eastern junction off Mexico: Paris, Académie des Sciences, Comptes Rendus, series II, 307, 617–626.

BRAUNMILLER, J., and NÁBĚLEK, J. (2008), Segmentation of the Blanco Transform fault zone from earthquake analysis: Complex tectonics of an oceanic transform fault, J. Geophys. Res. 113, B07108, doi:10.1029/2007JB005213.

BURKETT, E.R., and BILLEN, M.I. (2009), Dynamics and implications of slab detachment due to ridge-trench collision, J. Geophys. Res. 114, B12404, doi:10.1029/2009JB006402.

CANDE, S.C., and KENT, D.V. (1995), Revised calibration of the geomagnetic polarity timescale for the Late Cretaceous and Cenozoic, Geophys. Res. Letts. 100, 6093–6095.

DAUPHIN, J.P., and NESS, G.E. (1991), Bathymetry of the gulf and peninsular province of the Californias, In The Gulf and Peninsular Province of the Californias, AAPG Memoir 47 (eds. Dauphin, J.P. and Simoneit, B.R.T.) (AAPG, Tulsa, Oklahoma, 1991) pp. 21–24.

DAVISON, I. (1994), Linked fault systems; extensional, strike-slip and contractional, In Continental Deformation (ed. Hancock, P.L.) (Pergamon Press 1994) pp. 121–142.

DECHARON, A.V. (1989), Structure and tectonics of the Cascadia segment, central Blanco Transform fault zone, M.S. Thesis, Oregon State University, Corvallis, Oregon.

DEMETS, C., and STEIN, S. (1990), Present-day kinematics of the Rivera Plate and implications for tectonics of southwestern Mexico, J. Geophys. Res. 95, 21,931–21,948.

DEMETS, C., and WILSON, D.S. (1997), Relative motions of the Pacific, Rivera, North American, and Cocos plates since 0.78 Ma, J. Geophys. Res. 102, 2789–2806.

DEMETS, C., and TRAYLEN, S. (2000), Motion of the Rivera Plate since 10 Ma relative to the Pacific and North American plates and the mantle, Tectonophysics 318, 119–159.

EMBLEY, R.W., and WILSON, D.S. (1992), Morphology of the Blanco Transform fault zone-NE Pacific: Implications for its tectonic evolution, Mar. Geophys. Res. 14, 25–45.

GARFUNKEL, Z. (1986), Review of oceanic transform activity and development, J. Geol. Soc. London 143, 775–784.

GAZDAG, J. (1978), Wave-equation migration by phase-shift, Geophysics 43, 1342–1351.

GORDON, R.G. (1995), Plate motions, crustal and lithospheric mobility, and paleomagnetism: Prospective viewpoint, J. Geophys. Res., 100, 24,367–24,392.

HOLCOMBE, T.L., and SHARMAN, G.F. (1983), Post-Miocene Cayman Trough evolution: a speculative model, Geology 11, 714–717.

HOLCOMBE, T.L., VOGT, P.R., MATTHEWS, J.E., and MURCHISON, R.R. (1973), Evidence for sea-floor spreading in the Cayman Trough, Earth and Planetary Science Letters 20, 357–371.

KASTENS, K.A., MACDONALD, K.C., BECKER, K., and CRANE, K. (1979), The Tamayo Transform fault in the mouth of the Gulf of California, Mar. Geophys. Res. 4, 129–151.

KLITGORD, K.D., and MAMMERICKX, J. (1982), Northern East Pacific Rise: Magnetic anomaly and bathymetric framework, J. Geophys. Res. 87, 6726–6750.

LARSON, R.L. (1972), Bathymetry, magnetic anomalies, and plate tectonic history of the mouth of the Gulf of California, Geol. Soc. Am. Bull. 83, 3345–3360.

LIGI, M., BONATTI, E., GASPERINI, L., and POLIAKOV, A.N.B. (2002), Oceanic broad multifault transform plate boundaries, Geology 30, 11–14.

LONSDALE, P. (1995), Segmentation and disruption of the East Pacific Rise in the mouth of the Gulf of California, Mar. Geophys. Res. 17, 323–359.

LONSDALE, P. (1991), Structural patterns of the Pacific floor offshore of Peninsular California, In The Gulf and Peninsular Province of the Californias, AAPG Memoir 47 (eds. DAUPHIN, J.P. and SIMONEIT, B.R.T.) (AAPG, Tulsa, Oklahoma 1991) pp. 87–125.

LONSDALE, P. (1986), Tectonic and magmatic ridges in the Eltanin fault system, South Pacific, Mar. Geophys. Res. 8, 203–242.

MACDONALD, K.C., CASTILLO, D.A., MILLER, S.P., FOX, P., KASTENS, K.A., and BONATTI, E. (1986), Deep-Tow studies of the Vema Fracture Zone 1. Tectonics of a major slow slipping transform fault and its intersection with the Mid-Atlantic ridge, J. Geophys. Res. 91, 3334–3354.

MAMMERICKX, J. (1980), Neogene reorganization of spreading between the Tamayo and the Rivera fracture zone, Mar. Geophys. Res. 4, 305–318.

MAMMERICKX, J., and KLITGORD, K.D. (1982), Northern East Pacific Rise: Evolution from 25 m.y.B.P. to the present, J. Geophys. Res. 87, 6751–6759.

MANN, P., HEMPTON, M.R., BRADLEY, D.C., and BURKE, K. (1983), Development of pull-apart basins, Journal of Geology 91, 529–554.

MAUDUIT, T., and DAUTEUIL, O. (1996), Small-scale models of oceanic transform zones, J. Geophys. Res. 101, 20,195–20,209.

MENARD, H.W., and ATWATER, T. (1969), Origin of fracture zone topography, Nature 222, 1037–1040.

MERRILL, R.T., and McELHINNY, M.W. (1983), The Earth's Magnetic Field: Its History, Origin and Planetary Perspective, International Geophysical Series, vol. 32 (Academic Press, New York 1983).

MICHAUD, F., and BOURGOIS, J. (1995), Is the Rivera Fracture Zone a transform fault as currently accepted?, C.R. Acad. Sci. Paris, Série II, 321, 521–528.

MICHAUD, F., ROYER, J.-Y., BOURGOIS, J., DYMENT, J., CALMUS, T., BANDY, W., SOSSON, M., MORTERA-GUTIERREZ, C., SICHLER, B., REBOLLEDO-VIERA, M., and PONTOISE, B., (2006), Oceanic-ridge subduction vs. slab break off: Plate tectonic evolution along the Baja California Sur continental margin since 15 Ma, Geology 34, 13–16; doi:10.1130/G22050.1.

MICHAUD, F., ROYER, J.-Y., BOURGOIS, J., MERCIER DE LEPINAY, B., and LIAUDON, G.P. (1997), The Rivera Fracture Zone revisited, Marine Geology 137, 207–225.

MICHAUD, F., BOURGOIS, J., and AUBOUIN, J. (1990), Active fragmentation of the west Pacific Mexican coast: The Jalisco Block a future unsuspected terrain?, *In Tectonics of Circum-Pacific Continental Margins*, (eds. AUBOUIN, J., and BOURGOIS J.) (VSP-BV, The Netherlands 1990) pp. 51–76.

POCKALNY, R.A., FOX, P.J., FORNARI, D.J., MACDONALD, K.C., and PERFIT, M.R. (1997), *Tectonic reconstruction of the Clipperton and Siqueiros fracture zones: Evidence and consequences of plate motion change for the last 3 Myr*, J. Geophys. Res. 102, 3167–3181.

PROTHERO, W.A., and REID, I.D. (1982), *Microearthquakes on the East Pacific Rise at 21°N and the Rivera Fracture Zone*, J. Geophys. Res. 87, 8509–8518.

PROTHERO, W.A., REID, I., REICHLE, M.S., and BRUNE, J.N. (1976), *Ocean bottom seismic measurements on the East Pacific Rise and Rivera Fracture Zone*, Nature 262, 121–124.û

REID, I. (1976), *The Rivera plate: A study in seismology and plate tectonics*, Ph.D. thesis, University of California at San Diego, La Jolla, 288 p.

ROBERTS, A., and YIELDING, G. (1994), Continental extensional tectonics, In *Continental Deformation* (ed. HANCOCK, P.L.) (Pergamon Press, 1994) pp. 223–250.

SEARLE, R.C. (1983), *Multiple, closely spaced transform faults in fast-slipping fracture zones*, Geology 11, 607–610.

SCHOUTEN, H., and McCAMY, K. (1972), *Filtering marine magnetic anomalies*, J. Geophys. Res. 35, 7089–7099.

SINGH, S.K., and LERMO, J. (1985), *Mislocation of Mexican earthquakes as reported in international bulletins*, Geofis. Int. 24, 333–351.

SYLVESTER, A.G. (1988), *Strike-slip faults*, Geol. Soc. Am. Bull. 100, 1666–1703.

TEN BRINK, U.S., RYBAKOV, M., AL-ZOUBI, A.S., HASSOUNEH, M., FRIESLANDER, U., BATAYNEH, A.T., GOLDSCHMIDT, V., DAOUD, M.H., ROTSTEIN, Y., and HALL, J.K. (1999), *Anatomy of the Dead Sea transform: Does it reflect continuous changes in plate motion?*, Geology 27, 887–890.

TURCOTTE, D.L. (1974), *Are transform faults thermal contraction cracks?*, J. Geophys. Res. 79, 2573–2577.

WILSON, D.S., and DeMETS, C. (1998), *Reply*, J. Geophys. Res. 103, 24,251–24,253.

(Received February 23, 2010, revised August 3, 2010, accepted August 16, 2010, Published online November 9, 2010)

Pure Appl. Geophys. 168 (2011), 1415–1433
© 2010 Springer Basel AG
DOI 10.1007/s00024-010-0205-y

Overview of Recent Coastal Tectonic Deformation in the Mexican Subduction Zone

M. Teresa Ramírez-Herrera,[1] Vladimir Kostoglodov,[2] and Jaime Urrutia-Fucugauchi[3]

Abstract—Holocene and Pleistocene tectonic deformation of the coast in the Mexico subudction margin is recorded by geomorphic and stratigraphic markers. We document the spatial and temporal variability of active deformation on the coastal Mexican subduction margin. Pleistocene uplift rates are estimated using wave-cut platforms at ca. 0.7–0.9 m/ka on the Jalisco block coast, Rivera-North America tectonic plate boundary. We examine reported measurements from marine notches and shoreline angle elevations in conjunction with their radiocarbon ages that indicate surface uplift rates increasing during the Holocene up to ca. 3 ± 0.5 m/ka. In contrast, steady rates of uplift (ca. 0.5–1.0 m/ka) in the Pleistocene and Holocene characterize the Michoacan coastal sector, south of El Gordo graben and north of the Orozco Fracture Zone (OFZ), incorporated within the Cocos-North America plate boundary. Significantly higher rates of surface uplift (ca. 7 m/ka) across the OFZ subduction may reflect the roughness of subducting plate. Absence of preserved marine terraces on the coastal sector across El Gordo graben likely reflects slow uplift or coastal subsidence. Stratigraphic markers and their radiocarbon ages show late Holocene (ca. last 6 ka BP) coastal subsidence on the Guerrero gap sector in agreement with a landscape barren of marine terraces and with archeological evidence of coastal subsidence. Temporal and spatial variability in recent deformation rates on the Mexican Pacific coast may be due to differences in tectonic regimes and to localized processes related to subduction, such as crustal faults, subduction erosion and underplating of subducted materials under the southern Mexico continental margin.

Key words: Subduction zone, uplift, subsidence, marine terrace, holocene, pleistocene, southern Mexico.

1. Introduction

Coastal uplift and subsidence are crucial phenomena along subduction margins. Emergent coasts are commonly characterized by differential uplift within distinct segments that might be sustained as morphotectonic units on time scales of tens of thousands to millions of years (MELNICK *et al.*, 2006). These areas seem to be controlling and acting as quasi-independent rupture zones that drive deformation (OTA and YAMAGUCHI, 2004). Some examples of subduction zones indicate that crustal faults have controlled coastal deformation patterns and their subdivisions appear to rule seismotectonic segmentation, rupture propagation and local earthquake hazard (BRIGGS *et al.*, 2006; MELNICK *et al.*, 2006). This review focuses on the coastal sector of the Pacific subduction margin of Mexico, which is composed of distinct coastal segments. However, it is not known how tectonic deformation (uplift/subsidence) rates vary spatially and temporally, or how this long-term differential uplift and subsidence are manifested in an array of coastal landforms on the active Mexican Pacific margin coastal areas. Recent Pleistocene-Holocene interseismic and historic coseismic deformation along the Mexico subduction margin, considering changes in upper plate structures and Mexican subduction zone dynamics, is analyzed in this paper. We integrate reported geomorphic and stratigraphic markers indicative of tectonic deformation to constrain the spatial and temporal variability of the Mexican subduction margin deformation rates. Data analyzed integrate uplift rate over a 100-ka time scale (Pleistocene), periodicity of marine terrace formation, and the younger time scale (Holocene). Abandoned shorelines are reflected on the landscape as uplifted marine terraces, wave-cut surfaces, beach ridges and tidal notches (e.g., BRADLEY and GRIGGS, 1976; MERRITTS

[1] Geomorphology and Geohazards Group, Centro de Investigaciones en Geografía Ambiental, Universidad Nacional Autónoma de México, Antigua Carretera a Pátzcuaro No. 8701 Col. Ex-Hacienda de San José de La Huerta, 58190 Morelia, Michoacán, México. E-mail: mtramirez@ciga.unam.mx

[2] Departamento de Sismologia, Instituto de Geofísica, Universidad Nacional Autónoma de México, Ciudad Universitaria, Coyoacán, 04510 México, DF, México. E-mail: vladi@servidor.unam.mx

[3] Proyecto Universitario de Perforaciones en Oceanos y Continentes, Instituto de Geofísica, Universidad Nacional Autónoma de México, Ciudad Universitaria, Coyoacán, 04510 México, DF, México. E-mail: juf@geofisica.unam.mx

and BULL, 1989; ANDERSON et al., 1999; MUHS et al., 1990; LAJOIE et al., 1991. Landscape response to active tectonics has been observed in numerous parts of the world. The study of locations, rates and styles of Quaternary tectonic deformation using geomorphic markers on coastal areas has proved effective in providing information for patterns of tectonic deformation (e.g., CHAPPELL, 1983; LAJOIE et al., 1986; HANSON et al., 1994; OTA and CHAPPELL, 1996; CHAPELL et al. 1996; BURBANK et al., 2001).

The Mexican Pacific margin has mainly developed by subduction of the Rivera and Cocos plates along the Middle America trench (MAT). The study area extends from the northwestern end of the MAT where it intersects with the Tamayo fracture zone to the Tehuantepec isthmus in the southeast at the intersection with the Tehuantepec ridge (Fig. 1). We revise studies on morphotectonics of the southern Mexican Pacific coast (RAMIREZ-HERRERA and URRUTIA-FUCUGAUCHI, 1999; RAMIREZ-HERRERA et al., 1999), integrate studies of geomorphic, geodetic, seismologic and geochronologic data, and evaluate uplift/subsidence rate calculations over various time ranges. We present data on variability in space and time of coastal deformation rates and deliberate on plausible explanations for those changes.

2. Tectonic Setting and Seismicity

2.1. Tectonics

The area presently incorporating the Mexican subduction zone has experienced two major plate reorganizations over the past 25 Ma (KLITGORD and MAMMERICKX, 1982). Large Farallon plate, which initially subducted eastward beneath the North American plate, first evolved into the Cocos plate, which was split off, leaving the reduced Farallon plate to be renamed the Nazca plate, and initiating the Cocos-Nazca spreading at 23 Ma (LONSDALE, 2005); the Rivera plate fragmented from the Cocos plate at 10 Ma (DEMETS and TRAYLEN, 2000); they are now subducting along the Middle America. The MAT is a continuous feature off the southern Pacific margin of Mexico and Central America over a distance of more than 3,000 km (Fig. 1). MAT extends from the Gulf of California at the intersection with the Tamayo

fracture zone to the Osa Peninsula of Costa Rica in Central America. The MAT can be separated into two main zones, to the northwest and southeast of the intersection zone with the Tehuantepec ridge (FISHER, 1961; MOLNAR, 1977; AUBOIN et al., 1981). MAT maximum depths of about 6,000 m occur immediately southeast of the intersection zone with the Tehuantepec ridge. The Central America trench sector is characterized by an extended margin with a well-developed forearc basin. The Central American volcanic arc front develops some 150 km away from the trench. The Mexican segment shows a narrow, relatively steep trench inner wall and non- or incipiently developed forearc. The Trans-Mexican volcanic arc lies oblique to the trench axis (MOLNAR and SYKES, 1969; URRUTIA-FUCUGAUCHI and DEL CASTILLO, 1977; FERRARI and ROSAS-ELGUERA, 2000; RUIZ-MARTINEZ et al., 2010). Maximum trench depths in the Mexican segment exceed 4,400 m, except off Manzanillo and Zihuatanejo, and where submarine seamounts arise within the trench. Northwest of Acapulco, the trench is generally U-shaped in cross section, with a steeper inboard flank and a flat floor suggesting the presence of a sedimentary fill (MANEA et al., 2003). From Acapulco southeast to the western side of the Gulf of Tehuantepec, the trench deepens to 5,000 m in a series of basins. The southeast segment has an asymmetric V-shaped form in cross section with an irregular floor. A northeast-trending band of ridge-and-trough topography 111 km wide separates the ocean floor at a depth of 3,276–3,458 m adjacent to the trench off southern Mexico from the somewhat deeper (3,822–4,004 m depth) Guatemala Basin. This ridge-and-trough zone (Tehuantepec ridge) has been traced from an area several hundred kilometers offshore to an intersection with the MAT near the western side of the Gulf of Tehuantepec (FISHER, 1961). Cores recovered from eight sites along transects across the MAT off southwestern Mexico (Fig. 1) have revealed trench sediments ranging in age from Pleistocene to Miocene, indicating that the trench was in existence by the early Miocene (23–20 Ma). The sediments have been uplifted by as much as 2 to 3 km above the present-day trench floor by a combination of accretion and underplating (SHIPLEY et al., 1980; Moore et al., 1982; Watkins et al., 1982).

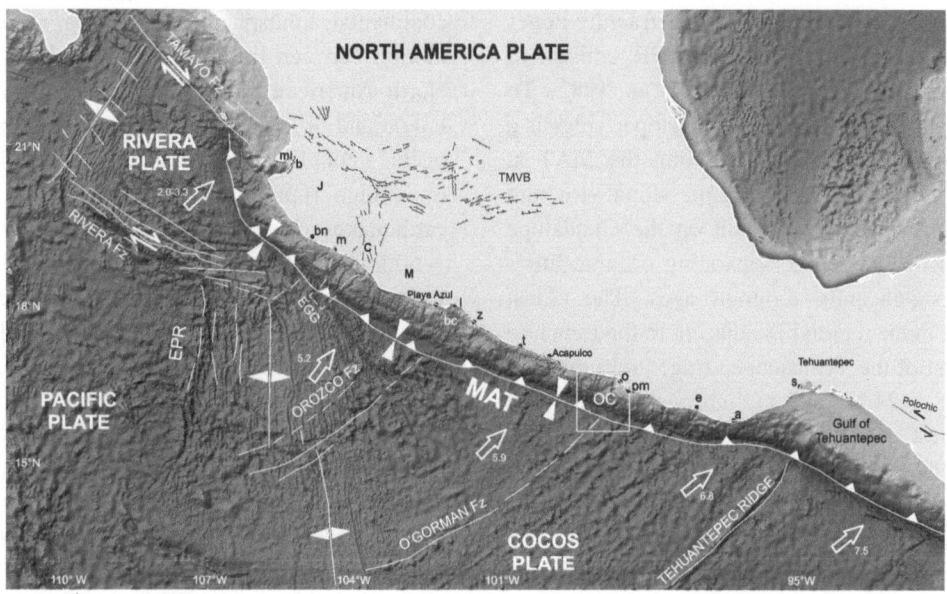

Figure 1

Simplified morphotectonic map and sectors of active tectonic deformation along the coast of the Mexican Pacific active margin. Symbols: *MAT* Middle American Trench; *EPR* East Pacific Rise; *EGG* El Gordo Graben; *Fz* fault zone; *J* Jalisco block; *C* Colima graben; *M* Michoacan; *TMVB* Trans-Mexican volcanic belt; *bc* balsas submarine canyon; *oc* ometepec submarine canyon. *Numbers* indicate convergence rate in cm/year (DEMETS and STEIN, 1990). *Square* shows location of Deep Sea Drilling project Leg 66 transects (MOORE *et al.*, 1982; WATKINS *et al.*, 1982). Site names: *mi* Punta Mita; *b* Bahia de Banderas; *bn* Barra de Navidad; *m* Manzanillo; *l* Lagunillas; *z* Zihuatanejo-Ixtapa; *t* Tecpan; *o* Ometepec; *pm* Punta Maldonado; *e* Puerto Escondido; *a* Puerto Angel; *s* Salina Cruz

The Rivera plate is bounded by the Rivera fracture zone, the East Pacific Rise (EPR), the Tamayo fracture zone, and the MAT (Fig. 1). The Rivera-Cocos plate boundary comprises three distinct morphotectonic units: an eastern zone characterized by lithospheric extension and delineated by the El Gordo graben; a structurally complex western zone marked by a regional bathymetric high and formed as a result of rift propagation and convergence between the Rivera and Cocos plates; and a central zone of undisturbed lithosphere, characterized by well-defined magnetic lineations, normal ocean depths and continuous seafloor spreading (BANDY *et al.*, 1995). The Rivera plate moves to the N-NE with respect to the Rivera-North America plate boundary (BANDY *et al.*, 1997; DEMETS *et al.*, 1994). Two types of models have been proposed for the convergence rate between the Rivera and North American plates. The first predicts convergence rates varying 2.0 and 5.0 cm/year (BANDY *et al.*, 1995; KOSTOGLODOV and BANDY, 1995). The second model bears out the lower convergence rates of between 2.0 and 3.3 cm/year near the southern end of the Rivera-North America

subduction zone and between 0.6 and 1.7 cm/year at its northern termination (e.g., DEMETS *et al.*, 1994). Rivera plate subduction is characterized by a steeper angle and more northerly trajectory than the adjacent Cocos plate, with dip angle increasing at depths around 100 km. YANG *et al.* (2009) have proposed that the Rivera plate and westernmost Cocos plate have recently rolled back toward the trench. The Rivera plate at the trench is of late Miocene age (~9 Ma) (KLITGORD and MAMMERICKX, 1982).

Motion of the Cocos plate relative to the North America plate is directed north-northeastwards, slightly counterclockwise to a line normal to the MAT. Convergence rate increases eastward from 4.8 cm/year at 104.5°W to 7.5 cm/year at 94°W (DEMETS *et al.*, 1994). Age of the Cocos plate being subducted varies along the MAT, with some discontinuities across the fracture zones that extend eastwards from the flank of the East Pacific Rise (Fig. 1). The most prominent of these are the Orozco (OFZ) and O'Gorman fracture zones (OGFZ). The OFZ is a broad topographic feature consisting of several deep parallel troughs and large ridges, whereas the OGFZ, the largest of the fracture zones in the region that does not offset the EPR, consists of a deep trough (KLITGORD and MAMMERICKX, 1982). To the southeast of the OGFZ the Tehuantepec ridge is a major topographic feature intersecting the MAT at $\sim 95°W$, which represents a transpressional structure along the former transform fault on the Guadalupe plate (MANEA et al., 2005) separating oceanic lithosphere of significantly different ages. The oldest Cocos plate in the region lies adjacent to the trench to the southeast of the Tehuantepec ridge and ranges in age from 28 to 24 Ma (Late Oligocene). To the west and northwest the age of the Cocos plate is 16 to 12 Ma (Middle Miocene), while elsewhere in the region, i.e., along the trench, it is of late Miocene-Pliocene age (KLITGORD and MAMMERICKX, 1982; PONCE et al., 1992; Manea et al., 2005).

A significant change in the dip of the subducting Cocos plate lithosphere and an associated increase in the maximum depth of the Benioff zone occur near the intersection of the Tehuantepec ridge with the MAT at about longitude 96° west (PONCE et al., 1992; MANEA et al., 2005). The subduction angle is subhorizontal ($\sim 15°$) west of the Tehuantepec ridge, but $\sim 45°$ to the east. In the region of subhorizontal subduction (Guerrero-Oaxaca) the earthquake hypocenters are shallower than 80 km, whereas earthquakes with foci as deep as 200 km occur beneath Chiapas and Central America where the subducted slab dips much more steeply to the east. In the Guerrero area, Cocos plate presents subhorizontal subduction for about 250 km, and then the dip increases just before the TMVB front with an angle of $\sim 75°$. The plate is truncated at depths of 500 km, probably associated with an E-W propagating tear in the subducted slab (HUSKER and DAVIS, 2009). A change in subduction angle also occurs to the northwest from being subhorizontal in the region of Guerrero to $\sim 30°$ at depths greater than 40 km at the western extremity of the Cocos plate.

2.2. Seismicity

Large magnitude ($M_w > 7.5$), shallow thrust earthquakes occurring along the Mexican subduction zone rupture a low-angle (10–15°) plane dipping to the north-northeast (Fig. 2). Coseismic rebound of the continental lithosphere is reversed to the relative motion between the Cocos plate and the overriding North American plate (ASTIZ and KANAMORI, 1984; PARDO and SUÁREZ, 1995). The area of strong seismogenic coupling in Central Mexico is shallow (maximum rupture depth of the subduction thrust earthquakes is ~ 25 km) and narrow (width $< \sim 60$ km) in comparison with subduction zones elsewhere. For example, the dip of the seismogenic zone in Mexico ($\sim 12°$) is only half that in Chile, and the distance between the trench and the coast (~ 70 km) and the thickness of the continental crust in Mexico (35–40 km) are about half of those in Chile. Width of the seismogenic locked interplate area may be the controlling factor that defines the magnitude of the largest earthquakes in the Mexican subduction zone (SUÁREZ and SÁNCHEZ, 1996).

During the past two centuries, large shallow earthquakes ($M_w > 7.5$) with recurrence intervals of 30–50 years have been occurring along the Mexican part of the Middle America Trench (MAT) in discrete segments of 100–200 km (Fig. 2). Four possible seismic gaps were identified along the MAT: the Jalisco, Guerrero, Michoacan and Tehuantepec (ASTIZ and KANAMORI, 1984).

In Jalisco the largest subduction earthquake of the last century, the 3 June 1932 ($M_w = 8.2$) was followed by the smaller, $M_w = 7.8$, 18 June 1932 event (SINGH et al., 1981, 1985; ASTIZ and KANAMORI, 1984). The 9 October 1995 earthquake ($M_w = 7.8$) occurred near Manzanillo and ruptured part of the Rivera-North America plate boundary (MELBOURNE et al., 1997; BANDY et al., 1997). This possibly indicates that the recurrence interval of 77 years for 7.5 magnitude earthquakes estimated by SINGH et al. (1985) is generally correct. However, no large magnitude ($M_w > 8$) earthquake has occurred since 1932.

The Guerrero seismic gap stands out as a region of the highest seismic potential because major seismic events are known to have occurred in this area in the beginning of the last century. In the NW Guerrro gap just one $M_w > 6.0$ earthquake has occurred since the previous large $M_w = 7.5$ event of 1911. This $M_w = 6.7$ shallow subduction thrust earthquake occurred on 18 April 2002 (IGLESIAS et al., 2003), but its rupture area was not large enough to fill

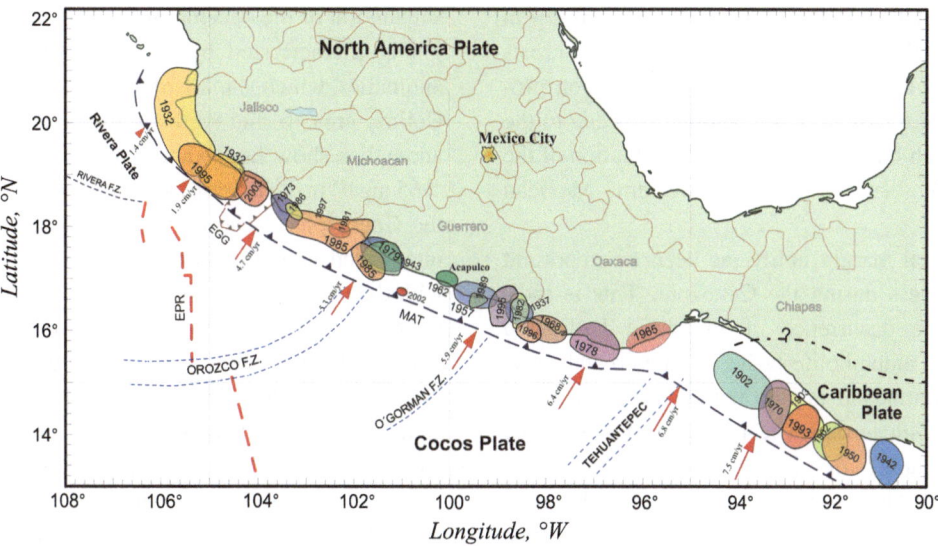

Figure 2
Seismicity along the Middle America Trench. *Map* shows the most important subduction thrust events of this century. *Shaded circles* show rupture zones, and numbers indicate the year of the event

the NW Guerrero seismic gap (Fig. 2). Significant thrust earthquakes on the SE Guerrero coast happened in 1957 ($M_w = 7.8$, Costa Chica) and in 1962 ($M_w = 7.0$, 7.1, Acapulco); thus, the SE Guerrero subduction zone should have accumulated a large amount of strain in the last ~50 years. Recently observed slow slip events (SSE) or "silent earthquakes" occurring within the Guerrero gap region (LARSON *et al.*, 2004) do not preclude this area from having a large subduction thrust event. Consequently, almost all the Guerrero coast is currently considered the most likely to experience a large severe earthquake or several smaller seismic events eventually (ANDERSON *et al.*, 1989; SONG *et al.*, 2009).

The area of the former Michoacan seismic gap is flanked on both sides by active seismic segments, and the OFZ intersects the MAT to the southeast of the gap. This gap was filled by a rupture area of a large ($M_w = 8.0$) earthquake on 19 September 1985. The region of maximum uplift during this earthquake is marked by elevated marine terraces, and these are often associated with a long-term coastal uplift resulting from periodic large magnitude earthquakes (MCNALLY and MINSTER, 1981; BODIN and KLINGER, 1986).

In the Tehuantepec sector of the subduction zone no significant shallow thrust earthquake has occurred

over the past 180 years (Fig. 2), and therefore it is considered to be either aseismic or seismic with anomalously long recurrence intervals for major earthquakes (ASTIZ and KANAMORI, 1984). The possibility that the Tehuantepec gap may be aseismic for large earthquakes has been explained by an influence of the subducting Tehuantepec ridge. This gap is located near the triple junction and is a transition zone with respect to the geometry of the Benioff zone (PONCE *et al.*, 1992; MANEA *et al.*, 2005).

3. Cenozoic Continental Margin Uplift

The western margin of North America has been the subject of a dominant process—plate convergence. Arc magmatism and plate tectonic paleoreconstructions provide evidence for long-term plate convergence in southern Mexico. The narrow forearc region and relatively small offscraped deep-sea sedimentary prism suggest removal of the forearc and accretionary prism (KARIG *et al.*, 1978). Calc-alkaline plutons are located at the margin close to the trench, and continental basement rocks extend to within some 35 km of the trench in the DSDP Leg 66 study area (e.g., LUNDBERG and MOORE, 1981). Structural trends in the continental basement display abrupt termination at the seaward edge, resulting in close juxtaposition of oceanic and continental crusts (Shipley, 1981). Continuous sediment accretion processes can be traced from the Middle Miocene to the present, which brackets initiation of subduction of the Cocos plate beneath the southwestern Mexican margin.

Continental margin uplift has been an important tectonic process during the Cenozoic. This is indicated by the occurrence of mylonitic belts and metamorphic amphibolite-facies middle crustal rocks along the continental margin within the Xolapa tectonostratigraphic terrane. Clear evidence for Neogene differential uplift of the continental margin of southwestern Mexico is given by the widespread exposure of Oligocene batholiths, which contrast with the presence of coeval volcanic rocks landward into the continental interior (Morán-Zenteno et al., 1996). The location of the Balsas river basin (second largest river in Mexico), which is oriented east-west parallel to the margin following the shape of the igneous and metamorphic bodies of the Xolapa terrane, may reflect the tectonic and uplift processes along the continental margin. Margin uplift and erosional processes have resulted in removal of a significant portion of the upper crust exposing the roots of the Xolapa complex magmatic arc.

Margin uplift and growth of the sedimentary accretionary prism along the southern Mexico continental margin have been attributed to underplating of material removed by subduction erosion (e.g., Shipley et al., 1980; Watkins et al., 1982). Morán-Zenteno et al. (1996) propose that progressive uplift along the continental margin was the result of thickening of the subducted slab at the junction of cylindrical flexures at the (trench-trench-transform) triple junction associated with the Chortis block, which was migrating towards the southeast. The left-lateral movement, according to paleogeographic reconstructions, led to progressive migration of the triple junction (e.g., Morán-Zenteno et al., 1996; Meschede et al., 1997).

Isostatic anomalies in central and southern Mexico feature continental margin characteristics (De la Fuente et al., 1994). The trench is characterized by an elongated negative isostatic anomaly (−45 to −55, up to <−75 mgals) associated with bending and subduction of the oceanic plate. The continental margin is characterized by a belt of positive isostatic anomalies, which is marked in the margin south of the Colima area to the south of El Gordo graben. The anomalies show larger amplitudes (45–60 mgals and >65 mgal) in the margin associated with subduction of the Cocos plate. The trend of isostatic anomalies is interrupted at the zones of intersection along the trench with the oceanic fracture zones. These features in the isostaic anomaly pattern can be observed associated with subduction of the Orozco fracture zone and the Tehuantepec ridge. The continental margin and volcanic arc are characterized by small isostatic anomalies. Crustal thickness increases from the coast to the continental interior, with thicker crust under the volcanic arc in the central and eastern TMVB segments (Urrutia-Fucugauchi and Flores-Ruiz, 1996; Urrutia-Fucugauchi et al., 1999). There is also a correlation with the characteristics of the trench shape and sediment infilling. The anomalies can be related to the dense subducted slab of oceanic lithosphere (e.g., Molnar, 1977; Bandy et al., 1999). Gravity models for areas studied along the margin such as the Tehuantepec Isthmus and the Jalisco and Colima sectors indicate that the subducted slab is responsible for the strong high-amplitude positive anomalies.

4. Coastal Uplift

An overview of field measurements coupled with the analysis of landforms, geology, stratigraphy and radiometric dating reveals significant variations in morphology, tectonic styles and rates of deformation along the coast of the Mexican subduction margin. Distinct areas experiencing short- and long-term deformation (Holocene and Pleistocene), and coseismic deformation have been identified in the region on the basis of their characteristic coastal morphology and modern geodetic measurements of tectonic deformation.

The coastal area presently incorporating the Jalisco block (Bourgois and Michaud, 1991) shows evidence of recent tectonic uplift (Fig. 1). The continental shelf here is relatively narrow, locally <1 km but reaching up to 2 km. The distance from the trench to the coast averages ∼73.4 km (standard deviation

9.6), and the predominant value is 55 km, but increases to about 126 km near Bahia de Banderas (Fig. 1). Bahia de Banderas shows evidence of active faulting. A submarine canyon has a half-graben structure where the associated Banderas fault extends westwards down to the MAT. Inland, the Banderas valley is identified as the extension of this active structure (ALVAREZ 2002, 2007; FERRARI and ROSAS-ELGUERA, 2000; URRUTIA-FUCUGAUCHI and GONZALEZ-MORAN, 2006). The coastal landscape here is characterized by cliffs and rocky promontories alternating with beaches, a wide alluvial plain at Bahia de Banderas and a small number of lagoons with narrow barriers further south. The inland landscape is typified by low plains with Pleistocene (?) marine terrace remnants, dissected hills and fault-block mountains. Clear evidence of recent tectonic uplift is provided by extensive marine terraces at 46 m (field measured with Abney level) and about 60 and 100 m (topographic map estimates; Fig. 3). Exposures within the lowest (46 m), seaward-dipping terrace (21°) uncover conglomerates overlain by carbonate rocks. The two higher terraces, which are the more extensive, are formed in tuff, volcanic breccia, and other volcanic rocks and conglomerates. No littoral deposits are found on the terrace surfaces (RAMIREZ-HERRERA and URRUTIA-FUCUGAUCHI, 1999). Correlations with eustatic sea-level curves (SIDALL et al., 2006) suggest that the higher terraces may correspond with oxygen isotope stage 3 and 5c highstands (60–100 ka).

Using wave cut platform shoreline angle averaged elevation correlated with sea-level highstand ages, plotted with an eustatic curve following LAJOIE (1986), SIDDALL et al. (2006) and SAILLARD et al. (2009), longer term uplift rates are estimated at ca. 0.7–0.9 m/ka (Fig. 4). We can preclude that our estimated rates are in the range of those of other similar subduction margins.

Field observations at Punta Mita, northern flank of the Jalisco block, reveal the presence of emerged beachrock and marine terraces at ca. 4 to 19 m above mean sea level (high precision geodetic survey). Marine terraces extend discontinuously along the coast from north of Puerto Vallarta to Punta de Mita, Nayarit, and their elevations can be correlated along 35 km of the coast. This stretch of coastline coincides with the

northernmost sector of the 3 June 1932 earthquake rupture (Fig. 2). Marine terraces show a thin layer of beach deposits. Mollusc shells from marine terrace deposits and beach rock were sampled and dated. The taxonomic analysis of these species (ex. *Chione* sp., *Anomalocardia* sp., *Trigoniocardia* sp. and *Trachycardium* sp.) indicate a moderately shallow water environment. Radiocarbon ages of marine shells collected from beach deposits at the marine terrace platforms provided dates of progressive emergence of these Holocene features. The radiocarbon ages ranging from 1 to 3 ka BP (G. Harper and A. Lefton, personal communication) for the lower terraces (ca. 4 m) indicate that these elevated landforms developed when sea level had reached or was close to its present level. Apparently, subsequent tectonic uplift raised these ancient shorelines above the present sea level. Marine terrace elevations following ANTONIOLI et al. (2009) were compared with the predicted value for sea level using the Lambeck model (LAMBECK et al., 2004) indicating ca. 0 to −0.5 m position and a corrected terrace elevation at ca. 3.5 to 4 m. Marine terrace elevation and the cosmogenic ^{14}C dates (1–3 ka BP) indicate an early Holocene age for these marine terraces and an estimated coastal uplift rate of ca. 0.8 to 3.3 ± 0.5 m/ka, which is in agreement with rates at active margins. It is however difficult to discriminate the coseseismic, if there is such, from the post- and interseismic component. It is inferred that the coast has been rising by interseismic uplift. In the light of historic coseimic subsidence and postseismic recovery reported for the southern part of Jalisco coast (CUMMING, 1933; MELBOURNE et al., 1997), the interseismic component could be of most importance in tectonic uplift and evolution of this coast.

The occurrence of a number of raised marine notches and wave-cut platforms indicate uplift of the Jalisco coast. Most of the notches are in volcanic breccia, some with wave cut platforms, and no deposits are present on the surface (RAMIREZ-HERRERA et al., 2004). These abrasion notches developed near high-tide level (the local tide level range is 1.3 m). Their heights range from ca. 1 to 4.5 m (−0.5 for the predicted sea level point using Lambeck model) above mean sea level. In situ intertidal organism (red algae and barnacles) radiocarbon dates of $1,262 \pm 51$ ^{14}C years BP (cal. AD 660–940; Table 1) indicate

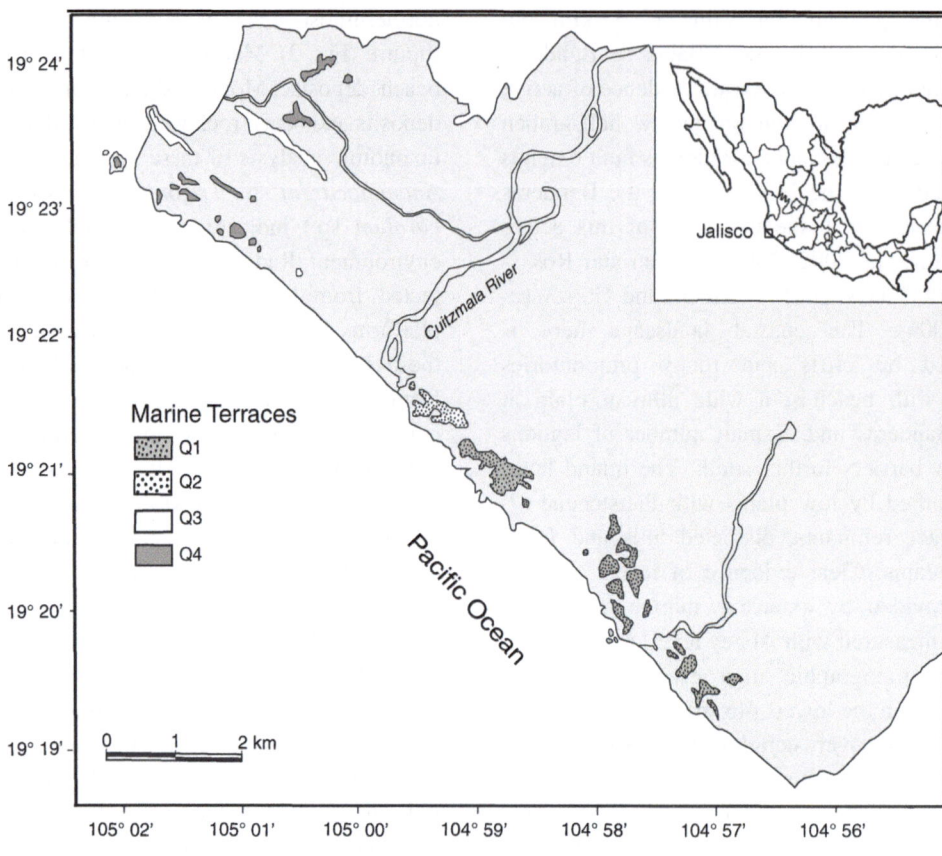

Figure 3
Map of the southern Jalisco coast showing the distribution of marine terraces. Shoreline angle elevations are estimated from topographic maps: Q1: 100–80 m, Q2: 60 m, Q3: 40 m, Q4: 20 m. Insert shows site location

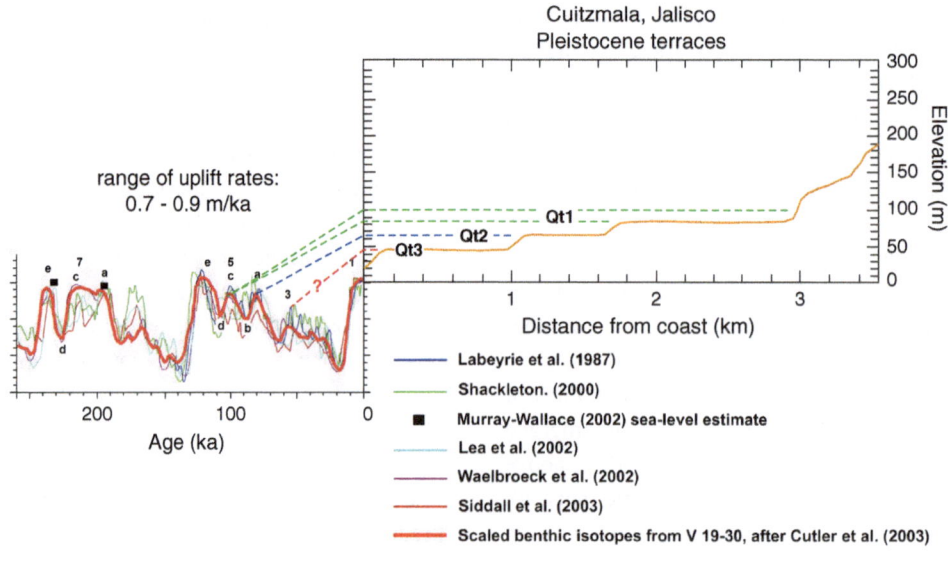

Figure 4
Pleistocene tectonic rates were computed from wave cut platform shoreline angle averaged elevations versus last sea-level highstand ages, plotted with eustatic curve from SIDDALL et al. (2006) and modified curve from SAILLARD et al. (2009)

tectonic uplift during at least the past 1,300 years BP at an average rate of about 3 ± 0.5 mm/year (RAMIREZ-HERRERA et al., 2004). This supports a model in which coseismic subsidence produced by offshore earthquakes is rapidly recovered during the postseismic and interseismic periods. There is no evidence of coastal interseismic and long-term subsidence along this sector of the Mexican Pacific active margin (RAMIREZ-HERRERA et al., 2004).

Further evidence of tectonic uplift is found on the coastal zone incorporated south of the El Gordo-Colima graben to where the onshore projection of the OFZ meets the coast in the vicinity of the Balsas submarine canyon (Fig. 1), incorporated in the Michoacan morphotectonic zone (RAMIREZ-HERRERA and URRUTIA-FUCUGAUCHI et al., 1999). The distance from the coast to the trench is greater towards the boundaries of the zone marked by the El Gordo graben and the OFZ to about 95 km, with an average distance of 79 km (standard deviation 9.6; Fig. 1). Here high (>600 m) fault-block mountains reach the sea and give rise to a coastal landscape characterized by steep cliffs and rocky promontories alternating with narrow pocket beaches. Inland the high fault-block and folded mountains have been deeply incised by coastal streams, suggesting the rivers response to tectonic uplift. Landforms provide evidence of uplift on this sector of the Mexican Pacific margin. A number of marine terraces, elevated wave-cut notches and river terraces are intermittently present between the Playa Azul and south of the Colima graben to the north (Fig. 1). At least two well-preserved marine and fluvial terraces are prominent and rise discontinuously through this coastline at elevations ranging from 15 to 120 m (the lowest measured with precise leveling). Gravels supporting well-developed clay-rich soils up to several meters thick are present on these terraces suggesting that they pre-date the Holocene. Correlations with eustatic sea-level curves (SIDALL et al., 2006) suggest that the terraces may correspond with oxygen isotope stage 3 and 5e highstands (60–120 ka) and support this interpretation. This indicates longer term uplift rates of 0.5–1.0 m/ka, following LAJOIE et al. (1986) and SAILLARD et al. (2009), within the central Michoacan coast (RAMIREZ-HERRERA et al., 1998). At the southern flank of the Michoacan morphotectonic zone

(RAMIREZ-HERRERA and URRUTIA-FUCUGAUCHI, 1999) and close to the OFZ, marine terraces range from 5–6 m to 8 m above mean sea level. Beach ridge deposits containing datable marine shells and pottery shards record progressive emergence during the Holocene. Cosmogenic date, 6.5 ± 0.5 ^{14}C ka (G. Harper and A. Lefton, personal communication), for uplifted, 3.5 ± 1-m-elevation beach ridges near the southern border of the Michoacan coast indicate an estimated uplift rate, using the Lambeck model, of 0.08 m/ka.

South of OFZ and in the area that incorporates the Guerrero morphotectonic zone (RAMIREZ-HERRERA and URRUTIA-FUCUGAUCHI, 1999; RAMIREZ-HERRERA et al., 1999), recent uplift is indicated by both marine and river terraces. Clear evidence comes from the area to the southeast of OFZ to about the Ometepec canyon. At least three extensive marine surfaces are present near Lagunillas (Fig. 1). The lowest is ca. 10 m above mean sea level, while the elevation of the intermediate terrace is ca. 20 m. This terrace, which is mantled by unconsolidated sands interbedded with pebble sand units, is the most extensive terrace along this sector of the coast. High river terraces are observed along all the largest rivers in the area, indicating uplift; for instance, the Rio Mezcala shows a series of terraces of up to 100 m above the present river level.

Evidence of uplift is more sporadic south of Lagunillas (Fig. 1). Two distinct surfaces are the only marine terraces observed and measured in this area, approximately 3 km north of Zihuatanejo-Ixtapa. The upper terrace, which is a little over 10 m high above the present mean sea level, is the most extensive, being approximately 2 km long and 1 km wide. It is formed of granite with a 2–3-m-thick mantle of gravel. The lower marine terrace is ca. 6 m above high tide. Scattered remnants of what appears to be an older higher marine terrace are present inland from the two lower terrace. OSL dating indicates a late Pleistocene age for the higher terraces (G. Harper and A. Lefton, personal communication). U-series dates of corals on the lowest uplifted beach ridge at La Saladita, north Guerrero, 4.5 m above mean sea level (compared with Lambeck model), ranging from ca. 584 ± 34 to 895 ± 29 year BP (G. Harper and A. Lefton, personal communication), suggest that uplift rates could be as high as 6.6 ± 2 m/ka. A notch

Table 1

Radiocarbon dates from cores one (ACA-04-01) and three (ACA-04-06), Laguna Mitla site

Sample lab ID[d,e]	Site	Core	Sample depth cm	Averaged elevations (m)[a]	Elevation amsl (m)	Material dated	[14]C age year BP	Calibrated age (2 sigma range)	[14]C/[13]C	Reference
AA03760	Barra de Navidad notch	–	–	0.46 to 0.6[b]	0.6 to 0.9	Shell	Post 0[c]	ca. 1959 AD[c]	–	RAMIREZ-HERRERA et al. (2004)
AA22193	Cuitzmala notch	–	–	4.5 to 4.7	4.5. to 4.7	Coralline algae	1,262 ± 51	660–940 AD	–25	RAMIREZ-HERRERA et al. (2004)
AA59462	Laguna Mitla, Guerrero	ACA-04-01	155	–	–	Wood	3,166 ± 34	3,336–3,456	–27.15	RAMIREZ-HERRERA et al. (2007)
AA59469	Laguna Mitla, Guerrero	ACA-04-06	394		–	Charcoal/ wood	4,630 ± 37	5,297–5,467	–30.173	RAMIREZ-HERRERA et al. (2007)
AA59470	Laguna Mitla, Guerrero	ACA-04-06	387	–	–	Wood	4,547 ± 36	5,051–5,317	–24.991	RAMIREZ-HERRERA et al. (2007)
AA59471	Laguna Mitla, Guerrero	ACA-04-06	377	–	–	Wood	4,559 ± 36	5,053–5,437	–27.77	RAMIREZ-HERRERA et al. (2007)
AA59472	Laguna Mitla, Guerrero	ACA-04-06	371	–	–	Wood	4,626 ± 37	5,294–5,468	–28.26	RAMIREZ-HERRERA et al. (2007)
AA59474	Laguna Mitla, Guerrero	ACA-04-06	234	–	–	Wood	2,836 ± 34	2,859–3,063	–27.412	RAMIREZ-HERRERA et al. (2007)
AA54043[e]	Laguna Mitla, Guerrero	ACA-03-02	299–300	–	–	Charcoal	2,818 ± 42	2,842–3,043 Cal BP	–	RAMIREZ-HERRERA et al. (2009)
AA54044[e]	Laguna Mitla, Guerrero	ACA-03-02	560–564	–	–	Wood	6,161 ± 53	6,901–7,176 Cal BP	–	RAMIREZ-HERRERA et al. (2009)

[a] Elevation data corrected for tide level at time of measurement; measurement error estimates ±0.2 m. Mean tide range for locations near Manzanillo, Colima, ca. ±0.5 m, and ±0.8 m for locations near Puerto Vallarta

[b] El Niño correction (1998) +0.14 to 0.32 m

[c] 109% of modern (Beta-119880)

[d] NSF-Arizona AMS facility. Radiocarbon calibrations were performed using the program CALIB 5.0 (STUIVER and REIMER, 1993) and the calibration dataset IntCal04 (REIMER et al., 2004)

[e] NOSAMS AMS laboratory

in bedrock and a highly irregular wave-cut surface are uplifted 5 m above high-tide level near Zihuatanejo-Ixtapa; however, these features are barren of deposits from which the cutting of the surface could be dated. Three river terraces, indicative of tectonic uplift, are present near Petatlan (southeast of Zihuatanejo-Ixtapa). Deposits on the highest, 25-m-elevation terrace and a thick soil cover suggest an old age (Pleistocene?) for this feature.

The Mexican Pacific coast is formed by rising mountains exhibiting triangular facets and V-shaped valleys indicating discrete faulting between the coastal plain and mountains inland, off the OGFZ and southeast of the Ometepec canyon (Fig. 1). The coastal landscape shows uplifted terraces in this area, near Punta Maldonado. A clear flight of marine and fluvial terraces is revealed on the Oaxaca coast along ca. 40 km, from Puerto Escondido to Puerto Angel

(Fig. 1). Two terraces rise at ca. 8 and 19 m elevation. Several streams dissect these terraces showing V-shaped valleys, indicative of active river incision. Moreover, marine terraces show a well-preserved and steep sea cliff. Active river incision and well-preserved sea cliffs on these terraces suggest active tectonic uplift in this area. Further work is required in this sector of the Pacific coast of Mexico to precisely determine rates of uplift.

5. Coastal Subsidence

The coastal area incorporated within the El Gordo-Colima graben coastal sector, which corresponds to the boundary between the Rivera and Cocos plates, is marked by the lack of evidence of elevated surfaces. Here the distance from the coast to the trench averages 77.5 km (5.7 km SD) and reaches up to ca. 88 km. The coastal landscape displays a wide low deltaic plain, sand bars, berms, estuaries and large coastal lagoons. Inland, fold mountains and Quaternary active volcanoes fringe the area. Active faults of the Colima graben show triangular facets and a remarkable depression further inland. Active extension has been suggested by other authors (BANDY et al., 2000, 2005; SUAREZ et al., 1994). A large, $M_w = 5.3$, normal event recently occurred (7 March 2000) in South-Colima graben (ZOBIN et al., 2000; PACHECO et al., 2003a, b). This event conclusively demonstrates that a NW-SE oriented extension is occurring in the South-Colima graben (BANDY et al., 2005). Depositional landforms on the coast suggest that the coast is either subsiding or perhaps slowly rising. Evidence of historical coseismic subsidence has been reported on this coast (CUMMING, 1933; MELBOURNE et al., 1997). We discuss this further below.

Stratigraphic, morphologic and archeological evidence of Late Holocene coastal subsidence has been reported for the Guerrero central coast (RAMIREZ-HERRERA et al., 2007, 2009). The sedimentary record of the Mitla and Coyuca Lagoon indicates a rapid shift from freshwater/brackish, i.e., a marginal lagoon, to marine conditions at ca. 3,400 year BP (Fig. 5). This rapid change in depositional environment is explained by a local earthquake that triggered a tsunami (RAMIREZ-HERRERA et al., 2007). The latter is inferred from the presence of discrete sand layers above the brackish mangrove peat and associated with a rapid marine inundation by a tsunami (RAMIREZ-HERRERA et al., 2007). The paleoseismic record thus suggests that subsidence of the coast occurred with a large thrust event that triggered a tsunami at the central Guerrero coast near Acapulco. Further indirect evidence for coastal subsidence along this coast is also revealed by local archeological sites, in the form of submerged shell middens at Laguna Coyuca (KENNETT et al., 2004; RAMIREZ-HERRERA et al., 2005) and marine microfossil proxies (pollen, diatoms and marine plankton) at the adjacent Laguna Tetitlan in the central Guerrero coast (GONZALEZ-QUINTERO, 1980).

Recent geodetic observations: leveling from 1995 to 1998 (KOSTOGLODOV et al., 2001) and continuous GPS records from 1997 up to 2008 suggest a short-term coastal subsidence of 1.1–1.4 cm/year (VERGNOLLE et al., 2010) at a period of ~4 years between the occurrence of aseismic slow slip events (SSE; YOSHIOKA et al., 2004; KOSTOGLODOV et al., 2003; COTTE et al., 2009). Tide gauge sea-level records available from 1953 reveal a long-term interseismic subsidence of ca. 0.3–0.4 cm/year on the Guerrero coast near Acapulco (ALVA and KOSTOGLODOV, 2007). A large subduction thrust earthquake should release the elastic strain accumulated during the interseismic epoch, which will result in a notable coseismic uplift. For example, an uplift of 22 cm was observed in 1962 during the M_w 7, 7.1 earthquake doublet that ruptured the plate interface right below the Acapulco coast (ORTIZ, 2000).

An apparent inconsistency exists between the expected coseismic "uplift" and a notable subsidence of the coastal area derived from the sedimentary record of a large tsunamigenic earthquake by ca. 3,400–3,500 year BP (Table 1) (RAMIREZ-HERRERA et al., 2009). A plausible explanation proposed to reconcile this inconsistency is that while the recent (100–150 years) large earthquakes in the central Mexican subduction zone have a limited width and rupture only strongly coupled interplate patches (within only 40-60 km and located below the coast), some prehistoric earthquakes most likely ruptured the entire coupled plate interface almost up to the trench.

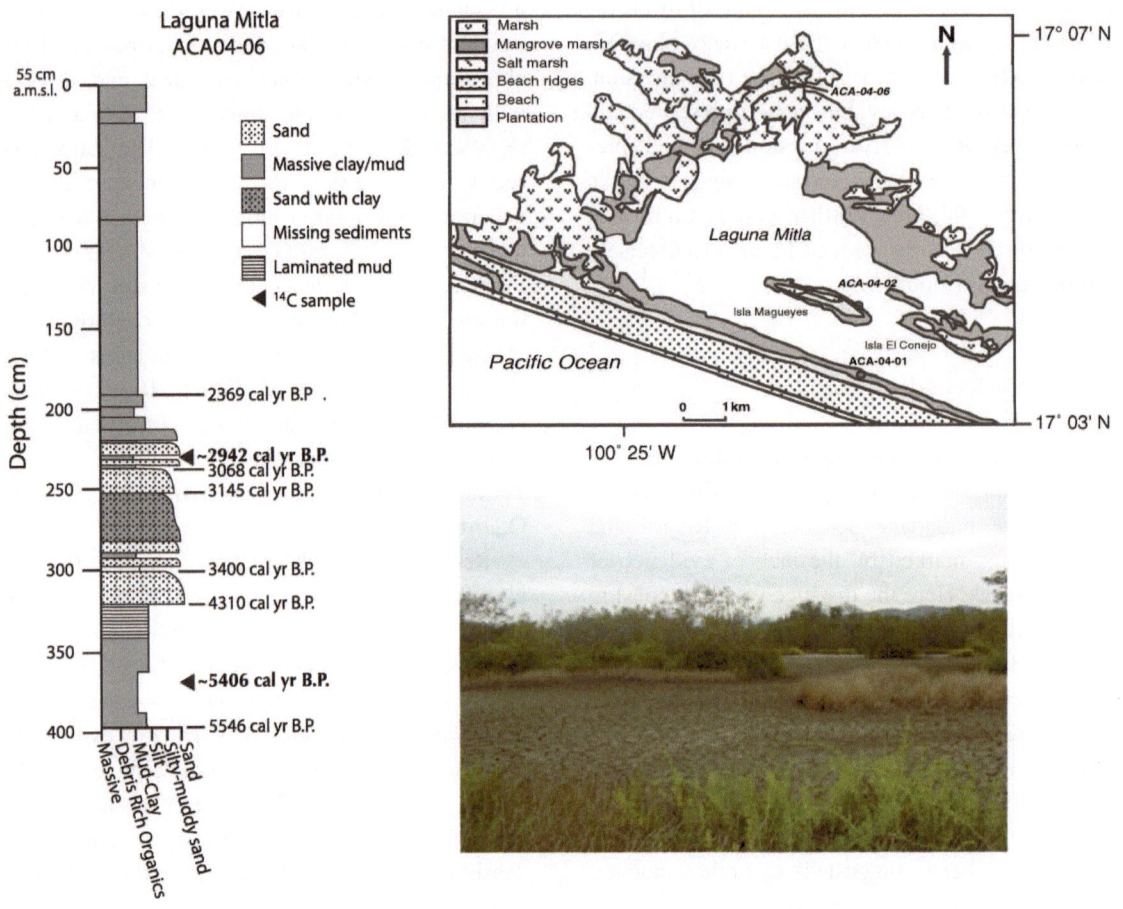

Figure 5

Sediment record from lagoonal marsh at Laguna Mitla, Guerrero. Cored sediments show a buried soil cover by sand (marine incursion) indicative of coastal subsidence. Insert shows location map and photograph of lagoon marshes (modified from RAMÍREZ-HERRERA et al., 2007)

Such a combined megathrust event should produce a considerable coastal area subsidence (RAMIREZ-HERRERA et al., 2009). Figure 6 illustrates possible models to explain the difference in uplift produced by recent subduction earthquakes and a hypothetical mega-event. An example of a near trench event is the M_w 6.7, 18 April 2002 thrust earthquake in front of the Guerrero coast, which ruptured a weakly coupled plate contact close to the trench and generated a small tsunami (Fig. 2; IGLESIAS et al., 2003). The recurrence period of mega-events such as the one that occurred by ca. 3,400–3,500 years BP (RAMIREZ-HERRERA et al., 2009) may be very long (10^2–10^3 years). Nevertheless, more detailed studies on the coseismic deformation of historical and prehistorical earthquakes and modeling is required to resolve this inconsistency between interseismic and expected coseismic coastal deformation on the Guerrero coast.

Further south and southeast of the Tehuantepec ridge offshore, the coastal landscape shows a remarkable change. In this area, low accumulative marine plains with eolian, marine and alluvial deposits, wide beaches, sandbars and lagoons characterize most of the coast. No evidence of elevated surfaces is found on this sector of the Mexican Pacific margin. However, high fault-block mountains border the area inland (RAMIREZ-HERRERA and URRUTIA-FUCUGAUCHI, 1999). The coast has been "apparently stable" in recent times. Offshore, there is a dramatic change with the presence of the Tehuantepec ridge and an increase in the width of the continental margin into a well-developed fore-arc basin, and no observed accretion. Here changes in convergence rates, geometry of the subduction and depth of earthquakes mark a seismotectonic transition to a different domain (SHIPLEY *et al.*, 1980; AUBOIN *et al.*, 1981; RAMIREZ-HERRERA and URRUTIA-FUCUGAUCHI, 1999; MANEA *et al.*, 2005). The palaeoecology and stratigraphy of the sequence on marsh and lagoonal sediments should help to distinguish evidence of sudden land elevation changes. Beach ridge sequences at this coast, if these exist, may reveal possible coseismic steps. However, further studies should be done to reconcile the Recent (Pleistocene-Holocene) coastal deformation of this region and to check a hypothesis that the coastal deformation here has been driven by a gradual, interseismic process.

5.1. Historic Coseismic Deformation

There is limited evidence of local coseismic deformation measured after the occurrence of large earthquakes on the Pacific Mexican margin. The 1985 earthquake, 8.1 (M_w), produced up to 1.0 m of coastal uplift in the Michoacan coastal segment (BODIN and

KLINGER, 1986). A survey of the vertical distribution of intertidal organisms documents local deformation through the coastal area enclosed in the 19 and 21 September 1985 earthquake rupture zones. These data however represent a conservative estimate of the actual magnitude of vertical displacement associated with the earthquakes. The region of greatest reported uplift coincides with an area marked by flights of uplifted marine terraces, mentioned above, associated with long-term coastal uplift and may represent residual uplift resulting from episodic large earthquakes on the underlying megathrust (BODIN and KLINGER, 1986).

Vertical deformation that produced coastal uplift was recorded by the extent of coralline algae mortality after the M_w 6.3, 2 February 1998 earthquake in Puerto Angel, Oaxaca (RAMIREZ-HERRERA and ZAMORANO, 2002). The extent of the coastal uplift reached up to 0.5 m near Agua Blanca. Coseismic uplift was recorded for ca. 70 km along the Oaxaca coast (RAMIREZ-HERRERA and ZAMORANO, 2002). The region of reported maximum coseismic coastal uplift also coincides with an area of clear uplifted marine

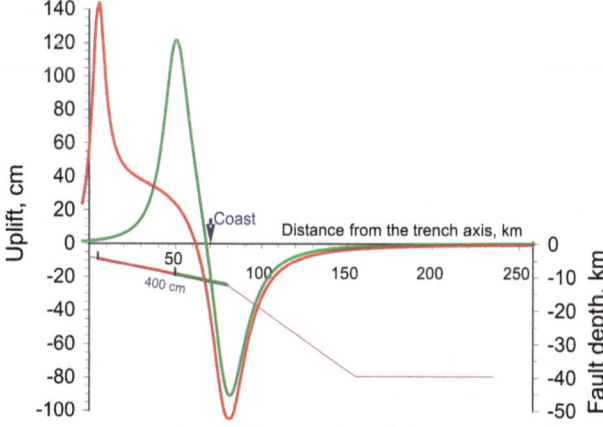

Figure 6

The model of tectonic uplift (dislocation in a homogeneous elastic half-space, SAVAGE, 1983). *Green line* is modeled uplift produced by recent subduction thrust events on the coast of Guerrero (1962 M_w 7.1, 7.0 events). The ruptured patch of the coupled plate interface with a coseismic displacement of 400 cm is shown as a *green line*. *Straight line* segments represent the subduction plate interface geometry (KOSTOGLODOV *et al.*, 1996; KIM *et al.*, 2010). While the tsunamis produced by these seismic events were only ~22 cm (ORTIZ *et al.*, 2000), their combined effect was negligible, in particular because the subsidence of the coast was <10 cm. The main subsidence occurred over the area of highland and was not affected by the tsunami at all. *Red line* is the uplift model for the case when the near trench coupled patch rupture together with a downdip coupled plate interface (*red + green* patches, with the same fault slip of 400 cm as during the 1962 events. Actually the fault slip for the higher magnitude earthquake may have been about two times larger than that in 1962. The uplift for this hypothetical earthquake (*red curve*) shows that the coastal area will subside noticeably, and the combined effect of tsunami amplitude and its runup should be very large. Later interseismic subsidence of the coast will be produced mostly by a strongly coupled downdip seismogenic patch (*green*). Recurrence period of the "combined" mega-thrust earthquake may be of the order of 10^3 years

terraces, mentioned above, and associated to long-term coastal uplift. The area is also enclosed within the rupture area of the 29 November 1978, $M = 7.6$ earthquake. This suggests that it may as well represent residual uplift produced by periodical large earthquakes on the underlying megathrust in southern Mexico.

The June 3 and 18 1932 earthquakes, $M_s = 8.1$ and 7.8, respectively, broke the Rivera-North America Plate boundary (SINGH et al., 1985) and produced coastal subsidence estimated at 40–75 cm on the stretch of the coast from Barra de Navidad and south of the Tecoman, southern Jalisco and Colima coasts (CUMMING, 1933). Recently, the 9 October 1995 earthquake ($M_w = 8.0$), located near Manzanillo, ruptured part of the Rivera-North America plate boundary (MELBOURNE et al., 1997; BANDY et al., 1997; ORTIZ et al., 1999). Coseismic subsidence of 80 ± 14 mm at Manzanillo, Colima, was estimated from GPS surveys performed before and after the earthquake (MELBOURNE et al., 1997). Pressured gauge records however indicated subsidence of 44 cm in Barra de Navidad and 11 cm in Manzanillo, Colima (KOSTOGLODOV et al., 1997).

6. Discussion

6.1. How Tectonic Deformation (Uplift/Subsidence) Rates Vary Spatially and Temporally on the Active Mexican Pacific Margin Coastal Areas

The Mexican Pacific margin shows spatial and temporal variability in the style and rates of tectonic deformation along-strike the subduction zone (Fig. 7). Non-steady long-term uplift rates are recorded from wave-cut platforms resulting from marine erosion during sea-level highstands, from beach ridges and tidal notches. Pleistocene uplift rates are estimated at ca. 0.7–0.9 m/ka on the Jalisco block active margin, within the Rivera-North America tectonic plate boundary, while uplift rates show an increase in the Holocene up to ca. 3 ± 0.5 m/ka. On the contrary, steady rates of uplift (ca. 0.5–1.0 m/ka), in the Pleistocene and Holocene, characterize the coastal sector south of El Gordo graben to the OFZ, Michoacan coast, which is

incorporated within the Cocos-North America plate boundary. Considerable high rates of uplift were measured in the margin sector across the OFZ subduction. Long-term, late Holocene subsidence was recorded in the Guerrero coastal sector near Acapulco, where the geometry of the subduction is very shallow. This is in agreement with a landscape and archeological evidence of coastal subsidence.

The difference in the deformation records from the Jalisco coastal sector and those from the Michoacan coast, particularly at the OFZ coastal sector, suggests that the uplift rate varies not only temporally (non-steady rates in Jalisco), but also spatially along-strike. The possible causes of uplift rate variability, temporal and spatial, are likely due to localized coastal tectonic processes linked to subduction. Marine terraces in the studied areas were developed indifferently on a heterogeneous substratum (i.e., volcanic breccia, granite). Therefore, lithologic variations in bedrock do not play a major role in the morphology of these terraces. Faulting in the Punta de Mita coastal area, Jalisco block, possibly the "Marieta-Punta de Mita Fault" (ALVAREZ, 2007), might have occurred after the formation of the Holocene marine terrace, which might explain the accelerated rates during the Holocene. Another plausible explanation for those changes in coastal morphology and surface deformation rates, i.e., uplift (Jalisco and Michoacan) and subsidence (Guerrero), is that slab subduction processes strongly influence the forearc tectonic deformation. The roughness of the subducted plate, such as irregularities, seamounts and ridges, can certainly lead to subduction erosion and underplating of material (e.g. HSU, 1992; HAMPEL et al., 2004; CLIFF and HARTLEY, 2007; SACK et al., 2009). These processes are even more evident on the coastal sector where the OFZ produces topographic irregularities on the subducting plate or material in the trench leading to rapid and episodic uplift rates. Thus, it is probable that the Mexican Pacific margin similarly to the Chilean forearc has experienced periods of lower uplift rates that might alternate with periods of accelerated uplift rates over time scales $\ll$ 1 Ma (SAILLARD et al., 2009), and also may be due to active faulting. Another plausible explanation for the variability of uplift rates of marine terraces above different tectonic settings along the coast of the

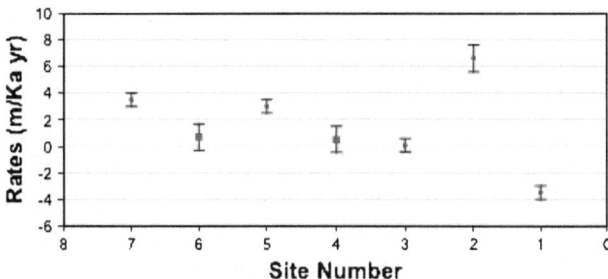

Figure 7

Tectonic rates with *error bars*. Temporal and spatial variability of coastal deformation rates on the Mexican Pacific margin. *Circles* with *error bars* in *black* show Holocene rates, *squares* with *error bars* in *gray* show Pleistocene rates. Sites: *1* Guerrero (Mitla), *2* Guerrero (Saladita), *3* Michoacan (Bufadero), *4* Michoacan (Neixpa), *5* Jalisco (Cuitzmala), *6* Jalisco (Farallon), *7* Jalisco-Nayarit (Punta Mita). Distance between sites is not to scale

Mexican Pacific margin could be related to the change in downdip position of the locked zone along the plate interface with respect to the continental forearc. Nevertheless, further data are required to determine which mechanism is responsible for the strong variation in uplift rates and style of deformation (uplift vs. subsidence).

6.2. Interseismic and Coseismic Deformation

Coseismic deformation produced during large thrust events results in coastal uplift and coastal subsidence along the Mexican Pacific margin. Coseismic deformation rates show that considerable deformation accompanies large thrust events ($M_w > 6.0$) on the Mexican Pacific margin. Coseismic uplift was measured after the 1985 event on the Michoacan coast reaching up to 1 m. Coastal uplift was also produced by the 1998 event on the Oaxaca coast that reached up to 0.5 m. These two sectors of the coast, Michoacan and Oaxaca, show marine terraces that indicate that uplift also takes place during interseismic periods. However, south of the Jalisco block, coastal coseismic subsidence of up to 0.4–0.75 m was measured after the 1932 earthquake at El Gordo-Colima graben coastal sector. Similarly, the Colima 1995 earthquake produced coastal subsidence of up to 0.08 m (measured with GPS) and 0.11–0.44 cm (tide gauge records) in the Manzanillo, Colima coast. The coastal landscape of this sector of Mexican Pacific margin is barren of morphologic evidence of long-term tectonic uplift, suggesting that either these coastal features have been eroded and/or

this coast is subject to interseismic slow uplift or to slow subsidence. Further work is required to support this hypothesis.

Coastal Holocene subsidence has occurred on the central Guerrero coast during great earthquakes that have ruptured the entire coupled plate interface almost up to the trench (RAMIREZ-HERRERA *et al.*, 2009). Based on models from registered seismic events, the expected coseismic coastal response is coseismic uplift in this coastal sector during events with limited width of the rupture area (within only 60 km and located below the coast).

Coastal rate estimates might have large uncertainties but show a consistent pattern of uplift in the Michoacán and Oaxaca coastal sectors, and subsidence in the Colima-south Jalisco coastal sectors, respectively. Elastic dislocation models constrained by interseismic geodetic or/and thermal data used to predict the coseismic pattern of deformation for the likely strain accumulation periods of plate convergence and for uniform megathrust slip are scarce for the Mexican megathrust fault. These models would provide a better understanding and help determine if predicted uplift/subsidence is in broad agreement with observed coseismic deformation.

7. Conclusions

Systematic measurements of marine wave-cut platforms, marine terraces, beach ridges and tidal notches suggest a pattern of non-steady uplift in the northwestern segment of the Mexican Pacific margin

(Jalisco block). Accelerated uplift rates in the late Holocene on the coastal stretch of the Jalisco block and Rivera plate boundary might be related to local faulting. The coastal segment incorporated in the Jalisco block-Rivera plate boundary has been rising by interseismic non-steady tectonic uplift. In light of historic coseismic subsidence and rapid postseismic recovery reported for the southern part of Jalisco coast (CUMMING, 1933; MELBOURNE et al., 1997), the interseismic component must be the most important process for tectonic uplift and the evolution of this coast. This supports a model in which coseismic subsidence produced by offshore earthquakes is rapidly recovered during the postseismic and interseismic periods. We show that recent uplift/subsidence rates in the Mexican Pacific coast have been highly variable, spatially and temporally, during the last 120 ka year, at time scales intermediate between those obtained using geodetic methods (~ 10 year) and those based on geological data (10^6 year). Variation in the uplift/subsidence rates is likely due to differences in tectonic regimes and to localized processes related to subduction, such as crustal faults, subduction erosion and underplating of subducted materials under the coastal zone (e.g., coast across the OFZ). This highlights the importance of identifying several datable geomorphic markers over the timescale of interest when attempting to assess regional or local uplift/subsidence rates. This is of most importance on the Mexican Pacific margin in light of the scarcity of such data and of elastic dislocation models.

Acknowledgments

We appreciate the kind permission of Dr. G. Harper and A. Lefton to allow access to their unpublished C14, U-Series and OSL dates and to reference them, and for providing insightful discussion vital to the success of this review. We thank J. A. Navarrete for help with figure design and GIS data management. Ramirez-Herrera acknowledges funding provided by DGAPA-PAPIIT IN123609. Constructive review by Associate Editor Eugenio Carminati greatly helped clarify the ideas presented in this work.

REFERENCES

ALVA, V.A., KOSTOGLODOV, V. (2007), Aseismic slow slip events in Mexico from tide gauge records. Geos 27(1), 119.

ALVAREZ, R. (2002), Banderas rift zone: a plausible NW limit of the Jalisco block. Geophys Res Lett 29(20), 1994. doi:10.1029/2002GL016089.

ALVAREZ, R. (2007), Submarine topography and faulting in Bahía de Banderas, Mexico. Geofísica Int 46(2), 93–116.

ANDERSON, J.G., SINGH, S.K., ESPINDOLA, M., and YAMAMOTO J. (1989), Seismic strain release in the Mexican subduction thrust, Phys Earth Planetary In 58(4), 307–322.

ANDERSON, R.S., DENSMORE, A.L., ELLIS, M.A. (1999), The generation and degradation of marine terraces, Basin Research 11, 7–19.

ANTONIOLI, F., FERRANTI, L., FONTANA, A., AMOROSI, A., BONDESAN, A., BRAITENBERG, C., DUTTON, A., FONTOLAN, G., FURLANI, S., LAMBECK, K., MASTRONUZZI, G., MONACO, C., SPADA G., STOCCHI, P. (2009), Holocene relative sea-level changes and vertical movements along the Italian and Istrian coastlines. Quaternary Int 206, 102–133. doi:10.1016/j.quaint.2008.11.008.

ASTIZ, L., and KANAMORI, H. (1984), An earthquake doublet in Ometepec, Guerrero, Mexico. Phys Earth Planetary Int 34, 24–45.

AUBOIN, J., STEPHAN, J.F., RENARD, V., ROUMP, J., and LOUSDALE, P. (1981), Subduction of the Cocos plate in the mid America trench, Nature 294, 147–150.

BANDY, W.L., MORTERA, C., URRUTIA-FUCUGAUCHI, J., and HILDE, T.W.C. (1995), The subducted Rivera-Cocos plate boundary: where is it, what is it, and what is its relationship to the Colima rift? Geophysical Research Letters 22, 3075–3078.

BANDY, W., KOSTOGLODOV V., SINGH S.K., PARDO M., PACHECO J., and URRUTIA-FUCUGAUCHI J. (1997), Implications of the October 1995 Colima-Jalisco Mexico earthquakes on the Rivera-North America Euler Vector, Geophys Res Lett 24(4), 485–488.

BANDY, W., KOSTOGLODOV, V., HURTADO-DIAZ, A., and MENA, M. (1999), Structure of the southern Jalisco subduction zone, Mexico, as inferred from gravity and seismicity, Geofis Int, 38(3), 127–136.

BANDY, W.L., HILDE, T.W.C., YAN, C.Y. (2000), The Rivera-Cocos plate boundary; implications for Rivera-Cocos relative motion and plate fragmentation. In Cenozoic tectonics and volcanism of Mexico (DELGADO-GRANADOS, H., AGUIRRE-DÍAZ, G., and STOCK, J.M., (eds.), Boulder, Colorado). Geological Soc. Am. Special Paper 334, 1–28.

BANDY, W.L., MICHAUD, F., BOURGOIS, J., CALMUS, T., DYMENT, J., MORTERA-GUTIERREZ, C.A., ORTEGA-RAMIREZ, J., PONTOISE, B., ROYER, J.-Y., SICHLER, B., SOSSON, M., REBOLLEDO-VIEYRA, M., BIGOT-COMIER, F., DIAZ-MOLINA, O., HURTADO-ARTUNDUAGA, A.D., PARDO-CASTRO, G., and TROUILLARD-PERROT, C. (2005), Subsidence and strike-slip tectonism of the upper continental slope off Manzanillo, Mexico, Tectonophysics 398(3–4), 115–140.

BODIN, P., and KLINGER, T. (1986), Coastal uplift and mortality of intertidal organisms caused by the September 1985 Mexico earthquakes, Science 233, 1071–1073.

BRADLEY, W.C., GRIGGS, G.B. (1976), Form, genesis, and deformation of central California wave-cut platforms, Geol Soc Am Bull 87, 433–449.

BRIGGS, R.W., SIEH, K., MELTZNER, A.J., NATAWIDJAJA, D., GALE-TZKA, J., SUWARGADI, B., HSU, Y.-J., SIMONS, M., HANANTO, N., SUPRIHANTO, I., PRAYUDI, D., AVOUAC, J.-P., PRAWIRODIRDJO, L., and BOCK, Y. (2006), *Deformation and slip along the Sunda megathrust in the great 2005 Nias-Simeulue earthquake*. Science 311, 1897–1901. doi:10.1126/science.1122602.

BURBANK, D.W., and ANDERSON, R.S., *Tectonic Geomorphology* (Blackwell Science, Burbank, D.W., and Anderson, R.S. 2001).

BOURGOIS J., and MICHAUD F. (1991), *Active fragmentation of the North America plate at the Mexican triple junction area off Manzanillo*, Geo-Marine Letters 11(2), 59–65.

CHAPELL, J., OTA, Y., and BERRYMAN, K. (1996), *Late Quaternary coseismic uplift history of Huon Peninsula, Papua New Guinea*, Quaternary Sci Rev 15, 7–22.

CHAPPELL, J.M. (1983), *A revised sea-level record for the last 300,000 years from Papua New Guinea*, Search 14, 99–101.

CLIFT, P.D., HARTLEY, A.J. (2007), *Slow rates of subduction erosion and coastal underplating along the Andean margin of Chile and Peru*. Geology 35, 503–506.

COTTE, N., WALPERSDORF, A., KOSTOGLODOV, V., VERGNOLLE, M., SANTIAGO, J.-A., MANIGHETTI, I., and CAMPILLO, M. (2009), *Anticipating the next large silent earthquake in Mexico*, Eos Trans AGU 90(21).

CUMMING, J.L. (1933), *Los terremotos de junio de 1932 en los estados de Colima y Jalisco*, Universidad de México, VI, 31–32, 68–104.

DE LA FUENTE, M., AITKEN, C.L.V., MENA, M., and SIMPSON, R.W. (1994), Cartas Gravimétricas de la Republica Mexicana, III. Carta de Anomalía Residual Isostática. Publicación del Instituto de Geofísica, UNAM, Mexico, scale 1:3,000,000.

DEMETS, C., and STEIN, S. (1990), *Present-day kinematics of the Rivera Plate and implications for the tectonics in southwestern Mexico*. J Geophys Res 95, 21931–21948.

DEMETS, C., GORDON, R.G., ARGUS, D.F., and STEIN, S. (1994), *Effect of recent revisions to the geomagnetic reversal time scale on estimates of current plate motions*, Geophys Res Lett 21, 2191–2194.

DEMETS, C., and TRAYLEN, S. (2000), *Motion of the Rivera Plate since 10 Ma relative to the Pacific and North American plates and the mantle*, Tectonophysics 318(1–4):119–159.

FERRARI, L., and ROSAS-ELGUERA, J. (2000). Late Miocene to quaternary extension at the northern boundary of the Jalisco block, western Mexico: the tepic-zacoalco rift revisited. In DELGADO-GRANADOS, H., AGUIRRE-DÍAZ, G., and STOCK, J.M. (eds.), *Cenozoic Tectonics and Volcanism of Mexico*. Boulder, Colorado, Geological Soc. Am. Special Paper 334, 41–63.

FISHER, R.L. (1961), *Middle America trench: topography and structure*. Geol Soc Am Bull 72, 703–720.

GONZÁLEZ-QUINTERO, L. (1980), Paleoecología de un sector costero de Guerrero, México (3000 años), Coloquio sobre paleo-botánica y palinología. Memorias, *Coleccion Científica Prehistoria* 86, pp. 133–157.

HAMPEL, A., ADAM J., and KUKOWSKI N. (2004), *Response of the tectonically erosive south Peruvian forearc to subduction of the Nazca ridge: analysis of three-dimensional analogue experiments*, Tectonics 23, TC5003. doi:10.1029/2003TC001585.

HANSON, K.L., WESLING, J.R., LETTIS, W.R., KELSON, K.I., and MEZGER, L. (1994), Correlation, ages, and uplift rates of Quaternary marine terraces: south-central coastal California, In *Seismotectonics of the Central California Coast Ranges* (ALTERMAN, MCMULLEN R.B., CLUFF, L.S., and SLEMMONS, D.B. (eds.) Boulder, Colorado, 1994) Geol S Am S, 292, 45–71.

HSU, J.T. (1992), *Quaternary uplift of the Peruvian coast related to the subduction of the Nazca ridge: 13.5 to 15.6 degrees south latitude*, Quatern Int 15/16:87–97.

HUSKER, A., and DAVIS, P.M. (2009), Tomography *and thermal state of the Cocos plate subduction beneath Mexico City*, J Geophys Res 114, B04306. doi:10.1029/2008JB006039.

IGLESIAS, A., SINGH, S.K., PACHECO, J.F., ALCANTARA, L., ORTIZ, M., and ORDAZ, M. (2003), *Near-trench Mexican earthquakes have anomalously low peak accelerations*, B Seismol Soc Am 93, 953–959.

KARIG, D.E., CALDWELL, R.K., MOORE, G.F., and MOORE, D.G. (1978), *Late Cenozoic subduction and continental margin truncation along the northern Middle America trench*, Geol Soc Am Bull 89, 265–276.

KENNETT D.J., JONES J.G., NEFF H., RAMIREZ-HERRERA T., VOORHIES, B. (2004), *Proyecto Costero Arcaico-Formativo: La costa de Guerrero II*, Instituto Nacional de Antropología e Historia, p. 41.

KIM, Y., CLAYTON, R.W., and JACKSON, J.M. (2010), *Geometry and seismic properties of the subducting Cocos plate in central Mexico*, J Geophys Res 115(B6), B06310.

KLITGORD, K., and MAMMERICKX, J. (1982), *Northern East Pacific rise: magnetic anomaly and bathymetric framework*, J Geophys Res 87(B8), 6725–6750.

KOSTOGLODOV, V., and BANDY, W. (1995), *Seismotectonic constraints on the convergence rate between the Rivera and North-American Plates*, J Geophys Res-Sol Ea, 100(B9), 17977–17989.

KOSTOGLODOV, V., BANDY, W., DOMINGUEZ, J., and MENA, M. (1996), *Gravity and seismicity over the Guerrero seismic gap, Mexico*, Geophys Res Lett 23(23), 3385–3388.

KOSTOGLODOV, V., SINGH, S.K., GORBATOV, A., PACHECO, J.F., FILONOV, A., and FIGUEROA, M.O. (1997), October 9, 1995, Mw 8.0 Colima-Jalisco earthquake: new constraints on the coseismic slip distribution. Resúmenes y Programa de la Reunión Anual de la Unión Geofísica Mexicana, 1997. GEOS 17(4), 259–260.

KOSTOGLODOV, V., VALENZUELA, R.W., GORBATOV, A., MIMIAGA, J., FRANCO, S.I., ALVARADO, J.A., Peláez, R. (2001), *Deformation in the Guerrero seismic gap, Mexico, from leveling observations*. J Geodesy 75(1), 19–32.

KOSTOGLODOV, V., SINGH, S.K., SANTIAGO, J.A., FRANCO, S.I., LARSON, K.M., LOWRY, A.R., and BILHAM, R., (2003), *A large silent earthquake in the Guerrero seismic gap, Mexico*, Geophys Res Lett 301(5), 1807. doi:10.1029/2003GL017219.

LAJOIE, K.R., *Coastal tectonics, in active tectonics* (National Academy Press, Washington, DC 1986), pp. 95–124.

LAJOIE, K.R., PONTI, D.J., POWELL II, C.L., MATHIESON, S.A., and SARNA-WOJCIKI, A.M., Emergent marine strandlines and associated sediments, coastal California: a record of Quaternary sea-level fluctuations, vertical tectonic movements, climatic changes, and coastal processes, In Quaternary Nonglacial Geology: Conterminous US, DNAG K-2. Geological Society of America (ed. R.B. Morrison) (Boulder, CO. 1991).

LARSON, K.M., LOWRY, A.R., KOSTOGLODOV, V., HUTTON, W., SAN-CHEZ, O., HUDNUT, K., and SUAREZ, G. (2004), *Crustal deformation measurements in Guerrero, Mexico*, J Geophys Res-Sol Ea, 109(B4), 1–19.

LAMBECK, K., ANTONIOLI, F., PURCELL, A., SILENZI, S. (2004), *Sea level change along the Italian coast for the past 10,000 yrs*. Quaternary Sci Rev 23, 1567–1598.

Reprinted from the journal

LONDSDALE, P. (2005), *Creation of the Cocos and Nazca plates by fission of the Farallon Plate*, Tectonophysics, 404(3–4), 237–264.

LUNDBERG, N., and MOORE, J.C. (1981), Structural features of the Middle America Trench slope off southern Mexico, deep sea drilling Project Leg 66. In Initial report DSDP 66, 793–805.

MANEA, M., MANEA V.C., and KOSTOGLODOV, V. (2003), *Sediment fill in the Middle America Trench inferred from gravity anomalies*, Geofis Int 42(4), 603–612.

MANEA, M., MANEA V.C., FERRARI, L., KOSTOGLODOV, V., and BANDY, W.L. (2005), *Tectonic evolution of the Tehuantepec ridge*, Earth Planet Sc Lett 238(1–2), 64–77.

MCNALLY, K.C., and MINSTER, B. (1981), *Nonuniform seismic slip rates along the Middle America Trench*, J Geophys Res 86 (NB6), 4949–4559.

MELBOURNE, T., CARMICHAEL, I., DEMETS, C., HUDNUT, K., SANCHEZ, O., STOCK, J., SUÁREZ, G., and WEBB, F. (1997), *The geodetic signature of the M8.0 October 9, 1995, Jalisco subduction earthquake*, Geophys Res Lett 24(6), 715–718.

MELNICK, D., BOOKHAGEN B., ECHTLER, H.P., STRECKER, M.R. (2006), *Coastal deformation and great subduction earthquakes, Isla Santa María, Chile (37°S)*. GSA Bull 118(11/12), 1463–1480. doi:10.1130/B25865.1.

MERRITTS, D.J., BULL, W.B. (1989), *Interpreting Quaternary uplift rates at the Mendocino triple junction, Northern California, from uplifted marine terraces*. Geology 17, 1020–1024.

MESCHEDE, M., FRISCH, W., HERRMANN U.R., RATSCHBACHER, L. (1997), *Stress transmission across an active plate boundary: an example from southern Mexico*, Tectonophysics 266, 81–100.

MOLNAR, P. (1977), *Gravity anomalies and the origin of the Puerto Rico trench*, Geophys J Roy Astr S 51, 701–708.

MOLNAR, P. and SYKES L.R. (1969), *Tectonics of the Caribbean and Middle America regions from focal mechanisms and seismicity*, Geol Soc Am Bull 80, 1639–1684.

MOORE, J.C., WATKINS, J.S., and SHIPLEY, T.H., Summary of accretionary process, deep sea drilling project leg 66: offscraping, underplating, and deformation of the slope apron. In *Initial reports of the deep sea drilling project, LXVI, Mazatlan, Mexico to Manzanillo, Mexico* (ed. Watkins, J.S., Moore, J.C.) (Washington, DC, 1982), US Government Printing Office, 827–836.

MORÁN-ZENTENO, D.J., CORONA-CHAVEZ, P., and TOLSON, G. (1996), *Uplift and subduction erosion in southwestern Mexico since the Oligocene: pluton geobarometry constraints*, Earth Planet Sci Lett 141, 51–65.

MUHS, D.R., KELSEY, H.M., MILLER, G.H., KENNEDY, G.L., WHELAN, J.F., MCINELLY, G.W. (1990), *Age estimates and uplift rates for Late Pleistocene marine terraces: southern Oregon portion of the Cascadia Forearc*, J Geophys Res 95, 6685–6688.

ORTIZ, M., SINGH, S.K., KOSTOGLODOV, V., and PACHECO, J. (2000), *Source areas of the Acapulco-San Marcos, Mexico earthquakes of 1962 (M 7.1; 7.0) and 1957 (M 7.7), as constrained by tsunami and uplift records*, Geofis Int 39(4), 337–348.

ORTIZ, M., SINGH, S.K., PACHECO, J., KOSTOGLODOV, V. (1999), *Rupture length of October 9, 1995 Colima-Jalisco earthquake (Mw 8) estimated from tsunami data*, Geophys Res Lett 25(15), 2857–2860.

OTA, Y., and CHAPPELL, J. (1996), *Late Quaternary coseismic uplift events on the Huon Peninsula, Papua New Guinea, deduces from coral terrace data*. J Geophys Res 101(B3), 6071–6082.

OTA, Y., YAMAGUCHI, M. (2004), *Holocene coastal uplift in the western Pacic Rim in the context of late Quaternary uplift*. Quatern Int 120, 105–117. doi:10.1016/j.quaint.2004.01.010.

PACHECO, J.F., BANDY, W., REYES-DÁVILA, G.A., NÚÑEZ-CORNÚ, F.J., RAMÍREZ-VÁZQUEZ, C.A., and BARRÓN. (2003a), *The Colima, Mexico, Earthquake (MW 5.3) of 7 March 2000: seismic activity along the southern Colima Rift*, Bull Seism Soc Am 93(4), 1458–1467. 10.1785/0120020193.

PACHECO, J.F., JIMÉNEZ, C., IGLESIAS, A., PÉREZ SANTANA, J., FRANCO, S.I., ESTRADA, J.A., CRUZ, J. L., CÁRDENAS, A., LI YI, T., GUTIÉRREZ, M.A., RUBÍ, B., and SANTIAGO, J.A. (2003b), *Sismicidad del Centro y Sur de México (Periodo Enero a Junio de 2003)*, GEOS 23(1), 28–36.

PARDO, M., and SUÁREZ, G. (1995), *Shape of the subducted Rivera and Cocos plates in southern Mexico: seismic and tectonic implications*, J Geophys Res 100, 12,357–312,372.

PONCE, L., GAULON, R., SUÁREZ, G., and LOMAS, E. (1992), *Geometry and state of stress of the downgoing Cocos plate in the Isthmus of Tehuantepec, Mexico*. Geophys Res Lett 19(8), 773–776.

RAMÍREZ-HERRERA, M.T., MARSHALL, J.S., and ZAMORANO OROZCO, J.J. (1998), Uplifted Holocene terraces and tectonic deformation of the Michoacan coast, southern Mexico, *GEOS*, Unión Geofísica Mexicana, Boletín Informativo, 18(4), 306–307.

RAMÍREZ-HERRERA, M. T., and URRUTIA-FUCUGAUCHI, J. (1999), *Morphotectonic zones along the coast of the Pacific continental margin, southern Mexico*. Geomorphology 28, pp. 237–250.

RAMÍREZ-HERRERA, M.T., KOSTOGLODOV, V., SUMMERFIELD, M., URRUTIA-FUCUGAUCHI, J., ZAMORANO, J. (1999), *A reconnaissance study of the morphotectonics of the Mexican subduction zone*, Ann Geomorphol 118, 207–226.

RAMÍREZ-HERRERA, M.T., and ZAMORANO, O.J. (2002), *Coastal uplift and mortality of coralline algae caused by a 6.3 Mw earthquake, Oaxaca, Mexico*, J Coastal Res 18(1), 75–81.

RAMÍREZ-HERRERA, M.T., URRUTIA-FUCUGAUCHI, J., KOSTOGLODOV, V. (2004), *Holocene Emerged Notches And Tectonic Uplift Along The Jalisco Coast, Southwest Mexico*, Geomorphology 58, 291–304.

RAMÍREZ-HERRERA, M.T., CUNDY A., KOSTOGLODOV, V. (2005), Evidence of Prehistoric Earthquakes and Tsunamis during the last 5000 years along the Guerrero Seismic Gap, Mexico: XV CNIS, Mexican Society of Seismic Engineering, I-07: 1–17.

RAMÍREZ-HERRERA, M.T., CUNDY, A., KOSTOGLODOV, V., CARRANZA-EDWARDS A., MORALES E., and METCALFE S. (2007), *Sedimentary record of late Holocen relative sea-level change and tectonic deformation from the Guerrero Seismic Gap, Mexican Pacific Coast*. The Holocene 17/8, 1211–1220.

RAMÍREZ-HERRERA, M.T., CUNDY, A., KOSTOGLODOV, V., and ORTIZ M. (2009), *Late Holocene tectonic land-level changes and tsunamis at Mitla lagoon, Guerrero, México*. Geofis Int 48(2), 195–209.

REIMER, P. J., BAILLIE, M. G. L., BARD, E., BAYLISS, A., BECK, J. W., BERTRAND, C. J. H., BLACKWELL, P. G., BUCK, C. E., BURR, G. S., CUTLER, K. B., DAMON, P. E., EDWARDS, R. L., FAIRBANKS, R. G., FRIEDRICH, M., GUILDERSON, T. P., HOGG, A. G., HUGHEN, K. A., KROMER, B., MCCORMAC, F. G., MANNING, S. W., RAMSEY, C. B., REIMER, R. W., REMMELE, S., SOUTHON, J. R., STUIVER, M., TALAMO, S., TAYLOR, F. W., VAN DER PLICHT, J., and WEYHENMEYER, C. E. 2004. IntCal04 Terrestrial radiocarbon age calibration, 26 - 0 ka BP. Radiocarbon 46, 1029–1058.

RUIZ-MARTINEZ, V.C., URRUTIA-FUCUGAUCHI, J., OSETE, M. (2010), *Paleomagnetism of the western and central sectors of the Trans-Mexican volcanic belt-implications for tectonic rotations and paleosecular variation in the past 11 Ma*. Geophys Jour Int. doi: 10.1111/j.1365-246X.2009.04447.x.

SACK, P.B., FISHER, D.M., GARDNER, T.W., LAFEMINA, P.C. (2009), *Rough crust subduction, forearc kinematics, and Quaternary uplift rates, Costa Rican segment of the Middle American Trench,* Geol Soc Am Bull 121(7–8), 992–1012.

SAILLARD, M., HALL, S.R., AUDIN, L., FARBER, D.L., HÉRAIL, G., MARTINOD, J., REGARD, V., FINKEL, R.C., BONDOUX, F. (2009), *Non-steady long-term uplift rates and Pleistocene marine terrace development along the Andean margin of Chile (31°S) inferred from 10Be dating,* Earth Planet Sci Lett 277(1–2): 50–63.

SAVAGE, J.C. (1983), *A dislocation model of strain accumulation and release at a subduction zone,* J Geophys Res 88(B6), 4984–4996.

SHIPLEY, T.H. (1981), Seismic facies and structural framework of the southern Mexico continental margin. In Initial report DSDP 66, 775–790.

SHIPLEY, T.H., MCMILLEN, K.J., WATKINS, J.S. (1980), *Continental margin and lower slope structures of the Middle America Trench near Acapulco (Mexico),* Mar Geol 35, 65–82.

SIDDALL, M., CHAPPELL, J., POTTER, E.-K., (2006). Eustatic sea-level during past interglacials. In *The climate of past interglacials* (SIROCKO, F., LITT, T., CLAUSSEN, M., SANCHEZ-GONI, M.-F. (eds.), Elsevier, Amsterdam), pp. 75–92.

SINGH, S.K., ASTIZ, L., and HAVSKOV, J.H. (1981), *Seismic gaps and recurrence periods of large earthquakes in the Mexican subduction zone: are-examination.* B Seismol Soc Am 71, 827–843.

SINGH, S.K., PONCE, L., and NISHENKO, S.P. (1985), *The great Jalisco, Mexico, earthquakes of 1932: subduction of the Rivera plate,* B Seismol Soc Am 75 (5), 1301–1313.

SONG, T.-R.A., HELMBERGER, D.V., BRUDZINSKI, M.R., CLAYTON, R.W., DAVIS, P., PEREZ-CAMPOS, X., SINGH, S.K. (2009), *Subducting slab ultra-slow velocity layer coincident with silent earthquakes in southern Mexico.* Science 324, 502–506. doi:10.1126/science.1167595.

STUIVER, M. and REIMER, P.J., (1993), Extended 14C data base and revised CALIB 3.0 14C age calibration program, Radiocarbon 35, 215–230.

SUAREZ, G., GARCIA-ACOSTA, V., GAULON, R., (1994), *Active crustal deformation in the Jalisco block, Mexico: evidence for a great historical earthquake in the 16th century.* Tectonophysics 234, 117–127.

SUÁREZ, G., and SÁNCHEZ, O. (1996), *Shallow depth of seismogenic coupling in southern Mexico: implications for the maximum size of earthquakes in the subduction zone,* Phys Earth Planet Int 93, 53–61.

URRUTIA-FUCUGAUCHI, J., and DEL CASTILLO, G.L. (1977), Un modelo del eje volcánico mexicano Traducido, *Boletin de la Sociedad Geologica Mexicana* 38, 18–28.

URRUTIA-FUCUGAUCHI, J., and FLORES-RUIZ, J. (1996), *Bouguer gravity anomalies and regional crustal structure in central Mexico,* Int Geol Rev 38(2), 176–194. doi:10.1080/0020681970 9465330.

URRUTIA-FUCUGAUCHI, J., FLORES-RUIZ, J.H., BANDY, W.L., and MORTERA, C. (1999), *Crustal structure of the Colima rift, western Mexico: gravity models revisited.* Geofís Int 38, 205–216.

URRUTIA FUCUGAUCHI, J., GONZALEZ MORAN, T. (2006), *Structural pattern at the northwestern sector of the Tepic-Zacoalco rift and tectonic implications for the Jalisco block, western Mexico.* Earth Planets Space 58, 1303–1308.

VERGNOLLE, M., WALPERSDORF, A., KOSTOGLODOV, V., TREGONING, P., SANTIAGO, J.A., COTTE, N., and FRANCO, S.I. (2010), *Slow slip events in Mexico revised from the processing of 11 year GPS observations,* J Geophys Res 115(B8), B08403.

WATKINS, J.S., MCMILLEN, K.J., BACHMAN, S.B., SHIPLEY, T.H., MOORE, J.C., and ANGEVINE, C., Tectonic synthesis, Leg 66: Transect and Vicinity. In *Initial reports of the deep sea drilling project, LXVI, Mazatlan, Mexico to Manzanillo, Mexico* (eds. Watkins, J.S., Moore, J.C., et al.,) (Washington, DC, 1982) US Government Printing Office, 837–849.

YANG, T., GRAND, S.P., WILSON, D., GUZMAN-SPEZIALE, M., GOMEZ-GONZALEZ, J.M., DOMINGUEZ-REYES, T., and NI, J. (2009), *Seismic structure beneath the Rivera subduction zone from finite-frequency seismic tomography,* J Geophys Res 114, B01302. doi:10.1029/2008JB005830.

YOSHIOKA, S., MIKUMO, T., KOSTOGLODOV, V., LARSON, K.M., LOWRY, A.R., and SINGH, S.K. (2004), *Interplate coupling and a recent aseismic slow slip event in the Guerrero seismic gap of the Mexican subduction zone, as deduced from GPS data inversion using a Bayesian information criterion,* Phys Earth Planet Int 146, 513–530.

ZOBIN, V.M., REYES-DÁVILA, G.A., PÉREZ-SANTA ANA, L.U., RAMÍREZ-VÁZQUEZ, C.A., VENTURA-RAMÍREZ, J. (2000), Estudio macrosísmico del temblor de Colima (Mw 5.3) del 6 de marzo del 2000, GEOS, 20 (4): 414–417.

(Received February 14, 2010, revised September 8, 2010, accepted September 10, 2010, Published online January 11, 2011)

Reprinted from the journal

Pure Appl. Geophys. 168 (2011), 1435–1447
© 2010 Springer Basel AG
DOI 10.1007/s00024-010-0201-2

Constraints on Jalisco Block Motion and Tectonics of the Guadalajara Triple Junction from 1998–2001 Campaign GPS Data

MICHELLE M. SELVANS,[1] JOANN M. STOCK,[1] CHARLES DEMETS,[2] OSVALDO SANCHEZ,[3] and BERTHA MARQUEZ-AZUA[4]

Abstract—A GPS campaign network in the state of Jalisco was occupied for ∼36 h per station most years between 1995 and 2005; we use data from 1998–2001 to investigate tectonic motion and interseismic deformation in the Jalisco area with respect to the North America plate. The twelve stations used in this analysis provide coverage of the Jalisco Block and adjacent North America plate, and show a pattern of motion that implies some contribution to Jalisco Block boundary deformation from both tectonic motion and interseismic deformation due to the offshore 1995 earthquake. The consistent direction and magnitude of station motion on the Jalisco Block with respect to the North America reference frame, ∼2 mm/year to the southwest (95% confidence level), perhaps can be attributed to tectonic motion. However, some station velocities within and across the boundaries of the Jalisco Block are also non-zero (95% confidence level), and the overall pattern of station velocities indicates both viscoelastic response to the 1995 earthquake and partial coupling of the subduction interface (together termed "interseismic deformation"). Our results show motion across the northern Colima rift, the eastern boundary of the Jalisco Block, which is likely to be sinistral oblique extension rather than pure extension. We constrain extension across both the Colima rift and the northeastern boundary of the Jalisco Block, the Tepic-Zacoalco rift, to ≤8 mm/year (95% confidence level), slow compared to relative rates of motion at nearby plate boundaries.

Key words: Colima rift, GPS, interseismic, Jalisco Block, Tepic-Zacoalco rift, triple junction.

1. Introduction

Jalisco is an interesting region for geodetic study for two main reasons. First, geologic evidence points to concentrations of tectonic deformation in two bounding rifts inland from the Rivera plate, the Tepic–Zacoalco and Colima rifts (see Fig. 1). Second, following the 9 Oct 1995 ($M_w = 8.0$) Colima-Jalisco earthquake, the hinge of deformation (between subsidence and uplift) quickly moved onshore (MELBOURNE *et al.*, 2002), providing better Global Positioning System (GPS) coverage of overall deformation than exists for most subduction megathrusts. Neither the deformation due to the earthquake cycle nor the local tectonics is fully understood.

This study seeks to constrain the tectonic motion of the Jalisco Block with respect to North America. We use average velocities during 1998–2001 for a network of twelve GPS stations to investigate rifting rates across the Tepic–Zacoalco and Colima rifts, as well as motion across the Chapala rift. We also investigate the contribution to station motion from earthquake cycle effects. The 1995 $M_w = 8.0$ event offshore from Jalisco may contribute significantly to the velocities of the network stations, so careful consideration of both tectonic and earthquake cycle signatures in the GPS data is necessary.

Constraining the current rate of tectonic motion in the Jalisco area will narrow down the likely scenarios for ongoing deformation in a region of active plate rearrangement (e.g. LUHR *et al.*, 1985; JOHNSON and HARRISON, 1990; ALLAN *et al.*, 1991; FERRARI, 1995; ROSAS-ELGUERA *et al.*, 1996; DEMETS and TRAYLEN, 2000). The Jalisco Block lies onshore from the northernmost section of the Middle America Trench, above the subducting Rivera plate, just south of rifting in the Gulf of California, and has often been cited as an example of continental rifting (e.g. LUHR *et al.*, 1985).

Characterizing the pattern and magnitude of earthquake cycle effects in the dataset is essential to

---

[1] Seismological Laboratory, California Institute of Technology, Pasadena, CA 91125, USA. E-mail: selvans@gps.caltech.edu

[2] Department of Geoscience, University of Wisconsin-Madison, Madison, WI 53706, USA.

[3] Instituto de Geofisica, Universidad Nacional Autonoma de Mexico, 04510 Mexico D.F, Mexico.

[4] DGOT/SisVoc, Universisdad de Guadalajara, Av. Maestros y Mariano Barcenas, 93106 Guadalajara, Jalisco, Mexico.

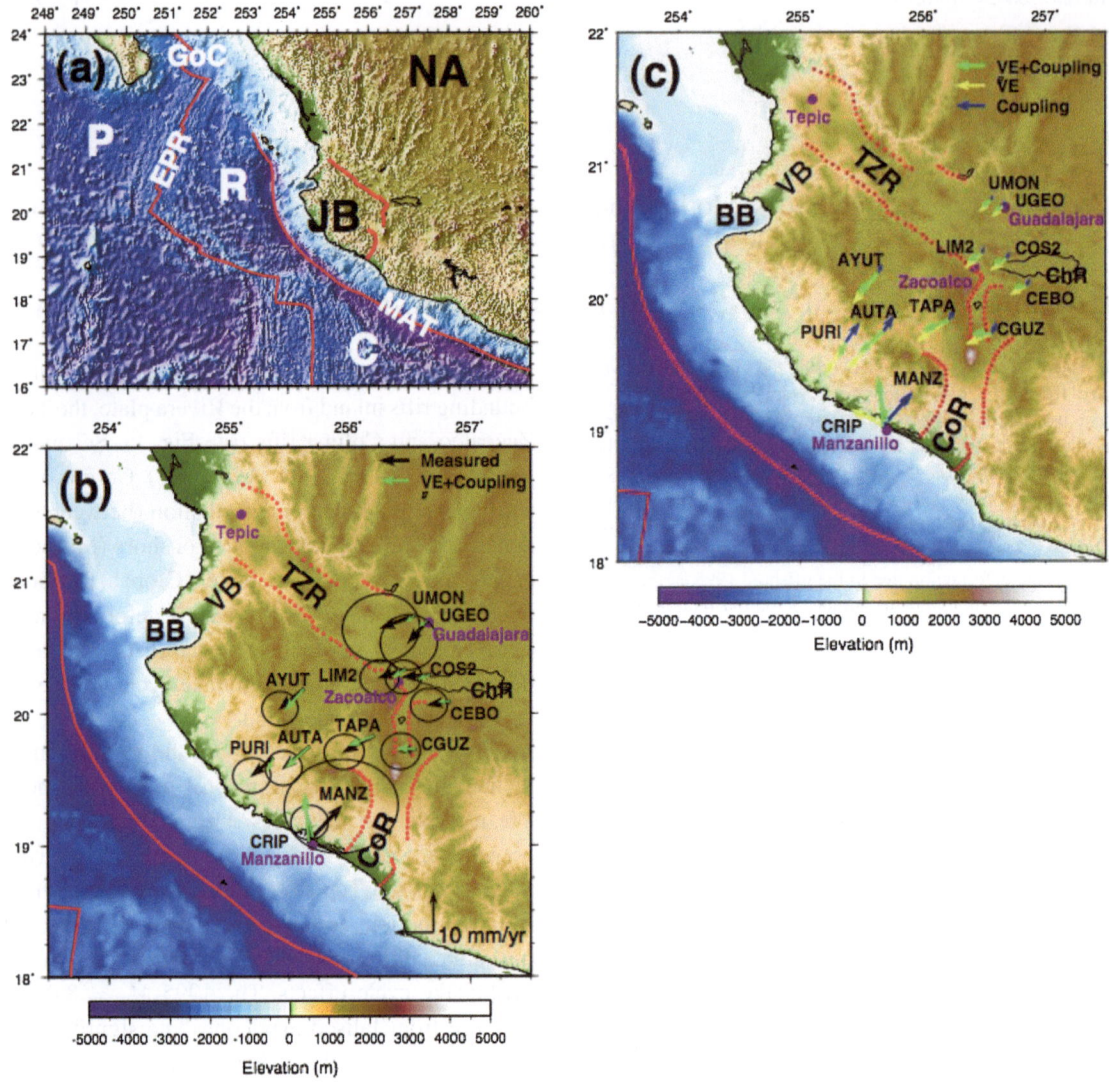

Figure 1

a Topography and bathymetry of western Mexico shows the plate tectonic context for the Jalisco Block (JB), with plate boundaries and Jalisco Block boundaries in red. The Jalisco Block is just southeast of the Gulf of California (GoC), and may move rigidly with respect to North America (NA). The Pacific (P) and Rivera (R) plates diverge at the East Pacific Rise (EPR). The Rivera and Cocos (C) subduct along the Middle America Trench (MAT). **b** Twelve GPS sites used in this study lie on the Jalisco Block and surround the rift-rift-rift triple junction, where the Tepic–Zacoalco, Colima, and Chapala rifts meet [TZR, CoR, and ChR, respectively; bounding faults for the Tepic–Zacoalco and Colima rifts are shown in red, after ALLAN (1986) and FERRARI et al. (1994)]. Bahía de Banderas (BB) and the Valle de Banderas (VB) are the proposed northwest boundary of the Jalisco Block (e.g. JOHNSON and HARRISON, 1990). Rivera plate boundaries are after DEMETS and WILSON (1997). Station velocities with respect to North America are plotted in black (see Appendix 2, Table 1), with north and east errors displayed as 2D 95% confidence intervals. For comparison, modeled interseismic velocity vectors are shown in green. **c** Modeled vectors include partial coupling along the subduction interface (50% coupling shown in blue) and viscoelastic (VE) deformation of the overriding plate (yellow) (MASTERLARK et al., 2001; MARQUEZ-AZUA et al., 2002). All measured and predicted station velocities are scaled to the reference vectors in the lower right. Shuttle Radar Topography Mission data and estimated seafloor topography are ~1 km resolution (BECKER and SANDWELL, 2006)

its reliable interpretation. GPS analysis and modeling of the coseismic and postseismic (transient) effects of the 1995 earthquake (HUTTON et al., 2001; MASTER-LARK et al., 2001; MARQUEZ-AZUA et al., 2002),

indicate that some signature of the earthquake cycle is expected during our study period, likely due to viscoelastic deformation of the mantle beneath our study area and partial to full coupling of the subduction interface (together termed "interseismic deformation").

We assess our network velocities with respect to forward-modeling predictions of earthquake cycle phenomena and predictions of multiple hypotheses for tectonic motion of the Jalisco Block. These phenomena predict different station velocity patterns: (1) stations on the Jalisco Block move together to the west or southwest relative to North America, due to rifting of the Jalisco Block from North America, (2) all stations move toward the 1995 rupture zone, due to viscoelastic response of North America to the earthquake, and (3) all stations move away from the 1995 rupture zone, due to partial coupling of the subduction interface. The first prediction has three variants (detailed below), based on alternative hypotheses for the formation and motion of the Jalisco Block and its bounding rifts. We also examine the smaller region around the "Guadalajara triple junction" (where three rifts—the Tepic–Zacoalco, Chapala, and Colima rifts—meet each other). For this region, we assume local tectonic rates dominate over any viscoelastic gradients, permitting us to place constraints on the velocity triangle that includes North America and the Jalisco and Michoacan Blocks, which surround this triple junction (e.g. JOHNSON and HARRISON, 1990).

2. Tectonic Setting

2.1. Regional Tectonics

Jalisco is a coastal state of western mainland Mexico, located southeast of the Gulf of California (Fig. 1a). The surrounding region is shaped by a series of recent tectonic events, including subduction along the west coast of North America, an eastward jump of the East Pacific Rise to produce rifting in the Gulf of California, and separation of the Rivera and Cocos plates, resulting in the current differential subduction beneath Jalisco and the rest of western Mexico to the southeast. Convergence is slower between the Rivera and North America plates than between the Cocos and North America plates, and is increasingly oblique from south to north along the Rivera-North America trench (e.g. KOSTOGLODOV and BANDY 1995; DEMETS and TRAYLEN 2000).

2.2. Boundaries of the Jalisco Block

The Tepic–Zacoalco rift consists of several tectonic depressions bounding the northeastern extent of a topographically high portion of the state of Jalisco (Fig. 1b). Structural mapping of the Tepic–Zacoalco rift indicates some extension and right-lateral motion occurred ~12–8.5 Ma, with extension continuing to the present day, probably related to opening of the Gulf of California (FERRARI, 1995; FREY et al., 2007). Recent rates of motion are small, with average minimum deformation rates that decrease from 0.75 mm/year in the late Miocene to 0.1 mm/year in the Quaternary (FERRARI and ROSAS-ELGUERA, 2000). Beginning at 4.7 Ma, alkaline and calc-alkaline volcanic lavas were concentrated within the Tepic–Zacoalco rift, some with compositions commonly found in ocean islands and intraplate rifts (e.g. ALLAN et al., 1991). Furthermore, rhyolitic ignimbrites were embedded in this rift on an order of magnitude greater volume during the interval of 5–3 Ma than is documented for the volcanism of the last ~1 Ma, indicating significant lithospheric extension occurred during that time (FREY et al., 2007). Seismicity to a depth of ~35 km within the Tepic–Zacoalco rift also indicates deep crustal faulting between the Jalisco Block and North America (NUÑEZ-CORNÚ et al., 2002). Additionally, a tomographic study of the crust and upper mantle in Jalisco and adjacent states reveals distinct low velocity lineaments beneath both the Tepic–Zacoalco and Colima rifts (WANG et al., 2008); in the upper mantle, these features are associated with tearing of the subducted slab (YANG et al., 2009).

The Colima rift bounds the eastern edge of the Jalisco highlands (Fig. 1b). Except for the massive deposition of rhyolitic ignimbrites (FREY et al., 2007), volcanism within this Jalisco Block boundary is similar in composition and duration to that of the Tepic–Zacoalco rift (e.g. ALLAN et al., 1991). Since ~5 Ma, rocks of the southern Colima rift have been faulted, both onshore (e.g. GARDUÑO-MONROY et al., 1998) and offshore (BOURGOIS et al., 1988; KHUTORSKOY et al., 1994; BANDY et al., 2005), and the northern Colima rift has subsided 0.07–0.7 mm/year (ROSAS-ELGUERA et al., 1996). Other crustal faults, such as the Tamazula fault to the west of the southern

Colima rift, may now form the southeastern boundary of the Jalisco Block (GARDUÑO-MONROY et al., 1998). These faults have had recent seismic activity (GARDUÑO-MONROY et al., 1998; PACHECO et al., 2003; ANDREWS et al., in press). Farther north, the eastern edge of the Jalisco block is roughly aligned with a sharp change in slab dip just east of the Colima rift (PARDO and SUAREZ, 1995), visible in seismic tomography (GRAND et al., 2007; YANG et al., 2009).

The Zacoalco half-graben lies at the continental rift-rift-rift triple junction (termed the Guadalajara triple junction) where the Tepic–Zacoalco, Colima, and Chapala rifts come together. A sequence of magnitude 1.5–3.5 earthquakes in 1997 on shallow normal faults (PACHECO et al., 1999) confirms formation of the half-graben as tilt blocks overlying listric faults (ROSAS-ELGUERA et al., 1997); additionally, the composite focal mechanism indicates possible right-lateral slip along a northwest-southeast oriented nodal plane within the Zacoalco graben near the triple junction (PACHECO et al., 1999). Historical records indicate the potential for much larger earthquakes at this triple junction, such as the >7.0 magnitude earthquake of 27 December 1568 (SUAREZ et al., 1994).

Receiver functions reveal Moho depths of 25–45 km in the continental interior of the Jalisco Block (SUHARDJA et al., 2007). To fully delineate the inland boundaries of the Jalisco Block, its northwestern corner must be defined. Seismicity and structural mapping suggest Valle de Banderas, trending northeast from Bahía de Banderas to the Tepic–Zacoalco rift, has been the northwestern limit of the Jalisco Block since ∼5 Ma (e.g. JOHNSON and HARRISON, 1990; NUÑEZ-CORNÚ et al., 2002); these lines of evidence are corroborated by gravity and magnetics data (ARZATE et al., 2006). An alternative interpretation of the geologic and magnetic data is that the northwest boundary of the Jalisco Block follows this same trend, but is located just to the northwest of Valle de Banderas (URRUTIA-FUCUGAUCHI and GONZALEZ-MORAN, 2006). While deformation is possible within and along all boundaries of the Jalisco Block, this study focuses on characterizing motion across the two most prominent boundaries between the Jalisco Block and neighboring continental material, the Tepic–Zacoalco and Colima rifts.

3. Predictions of Current Deformation

3.1. Hypotheses for Block Motion

Three hypothetical scenarios for the formation and motion of the Jalisco Block could explain the current morphologies of the Tepic–Zacoalco and Colima rifts. An early hypothesis for Jalisco Block formation and motion, based on regional tectonics, the clear inland delineation of the Jalisco Block by the Tepic–Zacoalco and Colima rifts, and the composition of volcanism in the rifts, was an imminent eastward jump of the East Pacific Rise to the Colima rift (e.g. LUHR et al., 1985). This hypothesis suggests the eventual attachment of the Jalisco Block to the Pacific plate (i.e. northwestward motion with respect to North America), and predicts opening in the Colima rift and primarily right-lateral strike slip along the Tepic–Zacoalco rift. A variant of this hypothesis, based on similarities between volcanism in the Tepic–Zacoalco rift and that of the Gulf of California 12–6 Ma, is that recent Tepic–Zacoalco volcanism is a precursor to rifting of the Jalisco Block from North America (Frey et al., 2007).

Alternatively, the Tepic–Zacoalco and Colima rifts are explained as passive responses of North America to tearing of the subducting slab, which stresses the continental crust (e.g. FERRARI, 1995, 2004). The Colima rift approximately overlies the sharp change in dip between the Rivera and Cocos slabs (Pardo and Suarez, 1995) and, as with the Tepic–Zacoalco rift, overlies a region of low seismic velocities; respectively, these low velocities may be due to differential motion between the subducting slabs (e.g. STOCK, 1993) and a lateral tear in the Rivera slab (e.g. NIXON, 1982). This hypothesis for Jalisco Block motion predicts opening along both the Colima and Tepic–Zacoalco rifts (i.e. southwestward motion with respect to North America), with the possibility of motion being dominantly trenchward (southward) (FERRARI et al., 1994; ROSAS-ELGUERA et al., 1996).

A third hypothesis for Jalisco Block formation and motion (which is potentially compatible with the preceding hypothesis) is that its inland boundaries accommodate little to no motion today, in keeping with geologic evidence for slow rates of opening

(average minimums of <1 mm/year) across the Tepic–Zacoalco and Colima rifts from the late Miocene
through the Quaternary (ROSAS-ELGUERA et al., 1996;
FERRARI and ROSAS-ELGUERA, 2000). This hypothesis
predicts opening of up to a few millimeters per year
across the inland Jalisco Block boundaries.

3.2. Earthquake Cycle Effects

The shallow portion of the slab interface ruptured
in a pair of large earthquakes ($M_w = 8.2$ and
$M_w = 7.8$) in 1932 (SINGH et al., 1985), after which
no large subduction-related earthquakes ruptured
the Rivera plate subduction interface until 1995
($M_w = 8.0$) and 2003 ($M_w = 7.2$) (e.g. MELBOURNE
et al., 1997; PACHECO et al., 1997; YAGI et al., 2004).
After the 1995 earthquake, GPS stations within
200 km of the rupture zone exhibited rapidly decaying
transient deformation attributable to a combination of
afterslip focused along areas of the subduction interface downdip from the rupture zone and viscoelastic
flow of the upper mantle due to the elevated stresses
from the 1995 earthquake (HUTTON et al., 2001;
MARQUEZ-AZUA et al., 2002; MELBOURNE et al.,
2002). Finite element modeling of the expected steady
deformation from frictional coupling of the subduction
interface and the transient, viscoelastically induced
deformation of the overriding North America plate
shows that these two processes cannot by themselves
match deformation recorded between 1993 and 2001
at a continuous GPS station directly onshore from
the 1995 rupture zone (MASTERLARK et al., 2001;
MARQUEZ-AZUA et al., 2002), in accord with the
aforementioned studies which conclude that fault
afterslip contributed significantly to the deformation
after the 1995 earthquake. Rapid transient postseismic
deformation after the 1995 earthquake concluded by
mid-1997, after which station motions were linear or
nearly linear until the 22 Jan 2003 Tecoman $M_w = 7.2$
earthquake offshore from the study area triggered
additional postseismic deformation consisting in part
of aseismic fault afterslip (SCHMITT et al., 2007).

By limiting the present analysis to GPS data
collected from 1998 to 2001, we exclude the years
when coseismic and postseismic signals (i.e. obvious
deviations from strictly linear motion) associated
with the 1995 and 2003 earthquakes dominated the

station velocities (2002 is excluded because no
campaign GPS data were collected that year). When
interpreting our data, we assume these 4 years are
representative of ongoing tectonic motion. However,
motion of the Jalisco Block may have varied over the
last few million years, and interseismic earthquake
cycle effects may still contribute significantly to
motion of GPS sites during the time interval of our
study.

Viscoelastic response of North America to the
1995 event would cause stations in our study to move
southwestward toward the earthquake rupture zone,
with the largest velocities closest to the epicenter, and
similar directionality but decreasing magnitude at
stations further inland (yellow vectors in Fig. 1c).
This effect would produce motion in generally the
same direction as predicted by the second hypothesis
for block motion; although, in that case no strain
gradient is expected. Partial coupling of the subduction interface would also result in a strain gradient,
again with the largest magnitudes at the coast, but of
generally northeastward motion (blue vectors in
Fig. 1c).

Differences in predicted strain patterns allow us
to assess the relative contributions of viscoelastic
response, partial coupling on the subduction interface,
and tectonic motion. We do not expect other large-
scale contributions to the station motion in the Jalisco
area. Because GPS stations farther inland in northern
Mexico do not move significantly with respect to
North America (MARQUEZ-AZUA and DEMETS, 2003;
2009), motion related to the Basin and Range region is
not expected to influence our study area.

4. Data and Methods

4.1. Data Collection and Processing

We use ten GPS stations from a Jalisco campaign
network, occupied for ~36 h per station, with up to
four stations simultaneously operating (HUTTON et al.,
2001), as well as two continuous sites that ran for
months at a time (UGEO and MANZ). We select
these twelve stations to provide good coverage of the
study area, including multiple data points on the
Jalisco Block and baselines across the Tepic–Zacoalco and Colima rifts (Fig. 1b), so that we can

investigate internal deformation as well as motion concentrated in the Jalisco Block boundaries.

Our analysis includes 62 sessions (days) over the 4-year time span. Fifteen GPS stations in North America (east of the Mojave desert) are used to define the reference frame relative to ITRF2000 (ALTAMIMI et al., 2002). Final orbits from the Jet Propulsion Laboratory are used for satellite positions.

For all sessions, loosely constrained least-squares solutions of position and velocity components and their correlation matrices are obtained for each station using GAMIT (HERRING et al., 1990). Using GLOBK (DONG et al., 1998), these quasi-observations for each session are processed with constraints on station position and velocity, reference frame motion, and orbital and Earth orientation parameter values, in order to determine station coordinate time series with respect to North America. Further processing with the Markov process (white noise) in a recursive, time-domain Kalman filter allows us to obtain station velocities for 1998–2001. See Appendix 1 for more detail on the data processing.

4.2. GPS Velocity Estimation

Seven GPS stations show small amounts of significant motion with respect to North America (AUTA, AYUT, CEBO, COS2, LIM2, PURI, TAPA; see Appendix 2, Table 1), with an overall pattern of southwestward and west-southwestward motion for all ten inland stations (black vectors in Fig. 1b). Sites in the Jalisco Block interior (AUTA, AYUT, PURI, TAPA) move most similarly to each other compared with any other group of stations in this study, at 2–3 mm/year to the southwest with respect to North America (95% confidence level). Stations closest to the Guadalajara triple junction (LIM2, COS2) also move ~2 mm/year with respect to North America (95% confidence level), in a more westward direction than those of the Jalisco Block interior. Since our uncertainties on vertical motion are so large, the contribution to LIM2 and COS2 motion from slip on high-angle normal faults and/or listric normal faults in the area (e.g. ROSAS-ELGUERA et al., 1997) cannot be determined. This is also true for the stations on the Michoacan Block (CGUZ, CEBO), both of which are at the edges of the rifts they bound and so may also

move in part due to local normal faulting. Five GPS stations (UMON, UGEO, CGUZ, CRIP, MANZ) do not move with respect to North America (95% confidence level), although we note that the coastal stations (CRIP and MANZ) do uniquely move inland. The direction of motion at stations CRIP and MANZ is likely due to interseismic strain accumulation (MAR-QUEZ-AZUA et al., 2002; SCHMITT et al., 2007); the relatively large error ellipse on the MANZ station is due to only having 2 years of data during the 1998–2001 interval, and our analyzing that data only on days when at least one campaign station was active.

5. Velocity Field Analysis Results

We use station TAPA as a reference point for station velocities (Appendix 2, Table 2) because it is the station on the Jalisco Block closest to the Guadalajara triple junction. With respect to TAPA, we see no significant motion (lower end of the 95% confidence level) within the Jalisco Block or across the Colima and Tepic–Zacoalco rifts. Directly across the Colima rift from TAPA, the CGUZ station shows at most ~8 mm/year eastward motion and ~7 mm/year northward motion with respect to TAPA (upper end of the 95% confidence level), which provides an upper bound on the opening rate of the northern Colima rift during the years 1998 to 2001. Similarly with respect to TAPA, LIM2 limits the opening rate across the Zacoalco half-graben to a maximum of ~5 mm/year east and ~6 mm/year north.

We can further limit the range of possible motions by considering velocity constraints on the triple junction, using a flat-earth assumption involving only the stations closest to the triple junction (TAPA, CGUZ, CEBO, UMON, UGEO; see Fig. 2). Our velocity diagram has northward velocity on the y axis and eastward velocity on the x axis. This diagram is centered on a velocity of zero, which corresponds to station TAPA, which we use as our local reference frame. The resulting velocities are minimum constraints on block motions.

We fix station TAPA, on the Jalisco Block (Fig. 2a), and plot the best fit velocity vectors and their 95% confidence limits of the other four stations (values from Appendix 2, Table 2 using the

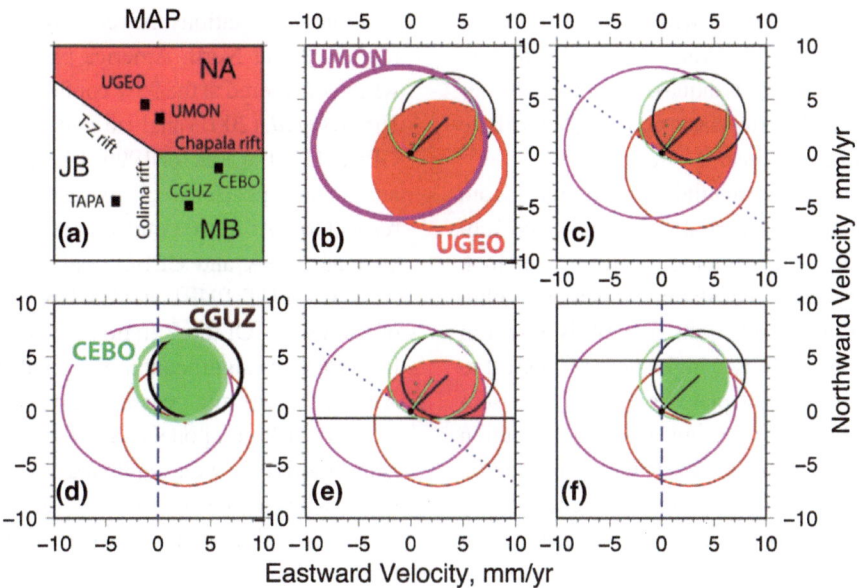

Figure 2
Velocity diagram analysis for a simplified Guadalajara triple junction constrains the sense of motion along the inland boundaries of the Jalisco Block (JB). **a** Schematic map showing the geometry of the triple junction and the location of the five stations. North America (NA) is in *red*; the Michoacan Block (MB) in *green*. Block JB (*white*) is the fixed block. Ellipses in *panels* (**b**) through (**f**) show the 95% confidence limits of station velocities relative to TAPA, whose velocity lies at the center of the diagram (*black dot* at coordinates 0,0). East–west velocity is on the *horizontal axis*; north–south velocity is on the *vertical axis*. *Straight lines* extending from the dot (best visible in *panels* **b** and **c**) indicate the best-fit velocity of each station with respect to TAPA, colored as follows: CEBO *green*, CGUZ *black*, UMON *pink*, UGEO *red*. **b** Intersection of UMON and UGEO ellipses shaded *red* to show allowable velocity values for NA motion relative to the JB. **c** Allowable velocities of NA relative to TAPA if no compression is occurring across the Tepic–Zacoalco (T-Z) rift. **d** Allowable velocities of the MB relative to TAPA if no compression is allowed across the northern Colima rift. **e** and **f** Allowable velocities of NA and the MB relative to TAPA if no compression is allowed across the Chapala rift. See Appendix 3 for a more detailed explanation

mathematical calculations of 2D Gaussian distributions (e.g., MOLNAR & STOCK, 1985)). We then constrain the possibilities for block velocities further using the assumptions detailed in Appendix 3.

For the N–S trending northern Colima rift, this indicates an upper limit on opening of ~8 mm/year, with a significant component of left-lateral strike-slip possible (green shaded velocity field, Fig. 2f); the lower limit of motion rate across the northern Colima rift is nearly 0 mm/year (lower end of the 95% confidence level). These velocity constraints allow for ~5 mm/year of pure extension across the northern Colima rift (i.e. normal to its strike). Within the uncertainties they also permit alternatives, such as up to 6 mm/year of left-lateral motion or oblique sinistral transtension. Across the N56°W trending Tepic–Zacoalco rift, an upper limit of ~8 mm/year opening between the Jalisco Block and North America is possible, with ~6 mm/year of pure extension possible.

Some amount of either pure left slip or pure right slip is possible (red shaded velocity field, Fig. 2e).

These constraints on opening rates across the Tepic–Zacoalco and Colima rifts (in the vicinity of the Guadalajara triple junction) permit more movement of the Jalisco Block with respect to North America than is observed geologically (average minimum values of ~0.1 mm/year since 5 Ma (ROSAS-ELGUERA *et al.*, 1996; FERRARI and ROSAS-ELGUERA, 2000; FREY *et al.*, 2007), and yet are consistent with the geology within the 95% confidence level of the velocity estimates.

Although the above analyses are based on holding station TAPA fixed, it is important to note that all four of our stations on the Jalisco highlands (TAPA, PURI, AUTA, AYUT) have similar velocities with respect to North America. The significant and coherent motion of these four stations, ~2 mm/year to the southwest with respect to North America, may be representative of the Jalisco Block rifting with respect to North America, with the caveat that viscoelastic deformation in response to the 1995 earthquake and partial coupling of the subduction interface may also contribute to the motions of these sites.

To understand the potential contributions from viscoelastic deformation, partial coupling of the subduction interface, and tectonic motion, we look qualitatively at the pattern of station velocities. Within the uncertainties, a distinct strain gradient is absent in the four stations on the Jalisco highlands (in contrast to predictions of both partial plate coupling and viscoelastic deformation), suggesting some contribution from Jalisco Block tectonic motion to the overall station velocities for 1998–2001. However, the overall pattern of estimated station velocities compares favorably with the modeled combination of viscoelastic response to the 1995 earthquake and 50% coupling along the subduction interface of previous researchers (yellow and blue arrows, respectively, in Fig. 1c) (MASTERLARK et al., 2001; MARQUEZ-AZUA et al., 2002).

The strain caused by coupling on the subduction interface consists of shortening, normal to the offshore subduction trench, counter to any extension across the Tepic–Zacoalco and Colima rifts. In contrast, the viscoelastic strain-rate gradient is extensional toward the rupture area of the 1995 earthquake, nearly opposite the sense of the gradient due to coupling, and so adds (temporarily) to any ongoing extension across the Tepic–Zacoalco or Colima rifts. The modeled and observed velocity vectors agree within the 2σ uncertainty ellipses, particularly with respect to the overall pattern of station motion (compare green and black vectors in Fig. 1b). While this similarity is suggestive, the relative role of off-fault and fault rheologies is still an open question (WANG, 2007), and the closest station to the 1995 rupture zone (PURI) does not fit well into the pattern of station motion predicted by the model of interseismic deformation.

6. Discussion

Although no previous analyses of GPS data have focused on motion across the Tepic–Zacoalco and Colima rifts, it is encouraging that similar earthquake cycle studies agree with our station velocity results. We find that GPS stations moved only ~2 mm/year with respect to the North America plate reference frame, when considered at the 95% confidence level (Fig. 1b). HUTTON et al. (2001) analyze campaign and continuous GPS data in the area for 1995–1999, and report varying amounts of postseismic motion for 1998–1999 with respect to NA: from 0 mm/year at CRIP, to ~10 mm/year at TAPA and CEBO, and to ~20 mm/year at AUTA, AYUT, PURI, and UMON, all to the southeast or southwest. Of these stations, they find that only the last four have significant motion with respect to North America (~10 mm/year at the 2σ level), and then only for the north component. Since this transient postseismic motion is well explained with a rate-and-state friction law model of the 1995 earthquake (HUTTON et al., 2001), it is an upper bound on annual velocity at these stations, and so is consistent with the significant motion of ~2 mm/year with respect to North America that we find at stations on the Jalisco highlands for 1998–2001 (at the 95% confidence level).

SCHMITT et al. (2007) analyze 1996–2003 GPS data in the Jalisco area to model interseismic and postseismic deformation due to the 2003 earthquake. They find that for the 1998–2001 time period, CRIP moved northeast with respect to the North America reference frame, and that TAPA, AUTA, AYUT, CGUZ, CEBO, and LIM2 moved southwest, all by ~20 mm or less over the study period, or >7 mm/year (in agreement with our velocity estimates, within 2σ error ellipses). Schmitt et al. (2007) do not report position component errors for each year, but do report them for coseismic offsets, so we use the latter for comparison. The 1σ error on coseismic position estimates for TAPA, PURI, LIM2, and UGEO is 2–10 times larger than the measurement itself, and similarly the 2σ error for CGUZ, AYUT, and CEBO is larger than the measurements (Schmitt et al., 2007). These relatively large errors suggest unresolved, although seemingly systematic, geodetic motion for 1998–2001 in the Jalisco area, as observed in our analysis of the data as well.

In both of the above analyses of Jalisco GPS data (SCHMITT et al., 2007; HUTTON et al., 2001), the intention was to model effects of the earthquake cycle on geodetic measurements in the area, and so neither focused on putting a robust constraint on the contribution from tectonic motion. We present a focused analysis of GPS data on and near the Jalisco Block that only covers the interseismic portion of the earthquake cycle, and obtain hard upper limits of 8 mm/year for opening along both the Tepic–Zacalco and Colima rifts (5 and 6 mm/year of pure extension, respectively, near the Guadalajara triple junction). Interestingly, for the northern Colima rift there also may be significant left-lateral strike-slip motion (up to 5 mm/year of pure left-lateral strike-slip) and, in fact, some left-lateral strike-slip motion is required for the minimum allowed velocity based on our velocity diagram analysis (Fig. 2). These constraints may include Jalisco Block motion with respect to North America, as well as partial coupling of the subduction zone and viscoelastic response of the overriding North America plate (i.e. interseismic deformation).

In terms of the hypotheses for Jalisco Block motion, the data do not support an eastward jump of the East Pacific Rise, because we do not find the motion across the Tepic–Zacalco rift to be dominantly right-lateral strike-slip. However, although only 1 mm/year of pure right-lateral strike-slip is allowed in our velocity diagram analysis (Fig. 2), the Tepic–Zacalco rift is comprised of many faults with a range of orientations (N56°W being the overall orientation of the rift), the more E–W oriented of which may have slightly more pure right-lateral strike-slip motion. The second and third hypotheses for Jalisco Block motion, either some opening at both rifts due to tearing of the subducted slab or no current Jalisco Block motion with respect to North America, are both consistent with our upper limit of 8 mm/year opening across the Tepic–Zacalco and Colima rifts, and the lower limit of only very slow motion across these boundaries (within the 95% confidence interval). These constraints confirm that relative motion between the Jalisco Block and North America (if present) is small compared to relative rates of motion at nearby plate boundaries (e.g. BANDY and PARDO, 1994), and is consistent with Quaternary geology.

With respect to earthquake cycle behavior of subduction megathrusts, we see station velocities following the 1995 earthquake that are largely consistent with modeled interseismic station velocities, suggesting homogenous response of the overriding North America plate and rapid resumption of coupling on the subduction interface. This indication of homogeneous earthquake cycle deformation onshore

from the Rivera plate is in contrast to the heterogeneous response to earthquakes beneath Oaxaca to the south (CORREA-MORA et al., 2008), and suggests the Jalisco area as a promising location for better understanding earthquake cycle behavior.

Acknowledgments

We would like to thank Jeff Genrich and Tom Herring for helpful discussions. Support for this research was provided by National Science Foundation grants EAR-0510395 (J. Stock) and EAR-0510553 (C. DeMets).

Appendix 1: Methods

Data are processed using the software packages GAMIT and GLOBK, developed at the Massachusetts Institute of Technology by T. A. Herring and D. Dong (HERRING et al., 1990; FEIGL et al., 1993; ZUMBERGE et al., 1997; DONG et al., 1998). Inputs into GAMIT are the data files, session specifications (year and day, receiver and antenna types, and antenna heights), and good initial station coordinates. Outputs are the loosely constrained solution files, which are passed to GLOBK for multi-session processing. A least-squares analysis is used to obtain the GAMIT solution files, and a combination of well-defined reference frame and Kalman filter with white noise are used in GLOBK to obtain the station coordinate time series and velocities.

The recursive, time-domain Kalman filter (run eight times) estimates the state of a dynamic system from a series of incomplete and noisy measurements, such as campaign GPS data (MAO et al., 1999); only the previous time step and the current measurement are needed to estimate the current state, with a linear relation used in the calculation, making it computationally efficient (HERRING et al., 1990). The Kalman filter uses a multivariate normal distribution for the process noise, which is independent of past process noise for every time step (i.e. the Markov process). In the simplest case, the Markov process allows separate noise levels for the north, east, and vertical components of position (we use 2 mm/year in the horizontal directions, 5 mm/year in the vertical direction).

The same a priori position and velocity constraints for each station are used in GAMIT and GLOBK, and are obtained from standard ITRF2000 files for the North America reference stations, and from the online Scripps Coordinate Update Tool for the campaign stations. Rotation and translation of three components of position and their rates are permitted when determining the reference frame in GLOBK. We iterate the reference frame solution eight times in order to stabilize the coordinate system, with 75% weighting on the coordinate sigmas of the previous iteration, and a 4-sigma cutoff for sites that are discordant with a priori values. Height residuals allowed in the stabilization are limited to 5 mm between the best and median for position (and 5 mm/year for the related rate), and 3 mm for the rms position (and 3 mm/year for the related rate).

Earth orientation parameters are tightly constrained in GLOBK by Markov process values of 0.25 mas/day in orientation and 0.1 mas/day in its rate of change. Since final orbits are used, a priori GPS satellite orbital parameters are also tightly constrained, with correspondingly tightly constrained random walk variation allowed while processing multiple sessions. Changes to orbital parameters due to random noise are constrained to 10 cm/day in XYZ, 0.01 mm/s/day for the XYZ time derivatives, 1%/day in direct and y-bias non-gravitational parameters, 0.1%/day in b axis bias and once-per-rev parameters, and 1 cm/day for SV antenna offsets.

Appendix 2: Results

See Tables 1 and 2.

Table 1

*Global [North America (NA)] and campaign GPS sites, their velocity components and 1σ errors (relative to North America, as defined by the stations with a *), and the cross-correlation (ρ) between north (N) and east (E) rates*

Long. (deg)	Lat. (deg)	E rate (mm/year)	N rate (mm/year)	Eσ	Nσ	ρ	H rate (mm/year)	Hσ	Site
284.912	38.777	0.06	0.44	1.29	1.14	0.028	−3.48	1.79	CHL1*
284.476	39.160	−0.07	0.61	1.97	1.80	0.042	1.05	4.13	DNRC*
284.430	39.561	−0.12	0.79	1.58	1.43	0.055	−1.17	2.27	RED1*
284.430	39.562	−2.99	−1.41	3.55	3.49	0.004	−3.57	10.83	RED2*
280.157	32.758	−0.04	0.75	1.19	1.14	−0.024	0.32	1.40	CHA1*
278.347	24.582	−0.69	0.59	1.15	1.18	0.031	3.96	1.48	KYW1*
273.910	36.358	1.59	−1.15	1.72	1.68	0.005	−0.25	2.50	HTV1*
265.183	35.367	1.28	−1.71	1.17	1.12	0.012	0.49	1.35	SAL1*
264.598	39.126	−0.02	−0.08	1.18	1.13	−0.005	−1.47	1.46	KAN1*
264.089	41.778	1.66	−1.02	1.62	1.57	−0.012	4.54	2.68	OMH1*
262.244	30.312	−0.24	0.75	1.22	1.16	0.070	−4.07	1.65	AUS5*
257.685	31.874	−0.88	0.09	1.25	1.12	0.098	−1.62	1.72	ODS5*
256.839	20.090	−5.30	−0.85	1.89	1.67	0.023	−2.09	4.02	CEBO
256.675	20.293	−6.79	−0.57	1.89	1.67	0.018	2.71	3.88	COS2
256.650	20.694	−4.66	−5.11	2.97	2.73	0.017	2.52	7.87	UGEO
256.554	19.730	−3.83	−0.57	2.05	1.80	0.031	0.87	4.68	CGUZ
256.547	20.737	−8.48	−2.98	3.99	3.39	0.027	4.38	12.83	UMON
256.472	20.335	−6.67	−2.03	2.02	1.75	0.030	3.99	5.09	LIM2
256.203	19.831	−7.42	−3.91	2.03	1.75	0.030	−4.29	4.50	TAPA
255.985	30.681	−1.94	0.13	1.43	1.37	0.006	0.97	1.73	MDO1
255.702	19.064	7.10	7.43	5.90	4.67	0.030	−8.17	15.94	MANZ
255.671	19.748	−6.62	−5.18	1.97	1.72	0.018	2.05	4.04	AUTA
255.667	19.031	0.19	4.60	1.95	1.71	0.017	62.74	3.67	CRIP
255.626	20.188	−6.19	−4.78	1.94	1.71	0.015	17.50	3.93	AYUT
255.363	19.665	−5.50	−4.28	1.98	1.74	0.016	12.42	4.09	PURI
251.881	34.302	−1.35	−2.36	1.45	1.43	0.019	0.01	1.91	PIE1
249.028	32.224	−1.51	0.48	2.49	2.01	0.162	3.42	4.48	COT1*

Uncertainties on vertical rates (H) are too large to constrain that component of motion. Stations are ordered by longitude

Table 2

GPS sites in the Jalisco region, their velocity components and 1σ errors (calculated relative to North America, as defined in Table 1, and presented relative to TAPA, a campaign site on the Jalisco Block), and the cross-correlation (ρ) between N and E rates

Long. (deg)	Lat. (deg)	E rate (mm/year)	N rate (mm/year)	Eσ	Nσ	ρ	H rate (mm/year)	Hσ	Site
256.203	19.831	0	0	0	0	0	0	0	TAPA
256.839	20.090	2.12	3.06	2.13	1.95	0.028	2.20	5.21	CEBO
256.675	20.293	0.63	3.34	2.13	1.96	0.026	7.00	5.10	COS2
256.650	20.694	2.75	−1.20	3.14	2.92	0.018	6.81	8.56	UGEO
256.554	19.730	3.58	3.35	2.21	2.03	0.031	5.17	5.38	CGUZ
256.547	20.737	−1.06	0.94	4.11	3.53	0.032	8.67	13.18	UMON
256.472	20.335	0.74	1.89	2.23	2.02	0.034	8.28	6.04	LIM2
255.702	19.064	14.52	11.35	5.96	4.76	0.030	−3.87	16.33	MANZ
255.671	19.748	0.80	−1.27	2.16	1.98	0.028	6.34	5.15	AUTA
255.667	19.031	7.61	8.52	2.11	1.95	0.025	67.03	4.78	CRIP
255.626	20.188	1.23	−0.87	2.14	1.97	0.025	21.79	5.05	AYUT
255.363	19.665	1.92	−0.37	2.16	1.99	0.027	16.71	5.18	PURI

Uncertainties on vertical rates (H) are too large to constrain that component of motion. UGEO, CRIP, and MANZ are continuous sites, while all other stations are part of the campaign. Stations are ordered by longitude

Appendix 3: Assumptions used in triple junction constraints

We assume a simplified geometry: three blocks (the Jalisco Block (JB), North America (NA), and the Michoacan Block (MB) (e.g. JOHNSON and HARRISON, 1990) meet at a continental triple junction formed by the Tepic–Zacoalco rift, the northern Colima rift, and the Chapala rift. GPS sites UGEO and UMON are on NA; GPS sites CEBO and CGUZ are on the MB; and GPS site TAPA is on JB. We assume a flat-earth geometry because of the close spacing of these stations (<100 km separation). We use the results from Appendix 2, Table 2 to constrain the velocity of NA and the MB relative to the JB, assuming no compression across any of the boundaries, as follows.

1. UGEO and UMON lie on NA, and should move together with respect to TAPA. Thus, the velocity of NA must lie within the intersection of the 95% confidence limits of the UMON and UGEO velocities relative to TAPA (red region in velocity diagram in Fig. 2b). Similarly, the velocity of the MB must lie within the intersection of the 95% confidence regions of the CEBO and CGUZ velocities.
2. We assume no compression across the Tepic–Zacoalco rift, which trends N56°W. Therefore, the velocity of NA with respect to TAPA must lie

northeast of a line with an azimuth of N56°W. This confines the allowable velocities for NA to points within the region shown in red in Fig. 2c.
3. We assume no compression in the northern Colima rift, which trends N–S. Thus, the velocities of stations on the MB (CEBO and CGUZ) must lie east of a line trending N–S from the origin. This requires their velocities to lie within the green region shown in Fig. 2d.
4. We assume no compression across the east–west trending Chapala rift. This eliminates velocities of NA from Fig. 2c that lie south of the southernmost point in the green velocity field in Fig. 2d, yielding the possible velocities of NA relative to block JB, shown in red in Fig. 2e. Similarly, we eliminate velocities of stations CGUZ and CEBO from Fig. 2d that lie north of an E-W line that passes through the northernmost point of the allowed velocity field for NA in Fig. 2e. The velocity for the MB then is restricted to the green field of Fig. 2f.

This yields the following constraints on the velocity triangle at the Guadalajara triple junction. At 95% confidence, the velocity of NA relative to the JB can lie anywhere in the red shaded region of Fig. 2e. The velocity of the MB relative to the JB can lie anywhere in the green-shaded region on Fig. 2f. However, the combination of velocities (one point from the red field and one point from the green field) must further satisfy two additional constraints. First, the red point cannot lie south of the green one (otherwise there would be compression across the Chapala rift). Second, the points on the velocity triangle must have the same topology as the blocks in map view (i.e. the JB, NA, and the MB must be encountered in clockwise order going around the triangle). A velocity triangle with the JB, NA, and the MB in counterclockwise order would imply that at least one of the boundaries is compressional.

References

ALLAN, J. F. (1986), *Geology of the Northern Colima and Zacoalco Grabens, southwest Mexico: Late Cenozoic rifting in the Mexican Volcanic Belt*, Geol. Soc. of Am. Bull. *97*, 473–485.

ALLAN, J. F., S. A NELSON, J. F. LUHR, I. S. E. CARMICHAEL, M. WOPAT, and P. J. WALLACE (1991), *Pliocene-Holocene Rifting and Associated Volcanism in Southwest Mexico: An Exotic Terrane in the Making*, in DAUPHIN, J. P. & SIMONEIT, B. R. T., eds., *The Gulf and Peninsular Province of the Californias*: Boulder, Colorado, AAPG Memoir 47, 425–445.

ALTAMIMI, Z., P. SILLARD, and C. BOUCHER (2002), *ITRF2000: A new release of the International Terrestrial Reference Frame for earth science applications*, J. of Geophys. Res., *107*(B10), 2214, doi:10.1029/2001JB000561.

ANDREWS, V., J. M. STOCK, G. REYES-DÁVILA and C. RAMÍREZ-VAZQUEZ, Double-difference relocation of the aftershocks of the Tecomán, Colima, Mexico earthquake of 22 January 2003, *PAGEOPH, in press*.

ARZATE, J. A., R. ALVAREZ, V. YUTSIS, J. PACHECO, and H. LOPEZ-LOERA (2006), *Geophysical modeling of Valle de Banderas Graben and its structural relation to Bahia de Banderas, Mexico*, Revista Mexicana de Ciencias Geologicas 23 (2), 184–198.

BANDY, W. and M. PARDO (1994), *Statistical examination of the existence and relative motion of the Jalisco and Southern Mexico Blocks*, Tectonics 13 (4), 755–768.

BANDY, B. L., F. MICHAUD, J. BOURGOIS, T. CALMUS, J. DYMENT, C. A. MORTERA-GUTIERREZ, J. ORTEGA-RAMIREZ, B. PONTOISE, J.-Y. ROYER, B. SICHLER, M. SOSSON, M. REBOLLEDO-VIEYRA, F. BIGOT-CORMIER, O. DIAZ-MOLINA, A. D. HURTADO-ARTUNDUAGA, G. PARDO-CASTRO, C. TROUILLARD-PERROT (2005), *Subsidence and strike-slip tectonism of the upper continental slope off Manzanillo, Mexico*, Tectonophysics 398, 115–140.

BECKER, J. J. and D. T. SANDWELL (2006), *SRTM30_PLUS V2.0: Data fusion of SRTM land topography with measured and estimated seafloor topography*, 29 July 2006, ftp://edcsgs9.cr.usgs. gov/pub/data/srtm/SRTM30.

BOURGOIS, J., V. RENARD, J. AUBOUIN, W. BANDY, E. BARRIER, T. CALMUS, J.-L. CARFANTAN, J. GUERRERO, J. MAMMERICKX, B. MERCIER DE LEPINAY, F. MICHAUD, M. SOSSON (1988), *Active fragmentation of the North American Plate: offshore boundary of the Jalisco Block off Manzanillo*, C. R. Acad., Sci. Paris 307 (II), 1121–1130.

CORREA-MORA, F., C. DEMETS, E. CABRAL-CANO, B. MARQUEZ-AZUA, and O. DIAZ-MOLINA (2008), *Interplate coupling and transient slip along the subduction interface beneath Oaxaca, Mexico*, Geophys. J. Int. *175*, 269–290.

DEMETS, C. and D. S. WILSON (1997), *Relative motions of the Pacific, Rivera, North American, and Cocos plates since 0.78 Ma*, J. of Geophys. Res. *102* (B2), 2789–2896.

DEMETS, C. and S. TRAYLEN (2000), *Motion of the Rivera plate since 10 Ma relative to the Pacific and North American plates and the mantle*, Tectonophysics *318*, 119–159.

DONG, D., T. A. HERRING, and R. W. KING (1998), *Estimating regional deformation from a combination of space and terrestrial geodetic data*, J. of Geodesy 72, 200 –214.

FEIGL, K. L., D. C. AGNEW, Y. BOCK, D. DONG, A. DONNELLAN, B. H. HAGER, T. A. HERRING, D. D. JACKSON, T. H. JORDAN, R. W. KING, S. LARSEN, K. M. LARSON, M. H. MURRAY, Z. SHEN, and F. H. WEBB (1993), *Space Geodetic Measurement of Crustal Deformation in Central and Southern California, 1984–1992*, J. of Geophys. Res. *98* (B12), 21,677–21,712.

FERRARI, L. (1995), *Miocene shearing along the northern boundary of the Jalisco block and the opening of the southern Gulf of California*, Geology 23 (8), 751–754.

FERRARI, L. (2004), *Slab detachment control on mafic volcanic pulse and mantle heterogeneity in central Mexico*, Geology 32 (1), 77–80.

FERRARI, L. and J. ROSAS-ELGUERA (2000), *Late Miocene to Quaternary extension at the northern boundary of the Jalisco block, western Mexico: The Tepic-Zacoalco rift revised*, in DELGADO-GRANADOS, H., G. AGUIRRE, and J. M. STOCK, eds., Cenozioc Tectonics and Volcanism of Mexico: Boulder, Colorado, GSA Special Paper 334, 41–63.

FERRARI, L., G. PASQUARE, S. VENEGAS, D. CASTILLO, and F. ROMERO (1994), *Regional tectonics of western Mexico and its implications for the northern boundary of the Jalisco Block*, Geofisica Internacional 33 (1), 139–151.

FREY, H. M., R. A. LANGE, C. M. HALL, H. DELGADO-GRANADOS, I. S. E. CARMICHAEL (2007), *A Pliocene ignimbrite flare-up along the Tepic-Zacoalco rift: Evidence for the initial stages of rifting between the Jalisco block (Mexico) and North America*, GSA Bulletin 119 (1/2), 49–64.

GARDUÑO-MONROY, V. H., R. SAUCEDO-GIRÓN, Z. JIMÉNEZ, J. C. GAVILANES-RUIZ, A. CORTÉS-CORTÉS, and R. M. URIBE-CIFUENTES (1998), *La falla Tamazula, Límite suroriental del Bloque Jalisco, y sus relaciones con el complejo volcánico de Colima, México*, Rev. Mex. Cienc. Geol. 15, 132–144.

GRAND, S. P., T. YANG, S. SUHARDJA, D. WILSON, M. G. SPEZIALE, J. GOMEZ GONZALEZ, G. LEON-SOTO, J. NI, and T. DOMINGUEZ REYES (2007), *Seismic structure of the Rivera subduction zone – the MARS experiment (abstract)*, Eos, Trans., AGU, T32A-02.

HERRING, T. A., J. L. DAVIS, and I. I. SHAPIRO (1990), *Geodesy by Radio Interferometry: The Application of Kalman Filtering to the Analysis of Very Long Baseline Interferometry Data*, J. of Geophys. Res. *95* (B8), 12,561–12,581.

HUTTON, W., C. DEMETS, O. SANCHEZ, G. SUAREZ, and J. STOCK (2001), *Slip kinematics and dynamics during and after the 1995 October 9 $M_w = 8.0$ Colima-Jalisco earthquake, Mexico, from GPS geodetic constraints*, Geophys. J. Int. *146*, 637–658.

JOHNSON, C. A. and C. G. A. HARRISON (1990), *Neotectonics in central Mexico*, Physics of the Earth and Planetary Interiors 64, 187–210.

KHUTORSKOY, M. D., L. A. DELGADO-ARGOTE, R. FERNANDEZ, V. I. KONONOV, B. G. POLYAK (1994), *Tectonics of the offshore*

Manzanillo and Tecpan basins, Mexican Pacific, from heat flow, bathymetric and seismic data, Geofis. Int. 33, 161–185.

KOSTOGLODOV, V. and W. BANDY (1995), Seismotectonic constraints on the convergence rate between the Rivera and North American plates, J. of Geophys. Res. 100 (B9), 17,977–17,989.

LUHR, J. F., S. A. NELSON, J. F. ALLAN, and I. S. E. CARMICHAEL (1985), Active rifting in southwestern Mexico: Manifestations of an incipient eastward spreading-ridge jump, Geology 13, 54–57.

MAO, A., C. G. A. HARRISON, and T. H. DIXON (1999), Noise in GPS coordinate time series, J. of Geophys. Res. 104 (B2), 2797–2816.

MARQUEZ-AZUA, B. M. and C. DEMETS (2003), Crustal velocity field of Mexico from continuous GPS measurements, 1993 to June 2001: Implications for the neotectonics of Mexico, J. of Geophys. Res. 108 (B9), 2450, doi:10.1029/2002/2002JB002241.

MARQUEZ-AZUA, B. M. and C. DEMETS (2009), Deformation of Mexico from continuous GPS from 1993 to 2008, Geochem. Geophys. Geosyst. 10, Q02003, doi:10.1029/2008GC002278.

MARQUEZ-AZUA, B. M., C. DEMETS, and T. MASTERLARK (2002), Strong interseismic coupling, fault afterslip, and viscoelastic flow before and after the Oct. 9, 1995 Colima-Jalisco earthquake: Continuous GPS measurements from Colima, Mexico; Geophysical Research Letters 29 (N.8), 122-1–122-4.

MASTERLARK, T., C. DEMETS, H. F. WANG, J. STOCK, and O. SANCHEZ (2001), Homogeneous vs. heterogeneous subduction zone models: Coseismic and postseismic deformation, Geophysical Review Letters 28, 4047–4050.

MELBOURNE, T., I. CARMICHAEL, C. DEMETS, K. HUDNUT, O. SANCHEZ, J. STOCK, G. SUAREZ, F. WEBB (1997), The geodetic signature of the M8.0 Oct. 9, 1995, Jalisco subduction earthquake, Geophys. Res. Let. 24 (6), 715–718.

MELBOURNE, T., F. WEBB, J. STOCK, and C. REIGBER (2002), Rapid postseismic transients in subduction zones from continuous GPS, J. Geophys. Res. 107 (B10), 2241, doi:10.1029/20001JB000555.

MOLNAR, P. and J. M. STOCK (1985), A method for bounding uncertainties in combined plate reconstructions, J. of Geophys. Res. 90 (B14), 13537–12544.

NIXON, G. T. (1982), The relationship between Quaternary volcanism in central Mexico and the seismicity and structure of subducted ocean lithosphere, Geol. Soc. of Am. Bull. 93, 514–523.

NUÑEZ-CORNÚ, F. J., R. L. MARTA, F. A. NAVA-P., G. REYES-DAVILA, and C. SUAREZ-PLASCENCIA (2002), Characteristics of seismicity in the coast and north of Jalisco Block, Mexico, Phys. of the Earth and Planet. Int. 132, 141-155.

PACHECO, J., S. K. SINGH, J. DOMINGUEZ, A. HURTADO, L. QUINTANAR, Z. JIMENEZ, J. YAMAMOTO, C. GUTIERREZ, M. SANTOYO, W. BANDY, M. GUZMAN, V. KOSTOGLODOV (1997), The October 9, 1995 Colima-Jalisco, Mexico earthquake (M_W 8): An aftershock study and a comparison of this earthquake with those of 1932, Geophys. Res. Let. 24 (17), 2223–2226.

PACHECO, J. F., C. A. MORTERA-GUTIERREZ, H. DELGADO, S. K. SINGH, R. W. VALENZUELA, N. M. SHAPIRO, M. A. SANTOYO, A. HURTADO, R. BARRON, and E. GUTIERREZ-MOGUEL (1999), Tectonic significance of an earthquake sequence in the Zacoalco half-graben, Jalisco, Mexico, Journal of South American Earth Sciences 12, 557–565.

PACHECO, J., W. BANDY, G. A. REYES-DAVILA, F. J. NUNEZ-CORNU, C. A. RAMIREZ-VAZQUEZ, J. R. BARRON (2003), The Colima, Mexico, earthquake (Mw 5.3) of 7 March 2000: seismic activity along the southern Colima rift, Bull. Seismol. Soc. Am. 93, 1458–1476.

PARDO, M. and G. SUAREZ (1995), Shape of the subducted Rivera and Cocos plates in southern Mexico: Seismic and tectonic implications, J. of Geophys. Res. 100 (B7), 12,357–12,373.

ROSAS-ELGUERA, J., L. FERRARI, V. H. GARDUÑO-MONROY, J. URRUTIA-FUCUGAUCHI (1996), Continental boundaries of the Jalisco block and their influence in the Pliocene-Quaternary kinematics of western Mexico, Geology 24 (10), 921–924.

ROSAS-ELGUERA, J., L. FERRARI, M. MARTINEZ, J. URRITIA-FUCAGAUCHI (1997), Stratigraphy and Tectonics of the Guadalajara Region and Triple-Junction Area, Western Mexico, International Geology Review 39, 125–140.

SCHMITT, S. V., C. DEMETS, J. STOCK, O. SANCHEZ, B. MARQUEZ-AZUA, and G. REYES (2007), A geodetic study of the 2003 January 22 Tecomán, Colima, Mexico earthquake, Geophys. J. Int. 169, 389–406.

SINGH, S. K., L. PONCE, S. P. NISHENKO (1985), The great Jalisco, Mexico, earthquakes of 1932: Subduction of the Rivera plate, Bull. Seis. Soc. Am. 75 (5), 1301–1313.

STOCK, J. M. (1993), Tectónica de placas y la evolución del bloque Jalisco, México, GEOS, Bol. Mex Geofis. Union 13 (3), 3–9.

SUAREZ, G., V. GARCIA-ACOSTA, and R. GAULON (1994), Active crustal deformation in the Jalisco block, Mexico: evidence for a great historical earthquake in the 16th century, Tectonophysics, 234, 117–127.

SUHARDJA, S., S. GRAND, D. WILSON, M. GUZMAN-SPEZIALE, J. GOMEZ-GONZALEZ, J. NI, and T. DOMINGUEZ-REYES (2007), Crustal structure beneath Southwestern Mexico (abstract), Eos, Trans., AGU, S33A-1052.

URRUTIA-FUCUGAUCHI, J. and T. GONZALEZ-MORAN (2006), Structural pattern at the northwestern sector of the Tepic-Zacoalco rift and tectonic implications for the Jalisco block, western Mexico, Earth Planets Space 58, 1303–1308.

WANG, K. (2007), Elastic and viscoelastic models of crustal deformation in subduction earthquake cycles, in The Seismogenic Zone of Subduction Thrust Faults, eds. T. H. Dixon and J. C. Moore, 540–575.

WANG, X., F. NIU, J. NI, and S. GRAND (2008), Crustal and Mantle Structure of the Jalisco Block of western Mexico from Surface Wave Tomography (abstract), Eos, Trans., AGU, S23A-1871.

YAGI, Y., T. MIKUMO, J. PACHECHO, and G. REYES (2004), Source Rupture Process of the Tecomán, Colima, Mexico Earthquake of 22 January 2003, Determined by Joint Inversion of Teleseismic Body-Wave and Near-Source Data, Bull. Seism. Soc. Am. 94, 1795–1807.

YANG, T., S. P. GRAND, D. WILSON, M. GUZMAN-SPEZIALE, J. M. GOMEZ-GONZALEZ, T. DOMINGUEZ-REYES, and J. NI (2009), Seismic structure beneath the Rivera subduction zone from finite-frequency seismic tomography, J. Geophys. Res. 114, B01302, doi:10.1029/2008JB005830.

ZUMBERGE, J. F., M. B. HEFLIN, D. C. JEFFERSON, M. M. WATKINS, and F. H. WEBB (1997), Precise point positioning for the efficient and robust analysis of GPS data from large networks, J. of Geophys. Res. 102 (B3), 5005–5017.

(Received February 10, 2010, revised July 10, 2010, accepted September 28, 2010, Published online November 9, 2010)

Reprinted from the journal

Pure Appl. Geophys. 168 (2011), 1449–1460
© 2010 Springer Basel AG
DOI 10.1007/s00024-010-0199-5

▮ Pure and Applied Geophysics

Imaging the Moho and Subducted Oceanic Crust at the Isthmus of Tehuantepec, Mexico, from Receiver Functions

DIEGO MELGAR[1,3] and XYOLI PÉREZ-CAMPOS[2]

Abstract—Using teleseismic data recorded along a transect, which we call VEOX (for Veracruz-Oaxaca seismic line), of 46 broadband stations installed across the Isthmus of Tehuantepec in southern Mexico, we obtained receiver functions and stacked them to study the Moho topography and back projected them to visualize the subducted slab geometry beneath the isthmus. We observed a back-azimuth dependent Moho thickness across the transect, particularly beneath the Los Tuxtlas Volcanic Field. Also, we observed the Cocos plate which subducts with an angle of 26° between 140 and 310 km from the trench. Comparison with regional seismicity indicates that it occurs below the oceanic crust.

Key words: Receiver functions, slab subduction, Moho depth.

1. Introduction

The Mexican subduction zone extends roughly from 106.5°W to 92°W, and the slab geometry across it is highly variable. PARDO and SUÁREZ (1995) inferred from hypocenter locations that the Cocos plate subducts at a steep angle ($\sim$50°) at the north, decreasing towards the south, eventually leading to a flat slab subduction beneath central Mexico. This structure was first reported by SUÁREZ et al., (1990), and verified by KOSTOGLODOV et al., (1996) by means of seismicity and gravity, and by PÉREZ-CAMPOS et al., (2008), by means of receiver functions. To the south,

the slab increases its dip again (PONCE et al., 1992) (Fig. 1).

The focus of this paper is to study the geometry of the subducted Cocos plate (and the Moho topography) at the Isthmus of Tehuantepec, a tectonically complex region. One of the most prominent tectonic features on the Cocos plate in this region is the Tehuantepec Ridge (TR, Fig. 1). According to KLITGORD and MAMMERICKX (1982), the TR separates the Cocos plate into two regions of different ages and tectonic regimes. On one hand, the age difference across its flanks is much greater than in other fracture zones that connect the East Pacific Rise with the Middle American Trench (MAT) (MANEA et al., 2005b); on the other, BRAVO et al., (2004) have suggested that the TR is a hinge fault that separates the portion of the Cocos plate that subducts beneath the North America plate from that portion that subducts beneath the Caribbean Plate.

Another notable feature of the region, this time on the North American plate, is the Los Tuxtlas Volcanic Field (LTVF, Fig. 1) which is isolated from other volcanic arcs in the region (the Trans-Mexican Volcanic Belt, TMVB, to the north, and the Modern Chiapanecan Volcanic Arc, MCVA, to the south; Fig. 1). Geodynamic modeling by MANEA and MANEA (2006) suggests that initially there was a traditional continuous volcanic arc across the whole of the subduction zone and that the transition to a variable dip slab across a smooth fold has produced the two non-parallel volcanic arcs (TMVB and MCVA) and the isolated LTVF between them. Meanwhile, FERRARI (2004) proposed a tear in the subducted slab to explain the mafic nature of a great part of the volcanism across this region. K–Ar dating by NELSON and GONZALEZ-CAVER (1992) sets the first episode of this mafic volcanism at 7–3.4 Ma.

---
[1] Facultad de Ingeniería, Universidad Nacional Autónoma de México, Coyoacán, 04510 Mexico, D.F., Mexico.
[2] Departamento de Sismología, Instituto de Geofísica, Universidad Nacional Autónoma de México, Circuito de la Investigación, s/n, Ciudad Universitaria, Coyoacán, 04510 Mexico, D.F., Mexico. E-mail: xyoli@geofisica.unam.mx
[3] Scripps Institution of Oceanography, University of California San Diego, 8800 Biological Grade, La Jolla, San Diego, CA 92093-0225, USA.

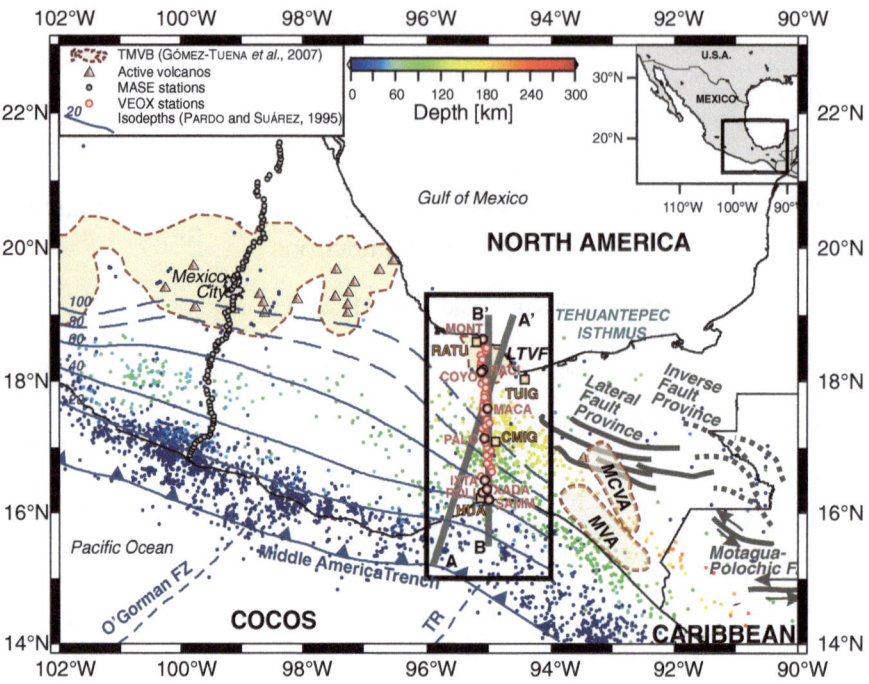

Figure 1

Main tectonic features of the region and location of stations. Los Tuxtlas Volcanic Field (LTVF), the Modern Chiapanecan Volcanic Arc (MCVA) and the Miocene Volcanic Arc (MVA) (from MANEA and MANEA, 2006) are shown in *yellow*. The main fault provinces in the region (from GUZMÁN-SPEZIALE and MENESES-ROCHA, 2000) are also shown. The regional seismicity during the operation of the VEOX network is denoted by *dots*, *color* coded by depth. Reference stations for Moho depth are indicated in squares. The *box* and profiles AA′ and BB′ are shown in Figs. 5, 6 and 7

The Meso-American Subduction Experiment (MASE) is a joint venture between the California Institute of Technology (Caltech), the Universidad Nacional Autónoma de México (UNAM) and the University of California Los Angeles; it aims to study and to model the geodynamics of the Mexican subduction zone. To this end, starting December, 2004, 100 stations were installed along a transect in central Mexico (Fig. 1); results from this transect are summarized in PÉREZ-CAMPOS et al., (2008). As a part of this continued effort, in July, 2007, Caltech and UNAM relocated the transect further south, where 46 stations were installed every 5 km along a new profile called VEOX (so named because it traverses the states of Veracruz and Oaxaca) in the Isthmus of Tehuantepec (Fig. 1). The focus of this new line is to study the processes taking place in a different part of the same subduction zone where the slab has clearly changed its behavior and geometry, but the Cocos Plate is still subducting North America, also avoiding triple junction complications.

Following the receiver function analysis for the MASE line (PÉREZ-CAMPOS et al., 2008), we identify the Moho and the subducted slab geometry along the VEOX transect. The slab can be traced between 140 and 310 km from the trench dipping at 26°. We compare these results with the isodepth curves obtained by PARDO and SUÁREZ (1995), with local seismicity obtained from the Mexican National Seismological Service (SSN) database and with double-difference relocated events from CASTRO ARTOLA (2010) in order to discuss their similarities and differences and the implications of the observed geometry.

2. Data and Processing

The dataset consists of 110 teleseismic events with $M_w > 5.8$ and angular distance between 30° and 90° (Fig. 2) registered between 12 July 2007 and 6 March 2009 on the VEOX network.

The VEOX network (Fig. 1) consists of 46 broadband stations installed along a profile between San Mateo del Mar, Oaxaca (station SAMM) on the Pacific coast and Monte Pío, Veracruz (station MONT) at the gulf coast. The station locations were chosen in the same manner as with the previous MASE transect, that is, installing them inside rural schools in towns along the profile (PÉREZ-CAMPOS, 2008). The remote locations of some of the stations meant sacrificing some of the profile geometry in exchange for station security. This approach proved to be successful, as only two solar panels and one GPS antenna suffered some sort of vandalism, and it permitted outreach efforts, which otherwise would have been impossible (PÉREZ-CAMPOS, 2008). Fortunately for the MASE transect, a main road perpendicular to the trench exists, making it easier to design a transect along the slab dipping orientation. This was not possible for the isthmus region; it would not have been possible to keep a 5 km separation among stations since there are only few roads and even fewer towns. Therefore, the main highway along the isthmus was chosen to define the orientation of the VEOX line.

Although a good number of earthquakes were adequately recorded along many stations of the profile, due to some problems (torrential rain which flooded some stations, less than adequate signal to noise ratios on some stations and occasional equipment failure), not all events were recorded at all stations throughout the transect (Fig. 3). Thus, only 38 stations yielded useful data for our analysis of

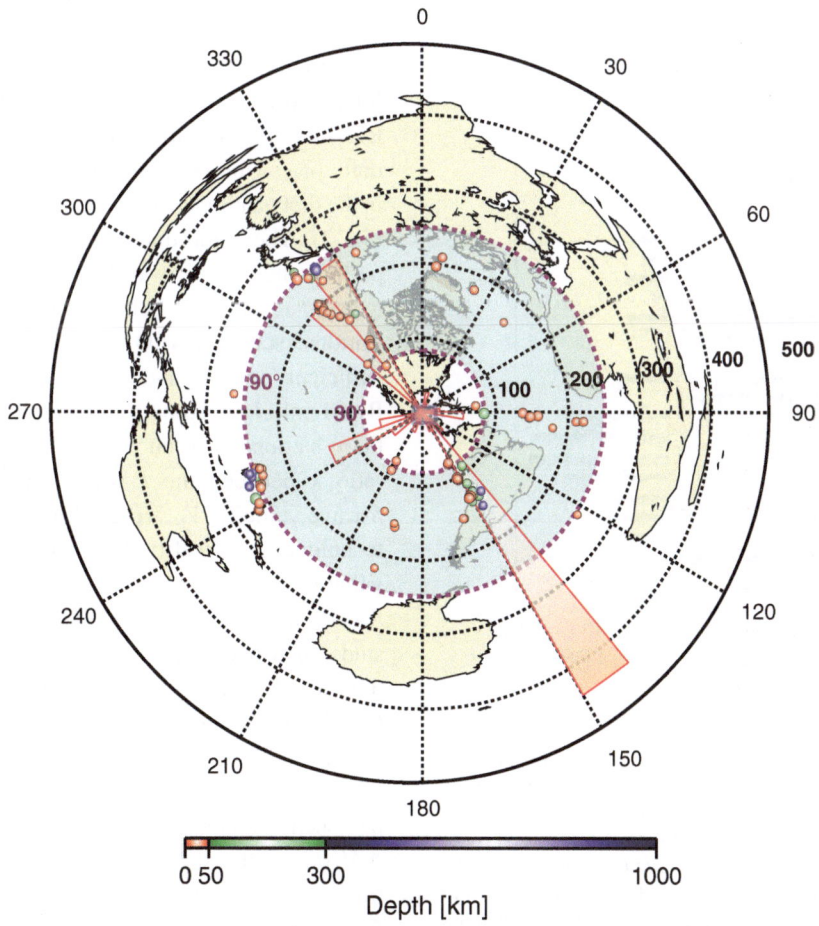

Figure 2

Teleseisms with $M_w > 5.8$ and angular distance between 30° and 90° (*blue shaded* region) from July 2007 to March 2009. The *rosette* indicates the RF azimuthal distribution

subducted oceanic crust geometry and only 35 were useful for our analysis of Moho topography.

Receiver functions (RFs) were first obtained using a frequency domain source equalization scheme (LANGSTON, 1979) with a low pass Gaussian filter and the water level technique (CLAYTON and WIGGINS, 1976). These were used for the stacking scheme used

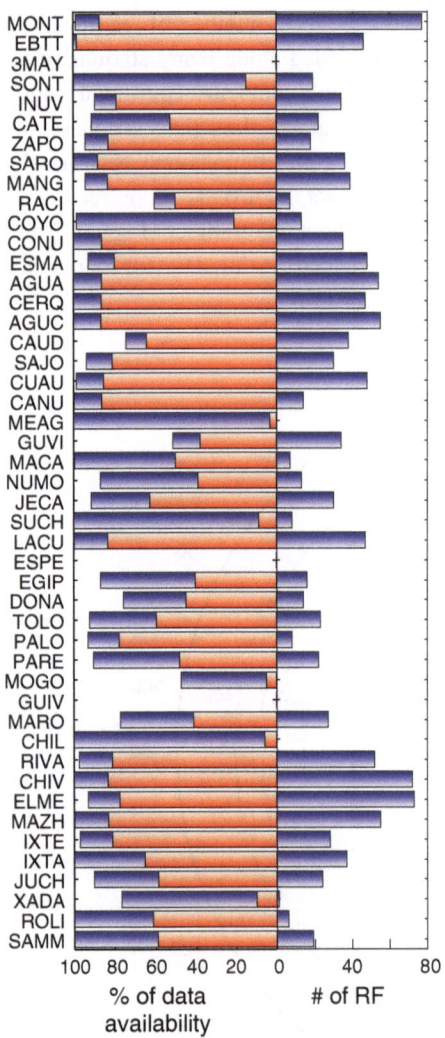

Figure 3

Data availability (in %, *left*) and number of receiver functions (RFs) (*right*) obtained at each station. The availability percentage is calculated both from the moment the station went into operation until the time it ceased (*blue*), and for the whole experiment (*red*), counting from the time the first station started recording until the last station stopped. Stations are displayed from north on the *top* to south on the *bottom*. Stations 3MAY, ELME and GUIV could not be installed because the sites were flooded before the installation campaign occurred. 19 stations had more than 80% data availability for the duration of the experiment

to analyze Moho topography. These RFs, however, were not of sufficient quality to analyze the subducted crust geometry; to improve the signal to noise ratio, an iterative deconvolution technique (KIKUCHI and KANAMORI, 1982) was employed instead, yielding better results for this analysis.

We obtained 1,279 RFs with a highly variable number of functions per station; station MONT had the most, 78 RFs, while station XADA had the fewest, 3 (Fig. 3). Figure 4 shows the stacked RFs without filtering for all stations and an example of the RFs obtained for station IXTA at the south of the transect; in this case they were filtered with a singular value decomposition (SVD) filter (CHEVROT and GIRARDIN, 2000) for visualization. Moho and the oceanic crust can be identified.

In order to obtain the Moho depth at each station and to verify its azimuthal variability (Table 1; Fig. 5), we performed a stacking scheme (ZHU and KANAMORI, 2000) for the RFs, after applying a SVD filter (CHEVROT and GIRARDIN, 2000), both, using all the RFs and grouping them according to their back azimuth (ϕ_B). Unfortunately, there were very few events in the first quadrant and not many more in the third (see Fig. 2). The minimum number of RFs required to perform the stacking was five. Thus, it was not possible to estimate the Moho depth for the first quadrant for all stations; station PALO produced only two RFs of sufficient quality, and stations ROLI, MACA and RACI only four, so no Moho depth was estimated for these stations. The uncertainty in the depth (H) and ratio κ between the P-wave velocity, V_P, and the S-wave velocity, V_S, were estimated with a bootstrap technique described by PERSAUD et al., (2007). A V_P value of 6.5 km/s was assumed for stations between SAMM to the south and COYO to the north, based on a study of Moho depths at SSN stations close to our transect (ESPÍNDOLA CASTRO, 2009); and a value of 6.35 km/s was used for the northern stations, based on a study of the Moho depths for an array in LTVF region (ZAMORA-CAMACHO et al., 2010). Once Moho depths had been obtained, they were projected along profiles AA' (along the slab dip) and BB' (along the station transect, see Fig. 1), and the depths for the second and fourth quadrant (90°–180° and 270°–360°) were interpolated using a loess function (CLEVELAND, 1993) with a first-order polynomial and a 0.25 smoothing

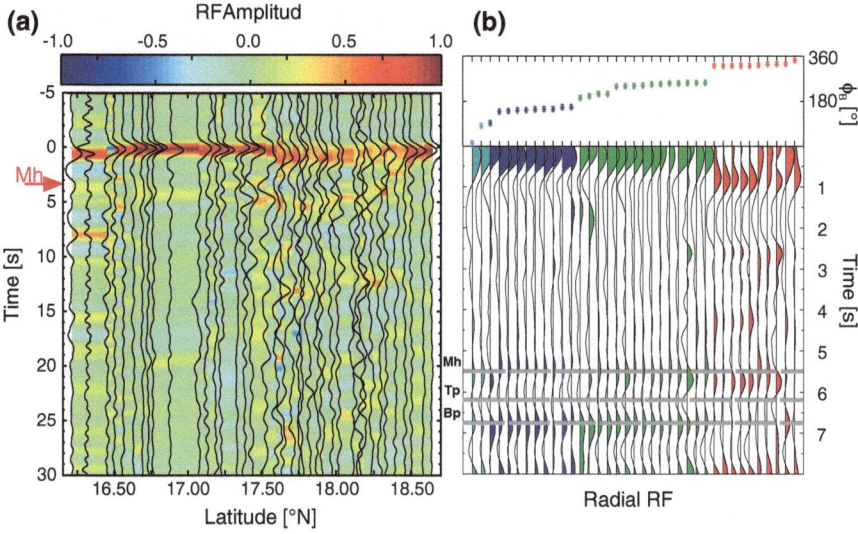

Figure 4
a Stacked RFs along the transect. **b** Receiver functions obtained at stations IXTA. They are ordered by back azimuth (ϕ_B), shown in the upper panel, and color coded according to their quadrant (*cyan* for $0° < \phi_B \leq 90°$, *blue* for $90° < \phi_B \leq 180°$, *green* for $180° < \phi_B \leq 270°$, and *red* for $270° < \phi_B \leq 360°$). The arrivals for the conversion at the continental Moho (Mh), and the *top* (Tp) and the *base* (Bp) of the oceanic crust of the Cocos plate are identified with the *dashed gray lines*

factor. Also, a loess function with the same characteristics was used to interpolate all the values (black line in Fig. 5).

The RFs obtained by the time deconvolution scheme were then transformed to the depth domain by a back-projection scheme (YAMAUCHI *et al.*, 2003; TONEGAWA *et al.*, 2005), assuming the IASPEI 91 velocity model (KENNET and ENGDAHL, 1991); next, they were projected to the AA′ profile along the slab dip proposed by PARDO and SUÁREZ (1995) (see Fig. 1) and averaged in 1×1 km cells (Fig. 6). The resulting back-projected profiles have an uncertainty of ~ 5 km (YAMAUCHI *et al.*, 2003).

3. Moho Geometry

There are subtle variations in Moho depth across the transect, from south to north (Fig. 6): the Moho starts at a 30 km average depth, increasing to 45 km around 18°N latitude (~ 275 km from the trench); later it becomes shallower (~ 28 km depth) at the coast of the Gulf of Mexico. Table 1 shows azimuthal variability. Moho profile plots by quadrant (Fig. 5), show a difference of up to 12 km in Moho depth, suggesting that Moho is not a horizontal interface.

The first quadrant has no Moho position owing to the lack of RFs. Although Moho depths change with back azimuth across the whole transect, the largest variations occur between the NW and SE Moho depth estimations close to LTVF.

A problem in identifying the Moho north of latitude 17°N is a layer of sedimentary rocks corresponding to the clastic units of the two main geological provinces in this part of the transect: the Veracruz and Tehuantepec basins (ORTEGA-GUTIEREZ *et al.*, 1992). These produce multiples of considerable amplitude arriving directly after the Moho conversion, which can be misleading in interpreting its depth (see Fig. 6).

4. Subducted Oceanic Crust Geometry

Because RFs from different back-azimuths are expected to sample different parts of the subducted crust and have different arrival times, we took the two most populated back-azimuth bins, 140°–150° and 310°–330°, (Fig. 2) with 430 and 425 RFs, respectively, and applied an SVD filter (CHEVROT and GIRARDIN, 2000) to each bin to enhance coherent arrivals. After this, a fourth order Butterworth filter

Table 1

Moho depths, H, in km, and $\kappa = V_P/V_S$ at each station for all receiver functions (numbers in parenthesis indicate the confidence interval), and for the three back azimuth (ϕ_B) ranges. Stations are listed from south to north

Station	$0° \leq \phi_B \leq 360°$			$90° \leq \phi_B \leq 180°$			$180° \leq \phi_B \leq 270°$			$270° \leq \phi_B \leq 360°$		
	n	H (H^-, H^+)	κ (κ^-, κ^+)	n	H (H^-, H^+)	κ (κ^-, κ^+)	n	H (H^-, H^+)	κ (κ^-, κ^+)	n	H (H^-, H^+)	κ (κ^-, κ^+)
SAMM	15	25.2 (24.4, 26.0) 28.5 (25.0, 32.0)[a]	1.88 (1.82, 1.90)							7	31.6 (30.2, 32.9)	1.80 (1.76, 1.85)
JUCH	14	27.3 (26.8, 27.9)								7	27.0 (26.2, 27.7)	1.67 (1.64, 1.71)
IXTA	22	30.5 (29.6, 31.3)	1.85 (1.81, 1.89)	5	34.5 (32.8, 35.0)	1.88 (1.80, 1.90)	7	30.5 (29.7, 31.5)	1.88 (1.83, 1.90)	8	23.9 (23.1, 24.5)	1.69 (1.66, 1.72)
IXTE	22	29.7 (28.4, 30.7)	1.86 (1.81, 1.90)	7	25.8 (24.9, 26.7)	1.89 (1.81, 1.90)				5	34.6 (33.5, 35.0)	1.86 (1.81, 1.90)
MAZH	44	27.9 (27.5, 28.5)	1.89 (1.86, 1.90)	21	27.3 (26.4, 28.0)	1.85 (1.79, 1.89)	5	37.6 (36.1, 38.9)	1.73 (1.69, 1.77)	12	34.8 (33.4, 36.5)	1.73 (1.69, 1.78)
ELME	54	33.8 (32.7, 34.7)	1.77 (1.70, 1.80)	18	30.5 (29.3, 31.8)	1.88 (1.84, 1.90)	6	36.5 (35.3, 38.4)	1.66 (1.62, 1.72)	19	29.6 (29.0, 30.5)	1.68 (1.64, 1.72)
CHIV	49	32.4 (31.4, 33.1)	1.85 (1.82, 1.89)	26	31.6 (31.2, 32.5)	1.89 (1.86, 1.90)	6	36.5 (34.8, 37.4)	1.64 (1.60, 1.68)	14	33.5 (32.2, 34.7)	1.78 (1.74, 1.82)
RIVA	35	28.8 (28.1, 30.1)	1.88 (1.84, 1.90)	16	30.7 (29.7, 31.9)	1.89 (1.86, 1.90)				16	28.3 (27.6, 29.0)	1.87 (1.84, 1.89)
MARO	23	33.8 (32.6, 34.8)	1.72 (1.68, 1.75)	6	35.2 (34.5, 35.9)	1.63 (1.61, 1.65)				10	32.8 (31.5, 33.9)	1.76 (1.72, 1.81)
PARE	18	29.3 (28.1, 30.9) 29.3 (28.8, 29.8)[b]	1.63 (1.60, 1.73)	7	38.0 (36.6, 38.1)	1.90 (1.78, 1.90)				8	33.1 (32.3, 34.0)	1.74 (1.71, 1.78)
TOLO	15	40.5 (38.6, 42.7)	1.89 (1.76, 1.90)	6	40.9 (39.2, 42.4)	1.89 (1.86, 1.90)				5	35.1 (33.0, 35.6)	1.62 (1.60, 1.66)
DONA	9	41.1 (39.6, 42.3)	1.80 (1.77, 1.84)	4	41.5 (39.5, 43.1)	1.80 (1.76, 1.84)						
LACU	36	33.3 (31.5, 33.9)	1.60 (1.60, 1.64)	13	37.7 (35.8, 39.3)	1.82 (1.79, 1.87)				14	33.9 (33.0, 34.7)	1.60 (1.60, 1.62)
SUCH	8	39.5 (38.5, 40.9)	1.73 (1.70, 1.76)	7	39.6 (38.3, 40.5)	1.73 (1.71, 1.76)						
JECA	24	37.1 (35.9, 38.2)	1.74 (1.71, 1.78)	15	37.4 (36.4, 38.3)	1.75 (1.72, 1.79)				9	32.0 (31.1, 32.8)	1.84 (1.81, 1.88)
NUMO	12	42.1 (41.0, 42.8)	1.69 (1.67, 1.71)	12	42.1 (41.0, 42.8)	1.69 (1.68, 1.72)						
GUVI	25	45.8 (44.3, 45.9)	1.78 (1.76, 1.90)	16	45.9 (45.8, 45.9)	1.74 (1.74, 1.75)				9	45.8 (45.0, 45.9)	1.90 (1.89, 1.90)
CANU	6	30.9 (29.7, 32.1)	1.82 (1.77, 1.88)									
CUAU	34	46.0 (45.4, 46.9)	1.73 (1.72, 1.76)	12	42.7 (41.2, 43.8)	1.89 (1.77, 1.90)				12	45.4 (45.0, 47.6)	1.76 (1.72, 1.77)
SAJO	16	35.5 (34.3, 36.5)	1.76 (1.74, 1.80)	6	35.8 (34.4, 37.0)	1.72 (1.67, 1.75)				7	37.6 (36.3, 38.7)	1.78 (1.73, 1.83)
CAUD	21	35.2 (35.1, 36.2)	1.83 (1.81, 1.85)	9	35.6 (35.2, 36.9)	1.85 (1.81, 1.87)				9	35.8 (35.2, 36.6)	1.81 (1.78, 1.84)
AGUC	34	35.2 (35.1, 36.0)	1.86 (1.84, 1.88)	17	35.1 (34.7, 35.7)	1.88 (1.86, 1.90)				11	35.2 (35.1, 37.3)	1.75 (1.73, 1.80)
CERQ	28	37.3 (36.9, 37.8)	1.82 (1.80, 1.83)	14	36.9 (36.4, 37.7)	1.82 (1.80, 1.84)				9	32.5 (30.1, 33.4)	1.88 (1.81, 1.90)
AGUA	39	39.5 (37.7, 40.4)	1.72 (1.70, 1.75)	16	35.8 (35.3, 36.4)	1.81 (1.78, 1.82)				18	40.0 (39.3, 40.4)	1.70 (1.69, 1.71)
ESMA	33	41.3 (41.0, 42.6)	1.90 (1.86, 1.90)	14	43.8 (42.8, 44.5)	1.82 (1.81, 1.85)				13	39.3 (37.7, 39.9)	1.71 (1.69, 1.81)
CONU	25	37.7 (37.3, 38.5)	1.70 (1.69, 1.72)	15	29.3 (28.9, 31.0)	1.90 (1.88, 1.90)				10	39.6 (38.6, 40.5)	1.69 (1.67, 1.75)
COYO	11	30.6 (29.9, 31.5)	1.72 (1.69, 1.77)	6	30.1 (28.9, 30.9)	1.79 (1.74, 1.85)						
MANG	28	28.0 (27.5, 28.5)	1.88 (1.85, 1.90)	12	27.8 (26.9, 28.8)	1.82 (1.79, 1.87)				14	29.4 (29.3, 29.9)	1.86 (1.84, 1.87)
SARO	19	27.4 (27.1, 27.8)	1.82 (1.80, 1.83)	6	31.5 (30.3, 32.4)	1.70 (1.66, 1.73)				14	27.6 (27.5, 28.2)	1.81 (1.79, 1.82)
ZAPO	9	27.9 (26.9, 29.0)	1.62 (1.60, 1.67)				6	25.6 (25.5, 27.3)	1.60 (1.60, 1.64)			
CATE	17	31.4 (31.0, 33.4)	1.90 (1.88, 1.90)	9	26.8 (25.9, 28.1)	1.70 (1.65, 1.75)				6	36.4 (35.1, 37.5)	1.85 (1.82, 1.89)
INUV	26	36.8 (36.2, 37.7)	1.88 (1.85, 1.90)	8	25.7 (25.5, 26.9)	1.61 (1.60, 1.65)	5	24.7 (23.8, 25.5)	1.88 (1.82, 1.90)	7	37.3 (36.5, 39.3)	1.89 (1.85, 1.90)
SONT	13	25.0 (24.0, 25.9) 25.7 (25.2, 26.2)[c]	1.70 (1.66, 1.74)	8	25.0 (24.2, 26.3)	1.72 (1.68, 1.76)				6	31.2 (30.6, 32.1)	1.89 (1.87, 1.90)

Table 1 *continued*

Station	0° ≤ ϕ_B ≤ 360°			90° ≤ ϕ_B ≤ 180°			180° ≤ ϕ_B ≤ 270°			270° ≤ ϕ_B ≤ 360°		
	n	H (H^-, H^+)	κ (κ^-, κ^+)	n	H (H^-, H^+)	κ (κ^-, κ^+)	n	H (H^-, H^+)	κ (κ^-, κ^+)	n	H (H^-, H^+)	κ (κ^-, κ^+)
EBTT	32	31.2 (30.0, 32.0)	1.75 (1.73, 1.80)	12	32.1 (30.3, 33.4)	1.61 (1.60, 1.69)	7	29.5 (28.4, 30.0)		13	32.0 (30.7, 33.1)	1.75 (1.71, 1.82)
MONT	61	23.5 (22.8, 24.1) 23.7 (23.2, 23.7)[d]	1.82 (1.79, 1.86)	24	24.0 (23.3, 25.0)	1.82 (1.77, 1.86)			1.89 (1.82, 1.90)	18	30.6 (29.7, 31.5)	1.64 (1.61, 1.67)

n = number of RFs for the stacking. The minimum allowed was five.

[a] Station SAMM was compared with station HUA, a temporary station operated by the Federal Electricity Commission, reported by Bravo et al. (2004)

[b] Station PARE was compared with station CMIG operated by the Mexican National Seismological Service (SSN), reported by Espíndola Castro (2009)

[c] Station SONT was compared with station RATU, reported by Zamora-Camacho et al. (2010)

[d] Station MONT was compared with station TUIG operated by the SSN, reported by Espíndola Castro (2009)

was applied to each bin with cutoff frequencies at 0.3 and 0.8 Hz. Then, the back-projected RFs were projected onto the along-dip profile AA'. The resulting section was then interpolated using a nearest-neighborhood scheme with a 25 km search radius; the Generic Mapping Tool software (Wessel and Smith, 1991) was used.

While there was some trial and error involved in determining the appropriate frequency band for filtering, it was conditioned by the realization that seismic waves will be affected by structures with sizes similar to their wavelengths. The IASP91 model (Kennet and Engdahl, 1991) shows crustal velocities of 5.8–6.5 km/s at 0–35 km depth, while a reflection study north of the study area (Valdés et al., 1986) has velocities of 5–7.1 km/s for the subducted crust at depths of 0–80 km. If we then assume a thickness of 8–10 km for the oceanic crust, a simple calculation shows that the frequency band selected is plausible and an adequate compromise.

From the resulting figure (Fig. 6b), we can observe a well aligned sequence of negative pulses that start at $\sim$140 km distance from the trench and $\sim$60 km depth. These denote a negative impedance contrast and thus a seismically slow medium underlying a fast medium. These pulses can be followed to $\sim$310 km distance from the trench and $\sim$145 km depth where the signal terminates abruptly. They are interpreted as corresponding to the top of the subducted oceanic crust (blue dashed line). Below it, we can observe a well aligned sequence of positive pulses which correspond to oceanic crust-mantle interface (red dashed line). These pulses give an average thickness of $\sim$9 km for the oceanic crust of the Cocos plate and a dip of 26°.

Below 150 km depth, the signature of the oceanic crust in the RFs is not evident. This coincides with the ending of the seismicity.

5. Discussion

Moho thicknesses for stations very close to the VEOX transect have been previously reported (Table 1) by Bravo et al., (2004) using local seismicity, and by Espíndola Castro (2009) and Zamora-Camacho et al., (2010) using receiver functions; they

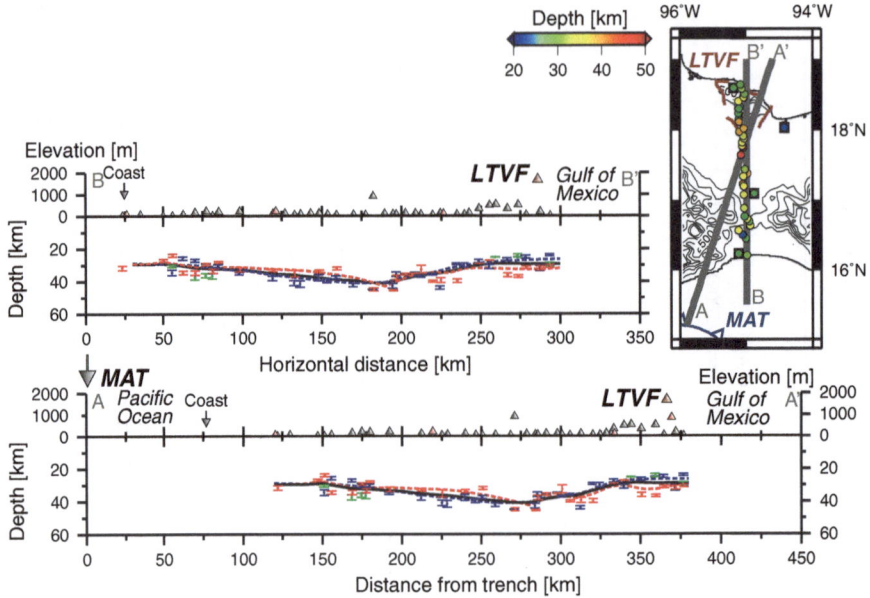

Figure 5

Moho geometry. *Right* Map with spatial distribution of the Moho depth estimations (*color coded circles*). The black squares correspond to stations from other studies (see text). Topography is shown with contour lines every 250 m. LTVF and MAT (*bottom panel*) are shown for reference. *Left* Profiles AA' (*bottom*) and BB' (*top*) with projected depths. The *color dots* denote the estimated depth (with *error bars*) for each station at each quadrant (*blue* for $90° < \phi_B \leq 180°$, *green* for $180° < \phi_B \leq 270°$, and *red* for $270° < \phi_B \leq 360°$). The *dashed lines* correspond to the interpolated Moho for the $90° < \phi_B \leq 180°$ quadrant (*blue*) and the $270° < \phi_B \leq 360°$ quadrant (*red*). The *solid black line* corresponds to the interpolated Moho using all the quadrants

estimated the Moho position at depths similar to those we obtain in this study if we stack the RFs over all back azimuths. However, as shown by Fig. 6, there are non-negligible azimuthal variations in the Moho depth estimates, and averaging out the calculations over all azimuths may lead to loss of detail in Moho topography.

A case in point is the depth estimate for the Moho in the vicinity of the LTVF: Moho positions estimated from RFs in the back azimuth range of 270°–360°, whose raypaths are through the center of the LTVF, are considerably deeper than the estimates from RFs on the 90°–180° quadrant whose raypaths would traverse the fringe of the LTVF. This could be explained because the LTVF consists of two main volcanic cones (San Martin and Santa Martha), both in excess of 1,700 m above mean sea level, surrounded by the essentially flat topography of the Veracruz basin. Thus, isostatic compensation of this load would result in something resembling the evident three-dimensional geometry; something roughly similar to the case of a point load on an elastic plate

(TURCOTTE and SCHUBERT, 2001). However, further inquiry is necessary.

Independent verification of the Moho depth estimates shown here by alternate geophysical techniques is desirable. However, gravity studies seem to focus on the seaward portion of the Tehuantepec Isthmus and the structure of Tehuantepec Ridge (e.g. COUCH and WOODCOCK, 1981; MANEA et al., 2003; MANEA et al., 2005a). URRUTIA-FUCUGAUCHI and FLORES-RUIZ (1996) estimate crustal thicknesses from gravity studies in central Mexico, but their study area is too far west of our profile and falls 1° of longitude short of the VEOX transect. However, they do obtain crustal thicknesses of 25 km for the continental shelf of the Gulf of Mexico immediately adjacent to the continent and of 30 km for the continental part of the coast, which is at least roughly consistent with Moho depth estimates shown here.

The final model for the geometry of the Moho and the subducted oceanic crust is shown in Fig 7. Comparing this final model with the one obtained by PARDO and SUÁREZ (1995) (Fig. 7), we can see that

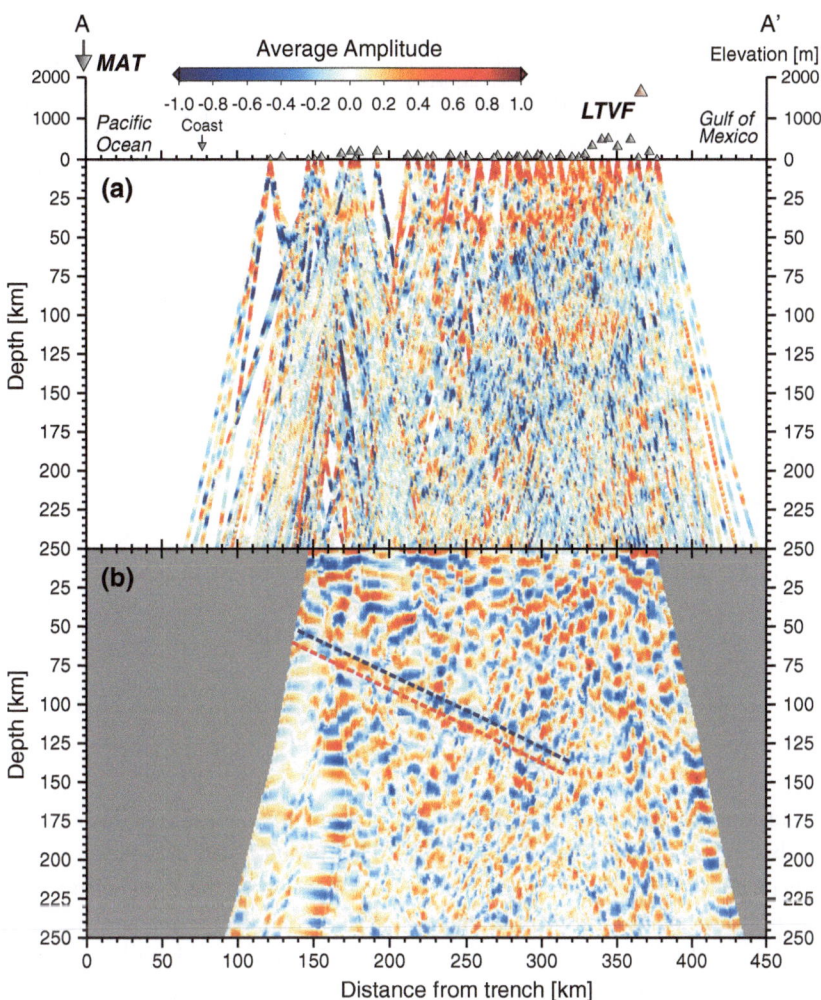

Figure 6

a Back-projected section AA' (see Fig. 1). **b** Interpolated projection (see text for details) with interpreted oceanic crust. The *blue dashed line* denotes the *top* of the slab and the *red dashed line* the *bottom* of the oceanic crust

our position for the slab is very similar to theirs, the difference being that they trace it all the way back to the trench, whereas our coverage does not allow us to do so. However, at 100 km depth the PARDO and SUÁREZ (1995) model was based on local seismicity terminates, whereas the RF analysis permits us to trace the slab to ~145 km depth.

The TMVB is not parallel to the trench (Fig. 1), and after some controversy regarding this, FERRARI (2004) put forth a subduction model where the tear in the Farallon slab, proposed by DICKINSON (1997) as having been produced by the arrival of the East Pacific Rise to the trench off Baja California, continued propagating laterally and parallel to the MAT,

where subduction was still active. According to this model, the tear would produce a transient thermal anomaly that has its surface reflection in the pulse of mafic volcanism observed along the TMVB. This mafic pulse, propagating at 100 km/Myr, would have reached 95°W longitude (roughly where profile AA' crosses the LTVF) at ~5.5 Myr.

Tomography studies (GORBATOV and FUKAO, 2005; HUSKER and DAVIS, 2009) and RF analyses (PÉREZ-CAMPOS et al., 2008) provide strong evidence for the detachment model. This, coupled with the fact that the model of FERRARI (2004) accurately explains the decrease in convergence rates from south to north along the MAT (DEMETS et al., 1990), as a natural

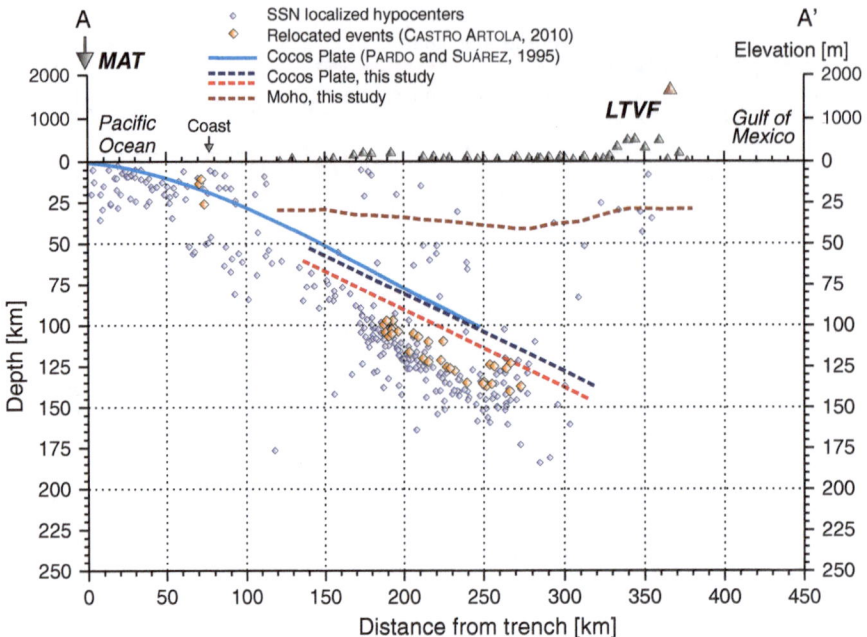

Figure 7
Geometry comparison between this study, PARDO and SUÁREZ (1995) and local seismicity recorded by the SSN and relocated by CASTRO-ARTOLA (2010)

consequence of the loss of the slab-pull force due to the tear, reinforces this model even further.

According to FERRARI (2004), if we take into account a convergence rate of 6.4 cm/year at the trench for 15°N latitude (DEMETS et al., 1990), then 358 km of slab have subducted since the tear reached 95°W longitude. If we add this to the length of slab already subducted when the tear reached the profile, then the detached slab would be barely visible at the edge of our profile, if at all.

In this context, it is difficult to affirm that the cessation of the subducted crustal signature at 145 km depth is evidence for propagation of this tear because, if so, it would be expected to happen at a greater depth. It is more likely that at this depth the thermal contrast between subducted crust and the surrounding mantle has diminished, together with a larger dipping angle, rendering the impedance contrast too small to be detected by our RF analysis.

Figure 7 also compares our results with local seismicity ($M > 3$) reported by the SSN between June 2007 and March 2009 and with double-difference relocated events using both VEOX and SSN

stations for the same period (CASTRO ARTOLA, 2010). The position of the oceanic crust is consistently shallower than the local seismicity obtained from the SSN. Taking into account our uncertainty of 5 km, the distance of the relocated events to the oceanic crust can be 5–15 km. Caution must be exercised in comparing these two sets of data, since the velocity model used by the SSN and CASTRO ARTOLA (2010) to locate hypocenters is the flat layer model of CAMPILLO et al., (1996), whereas we are using the IASPEI 91 model (KENNET and ENGDAHL, 1991). In spite of this bias, the general trend of the relocated seismicity is similar in dip angle to the geometry proposed here.

Then, assuming that bias is minimal (~ 5 km), the locations of the hypocenters would place some of the seismicity at the bottom of the oceanic crust or within the uppermost oceanic mantle. This phenomenon has been observed in Guerrero along the Mexican subduction zone (PACHECO and SINGH, 2010), where inslab earthquakes are at the bottom of the oceanic crust or at the top of the oceanic mantle. This has also been observed in the Cascadia subduction zone (PRESTON et al., 2003; ABERS et al. 2009); and has been associated with dehydration of the lithosphere.

6. Conclusions

We have shown with RF analyses that the Moho beneath southern Mexico is not a simple horizontal interface, revealing a three-dimensional structure; this is evident from the azimuthal dependence of depth estimates for single stations across the whole transect. The largest variations occur close to or beneath the LTVF and may be explained by isostatic compensation of the LTVF load on an elastic plate.

We have also identified the subducted oceanic crust as having a thickness of ~9 km and have shown that it has a dip of 26° between ~140 and ~310 km from the MAT. At this point, the signal corresponding to the subducted oceanic crust ceases; we cannot relate this to the slab tear hypothesis proposed by FERRARI (2004) because the detached slab is most likely at the northern edge of our profile, if at all. It is more likely that at 310 km from the trench the impedance contrast between crust and mantle is not high enough to show up in the receiver functions. No information is available for the segment between the trench and the coast.

Comparison with local seismicity has shown that intraslab seismicity occurs in the oceanic mantle. Even though SSN hypocenters show considerable scatter, a relocated subset of these (CASTRO ARTOLA, 2010) shows that seismicity occurs approximately 5–15 km beneath the subducted crust at the top of the oceanic mantle.

Acknowledgments

This work was supported by the Tectonics Observatory at Caltech and Conacyt project J51566-F. The VEOX experiment was funded by the Gordon and Betty Moore Foundation. We thank O. Castro Artola for providing his relocated hypocenter data and all the volunteers who contributed their time to the field work. We thank the editor and two anonymous reviewers for valuable comments that improved the article.

REFERENCES

ABERS, G. A., MACKENZIE, L. S., RONDENAY, S., ZHANG, Z., WECH, A. G., and CREAGER, K. (2009), *Imaging the source region of Cascadia tremor and intermediate-depth earthquakes*, Geology, *37*, 1119–1122.

BRAVO, H., REBOLLAR, C., URIBE, A., and JIMÉNEZ, O. (2004), *Geometry and state of stress of the Wadati-Benioff zone in the Gulf of Tehuantepec*, J. Geophys. Res., *109*. doi:10.1029/2003JB002854

CAMPILLO, M., SINGH, S. K., SHAPIRO, N., PACHECO, J., and HERMANN, R. B. (1996), *Crustal structure of the Mexican volcanic belt based on group velocity dispersion*, Geofísica Internacional, *35*, 361–370.

CASTRO-ARTOLA, O. A. (2010), *Caracterización de la geometría de la zona de Benioff con una red densa de banda ancha en el Istmo de Tehuantepec*, Bachelor's Thesis, Facultad de Ingeniería, Universidad Nacional Autónoma de México, México, 65 pp.

CHEVROT, S., and GIRARDIN, N. (2000), *On the detection and identification of converted and reflected phases from receiver functions*, Geophys. J. Int., *141*, 801–808.

CLAYTON, R. W., and WIGGINS, R. A. (1976), *Source shape estimation and deconvolution of teleseismic body waves*, Geophys, J. R. Astron. Soc., *47*, 151–177.

CLEVELAND, W. S., *Visualising Data*, (Hobart Press, 1993).

COUCH, R., and WOODCOCK, S. (1981), *Gravity and structure of the continental margins of Southwestern Mexico and Northern Guatemala*, J. Geophys. Res., *86*(B3), 1829–1840.

DEMETS, C., GORDON, R. G., ARGUS, D. F., and STEIN, S. (1990), *Current plate motions*, Geophys. J. Int., *101*, 425–478.

DICKINSON, W. (1997), *Tectonic implications of Cenozoic volcanism in coastal California*, Geol. Soc. Am. Bull., *109*, 936–954.

ESPÍNDOLA CASTRO, V. H. (2009), *Modelos de velocidad cortical utilizando funciones de receptor aplicado a estaciones de banda ancha del SSN, Mexico*, Ph. D. Thesis, Instituto de Geofísica, Universidad Nacional Autónoma de México, D.F., Mexico.

FERRARI, L. (2004), *Slab detachment control on mafic volcanic pulse and mantle heterogeneity in central Mexico*, Geology, *32*, 77–80. doi:10.1130/G19887.1

GÓMEZ-TUENA, A., LANGMUIR, C. H., GOLDSTEIN, S. L., STRAUB, S. M., and ORTEGA-GUTIÉRREZ, F. (2007), *Geochemical evidence for slab melting in the Trans-Mexican Volcanic Belt*, J. Petrol., *48*, 537–562.

GORBATOV, A., and FUKAO, Y. (2005), *Tomographic search for missing link between the ancient Farallon subduction and the present Cocos subduction*, Geophys. J. Int., *160*, 849–854.

GUZMÁN-SPEZIALE, M., and MENESES-ROCHA, J. J. (2000), *The North America – Caribbean plate boundary west of the Motagua-Polochic fault system: a fault jog in southeastern Mexico*, J. S. Am. Earth. Sci., *13*, 459–468.

HUSKER, A., and DAVIS, P. M. (2009), *Tomography and thermal state of the Cocos plate subduction beneath Mexico City*, J. Geophys. Res., *114*. doi:10.1029/2008JB006039

KENNET, B. L. N., and ENGDAHL, E. R. (1991), *Travel times for global earthquake location and phase identification*, Geophys. J. Int., *105*, 429–465.

KIKUCHI, M., and KANAMORI, H. (1982), *Inversion of complex body waves*, Bull. Seism. Soc. Am., *72*, 491–506.

KLITGORD, K. D., and MAMMERICKX, J. (1982), *Northern east Pacific rise: magnetic anomaly and bathymetric framework*, J. Geophys. Res., *87*, 6725–6750.

KOSTOGLODOV, V., BANDY, W., DOMINGUEZ, J., MENA, M. (1996), *Gravity and seismicity over the Guerrero seismic gap, Mexico*, Geophys. Res. Lett., *23*, 3385–3388.

LANGSTON, C. A. (1979), *Structure under Mount Rainier, Washington, inferred from teleseismic body waves*, J. Geophys. Res., *84*, 4749–4762.

MANEA, V. C., and MANEA, M. (2006), *Origin of modern Chiapanecan volcanic arc in southern Mexico inferred from thermal models*. In (Rose, W. I., Bluth, G. J. S., Carr, M. J., Ewert, W., Patiño, L. C., and Vallance, eds), Volcanic Hazards in Central America. Geol. Soc. Am., *411*, 27–38.

MANEA, M., MANEA, V.C., and KOSTOGLODOV, V. (2003), *Sediment fill in the Middle America trench inferred from gravity anomalies*, Geofísica Internacional, *42*(4), 603–612.

MANEA, M., MANEA, V.C., KOSTOGLODOV, V., and GUZMAN-SPEZIALE, M. (2005a), *Elastic thickness of the oceanic lithosphere beneath Tehuantepec Ridge*, Geofísica Internacional, *44*(2), 157–168.

MANEA, M., MANEA, V. C., FERRARI, L., KOSTOGLODOV, V., and BANDY, W. (2005b), *Tectonic evolution of the Tehuantepec ridge*, Earth Planet. Sci., *238*, 64–77.

NELSON S. A., and GONZALEZ-CAVER, E. (1992), *K-Ar dating of the Tuxtla volcanic field, Veracruz, Mexico*, Bull Volcanol, *55*, 85–96.

ORTEGA-GUTIÉRREZ, F., MITRE-SALAZAR, L. M., ROLDÁN-QUINTANA, J., ARANDA-GÓMEZ, J. J., MORÁN-ZENTENO, D., ALANIZ-ÁLVAREZ, S. A., and NIETO-SAMANIEGO, A. F. (1992), *Texto explicativo de la quinta edicion de la carta geologica de la republica Mexicana, escala 1:2,000,000*, Universidad Nacional Autónoma de México, Instituto de Geología, and Secretaría de Energía, Minas e Industria Paraestatal, Consejo de Recursos Minerales, Mexico DF.

PACHECO, J. F., and SINGH, S. K. (2010), *Seismicity and state of stress in Guerrero segment of the Mexican subduction zone*, J. Geophys. Res., *115*, B01303.

PARDO, M., and SUÁREZ, G. (1995), *Shape of the subducted Rivera and Cocos plates in southern Mexico, seismic and tectonic implications*, J. Geophys. Res., *100*, 12357–12373.

PÉREZ-CAMPOS, X. (2008), *MASE: Undergraduate research and outreach as part of a large project*, Seismol. Res. Lett., *79*, 232–236.

PÉREZ-CAMPOS, X., KIM, Y., HUSKER, A., DAVIS P.M., CLAYTON, R. W., IGLESIAS, A., PACHECO, J., SINGH, S. K., MANEA, V. C., and GURNIS, M. (2008), *Horizontal subduction and truncation of the Cocos plate beneath central Mexico*, Geophys Res. Lett., *35*. doi: 10.1029/2008GL035127

PERSAUD, P., PÉREZ-CAMPOS, X., and CLAYTON, R. W. (2007), *Crustal thickness variations in the margins of the Gulf of California from receiver functions*, Geophys. J. Int., *170*, 687–699.

PONCE, L., GAULON, R., SUÁREZ, G., and LOMAS, E. (1992), *Geometry and state of stress of the downgoing Cocos plate in the Isthmus of Tehuantepec, Mexico*, Geophys. Res. Lett., *19*, 773–776.

PRESTON, L. A., CREAGER, K. C., CROSSON, R. S., BROCHER, T. M., and TREHU, A. M. (2003), *Intraslab earthquakes: Dehydration of the Cascadia slab*, Science, *302*, 1197–1200.

SUÁREZ, G., MONFRET, T., WITTLINGER, G., and DAVID, C. (1990), *Geometry of subduction and depth of the seismogenic zone in the Guerrero gap, Mexico*, Nature, *345*, 336–338.

TONEGAWA, T., HIRAHARA, K., and SHIBUTANI, T. (2005), *Detailed structure of the upper mantle discontinuities around the Japan Subduction zone imaged by receiver function analyses*, Earth Planets Space, *57*, 5–14.

TURCOTTE, D. L., and SCHUBERT, G., *Geodynamics*, (Cambridge Univ. Press, 2001).

URRUTIA-FUCUGAUCHI, J., and FLORES-RUIZ, J. (1996), *Bouguer gravity anomalies and regional crustal structure in Central Mexico*, Int. Geol. Rev., *38*(2), 176–194.

VALDÉS, C. M., MOONEY, W. D., SINGH, S. K., MEYER, R. P., LOMNITZ, C., LUETGERT, J. H., HELSLEY, C. E., LEWIS, B. T. R., and MENA, M. (1986), *Crustal structure of Oaxaca, Mexico, from seismic refraction measurements*, Bull. Seism. Soc. Am., *76*(2), 574–563.

WESSEL, P., and SMITH, W. H. F. (1991), *Free software helps map and display data*, EOS, Trans. Am. Geophys. Un., *72*, 445–446.

YAMAUCHI, M., HIRAHARA K., and SHIBUTANI, T. (2003), *High resolution receiver function imaging of the seismic velocity discontinuities in the crust and uppermost mantle beneath southwest Japan*, Earth Planets Space, *55*, 59–64.

ZAMORA-CAMACHO A., ESPÍNDOLA V. H., PACHECO J. F., ESPÍNDOLA J. M., and GODÍNEZ M. L. (2010), *Crustal thickness at the Tuxtla Volcanic Field, (Veracruz, Mexico) from receiver functions*. Phys. Earth Planetary Int, *182*, 1–9.

ZHU, L., and KANAMORI, H. (2000), *Moho depth variations in southern California from Teleseismic Receiver Functions*. J. Geophys. Res., *105*(B2), 2969–2980.

(Received January 21, 2010, revised September 2, 2010, accepted September 3, 2010, Published online November 9, 2010)

Pure Appl. Geophys. 168 (2011), 1461–1474
© 2010 The Author(s)
This article is published with open access at Springerlink.com
DOI 10.1007/s00024-010-0200-3

▌Pure and Applied Geophysics

An Evaluation of Proposed Mechanisms of Slab Flattening in Central Mexico

STEVEN M. SKINNER[1] and ROBERT W. CLAYTON[1]

Abstract—Central Mexico is the site of an enigmatic zone of flat subduction. The general geometry of the subducting slab has been known for some time and is characterized by a horizontal zone bounded on either side by two moderately dipping sections. We systematically evaluate proposed hypotheses for shallow subduction in Mexico based on the spatial and temporal evidence, and we find no simple or obvious explanation for the shallow subduction in Mexico. We are unable to locate an oceanic lithosphere impactor, or the conjugate of an impactor, that is most often called upon to explain shallow subduction zones as in South America, Japan, and Laramide deformation in the US. The only bathymetric feature that is of the right age and in the correct position on the conjugate plate is a set of unnamed seamounts that are too small to have a significant effect on the buoyancy of the slab. The only candidate that we cannot dismiss is a change in the dynamics of subduction through a change in wedge viscosity, possibly caused by water brought in by the slab.

Key words: Mexico, flat slab, subduction.

1. Introduction

The major driving force of plate motion is slab buoyancy and the pull of subducting slabs descending into the mantle (BILLEN and HIRTH, 2007; CHAPPLE and TULLIS, 1977; FORSYTH and UYEDA, 1975). However, the current understanding of the initiation of subduction zones and the balance of forces controlling the 3D geometry and evolution of a subducting slab is not well understood (BILLEN, 2008). The angle of subduction influences the overall state of stress in the overriding slab, the resulting mode of deformation, and the location and type of arc volcanism.

Shallow or flat subduction occurs in 10% of the subduction zones present today (VAN HUNEN et al., 2002). The global variation of slab dips is shown in Fig. 1. Present day zones of shallow subduction include the Nankai trough of Japan, northern and southern Peru, Central Chile, East Aleutians in Alaska, and Mexico. A number of these are coincident with oceanic impactors, anomalously thick crust in the form of an aseismic ridge or plateau, that are presumed to be the cause of the shallow geometry. The Chilean flat slab coincides with the subduction of the Juan-Fernandez ridge (ANDERSON et al., 2007; KAY and ABBRUZZI, 1996; PILGER, 1981). The Peruvian flat slab is a combination of two adjacent flat segments resulting from subduction of the Nazca ridge and the Inca plateau (GUTSCHER et al., 1999b). There is a possible flat slab segment in Ecuador that correlates with the subduction of the Carnegie ridge (GUTSCHER et al., 1999a). Oceanic lithosphere of the Caribbean oceanic plateau might be causing a flat slab in northwestern Columbia (GUTSCHER et al., 2000a). Subduction of the Cocos ridge has led to a flat slab in Costa Rica (PROTTI et al., 1995; SAK et al., 2009), and the Yakutat terrane is subducting in the zone of the East Aleutian flat slab (BROCHER et al., 1994; FUIS et al., 2008). The flat slab of southwestern Japan has been linked to subduction of the Izu Bonin arc and the Palau–Kyushu ridge (GUTSCHER et al., 2000b), and western New Guinea has a flat segment linked to subduction of the Euripik ridge (GUTSCHER et al., 2000b). In northern Chile, the current dip of the slab is not flat but is actively flattening due to the subduction of the Iquique ridge (ESPURT et al., 2008).

However, in two cases there is no obvious impactor associated with the flat subduction. In the Cascadia subduction zone, for example, there is no evidence for thickened crust subducting along the

---
[1] Division of Geological and Planetary Sciences, California Institute of Technology, Pasadena, CA 91125, USA. E-mail: Skinner@caltech.edu

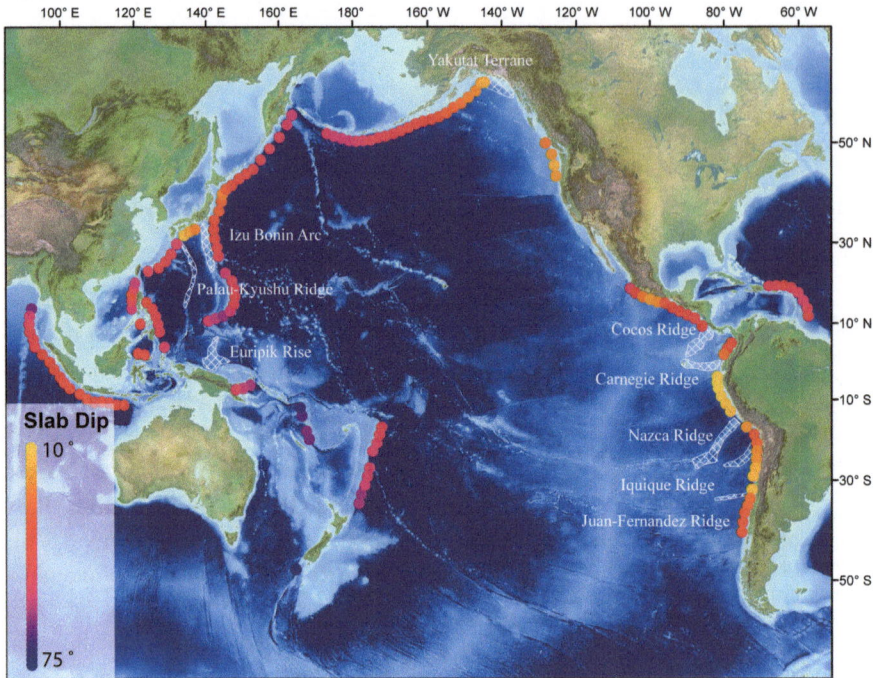

Figure 1
Map of the Pacific seafloor showing the dip of the shallow (<125 km depth) portion of subducting slabs (LALLEMAND *et al*., 2005), and subducting bathymetric highs that have been correlated with zones of shallow subduction (*white crosshatched pattern*)

shallow dipping Washington segment (GUTSCHER *et al*., 2000c). The flat slab in Mexico has been attributed to the Tehuantepec ridge (GUTSCHER *et al*., 2000c); however, the Tehuantepec ridge is being subducted at a point where the slab is dipping at 30° and appears to have little effect on the angle. In locations where an impactor has been identified, the spatial correlation between the impactor and zone of shallow subduction does not hold up when looked at in detail. In the Nankai trough, for example, the Palau–Kyushu ridge is entering the trench at the southern limit of the shallow zone. Figure 2 shows the anticorrelation between the shallow zone and where ridges are subducting. The fact that the impactor and zone of shallow subduction do not align suggests that it is not the buoyancy of the ridge itself that is holding up the slab, but a dynamic process that continues to operate in the wake trailing the impacting ridge.

As shown in Fig. 3, there are also cases where apparent buoyant impactors have little to no effect on the geometry of the subducting slab. The Emperor seamounts are subducting at the Kurile trench; the

Ogasawara plateau, Magellan seamounts, and Caroline ridge are all subducting at the Mariana trench; the Ozbourn–Louisville seamounts are subducting at the Kermadec trench; and the Chile rise is subducting at the Peru–Chile trench. These are just a sample of the largest thickness anomalies that are subducting without shallowing the dip of the downgoing slab.

2. Current State of Subduction in Central Mexico

The central Mexico subduction zone is of particular interest because it does not have an impacting ridge yet is one of the shallowest slabs that has been measured. Understanding the flat slab in Mexico is key to reevaluating the proposed mechanisms for shallow slabs around the globe. Along the western Mexico margin, the Cocos plate is subducting under the North America plate at a rate varying between 4.7 and 6.8 cm/year (DEMETS *et al*., 1994). As shown in Fig. 4, the subducted slab is shown by receiver function analysis to transition from a normal dip at

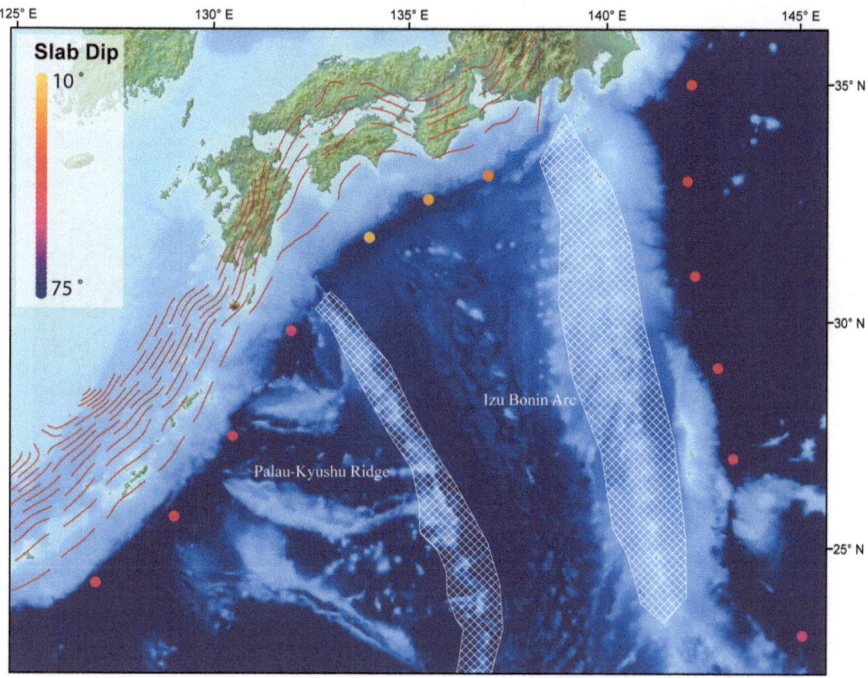

Figure 2
Detail view of the shallow slab segment of Japan. *Dashed red lines* are 20 km *contour lines* of slab depth from model of HAYES *et al.* (2009). *Colored dots* are slab dip from LALLEMAND *et al.* (2005). The shallow segment appears to correlate with the subduction of the Shikoku basin rather than the two ridges that flank it

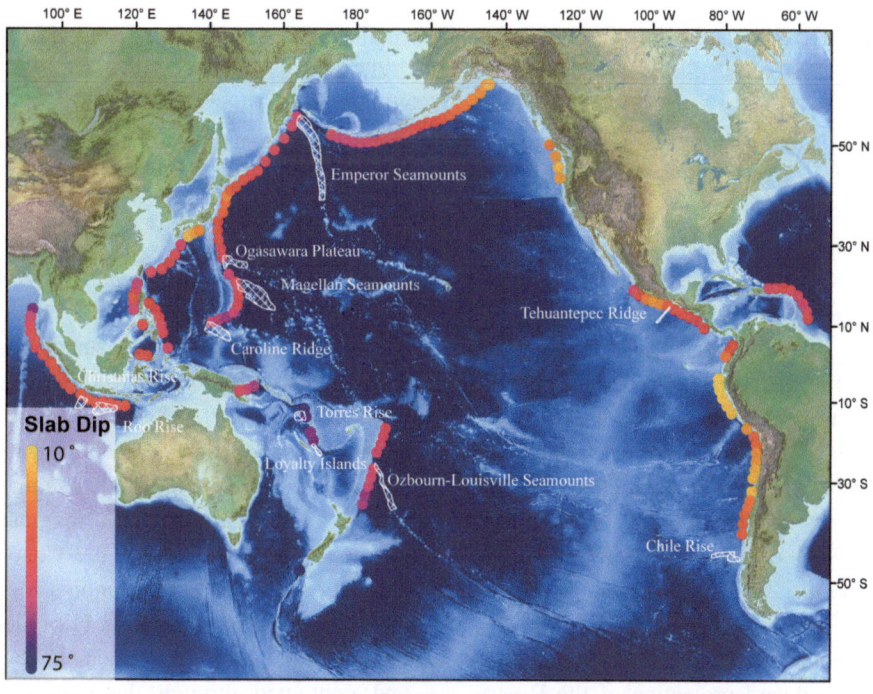

Figure 3
Map of the Pacific seafloor with labeled lithosphere anomalies (*white crosshatched regions*) that are subducting with no apparent effect on slab dip. *Colored dots* are slab dip from LALLEMAND *et al.* (2005)

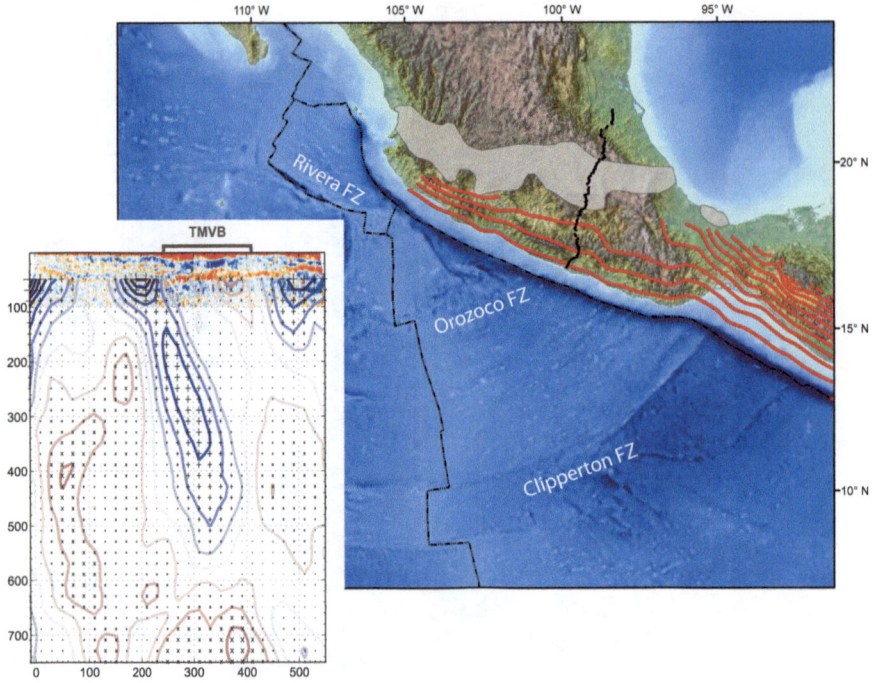

Figure 4
Combined receiver function and tomographic image from the Meso American Subduction Experiment (MASE) transect modified from Pérez-Campos et al. (2008). *Vertical axis* is kilometers below sea level, *horzontal axis* is distance along the MASE transect. The location map of Mexico shows the relative location of the MASE transect (*black dots*) to the trans-Mexican Volcanic Belt (TMVB) (*gray area*), offshore bathymetric features, and the dip of the subducted Cocos plate in 20 km contours (Hayes et al., 2009). Plate boundaries from Bird (2003)

the trench to sub-horizontal at 80 km from the trench (Kim et al., 2010; Pardo and Suárez, 1995; Pérez-Campos et al., 2008; Suárez et al., 1999). The horizontal slab persists to 250 km from the trench where it descends into the mantle with a 75° dip and is recognizable in tomographic images to a depth of 500 km (Husker and Davis, 2009; Kim et al., 2010; Pérez-Campos et al., 2008). An ultra low velocity layer, approximately 3 km thick is imaged on top of the slab from the trench through the horizontal section. The overriding plate appears to be in an overall state of extension rather than compression (Singh and Pardo, 1993), which is counterintuitive when considering the compressive forces associated with the subduction collision and the traction of an under-plated slab (De Franco et al., 2007; Keppie, 2009; Moran-Zenteno et al., 2007; Nieto-Samaniego et al., 2006).

The trans-Mexican Volcanic Belt has embayments along the landward projection of the Rivera, Orozoco, and Clipperton fracture zones suggesting that the Cocos plate is being further divided into

smaller plates by tearing of the slab (Blatter et al., 2007; Menard, 1978). The breakup of the Cocos plate allows the smaller fragments to rollback faster and results in the along trench dip variation (Billen, 2008).

3. History of Subduction in Mexico

The western Mexican margin has been a subduction margin for the past 160 million years (Keppie, 2004; Solari et al., 2007). The Sierra Madre Occidental, the subduction-related arc of western Mexico, initiated in the Jurassic and contains a continuous record of subduction-related magmatism from the Cretaceous and throughout the Cenozoic. The area has undergone moderate compressional deformation that correlates in time with Laramide deformation further north. Extension began in the early Eocene and continued through the Oligocene. Associated with the extension, is an ignimbrite flare-up that signals slab rollback or detachment of the slab

(FERRARI *et al.*, 2007). All of this early extension occurred while the margin was still under the compressive forces of subduction.

The details of the assembly of southwestern Mexico are complicated, but there are some aspects that can constrain the evolution of the slab geometry. The extent and migration of Cenozoic volcanism is related to the location of the subducted slab. Age data from the North American Volcanic Database (NAVDAT) and MORAN-ZENTENO *et al.*, (2007) are plotted in Fig. 5 against distance from the paleo-trench to show the space and time evolution of subduction related magmatic activity. At 20 Ma the locus of subduction magmatism jumps 200 km inland from the trench. At 10 Ma a rollback phase starts as the volcanism migrates toward the trench.

The migration of the arc needs to be viewed in relation to the reorganization of the oceanic plates offshore, namely the ridge jumps at 25, 12.5–11, and 6.5–3.5 Ma (KLITGORD and MAMMERICKX, 1982; MAMMERICKX and KLITGORD, 1982; MORAN-ZENTENO *et al.*, 2007). The southern Mexican margin has undergone a major reshaping in Tertiary time

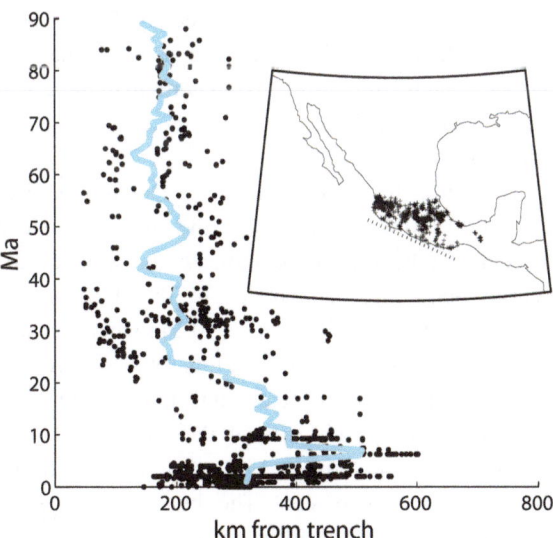

Figure 5

Distance of arc magmatism from the trench through time. The *blue line* is an 0.2 million-year moving average of the distance from the trench. There is a distinct change in the location of the arc starting at 25 Ma that shows the location of active volcanism migrating northward away from the trench, then starts a rollback to the south. The inset map shows the extent of the data used (*crosses*) and the *dotted line* approximating the trench. Data are from the North American Volcanic Database and MORAN-ZENTENO *et al.*, (2007)

(MORAN-ZENTENO *et al.*, 1996). The truncation of structural trends in addition to the juxtaposition of the modern trench and the Paleogene batholith suggests subsequent forearc removal (KARIG, 1978; MORAN-ZENTENO *et al.*, 2007, 1996; SCHAAF *et al.*, 1995). The Chortis block is often assumed to be the missing forearc, though this correlation is just as often called into question (KEPPIE and MORAN-ZENTENO, 2005; MORAN-ZENTENO *et al.*, 2009; ORTEGA-GUTIERREZ *et al.*, 2007; ORTEGA-OBREGON *et al.*, 2008). Recent studies evaluating the multiple reconstructions proposed for the Chortis block do not find much evidence to support the hypothesis that it represents the missing forearc and prefer a model of wholesale subduction erosion (KEPPIE, 2009).

4. Proposed Causes of Zones of Shallow Subduction

There are several factors that affect the geometry of subduction zones. A rapid convergence rate, trench-ward absolute motion of the upper plate, subduction of thickened oceanic crust, and young oceanic lithosphere are four factors that lead to shallowing of subducting plates (CROSS and PILGER, 1982). These factors are discussed specifically for Mexico.

4.1. Tehuantepec Ridge

The southern Mexico subduction zone, near the Isthmus of Tehuantepec, exhibits all of the four factors that would lead to a shallow slab geometry as described by CROSS and PILGER (1982): the convergence rate of the Cocos and North American plates is rapid (approximately 6 cm/year); the North American plate is overriding the Cocos plate in an absolute motion reference frame; the Tehuantepec ridge is currently being subducted; and the subducting lithosphere has been younger than 10 Ma for the past 40 Ma (CROSS and PILGER, 1982; MÜLLER *et al.*, 2008). These factors predict that the subducted Cocos plate in this region should have a very shallow dip, but it actually has a moderate dip of 30°.

One of the most obvious positive seafloor anomalies on the Cocos plate is the Tehuantepec ridge. The ridge is a compression structure that stretches for

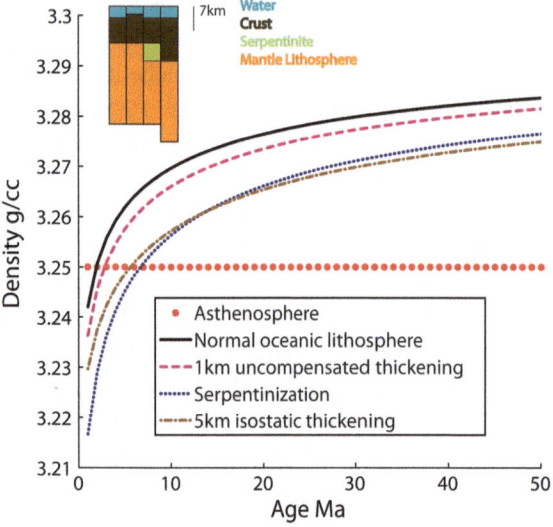

Figure 6

Average buoyancy at a given age of the crustal columns for four possible types of oceanic lithosphere subducted under Mexico, calculated using half space cooling model (TURCOTTE and SCHUBERT, 2002). *Red dotted line* is the density of asthenosphere for reference. The modeled lithosphere will resist subduction until it crosses above the asthenosphere line. The 5 Ma time slice of the four models of oceanic lithosphere used in calculating the density variation with age are shown. The models include unaltered normal oceanic lithosphere, 1 km of uncompensated thickening to represent the Tehuantepec ridge, 15% serpentinization if the upper 5 km of the oceanic mantle lithosphere, and a 5 km isostatic compensated thickening of the oceanic crust to represent seamounts formed on a spreading ridge

more than 200 km along the Clipperton fracture zone. The ridge marks the boundary of oceanic lithosphere that is, on average, 7 million years younger and 800 m shallower to the north (MANEA *et al.*, 2005). The ridge itself has a maximum relief of roughly 1 km relative to the seafloor to the north and is, on average, 10 km wide. Assuming the Tehuantepec ridge is simply a kilometer increase in oceanic crust, the resultant buoyancy increase is only 0.12% (see Fig. 6). The Tehuantepec ridge is thought to have formed as a transform fault on the Guadalupe plate at 15–20 Ma; in addition, it is currently encountering the trench at the transition zone of shallow to steep subduction and has no historic or kinematic link to the current zone of flat subduction (MANEA *et al.*, 2005). The Tehuantepec ridge has a trend perpendicular to the trench which reduces the effect of any positive buoyancy (MARTINOD *et al.*, 2005). The Tehuantepec ridge impacts in the wrong place

(500 km to the southeast of the zone of flat subduction) and has no history of lateral movement along the trench (MANEA *et al.*, 2005).

4.2. Seamounts

There is a seamount chain (the Moonless Mountains) on the Pacific plate between the Murray and Clarion fracture zones that may have had a correlative chain, the Chumbia seamount ridge, on the now subducted Farallon plate (KEPPIE and MORAN-ZENTENO, 2005). The seamounts in this chain do not have flexural or gravity moats around them, indicating that they were formed on or very near the spreading ridge (WATTS and RIBE, 1984). The lithosphere that surrounds the Moonless Mountains is roughly 40 million years old (MÜLLER *et al.*, 2008). When the Cocos plate started to shallow 30 million years ago, as evidenced by migration of volcanism, the lithosphere at the trench was 10 million years old (MÜLLER *et al.*, 2008) and would be neutrally buoyant. If a corollary to the Moonless Mountains did exist on the Cocos plate, it is of the right age to contribute to the flattening of the slab; however, reconstructions based on the rotation poles and the error analysis of DOUBROVINE and TARDUNO (2008) (see Fig. 7) show that the Moonless Mountains mirror image would intersect the Mexican margin further to the north than the extent of the zone of shallow subduction, and, hence, is not likely the cause.

By using the stage rotations of DOUBROVINE and TARDUNO (2008), a conjugate to the current Mexican margin can be rotated to indicate the area of the Pacific plate that corresponds to the area on the Farallon plate that subducted at 30 Ma when the slab shallowed. As shown in Fig. 8, this rotation reveals a set of small unnamed seamounts that would have intersected the margin around the latitude of Acapulco and can be correlated in space and time to the flat segment of the slab. The buoyancy of these seamounts alone is insufficient to cause a flat slab (CLOOS, 1993). We can use a simplified geometry to estimate the volumetric differences and resulting changes in buoyancy due to various forms of thickened oceanic lithosphere. From global bathymetry data we extract a representative width and height of the given bathymetric anomaly, then calculate the volume assuming a conical shape for a seamount or a

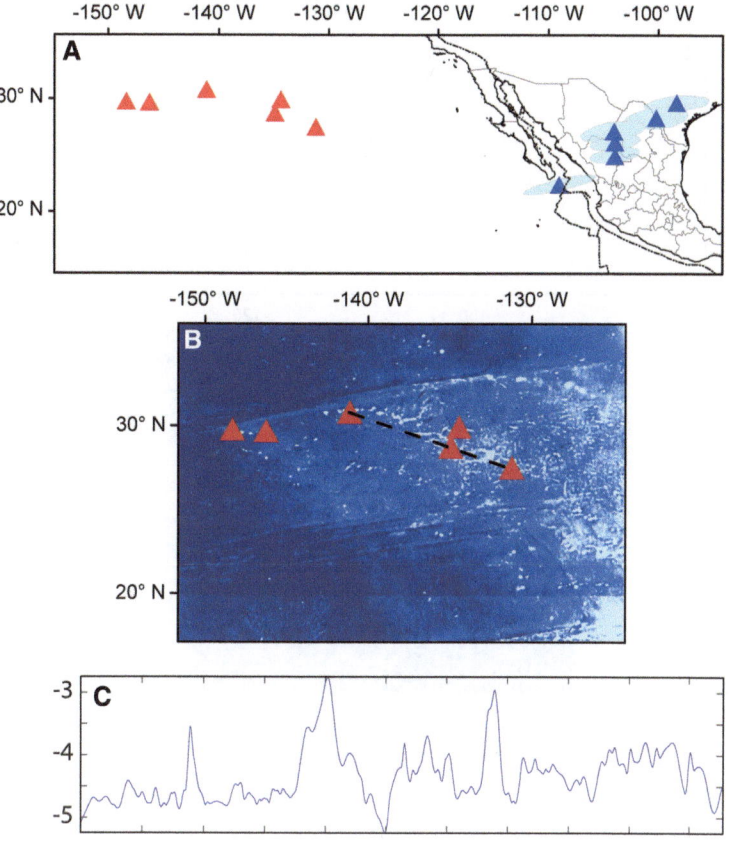

Figure 7

A tectonic reconstruction of the Moonless Mountains at 30 Ma. **a** Depicts the current location of the Moonless Mountains as *red triangles*. *Blue triangles* show the reconstructed location relative to North America of the hypothetical correlative chain of seamounts on the Farallon plate at 30 Ma in a fixed North America reference frame. *Light blue* areas are the error ellipses of the rotations given by DOUBROVINE and TARDUNO (2008). **b** Shows the bathymetry of the area around the moonless mountains and the location of the representative bathymetric profile shown to the right. **c** Is a representative profile along the *dashed line* in (**b**). *Vertical axis* is kilometers below sea level, vertical exaggeration is 100 times

triangular prism for an aseismic ridge. The estimated increase in crustal volume is then normalized by the aerial extent of the feature in order to compare thickening per unit area. Using this method the unnamed mountains are approximately 10% of the crustal volume increase associated with the Nazca or Juan Fernandez ridge.

4.3. Age of the Subducting Plate

One of the predictions of plate tectonics is that the angle of subduction is a function of the age of the subducting plate, because as a plate ages it cools and increases in density (BILLEN and HIRTH, 2007; PARSONS and SCLATER, 1977). The relationship between age and density is clearly seen in the half space cooling models of Fig. 6. However, when the angle of subduction and the age of actual subduction zones are analyzed, the correlation is quite weak (CRUCIANI *et al.*, 2005; JARRARD, 1986). This is evident in the case of central Mexico, where the Cocos plate exhibits steep subduction in the north where the subducting oceanic lithosphere is younger than the lithosphere of the flat segment to the south (MÜLLER *et al.*, 2008; PARDO and SUÁREZ, 1993, 1995) (see Fig. 4).

It is possible for an ephemeral spreading center to have existed between the Farallon and an unknown microplate. If this failed ridge was near the trench it could produce very young and buoyant lithosphere that would decrease the angle of subduction. This hypothetical ridge would be entirely contained within

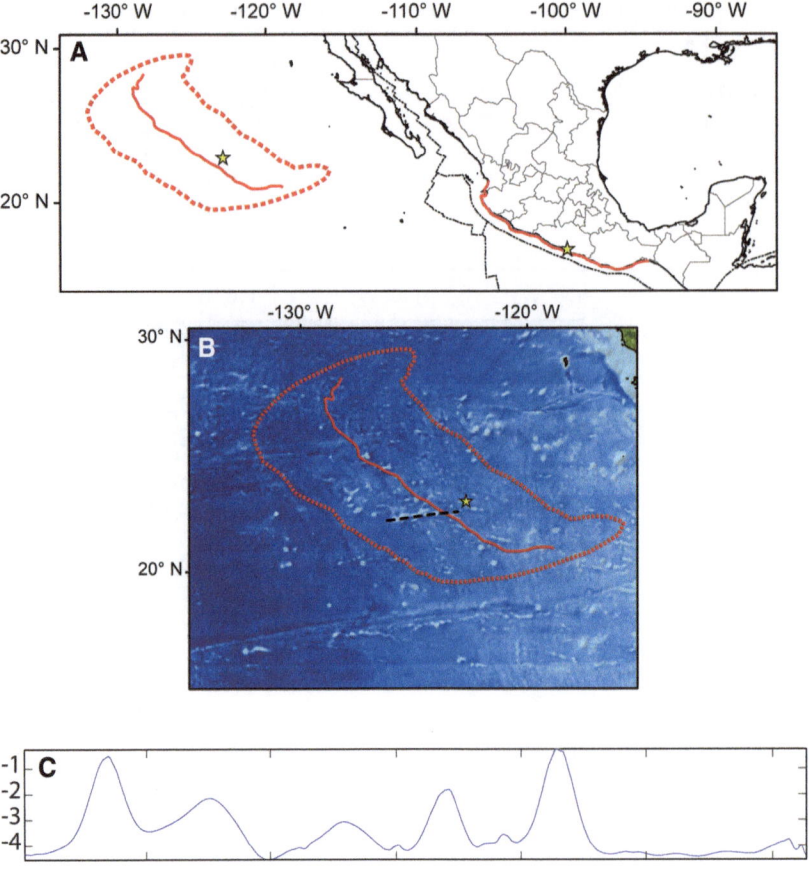

Figure 8

a Shows the Mexican coastline transformed by the rotations of DOUBROVINE and TARDUNO (2008) to show the area of the Pacific plate that is the corollary to the oceanic lithosphere that was subducting along the southern Mexican margin at 30 Ma. The *yellow stars* are the current and rotated location of Acapulco for reference. There is a small chain of seamounts near what would have been the latitude of Acapulco. The *dashed red line* is the total 95% confidence area of the error ellipses associated with the rotated points of the coast (*solid red line*). **b** Shows the bathymetry of the area around the unnamed seamounts and the location of the representative bathymetric profile. **c** Is a representative profile along the *dashed line* in (**b**). *Vertical axis* is kilometers below sea level, vertical exaggeration is 100 times

subducted Farallon plate, and the evidence for it completely subducted. Although the tectonic plates in the area underwent frequent reorganization around the time of the slab shallowing, there is no evidence for such a spreading ridge in the geologic or geochemical record of the upper plate.

4.4. Hydrothermal Alteration

Age alone may not be the sole cause for the angle of the Cocos slab, but could be a major component. The seafloor on both sides of the spreading ridge in the zone of flat subduction is extremely rough. The area is the site of numerous fracture zones and failed rifting events. One of the mapped failed rifts on the

Pacific plate was dredged as part of the Ocean Drilling Project and the recovered sample contained serpentinite (LONSDALE, 2005). The alteration or serpentinization of the oceanic lithosphere causes a decrease in the average density of the lithosphere and could increase the buoyancy of the slab, causing it to go flat. Hydrothermal alteration will likely increase with increased fracturing, although we have no way of knowing the fracture density of the plate that subducted at 20 Ma given the fact that fracture causing events such as ridge jumps and forearc bulges are not necessarily recorded symmetrically about the new spreading center. The bending of the Cocos plate preferentially induces the reactivation of faults and fractures, creating a horst and graben structure

(AUBOUIN et al., 1982; GREVEMEYER et al., 2005; RUFF, 1989). The faulting of the lithosphere allows water to penetrate into the young warm slab and alter the density. This is a process that occurs at all subduction zones; however, due to the consistently young lithosphere subducting in this area, the higher temperature of the slab will increase hydrothermal alteration independent of the degree of fracturing. Altering the top 5 km of the mantle lithosphere by 15% serpentinization doubles the length of time for which a slab is neutrally buoyant (see Fig. 6). Recent geophysical studies in Mexico have determined that there is a hydrous layer at the plate interface (KIM et al., 2010). Remobilization of fluids entrained with the downgoing slab by serpentinization may be the source of these hydrous phases.

4.5. Slab Detachment and Flexure

Tomographic images reveal the foundering segments of the Farallon slab beneath North America. The tomographic model of GORBATOV and FUKAO (2005) reveals a southward propagating tear in the slab at 600 km depth that is a result of the differential motion between the Cocos and subducted Farallon plates. They speculate that the tear and differential rotation buckles the Cocos plate and caused uplift of the slab in the region of the TMVB, producing the flat slab geometry. There are also large discrepancies between tomographic models of the region. The more detailed tomographic model of HUSKER and DAVIS (2009) places the truncated edge of the slab roughly 500 km to the south of where Gorbatov and Fukao locate it, which makes the uplift mechanism less likely. Other tomographic models locate a shallower gap in the slab under northern Central America (ROGERS et al., 2002), and it is not clear how truncation of the slab at a depth of 300 km beneath Guatemala, Honduras, and Nicaragua would relate to the model of Gorbatov and Fukao.

4.6. Chortis Block

The origin and location of the Chortis block (present day Nicaragua) through time is highly debated. One reconstruction places the Chortis block along the Acapulco trench at 50 Ma (PINDELL et al.,

1988; ROSS and SCOTESE, 1988). The block then migrates to the east with the Farallon-North America-Caribbean triple junction, which changes the margin from a North American-Caribbean to North American-Farallon plate boundary (MORAN-ZENTENO et al., 1996). The change in the plate pair exposes the southern Mexican margin to the faster Farallon-North America convergence rate, which may lead to the flattening of the slab, though it is unclear why the margin to the north with the same convergence rate would not also be flat. Other studies (KEPPIE, 2009; KEPPIE and MORAN-ZENTENO, 2005) propose models for the evolution of the Chortis block that make it unrelated to the flat slab in central Mexico. It is unlikely that the Chortis block is the cause of the flat slab, yet the knowledge of its location through time is needed for a complete model of the area.

4.7. Continental Root

Slab suction is an important force influencing the geometry at subduction zones. Viscously driven flow of the asthenosphere by the downgoing slab creates a zone of negative pressure in the mantle wedge (TOVISH et al., 1978). The suction force alone may not provide enough lift to drive slabs flat, but may prove more effective when combined with excessively buoyant lithosphere in the form of an oceanic plateau (VAN HUNEN et al., 2004). The suction force in the mantle wedge can be greatly increased by a continental root that penetrates the asthenosphere (O'DRISCOLL et al., 2009). The crustal root blocks flow perpendicular to the trench resulting in a higher negative pressure in the space between the trench and the root that can assist in pulling up the slab. This mechanism is proposed as a contributing factor for the Laramide, and has been suggested for central Mexico because the elevated TMVB may indicate the presence of the a crustal root (URRUTIA-FUCUGAUCHI and FLORES-RUIZ, 1996). However, as shown in PÉREZ-CAMPOS et al. (2008), the crust under the TMVB is only 45 km thick and, hence, there is no deep crustal root.

4.8. Hydration of the Mantle Wedge

The viscosity of the mantle wedge can be decreased by the addition of fluids released from the slab; this low viscosity wedge or channel can change the dip of the downgoing slab and has been modeled to create flat lying slabs as observed in Mexico (MANEA and GURNIS, 2007). There is some evidence in the attenuation study of CHEN and CLAYTON (2009) that zones of low Q in the mantle wedge may be due to fluids from the slab. Geochemical studies of the TMVB show that the sub-arc mantle is highly heterogeneous and have found locations with a magmatic water content in excess of 8 wt% (BLATTER and CARMICHAEL, 1998; JOHNSON et al., 2009). We know that excess hydration can cause a slab to flatten; however, the cause of excess water in the Mexican subduction zone has yet to be explained. Tectonic erosion is one way to subduct large amounts of water laden sediments (DOMINGUEZ et al., 2000). In Mexico there is evidence for extreme tectonic erosion, namely, the entire Oligocene forearc is missing and the associated batholith is sitting adjacent to the modern trench (KEPPIE et al., 2009a, b; MORAN-ZENTENO et al., 2007). The juxtaposition of the Oligocene arc with the modern trench and the truncation of other structural features reveals how much of the Mexican margin has been lost to tectonic erosion.

Seamounts may not have enough positive buoyancy to flatten the slab, but they do create a long lived period of subducting extreme relief that could lead to a prolonged period of subduction erosion (VON HUENE and SCHOLL, 1991). The subduction on individual seamounts has been shown, through analog models, to cause erosion of the overriding plate (DOMINGUEZ et al., 1998, 2000). The unnamed mountains range in age from roughly 35–25 Ma and stretch across 500 km (see Fig. 8). The convergence rate along the Middle America Trench varies widely though averaging in space and time the margin would be continually impacted for a span of 6 Ma assuming a perfect mirroring of the unnamed mountains (DOUBROVINE and TARDUNO, 2008; MÜLLER et al., 2008). The erosion of the margin corresponds with a 29–19 Ma gap in arc magmatism (KEPPIE et al., 2009b). Recent numerical modeling has shown the rapid removal of large blocks of continental forearc as one possible mode of subduction erosion that shaped the Mexican margin (KEPPIE et al., 2009a).

The eroded forearc would be highly fractured in this catastrophic event leading to an increase in pore space for fluids to be entrained with the downgoing plate. Modeling indicates that the eroded material could be underplated or transported deep into the mantle (KEPPIE et al., 2009a). The low viscosity channel that forms from the subducted material and fluid would also decouple the upper and lower plates and cause the lack of compression that we see in Mexico.

5. Discussion

Looking for a single cause of flat slab subduction reveals the complexity and multifaceted nature of subduction zone dynamics. Single trench correlations quickly break down when extended to the global scale. The often called upon correlation between the location of flat slabs and the presence of a subducting aseismic ridge or plateau is quite strong; however, this does not imply direct causation. Most flat slabs have an associated subducting ridge, but not all subducting ridges produce flat slabs. The fact that the correlation between ridges and shallow zones is not one-to-one means that it is not likely the sole cause of flat subduction. We have shown that, by adding a second variable, in this case age, we are able to explain some of the zones where a ridge is subducting yet fails to produce a shallow slab segment (see Fig. 9). This is just one example of the need for a comprehensive evaluation of the parameters that influence the dip of subducting plates. One proposed cause of the Laramide flat slab is the subduction of a conjugate oceanic plateau to the Shatsky Rise; though, as we have shown, subduction of thickened oceanic lithosphere is neither a sufficient or necessary condition for shallow subduction. We find hydration of the mantle wedge to be the only mechanism that there is no evidence against causing the flat slab in Mexico. Further study of the fluid budget of the downgoing slab and the change in mantle viscosity with the addition of fluids is needed to evaluate hydration as the cause of the Mexican flat slab and possibly the key mechanism for shallow slabs worldwide.

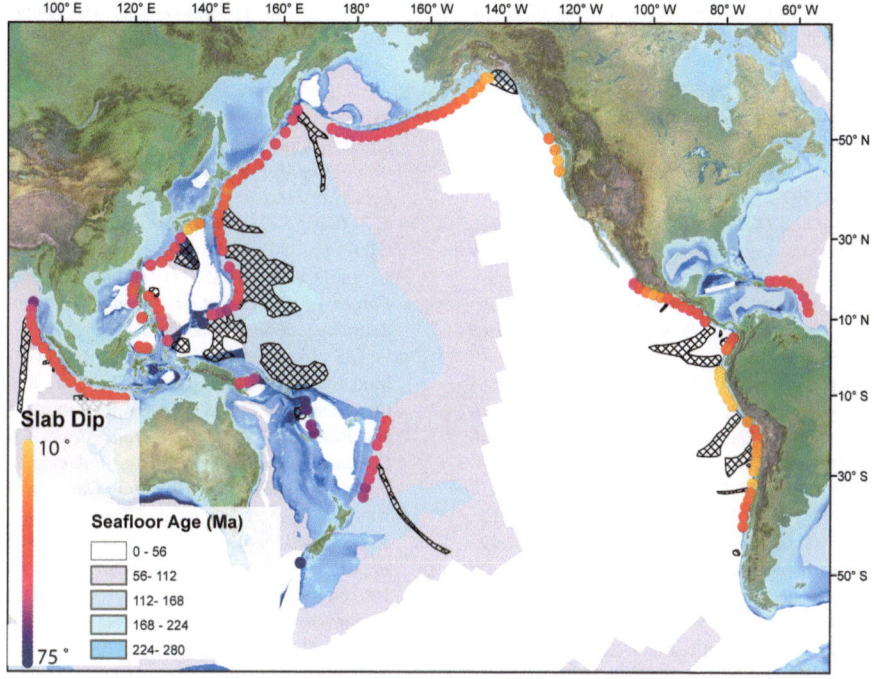

Figure 9

Map of the Pacific seafloor age (MÜLLER *et al.*, 2008), shallow slab segment dips (LALLEMAND *et al.*, 2005), and subducting bathymetric highs (*crosshatched pattern*). Not all bathymetric highs are correlated with a zone of shallow subduction. Although there is no direct correlation between the age of the subducting lithosphere and the dip of the slab, there appears to be a maximum plate age past which the slab cannot support a flat segment. This explains the subduction of ridges that do not form a shallow slab in the western Pacific

6. Conclusions

Subducting buoyant ridges, seamounts, and plateaus do not directly cause flat slabs but are, rather, a catalyst of other dynamic mantle processes. Determining the combination of forces that lead to flat slabs is important not only for our understanding of the current zones of flat subduction but also the geologic history of western North America and inferred periods of flat subduction. The geometry that we see in the present day Mexican flat slab appears to be the result of the dynamic response of subduction to hydration of the mantle wedge that occurred 30 million years ago. The direct evidence for the flattening mechanism has long been destroyed, and there is no suitable impactor on the conjugate plate. Hydration of the mantle wedge is the only feasible mechanism to change the slab geometry in Mexico, although the process is not completely understood. The cause of the intense subduction erosion that leads to the hydration has yet to be identified, yet appears to be the only viable explanation for the geometry of both the slab and the margin.

Acknowledgments

This study was supported by the Gordon and Betty Moore Foundation through the Tectonics Observatory at Caltech. This is contribution number 136 from the Caltech Tectonics Observatory.

REFERENCES

ANDERSON, M., ALVARADO, P., ZANDT, G. and BECK, S. (2007), *Geometry and Brittle Deformation of the Subducting Nazca Plate, Central Chile and Argentina*, Geophysical Journal International 171, 419–434.

AUBOUIN, J., STEPHAN, J.F., ROUMP, J. and RENARD, V. (1982), *The Middle America Trench as an Example of a Subduction Zone*, Tectonophysics 86, 113–132.

BILLEN, M.I. (2008), *Modeling the Dynamics of Subducting Slabs*, Annual Review of Earth and Planetary Sciences 36, 325–356.

BILLEN, M.I. and HIRTH, G. (2007), *Rheologic Controls on Slab Dynamics*, Geochemistry Geophysics Geosystems 8, Q08012, doi:10.1029/2007gc001597.

BIRD, P. (2003), *An Updated Digital Model of Plate Boundaries*, Geochemistry Geophysics Geosystems 4, 1027, doi:10.1029/2001gc000252.

BLATTER, D.L. and CARMICHAEL, I.S.E. (1998), *Hornblende Peridotite Xenoliths from Central Mexico Reveal the Highly Oxidized Nature of Subarc Upper Mantle*, Geology 26, 1035–1038.

BLATTER, D.L., FARMER, G.L. and CARMICHAEL, I.S.E. (2007), *A North-South Transect across the Central Mexican Volcanic Belt at Similar to 100 Degrees W: Spatial Distribution, Petrological, Geochemical, and Isotopic Characteristics of Quaternary Volcanism*, Journal of Petrology 48, 901–950.

BROCHER, T.M., FUIS, G.S., FISHER, M.A., PLAFKER, G., MOSES, M.J., TABER, J.J. and CHRISTENSEN, N.I. (1994), *Mapping the Megathrust beneath the Northern Gulf of Alaska Using Wide-Angle Seismic Data*, Journal of geophysical research 99, 11663–11685.

CHAPPLE, W.M. and TULLIS, T.E. (1977), *Evaluation of Forces That Drive Plates*, Journal of geophysical research 82, 1967–1984.

CHEN, T. and CLAYTON, R.W. (2009), *Seismic Attenuation Structure in Central Mexico: Image of a Focused High-Attenuation Zone in the Mantle Wedge*, Journal of geophysical research 114, B07304, doi:10.1029/2008jb005964.

CLOOS, M. (1993), *Lithospheric Buoyancy and Collisional Orogenesis; Subduction of Oceanic Plateaus, Continental Margins, Island Arcs, Spreading Ridges, and Seamounts*, Geological Society of America Bulletin 105, 715–737.

CROSS, T.A. and PILGER, R.H. (1982), *Controls of Subduction Geometry, Location of Magmatic Arcs, and Tectonics of Arc and Back-Arc Regions*, Geological Society of America Bulletin 93, 545–562.

CRUCIANI, C., CARMINATI, E. and DOGLIONI, C. (2005), *Slab Dip Vs. Lithosphere Age: No Direct Function*, Earth and Planetary Science Letters 238, 298–310.

DE FRANCO, R., GOVERS, R. and WORTEL, R. (2007), *Numerical Comparison of Different Convergent Plate Contacts: Subduction Channel and Subduction Fault*, Geophysical Journal International 171, 435–450.

DEMETS, C., GORDON, R.G., ARGUS, D.F. and STEIN, S. (1994), *Effect of Recent Revisions to the Geomagnetic Reversal Time-Scale on Estimates of Current Plate Motions*, Geophysical research letters 21, 2191–2194.

DOMINGUEZ, S., LALLEMAND, S.E., MALAVIEILLE, J. and VON HUENE, R. (1998), *Upper Plate Deformation Associated with Seamount Subduction*, Tectonophysics 293, 207–224.

DOMINGUEZ, S., MALAVIEILLE, J. and LALLEMAND, S.E. (2000), *Deformation of Accretionary Wedges in Response to Seamount Subduction: Insights from Sandbox Experiments*, Tectonics 19, 182–196.

DOUBROVINE, P.V. and TARDUNO, J.A. (2008), *A Revised Kinematic Model for the Relative Motion between Pacific Oceanic Plates and North America since the Late Cretaceous*, Journal of geophysical research 113, B12101, doi:10.1029/2008jb005585.

ESPURT, N., FUNICIELLO, F., MARTINOD, J., GUILLAUME, B., REGARD, V., FACCENNA, C. and BRUSSET, S. (2008), *Flat Subduction Dynamics and Deformation of the South American Plate: Insights from Analog Modeling*, Tectonics 27, TC3011, doi:10.1029/2007tc002175.

FERRARI, L., VALENCIA-MORENO, M. and BRYAN, S., Magmatism and Tectonics of the Sierra Madre Occidental and Its Relation with the Evolution of the Western Margin of North America, In *Geology of México: Celebrating the Centenary of the Geological Society of México: Geological Society of America Special Paper 422* (eds. S.A. Alaniz-Alvarez and A.F. Nieto-Samaniego) (2007) pp. 1–39.

FORSYTH, D. and UYEDA, S. (1975), *Relative Importance of Driving Forces of Plate Motion*, Geophysical Journal of the Royal Astronomical Society 43, 163–200.

FUIS, G.S., MOORE, T.E., PLAFKER, G., BROCHER, T.M., FISHER, M.A., MOONEY, W.D., NOKLEBERG, W.J., PAGE, R.A., BEAUDOIN, B.C., CHRISTENSEN, N.I., LEVANDER, A.R., LUTTER, W.J., SALTUS, R.W. and RUPPERT, N.A. (2008), *Trans-Alaska Crustal Transect and Continental Evolution Involving Subduction Underplating and Synchronous Foreland Thrusting*, Geology 36, 267–270.

GORBATOV, A. and FUKAO, Y. (2005), *Tomographic Search for Missing Link between the Ancient Farallon Subduction and the Present Cocos Subduction*, Geophysical Journal International 160, 849–854.

GREVEMEYER, I., KAUL, N., DIAZ-NAVEAS, J.L., VILLINGER, H.W., RANERO, C.R. and REICHERT, C. (2005), *Heat Flow and Bending-Related Faulting at Subduction Trenches: Case Studies Offshore of Nicaragua and Central Chile*, Earth and Planetary Science Letters 236, 238–248.

GUTSCHER, M.A., MALAVIEILLE, J., LALLEMAND, S. and COLLOT, J.Y. (1999a), *Tectonic Segmentation of the North Andean Margin: Impact of the Carnegie Ridge Collision*, Earth and Planetary Science Letters 168, 255–270.

GUTSCHER, M.A., OLIVET, J.L., ASLANIAN, D., EISSEN, J.P. and MAURY, R. (1999b), *The "Lost Inca Plateau": Cause of Flat Subduction beneath Peru?*, Earth and Planetary Science Letters 335–341.

GUTSCHER, M.-A., SPAKMAN, W., BIJWAARD, H. and ENGDAHL, E.R. (2000a), *Geodynamics of Flat Subduction: Seismicity and Tomographic Constraints from the Andean Margin*, Tectonics 19.

GUTSCHER, M.A., MAURY, R., EISSEN, J.-P. and BOURDON, E. (2000b), *Can Slab Melting Be Caused by Flat Subduction?*, Geology 28, 535–538.

GUTSCHER, M.A., SPAKMAN, W., BIJWAARD, H. and ENGDAHL, E.R. (2000c), *Geodynamics of Flat Subduction: Seismicity and Tomographic Constraints from the Andean Margin*, Tectonics 19, 814–833.

HAYES, G.P., WALD, D.J. and KERANEN, K. (2009), *Advancing Techniques to Constrain the Geometry of the Seismic Rupture Plane on Subduction Interfaces a Priori: Higher-Order Functional Fits*, Geochemistry Geophysics Geosystems 10, Q09006, doi:10.1029/2009gc002633.

HUSKER, A. and DAVIS, P.M. (2009), *Tomography and Thermal State of the Cocos Plate Subduction beneath Mexico City*, Journal of geophysical research 114, B04306, doi:10.1029/2008JB006039.

JARRARD, R.D. (1986), *Relations among Subduction Parameters*, Reviews of Geophysics 24, 217–284.

JOHNSON, E.R., WALLACE, P.J., DELGADO GRANADOS, H., MANEA, V.C., KENT, A.J.R., BINDEMAN, I.N. and DONEGAN, C.S. (2009), *Subduction-Related Volatile Recycling and Magma Generation* beneath Central Mexico: Insights from Melt Inclusions, Oxygen Isotopes and Geodynamic Models, Journal of Petrology 50, 1729–1764.

KARIG, D.E. (1978), Late Cenozoic Subduction and Continental Margin Truncation Along the Northern Middle America Trench, Geological Society of America Bulletin 89, 265–276.

KAY, S.M. and ABBRUZZI, J.M. (1996), Magmatic Evidence for Neogene Lithospheric Evolution of the Central Andean "Flat-Slab" between 30°S and 32°S, Tectonophysics 259, 15–28.

KEPPIE, J.D. (2004), Terranes of Mexico Revisited: A 1.3 Billion Year Odyssey, International Geology Review 46, 765–794.

KEPPIE, D.F., 2009. Subduction Erosion Processes with Application to Southern Mexico. PhD Thesis, McGill University, Montreal, 172 pp.

KEPPIE, J.D. and MORAN-ZENTENO, D.J. (2005), Tectonic Implications of Alternative Cenozoic Reconstructions for Southern Mexico and the Chortis Block, International Geology Review 47, 473–491.

KEPPIE, D.F., CURRIE, C.A. and WARREN, C. (2009a), Subduction Erosion Modes: Comparing Finite Element Numerical Models with the Geological Record, Earth and Planetary Science Letters 287, 241–254.

KEPPIE, J.D., MORAN-ZENTENO, D.J., MARTINY, B. and GONZALEZ-TORRES, E. (2009b), Synchronous 29-19 Ma Arc Hiatus, Exhumation and Subduction of Forearc in Southwestern Mexico, Geological Society, London, Special Publications 328, 169–179.

KIM, Y., CLAYTON, R.W. and JACKSON, J.M. (2010), Geometry and Seismic Properties of the Subducting Cocos Plate in Central Mexico, Journal of geophysical research 115, B06310, doi: 10.1029/2009jb006942.

KLITGORD, K.D. and MAMMERICKX, J. (1982), Northern East Pacific Rise—Magnetic Anomaly and Bathymetric Framework, Journal of geophysical research 87, 6725–6750.

LALLEMAND, S., HEURET, A. and BOUTELIER, D. (2005), On the Relationships between Slab Dip, Back-Arc Stress, Upper Plate Absolute Motion, and Crustal Nature in Subduction Zones, Geochemistry Geophysics Geosystems 6, Q09006, doi: 10.1029/2005GC000917.

LONSDALE, P. (2005), Creation of the Cocos and Nazca Plates by Fission of the Farallon Plate, Tectonophysics 404, 237–264.

MAMMERICKX, J. and KLITGORD, K.D. (1982), Northern East Pacific Rise—Evolution from 25 My Bp to the Present, Journal of geophysical research 87, 6751–6759.

MANEA, V. and GURNIS, M. (2007), Subduction Zone Evolution and Low Viscosity Wedges and Channels, Earth and Planetary Science Letters 264, 22–45.

MANEA, M., MANEA, V.C., FERRARI, L., KOSTOGLODOV, V. and BANDY, W.L. (2005), Tectonic Evolution of the Tehuantepec Ridge, Earth and Planetary Science Letters 238, 64–77.

MARTINOD, J., FUNICIELLO, F., FACCENNA, C., LABANIEH, S. and REGARD, V. (2005), Dynamical Effects of Subducting Ridges: Insights from 3-D Laboratory Models, Geophysical Journal International 163, 1137–1150.

MENARD, H.W. (1978), Fragmentation of Farallon Plate by Pivoting Subduction, Journal of Geology 86, 99–110.

MORAN-ZENTENO, D., CERCA, M. and KEPPIE, J.D. (2007), The Cenozoic Tectonic and Magmatic Evolution of Southwestern Mexico; Advances and Problems of Interpretation, In Geology of México: Celebrating the Centenary of the Geological Society of México: Geological Society of America Special Paper 422 (eds. S.A. Alaniz-Alvarez and A.F. Nieto-Samaniego) pp. 71–91.

MORAN-ZENTENO, D., CORONA-CHAVEZ, P. and TOLSON, G. (1996), Uplift and Subduction Erosion in Southwestern Mexico since the Oligocene: Pluton Geobarometry Constraints, Earth and Planetary Science Letters 141, 51–65.

MORAN-ZENTENO, D., KEPPIE, D.J., MARTINY, B. and GONZÁLEZ-TORRES, E. (2009), Reassessment of the Paleogene Position of the Chortis Block Relative to Southern Mexico: Hierarchical Ranking of Data and Features, Revista Mexicana De Ciencias Geologicas 26, 177–188.

MÜLLER, R.D., SDROLIAS, M., GAINA, C. and ROEST, W.R. (2008), Age, Spreading Rates, and Spreading Asymmetry of the World's Ocean Crust, Geochemistry Geophysics Geosystems 9, Q04006, doi:10.1029/2007gc001743.

NIETO-SAMANIEGO, A.F., ALANIZ-ALVAREZ, S.A., SILVA-ROMO, G., EGUIZA-CASTRO, M.H. and MENDOZA-ROSALES, C.C. (2006), Latest Cretaceous to Miocene Deformation Events in the Eastern Sierra Madre Del Sur, Mexico, Inferred from the Geometry and Age of Major Structures, Geological Society of America Bulletin 118, 238–252.

O'DRISCOLL, L.J., HUMPHREYS, E.D. and SAUCIER, F. (2009), Subduction Adjacent to Deep Continental Roots: Enhanced Negative Pressure in the Mantle Wedge, Mountain Building and Continental Motion, Earth and Planetary Science Letters 280, 61–70.

ORTEGA-GUTIERREZ, F., SOLARI, L.A., ORTEGA-OBREGON, C., ELIAS-HERRERA, M., MARTENS, U., MORAN-ICAL, S., CHIQUIN, M., KEPPIE, J.D., DE LEON, R.T. and SCHAAF, P. (2007), The Maya-Chortis Boundary: A Tectonostratigraphic Approach, International Geology Review 49, 996–1024.

ORTEGA-OBREGON, C., SOLARI, L.A., KEPPIE, J.D., ORTEGA-GUTIERREZ, F., SOLE, J. and MORAN-ICAL, S. (2008), Middle-Late Ordovician Magmatism and Late Cretaceous Collision in the Southern Maya Block, Rabinal-Salama Area, Central Guatemala: Implications for North America-Caribbean Plate Tectonics, Geological Society of America Bulletin 120, 556–570.

PARDO, M. and SUÁREZ, G. (1993), Steep Subduction Geometry of the Rivera Plate beneath the Jalisco Block in Western Mexico, Geophysical research letters 20, 2391–2394.

PARDO, M. and SUÁREZ, G. (1995), Shape of the Subducted Rivera and Cocos Plates in Southern Mexico: Seismic and Tectonic Implications, Journal of geophysical research 100, 12357–12373.

PARSONS, B. and SCLATER, J.G. (1977), Analysis of Variation of Ocean-Floor Bathymetry and Heat-Flow with Age, Journal of geophysical research 82, 803–827.

PÉREZ-CAMPOS, X., KIM, Y., HUSKER, A., DAVIS, P.M., CLAYTON, R.W., IGLESIAS, A., PACHECO, J.F., SINGH, S.K., MANEA, V.C. and GURNIS, M. (2008), Horizontal Subduction and Truncation of the Cocos Plate beneath Central Mexico, Geophysical research letters 35, L18303, doi:10.1029/2008gl035127.

PILGER, R.H. (1981), Plate Reconstructions, Aseismic Ridges, and Low-Angle Subduction beneath the Andes, Geological Society of America Bulletin 92, 448–456.

PINDELL, J.L., CANDE, S.C., PITMAN III, W.C., ROWLEY, D.B., DEWEY, J.F., LABRECQUE, J. and HAXBY, W. (1988), A Plate-Kinematic Framework for Models of Caribbean Evolution, Tectonophysics 155, 121–138.

PROTTI, M., GUENDEL, F. and MCNALLY, K. (1995), Correlation between the Age of the Subducting Cocos Plate and the Geometry of the Wadati-Benioff Zone under Nicaragua and Costa Rica, In Geologic and Tectonic Development of the Caribbean Plate Boundary in Southern Central America. Geological Society of America Special Paper 295 (ed' P. Mann) pp. 309–326.

Reprinted from the journal

ROGERS, R.D., KARASON, H. and VAN DER HILST, R.D. (2002), *Epeirogenic Uplift above a Detached Slab in Northern Central America*, Geology 30, 1031–1034.

ROSS, M.I. and SCOTESE, C.R. (1988), *A Hierarchical Tectonic Model of the Gulf of Mexico and Caribbean Region*, Tectonophysics 155, 139–168.

RUFF, L.J. (1989), *Do Trench Sediments Affect Great Earthquake Occurrence in Subduction Zones*, Pure and Applied Geophysics 129, 263–282.

SAK, P.B., FISHER, D.M., GARDNER, T.W., MARSHALL, J.S. and LAFEMINA, P.C. (2009), *Rough Crust Subduction, Forearc Kinematics, and Quaternary Uplift Rates, Costa Rican Segment of the Middle American Trench*, Geological Society of America Bulletin 121, 992–1012.

SCHAAF, P., MORAN-ZENTENO, D., HERNANDEZ-BERNAL, M.D., SOLIS-PICHARDO, G., TOLSON, G. and KOHLER, H. (1995), *Paleogene Continental Margin Truncation in Southwestern Mexico: Geochronological Evidence*, Tectonics 14, 1339–1350.

SINGH, S.K. and PARDO, M. (1993), *Geometry of the Benioff Zone and State of Stress in the Overriding Plate in Central Mexico*, Geophysical research letters 20, 1483–1486.

SOLARI, L.A., DE LEON, R.T., PINEDA, G.H., SOLE, J., SOLIS-PICHARDO, G. and HERNANDEZ-TREVINO, T. (2007), *Tectonic Significance of Cretaceous-Tertiary Magmatic and Structural Evolution of the Northern Margin of the Xolapa Complex, Tierra Colorada Area, Southern Mexico*, Geological Society of America Bulletin 119, 1265–1279.

SUÁREZ, G., ESCOBEDO, D., BANDY, W. and PACHECO, J.F. (1999), *The 11 December, 1995 Earthquake (Mw = 6.4): Implications for the Present-Day Relative Motion on the Rivera-Cocos Plate Boundary*, Geophysical research letters 26, 1957–1960.

TOVISH, A., SCHUBERT, G. and LUYENDYK, B.P. (1978), *Mantle Flow Pressure and the Angle of Subduction: Non-Newtonian Corner Flows*, Journal of geophysical research 83, 5892–5898.

TURCOTTE, D.L. and SCHUBERT, G., *Geodynamics* (Cambridge University Press, New York 2002).

URRUTIA-FUCUGAUCHI, J. and FLORES-RUIZ, J. (1996), *Bouguer Gravity Anomalies and Regional Crustal Structure in Central Mexico*, International Geology Review 38, 176–194.

VAN HUNEN, J., VAN DEN BERG, A.P. and VLAAR, N.J. (2002), *On the Role of Subducting Oceanic Plateaus in the Development of Shallow Flat Subduction*, Tectonophysics 352, 317–333.

VAN HUNEN, J., VAN DEN BERG, A.P. and VLAAR, N.J. (2004), *Various Mechanisms to Induce Present-Day Shallow Flat Subduction and Implications for the Younger Earth: A Numerical Parameter Study*, Physics of The Earth and Planetary Interiors 146, 179–194.

VON HUENE, R. and SCHOLL, D.W. (1991), *Observations at Convergent Margins Concerning Sediment Subduction, Subduction Erosion, and the Growth of Continental-Crust*, Reviews of Geophysics 29, 279–316.

WATTS, A.B. and RIBE, N.M. (1984), *On Geoid Heights and Flexure of the Lithosphere at Seamounts*, Journal of geophysical research 89, 1152–1170.

(Received February 5, 2010, revised June 28, 2010, accepted August 18, 2010, Published online November 16, 2010)

Pure Appl. Geophys. 168 (2011), 1475–1487
© 2010 Springer Basel AG
DOI 10.1007/s00024-010-0207-9

Flat-Slab Thermal Structure and Evolution Beneath Central Mexico

Vlad C. Manea[1] and Marina Manea[1]

Abstract—Recent seismic and magnetotelluric experiments, aimed at better characterizing the shape and state of the subducting slab and continental crust beneath Central Mexico, exposed significant differences with conclusions of previous studies. A new slab geometry is revealed in which the subducting Cocos slab is perfectly flat between 120 to 290 km from the trench, after which it plunges into the asthenosphere at a dip angle of $\sim 65°$, in sharp contrast with the previously proposed $\sim 20°$ dip angle. Seismic tomography studies show negative P-wave velocity anomalies (-2 to -4%) in the mantle wedge beneath the Mexican Volcanic Belt, and positive anomalies ($+2$ to $+3\%$) for the subducted Cocos slab. Magnetotelluric experiments exposed a very low-resistivity area ($1–10$ Ωm) located within the continental crust just below the Mexican Volcanic Arc. Finally, several spots of non-volcanic tremors (NVTs) have been recorded inside the continental crust above the flat-slab segment. While all these experiments provide a better picture of the subduction system beneath Central Mexico, several key processes need further investigation. In this study, we take advantage of these new observations to better constrain the thermal structure beneath Central Mexico. Two different thermal models are computed for a mantle potential temperature (T_p) of 1,350 and 1,450°C, respectively. The new thermal structures are then converted into P-wave velocity anomalies and compared with the observed V_p anomalies. We found that a T_p of 1,450°C produced larger V_p anomalies that do not fit the observations. However, using a T_p of only 1,350°C, our predicted V_p anomalies are positive ($+2$ to $+3\%$) for the cold slab and negative (-2 to -4%) in the mantle wedge. These V_p estimates are consistent with the observed seismic tomography from P-wave arrivals, and therefore we conclude that a T_p of 1,350°C is a better estimate for the mantle potential temperature beneath Central Mexico. The new thermal model, in conjunction with phase diagrams for sediments, hydrated basalt and lithospheric mantle, have been used to estimate the amount and location of fluids released from the subducting Cocos slab. Several dehydration pulses have been identified along the slab interface where most of the fluids stored in sediments and oceanic crust are released into the overlying continental crust above the flat-slab. We found a good correlation between the pattern of these dehydration pulses and the location of NVTs, suggesting that slab dehydration is responsible for triggering the tremors. We suggest that NVT bursts localized above the flat slab segment represent the manifestation of ongoing continental crust hydration

and weakening, a process that has been going on since 15 Ma ago when the Cocos slab entered into a flat-slab regime. Such continuous weakening would have reduced the suction forces that kept the slab in a flat regime in the last 15 Ma, allowing the slab to easily roll back. The continuous low-resistivity region recorded beneath the volcanic front in Central Mexico might represent the evidence of slab dehydration and crust weakening over time.

Key words: Flat-slab, thermal structure, Central Mexico, slab dehydration, slab rollback.

1. Introduction

Flat-slab subduction takes place at $\sim 10\%$ of the present-day convergent margins, in general where a correlation with buoyant plateaus and/or aseismic ridges exists (Cross and Pilger, 1982; McGeary et al., 1985; Gutscher et al., 2000). Nevertheless, such correlation does not exist in Central Mexico below the state of Guerrero, where a shallow subduction angle has been revealed at around 40–50 km depth and at distances up to ~ 250 km from the trench (Suarez et al., 1990). The slab geometry farther inland, beneath the Mexican Volcanic Arc, has been difficult to image because of the complete lack of intraslab seismicity. It was considered that from a distance of ~ 250 km from the trench the slab plunges into the asthenosphere at a shallow $\sim 20°$ angle, reaching a depth of around 100 km beneath the active Popocatepetl stratovolcano (Pardo and Suarez, 1995). This slab geometry was used by previous studies to infer and interpret the thermal structure in the area (Currie et al., 2002; Manea et al., 2004, 2005a). The period of flat-slab initiation in Mexico has been considered early-middle Miocene (Ferrari et al., 1999), but without upper plate deformation, as in the case of the Chilean flat-slab. Also, the initial flat-slab length is considered to have been larger in the past than it is today because the volcanic arc

---
[1] Computational Geodynamics Laboratory, Centro de Geociencias, Campus Juriquilla-Queretaro, Universidad Nacional Autonoma de Mexico, Mexico City, Mexico. E-mail: vlad@geociencias.unam.mx

migrated trenchward in the last 15 Ma at a rate of ~ 10 km/Ma (FERRARI et al., 2001).

In recent years, several studies and experiments have been carried out with the main purpose of exposing in great detail the subduction structure and state in this area. The Middle America Seismic Experiment (MASE) imaged the subducted Cocos plate beneath Central Mexico (CLAYTON et al., 2007). The results from this research revealed a longer flat slab segment that extends further inland to ~ 300 km from the trench, where then it sinks into the astheno-sphere at a steep angle of ~ 65° (Fig. 1). Also, an ultra slow velocity layer, ~ 3–5 km thick, interpreted as relict serpentinized mantle, was found on top of the flat-slab segment (PEREZ-CAMPOS et al., 2008; SONG et al., 2009). On the same profile, several regions of non-volcanic tremors (NVTs) have been identified above the flat-slab, with the majority of tremors con-centrated in an area located ~ 220–240 km from the trench within the overriding continental crust (PAYERO et al., 2008). NVTs are long-period low-frequency events with periods of several weeks, and are sus-pected to involve a chain reaction of small fractures caused by super-critical fluids (OBARA, 2002; KODAIRA et al., 2004). Although the processes that generate NVTs are not well understood, in southwest Japan the

good agreement of the NVTs with the shape and position of the seismogenic zone confirm a tectonic origin for these vibrations (OBARA, 2002; JULIAN, 2002). The places where NVTs have been positively identified include Nankai, Cascadia and Mexico sub-duction zones, where young plates produce abundant fluids from the dehydration of sediments and metaba-salt (DRAGERT et al., 2004). In the Nankai subduction zone, NVTs have been located near the seismic-aseismic transition zone (OBARA, 2002; SHELLY et al., 2006; NUGRAHA and MORI, 2006; MIYAZAWA et al., 2008). Also, apparently there is no kinetic delay in triggering NVTs; the speed at which these tremors move can reach 9 km/day, and might reflect the migration of fluids released by metamorphism (JULIAN, 2002). The fluid release from the subducting slab is the most common explanation for the origin of NVTs and this could be the case for Central Mexico, too. For this reason, a better-constrained thermal model is a useful tool to predict accurately where the slab dehydrates, and whether or not there is a correlation with the location of recorded NVTs.

The tomographic study of GORBATOV and FUKAO (2005) shows low-velocity regions (−2%) (probably mantle wedge) beneath the Mexican Volcanic Arc and high-velocity areas (+2%) corresponding to the subducted slab. Additionally, a low velocity area is revealed beneath the Cocos plate close to the trench, possibly attributed to the young (~13–14 Ma) incoming Cocos plate. A similar recent study (YANG et al., 2009), but farther west and with higher reso-lution, shows negative P-wave perturbations in the mantle wedge (−2 to 4%) and positive perturbations (+2 to +4%) interpreted as the subducted Cocos and Rivera plates. Since these seismic anomalies are often interpreted as thermal anomalies, they can be used to better constrain the thermal structure of a subduction zone (MANEA et al., 2005b).

A magnetotelluric (MT) experiment, performed along the same profile as MASE, exposed a contin-uous high-conductive low-resistivity region located beneath the volcanic arc and in the continental crust (JÖDICKE et al., 2006). These low-resistivity areas are interpreted as a consequence of fluid release from the subducting slab or partial melt, and therefore can provide insights about the past dynamics of the sub-duction system in Central Mexico.

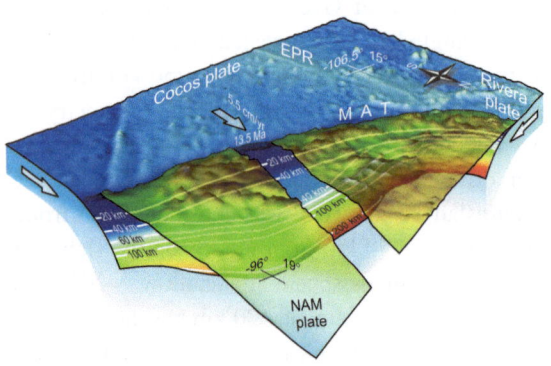

Figure 1

Three-dimensional visualization of the Mexican subduction zone from the northeast. Surface relief is shown as a semi-transparent layer. Labeled *white contours* and *color* gradation of the subducting plate indicate depths to the slab surface from the Earth's surface (PARDO and SUREZ, 1995; PEREZ-CAMPOS et al., 2008). *Bold arrow* shows the direction of the Cocos Plate movement relative to North America. *EPR* East Pacific Rise, *MAT* Middle America Trench, *NAM plate* North America plate. Band cut through the surface relief is located above the flat-slab in Central Mexico where we performed the 2D numeric modeling

In this paper, we provide a new 2D steady-state thermal structure for Central Mexico, constrained by the new slab geometry and recent P-wave seismic tomography. Also, we explore the implications of slab dehydration in light of recently discovered NVTs. Based on the MT results and the NVT patterns we propose a time–space evolutionary model of the flat-slab since the Miocene in Central Mexico.

2. Thermal Models, Mantle Potential Temperatures and Seismic Tomography

2.1. Thermal Models

Using the numeric procedure proposed by MANEA et al. (2004) and the new slab geometry constrained from MASE, we estimated the 2D temperature distribution along the MASE profile for two possible mantle potential temperatures, 1,350 and 1,450°C. Other parameters are kept the same as in previous studies (MANEA et al., 2004, 2005a), including Cocos plate age and convergence rate of 13.7 Myr and 5.5 cm/yr, respectively, and a high pore pressure ratio along the subduction interface of 0.98. The pore pressure ratio effect is considered only along the plate interface from the trench to the hinge point, where the slab plunges abruptly into the mantle. The high pore pressure ratio between the oceanic and continental plates is consistent with the ultraslow velocity, high pore fluid pressure (HPFP) layer found on top of the subducted Cocos plate by SONG et al. (2009). The pore pressure ratio (PPR) controls the position of the 450°C isotherm, which represents the transition from partially coupled zone to stable sliding (HYNDMAN and WANG, 1993). According to GPS studies in Central Mexico the transition zone is located 200–220 km from the trench (KOSTOGLODOV et al., 2003). In our modeling, we used a PPR value of 0.98, which positions the 450°C isotherm at ~200 km from the trench (Fig. 2). The mantle wedge above the slab has a temperature-dependent viscosity, with a reference viscosity of 10^{21} Pa s and an activation energy for olivine of 300 kJ/mol (MANEA et al., 2004). The subducting slab drives the mantle wedge flow and there is no additional flow induced by slab rollback.

2.2. Mantle Potential Temperatures

Mantle potential temperature (T_p) is the temperature of mantle volume that rises towards the Earth's surface along an adiabat without melting (MCKENZIE and BICKLE, 1988). Estimates of T_p range from 1,280 to 1,310°C (MCKENZIE and BICKLE, 1988; PRESNALL et al., 2002) to 1340–1475°C (HERZBERG and O'HARA, 1998; PUTIRKA, 1999; GREEN et al., 2001; PUTIRKA, 2005). Using petrological and geochemical characteristics of primitive basalts, picrites, and komatiites, PUTIRKA et al. (2007) show that ambient mantle temperatures at normal oceanic ridges are in the range of 1,280–1,400°C, and that they can be even as high as 1,460°C below present-day Iceland.

We modeled the thermal structure beneath Central Mexico using two different T_p values, 1,350 and 1,450°C. The modeling results show that the mantle wedge above the slab is the most influenced by the T_p value (Fig. 2a, b). However, in both cases we predict melting of wet peridotite beneath the volcanic arc, and it is uncertain which T_p represents a better fit for the mantle potential temperature. In this case, we used an indirect method to better constrain T_p in our models. We converted the temperature distribution into synthetic V_p anomalies (MANEA et al., 2005b), and then we compared the results with the observed V_p anomalies in Mexico (GORBATOV and FUKAO, 2005; YANG et al., 2009).

2.3. Seismic Tomography in Central Mexico

Negative velocity anomalies in the mantle wedge above subducting slabs are interpreted as thermally induced structures (TAMURA et al., 2002). GERYA et al. (2006) developed a coupled petrological–thermomechanical model that permits prediction of seismic velocity anomalies in subduction zones. In this study, we employ a similar, but simplified approach from MANEA et al. (2005b) to predict seismic velocity anomalies (V_p) beneath Central Mexico using only the temperature dependence of seismic wave velocity from KARATO (1993). The predicted V_p perturbations are calculated relative to the PREM model (DZIEWONSKI and ANDERSON, 1981). We compared the predicted V_p anomalies with the results from several recent tomographic studies in Mexico in order to

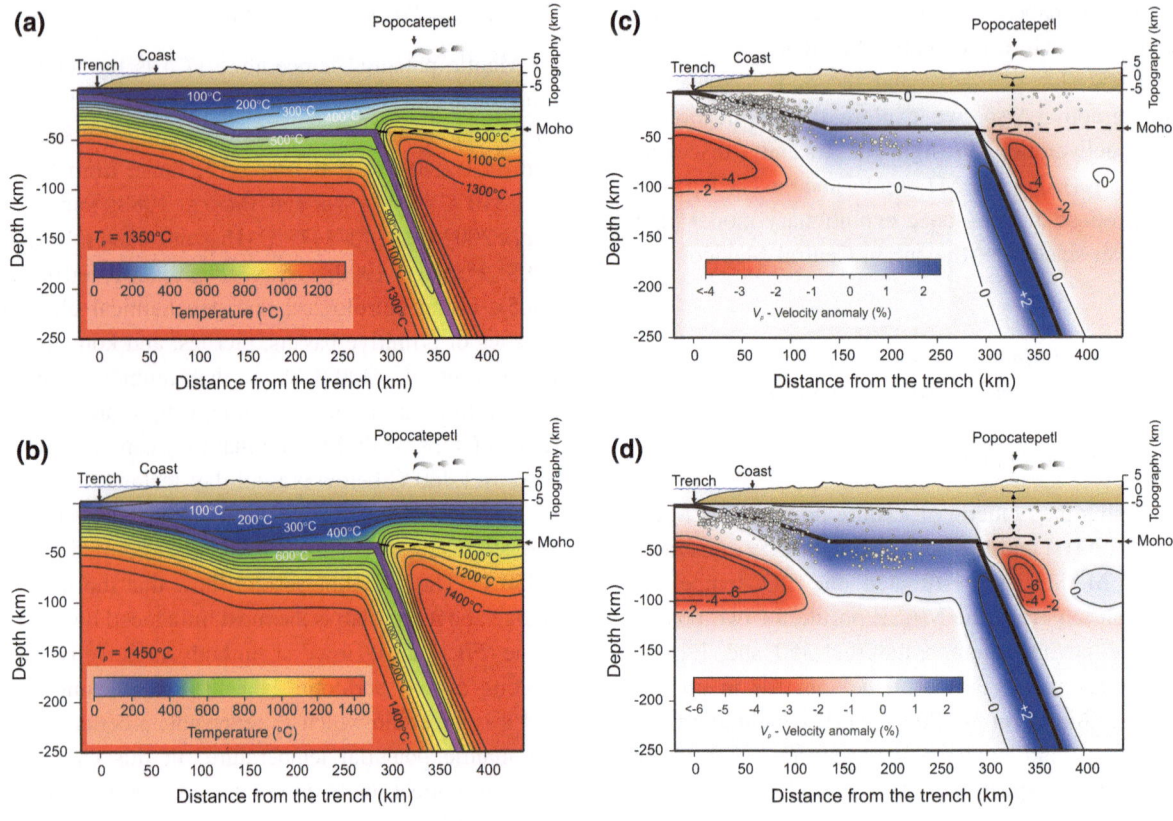

Figure 2
Predicted thermal models and V_p anomaly beneath Central Mexico. **a** Thermal model for $T_p = 1,350°C$. **b** Thermal model for $T_p = 1,450°C$. *Purple band* drawn over thermal model depicts the oceanic crust. **c** V_p velocity anomaly estimation from the temperature model with $T_p = 1,350°C$. **d** V_p velocity anomaly estimation from the temperature model with $T_p = 1,450°C$. *White circles* are the projection of earthquake hypocenters ($M > 4$) from SSN catalog for 2005–2006 epoch. *Solid* and *dashed black lines* illustrate the Cocos-North America plates interface and Moho from PEREZ-CAMPOS *et al.* (2008)

better constrain the thermal structure. As an input parameter, we used the thermal structure calculated for the two different T_p values, 1,450 and 1,350°C (Fig. 2a, b). The modeling results are presented in Fig. 2c, d and both show strong negative V_p anomalies in the mantle wedge and beneath the incoming Cocos plate in the range of -2 to -6%. Observed V_p anomalies in the wedge are in the range -2 to -4%, and therefore the best fitting model is obtained when we use a T_p of 1,350°C (Fig. 2c). Also, our modeling result (Fig. 2c) is consistent with the P-wave tomography of GORBATOV and FUKAO (2005), although the magnitude of the P-wave velocity anomaly estimated from thermal modeling (-4%) is higher than the observed one (-2%). However, a recent seismic experiment located few hundred km to the north from

Guerrero, (YANG *et al.*, 2009) shows P velocity perturbations in the mantle wedge up to -4%. In our modeling the cold slab induces a positive velocity anomaly of $\sim +2\%$, a value consistent with both tomographic studies mentioned above, and also with the recent tomographic image from PEREZ-CAMPOS *et al.* (2008).

The negative velocity anomaly associated with mid-ocean ridges and young plates, as the Cocos and Rivera plates offshore from Mexico, is a common feature in global or regional seismic tomography (GUNG and ROMANOWICZ, 2004). Such a negative anomaly can also be observed beneath the incoming Cocos plate (GORBATOV and FUKAO, 2005 and also in our predicted V_p tomography inferred from temperature as seen in Fig. 2c, d).

3. Slab Dehydration

The water carrier in active subduction systems is the oceanic plate composed of sediments, oceanic crust and mantle layers. The young Cocos plate has a thin sediment layer only ~ 200 m thick, but most of these hydrated sediments enter into the subduction system (MANEA et al., 2003). Based on the newly constrained thermal model (Fig. 2a) and phase diagrams for sediment-, basalt-, and peridotite-water systems from RÜPKE et al. (2004), we analyze the stability of hydrous phases and estimate the water content in the subducting Cocos plate. We also explore the locations where various hydrous phases break down, and compare them with the position of NVTs and low-resistivity areas.

3.1. Sediment-water System

Figure 3a presents the estimation of H_2O amount retained by minerals in sediments as function of $P–T$ along the top slab interface. Sediments start to dehydrate at ~ 35 km depth and ~ 100 km from the trench where they drop from 4% wt. H_2O to 3.5% wt. H_2O. Then there are several similar dehydration pulses located along the flat-slab. A second pulse of 0.5% H_2O is released at 160–170 km from the trench and a third and larger one ($\sim 1\%$ H_2O) at 230–240 km from the trench. The location of these two former dehydration pulses corresponds with the position of areas of intense NVTs activity. The last dehydration pulse (0.5% H_2O) is located at 270–280 km from the trench just before the slab starts plunging into the asthenosphere. At a distance of ~ 300 km from the trench the sedimentary layer dehydrates $\sim 75\%$, and a total of $\sim 3\%$ wt. H_2O is released mostly above the flat-slab area (Fig. 4, inset). The remaining 1% wt. H_2O is carried down into subduction zone, where it can contribute to the water content of the upper mantle (ONO, 1998).

3.2. Basalt-water System

To estimate the water content stored in the basaltic oceanic crust, we use the phase diagram for metabasalt (RÜPKE et al., 2004) and top and bottom oceanic crust isotherms from the best-fit thermal model (Fig. 2a). The results are presented in Fig. 3b, and show that no dehydration occurs before the slab enters into the flat regime. Even then, the slab runs for another ~ 100 km, when a large dehydration pulse goes off the slab at 230–240 km from the trench (Fig. 4). In this narrow band, the oceanic crust dehydrates $\sim 85\%$ and more than 3% wt. H_2O is released into the overriding continental crust. The location of this major dehydration pulse corresponds with the area where $\sim 80\%$ of the NVT bursts occur (Fig. 4). Also, this region overlaps with a smaller dehydration pulse ($\sim 1\%$ wt. H_2O) from sediments. The remaining 0.5% wt. H_2O stored in the basaltic crust is later released into the mantle wedge at a depth of ~ 90 km, below the active Popocatepetl volcano.

3.3. Peridotite-water System

The peridotite layer of oceanic plates represents one of the major water sources that are carried down in active subduction systems. We estimate the H_2O amount stored into the serpentinized subducting lithosphere beneath Central Mexico using two slab geotherms, located at the base of oceanic crust and 7 km below, and the phase diagram for serpentinized mantle (RÜPKE et al., 2004). The results presented in Fig. 3c show that a significant amount of water (6.5 wt%) is preserved in the serpentinized peridotite layer down to 60–70 km depth. The first dehydration occurs in the flat slab segment when for the same pressure the temperature increases above 500°C. Here only 0.5% is released and the dehydration front is distributed sub-horizontally for a distance of about 50 km beneath the first cluster of NVTs (Fig. 4). At greater depths, the slab crosses the choke point at 600–700°C at ~ 2 GPa (60–70 km depth) and a strong dehydration process occurs where $\sim 90\%$ of the fluid stored is released into the overlying mantle wedge through the oceanic crust. The slab depth beneath Popocatepetl is ~ 120 km, and the difference between this depth and the depth where the serpentinized oceanic lithosphere releases fluids ($\sim 60–70$ km), can be explained by the time necessary for the released water to cross the oceanic crust and sediments and to run off the slab into the overlying mantle at ~ 120 km depth. We conclude that this significant amount of H_2O released through

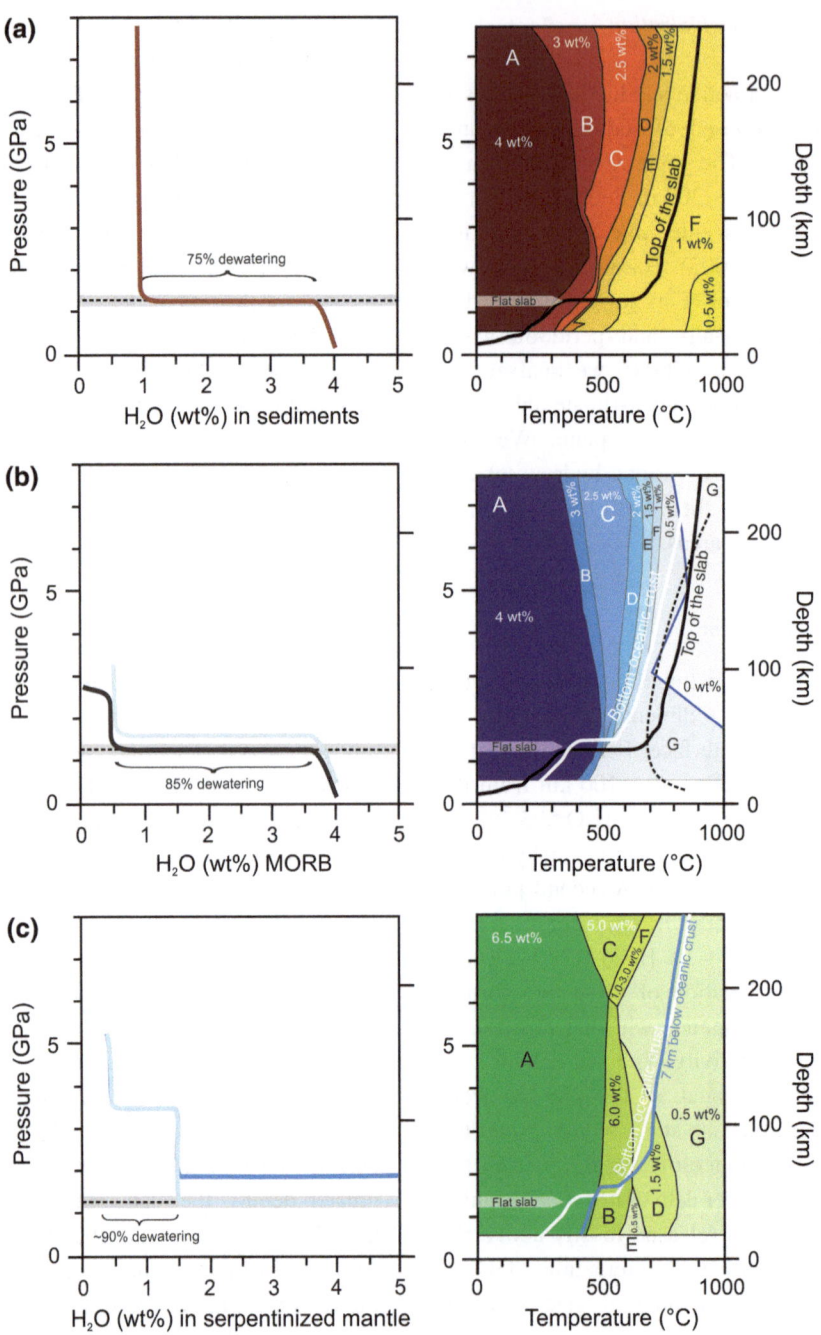

Figure 3

Computed phase equilibria and *P–T* water content for sediments (**a**), metabasalt (**b**), and serpentinized mantle (**c**). *Black*, *white* and *blue* *curves* represent the geotherms, all from the thermal model shown in Fig. 2a, at the *top* of the slab, the *bottom* of the oceanic crust, and 7 km below the oceanic crust, respectively. The plots on the *left* side illustrate how much H_2O (%) is released in each layer

slab deserpentinization can induce partial melting of the mantle above the slab and explain the volcanic productivity and the reasonably high H_2O found in the erupted mafic magmas, similar to magmas from other arcs (ROBERGE *et al.*, 2009; CERVANTES and WALLACE, 2003).

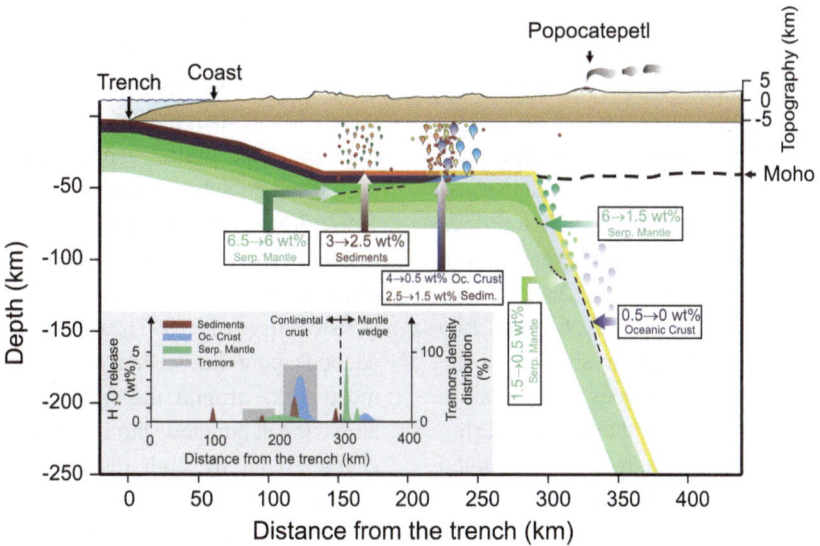

Figure 4

Sediments, oceanic crust, and serpentinized mantle dehydration along the subducting Cocos slab beneath Central Mexico. *Red* and *yellow dots* are the location of NVTs from PAYERO *et al.* (2008). *Colored drops* illustrate where the oceanic sediments, crust and serpentinized mantle dehydrate the most. *Inset plot* shows the location of the main dehydration pulses and NVTs along the slab interface. Note the good correlation between the position of dehydration episodes and the location of *NVT* bursts

4. Discussion and Conclusions

In this study, we combine our modeling results with results from recent experiments in Central Mexico to interpret key processes related to flat-slab subduction structure and dynamics. Here we computed a 2D steady-state thermal model; more realistic time-dependent 2D or 3D thermal models are proposed for future studies. However, despite the limitations of steady-state models, the new thermal structure presented here is well constrained by both GPS deformation studies (KOSTOGLODOV *et al.*, 2003) and seismic experiments (GORBATOV and FUKAO, 2005; PEREZ-CAMPOS *et al.*, 2008), and therefore it provides a reliable picture of the present-day thermal structure.

In the next sections, we provide a scenario of how slab dehydration could control and affect the resistivity distribution in the upper crust and could affect NVT patterns. Finally, we propose an evolution model for the flat-slab in Central Mexico for the last 15 m.y. based on present-day geophysical and past geological data.

4.1. Slab Dehydration, NVTs and Crust Resistivity

As shown in Sect. 3, the subducted oceanic crust and sediment layers are able to release significant fluids into the overlying continental crust. Additionally, the dehydration process occurs in several pulses that differ both in location and the amount of H_2O released (Fig. 4, inset). The major discharge is located in a narrow band, 230–250 km from the trench, where both sediments and oceanic crust strongly dehydrate, and a total amount of $\sim 4\%$ wt. H_2O is flushed into the continental crust. This narrow band corresponds with the area where most of the NVT activity occurs. Actually, $\sim 80\%$ of the NVT bursts are recorded in this spot, strongly suggesting that slab dehydration is the main controlling source of tremors distribution above the flat-slab in Central Mexico. The other region where a smaller number of NVTs were recorded is situated closer to the trench at ~ 150–180 km. In this area we found that only a small amount of fluids (0.5% wt. H_2O) is released from oceanic minerals breakdown reactions.

The triggering mechanism of NVTs depends on where and how fluids are distributed along the subducting slab interface, and several mechanisms have been proposed. For example, for southwest Japan, MIYAZAWA *et al.* (2008) suggested that normal stress reduction accompanied by horizontal compression can trigger these tremors. However, for the Cascadia subduction zone, shear stress changes from Love waves are proposed to induce NVTs (RUBINSTEIN *et al.*, 2010). These observations could indicate a difference in physical properties between the two subduction regions. MIYAZAWA *et al.* (2008) proposed that the fluid pattern on and off the fault plane actually controls the triggering mechanism. A homogenous fluid distribution along the slab interface would favor normal stress reduction, whereas a more heterogeneous fluid distribution could induce a shear strain change. In Central Mexico, we observed two distinct locations where NVTs occur and where the subducting Cocos plate undergoes dehydration (Fig. 4). This fluid pattern favors the first triggering mechanism where normal stress reduction coupled with horizontal compression could effectively induce NVTs.

In our interpretation, we take advantage of the magnetotelluric survey in Central Mexico carried out by JÖDICKE *et al.* (2006). High-conductive low-resistivity (<10 Ωm) areas imaged by MT surveys are often interpreted as zone rich in fluids (OGAWA *et al.*, 2001; RYBIN *et al.*, 2004; SOYER and UNSWORTH, 2006). In general, the overall resistivity of a rock is considered as the sum of both solid grains and pore

spaces saturated with fluids. Moreover, if the fluids contain dissolved ions, the resistivity of the fluid part decreases further because the ions can easily move. Also, low-resistivity of rocks could be due to interconnected grain boundary phases such as water, partial melts, sulphides or graphite (SCHILLING *et al.*, 1997; JONES, 1999). Magnetotelluric exploration of the San Andreas Fault has revealed that the fault is characterized by a low-resistivity wedge (~3 Ωm) (UNSWORTH *et al.*, 1997), which has been attributed to aqueous pore fluids that dominate the resistivity of most rocks around the fault. The continental crust above the subducted slab in Central Mexico is mostly characterized by high resistivity. However, an isolated low-resistivity (~50 Ωm) spot can be identified at ~100 km from the trench at a depth of 20–30 km. From our modeling results, this area corresponds to the first dehydration pulse from the sedimentary layer (Fig. 5), and we interpret this low-resistivity area as a result of sediment dehydration. The continental crust beneath the volcanic arc is characterized by low-resistivity (1–10 Ωm) distributed in a ~300-km long band. This LR band ends where the NVT area is located above the eastern end of the flat-slab segment

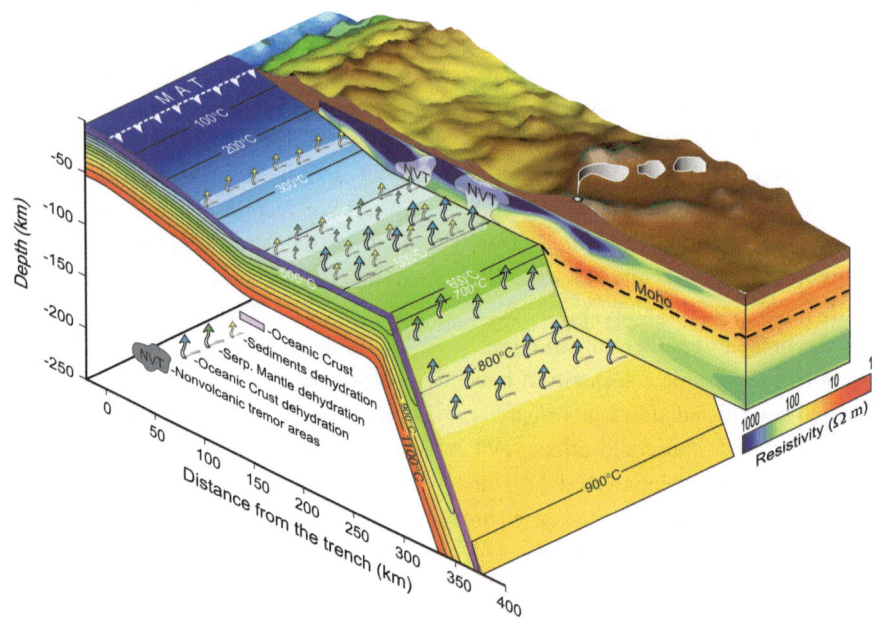

Figure 5

3D view of the flat-slab subduction in Central Mexico. The slab surface is shown in *color shades* that correspond to the temperature model from Fig. 2a. *Yellow, blue* and *green arrows* on the slab surface represent the locations where oceanic sediments, crust and serpentinized mantle dehydrate. The background image cut through the continental crust is the resistivity model of JÖDICKE *et al.* (2006)

(Fig. 5). We interpret this elongated low-resistivity region as a consequence of continuous subducting slab dehydration over the last 15–20 Ma. In fact, FERRARI (2004) shows that the Mexican Volcanic Belt has migrated trenchward at a rate of ~10 km/Myr, probably implying that the slab has rolled back and the length of the flat-slab segment has diminished since the Miocene. Similar low-resistivity in the overriding crust has been imaged above the flat slab region in central Chile, where an MT survey revealed a continuous conductive region in the continental crust at ~25 km depth and above the flat slab (BOOKER et al., 2004). On the other hand, the MT survey performed for the Cascadia subduction zone, which did not experienced flat-slab and rollback, shows high resistivity areas in the continental crust (100–1,000 Ωm) with only some small-scale low-resistivity spots above the Moho. These low-resistivity localized areas are interpreted as the consequence of aqueous fluids released from the subducting Juan de Fuca slab (SOYER and UNSWORTH, 2006).

The difference in the low resistivity pattern between the two types of subducting slabs suggests that the flat slab dewatering followed by rollback causes the MT anomaly recorded in central México beneath the volcanic arc (Fig. 5). However, the nature of this low resistivity area could also be interpreted in terms of partial melt. The temperatures (800–1,000°C) predicted for the lowermost continental crust beneath the volcanic arc (Fig. 2a, b) are well above the wet melting temperatures for a wide range of lower crustal rock types, therefore there could be partial crustal melting in that area. In the western US during the Laramide orogeny, a slab rollback process, similar to what we proposed here, likely led to voluminous crustal melting and created widespread rhyolitic eruptions. In the area of the Tepic-Zacoalco graben, the first episodes of voluminous rhyolithic volcanism started in the late Miocene (~7 Ma) and continued in the early Pliocene and late Pliocene-Quaternary (GÓMEZ-TUENA et al., 2007). Also, it is worth mentioning that this rhyolithic manifestation coincides with the trenchward migration of the volcanic front. However, the solidification of partial melt at the base of the crust could also induce the accumulation of large amounts of fluids that could explain the observed low-resistivity anomaly beneath the arc and backarc (SOYER and UNSWORTH, 2006). Although partial melt in the lower crust might somehow explain the low-resistivity area beneath the volcanic arc, part of this electric anomaly is located above the eastern end of the flat-slab segment where our thermal models predict temperatures as low as 400°C. A combination of interconnected partial melts and fluids in the lower crust would explain the exceptionally low-resistivity (~1 Ωm) recorded beneath the volcanic arc and above the flat-slab in Central Mexico (Fig. 5).

4.2. Crust Weakening, Suction Forces and Slab Rollback

In active subduction systems, the slab geometry reflects the balance of slab pull, elastic resistance and hydrodynamic forces (suction forces) in the mantle wedge (PEREZ-GUSSINYÉ et al., 2008; MANEA and GURNIS, 2007). Whereas the slab density excess, with respect to the surrounding mantle, controls slab pull force, the suction forces act as a counterbalance force which is mainly controlled by the mantle wedge viscosity. MANEA and GURNIS (2007) show how mantle wedge viscosity affects the slab dip evolution in an active subduction zone. A low viscosity wedge tends to increase the slab dip, whereas a high wedge viscosity decreases the subduction angle and can even produce flat-slabs. In the latter case, the suction forces in the mantle wedge are sufficiently high to compensate the slab positive buoyancy. Thus, elevated suction forces along the slab-mantle interface could sustain long-lived flat-slab systems. This is likely the case in Central Mexico, where flat-slab subduction has been going on since the Miocene (FERRARI et al., 1999). Geological evidence shows that since the Miocene the volcanic arc has moved slowly trenchward at a rate of ~10 km/Myr (FERRARI, 2004), suggesting slab rollback and flat-slab shortening mechanisms. Rollback of the subducted Cocos plate over the last 2 Myr has been proposed to explain high-Nb alkali basaltic magmas for the Michoacan-Guanajuato Volcanic Field located only several hundreds of km to the west of our study area (JOHNSON et al., 2009). Integrating these observations with the 2D thermo-mechanical model of the subduction zone beneath Central Mexico, we propose the following scenario for flat-slab evolution:

1. When flat-slab started in Mexico around 15 Ma
   ago, the flat segment length was larger than today,
   extending probably some 450 km from the trench
   (Fig. 6a). This assumption is supported by the age
   data for mafic rocks in Central Mexico (FERRARI,
   2004). The main reason for initiating such a long
   flat-slab is still unknown.

2. A continuous process of slab dehydration at the
   eastern end of the flat-slab segment weakens the
   base of the continental crust (Fig. 6b). Numerical
   simulations support the assumption that slab
   dehydration facilitates significant weakening of
   the overriding plate (ARCAY et al., 2006). In
   Central Mexico, a low-resistivity area can now

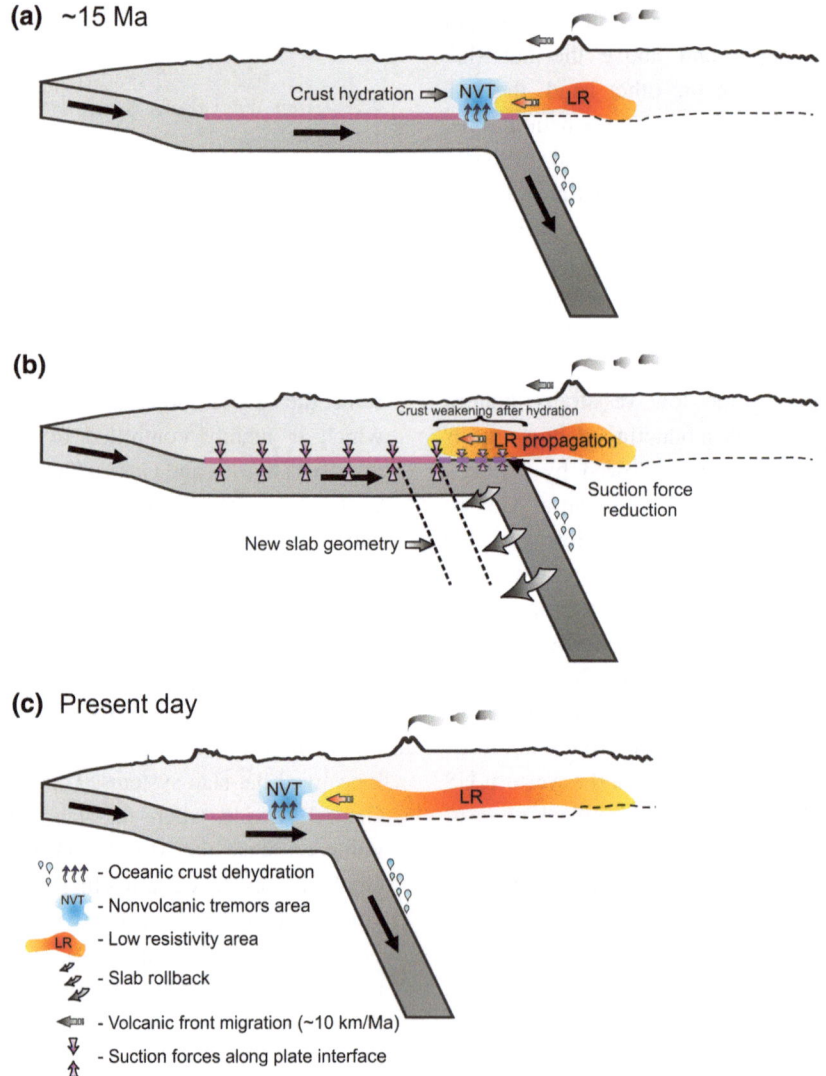

Figure 6

Theoretical evolutionary model proposed for the flat-slab subduction beneath Central Mexico. **a** A longer flat-slab segment could have existed ~15 Ma ago. Over time, slab dehydration weakens the overriding continental crust. *NVT*s can be an indication of the weakening process. **b** Once the crust is sufficiently weakened, the magnitude of the suction force that kept the subducting Cocos plate in flat-slab regime diminishes, allowing the slab to roll back. As a consequence, the volcanic arc and the low-resistivity area (*LR*) start propagating trenchward. **c** In time, the length of the flat-slab section reduces, the *LR* area propagates even further until the slab takes the present day geometry as revealed by MASE. *NVT* bursts still occur at present, suggesting that crust weakening is an ongoing process

propagate trenchward into the newly weakened and hydrated crust. We propose that such a prolonged flat-slab dehydration regime might have weakened the overriding continental crust enough to allow the slab to start rollback. The actual process behind the slab rollback and overlying crust weakening is still unknown. However, slab rollback permits the hot mantle wedge to flow back into the newly created space, explaining why the volcanism migrated trenchward. In our view, continuous weakening of the continental crust above the flat-slab segment leads to a significant reduction in viscosity, and therefore suction forces, allowing the slab to decouple from the continental plate (MANEA and Gurnis, 2007; PEREZ-GUSSINYÉ *et al.*, 2008).

3. We propose that slab rollback occurs in steps, and several episodes of crust weakening and slab rollback lead to the present day subduction geometry in Central Mexico (Fig. 6c). Present day strong NVTs above the flat-slab would be the manifestation of ongoing hydration and weakening processes, that will eventually promote another episode of slab rollback in the future.

Acknowledgments

Numerical simulations were performed on Computational Geodynamics Laboratory's Supercomputing facility (Horus), at GeoSciences Center, UNAM. This study was supported by PAPIIT IN110709, PAPIIT IN115810 and CONACyT 84035.

REFERENCES

ARCAY, D., DOIN, M.-P., TRIC, E., BOUSQUET, R., and DE CAPITANI, C. (2006), Overriding plate thinning in subduction zones: localized convection induced by slab dehydration, *Geochem. Geophys. Geosyst.* 7, Q02007. doi:10.1029/2005GC001061.

BOOKER, J., FAVETTO, A., and POMPOSIELLO C. (2004), Low electrical resistivity associated with plunging of the Nazca flat slab beneath Argentina. *Nature* 429, 399–403.

CERVANTES, P. and WALLACE, P. (2003), Role of H_2O in subduction zone magmatism: new insights from melt inclusions in high-Mg basalts form central Mexico. *Geology* 31, 235–238.

CLAYTON, R. W., DAVIS, P. M., and PEREZ-CAMPOS, X. (2007), Seismic structure of the subducted Cocos plate. *Eos Trans. AGU, Jt. Assem. Suppl.* 88(23), Abstract T32A-01.

CROSS, T. A. and PILGER R. H. Jr. (1982), Controls of subduction geometry, location of magmatic arcs, and tectonics of arc and back-arc regions. *Geol. Soc. Am. Bull.* 93, 545–562.

CURRIE, C. A., HYNDMAN, R. D., WANG, K. and KOSTOGLODOV, V. (2002), Thermal models of the Mexico subduction zone: Implications for the megathrust seismogenic zone, *J. Geophys. Res.* 107(B12), 2370. doi:10.1029/2001JB000886.

DRAGERT, H., WANG, K., and ROGERS, G. (2004), Geodetic and seismic signatures of episodic tremor and slip in the northern Cascadia subduction zone. *Earth Planets Space* 56, 1143–1150.

DZIEWONSKI, A. M. and ANDERSON, D. L. (1981), Preliminary reference Earth model. *Phys. Earth. Planet. Int.* 25, p 297–356.

FERRARI, L. (2004), Slab detachment control on mafic volcanic pulse and mantle heterogeneity in central Mexico, *Geology* 32, 77–80.

FERRARI L., LOPEZ-MARTINEZ M., AGUIRRE-DIAZ G., and CARRASCO-NUÑEZ, G. (1999), Space-time patterns of Cenozoic arc volcanism in central Mexico: from the Sierra Madre Occidental to the Mexican Volcanic Belt. *Geology* 27, 303–306.

FERRARI, L., PETRONE, C. M., and FRANCALANCI, L. (2001), Generation of oceanic-island basalt-type volcanism in the western Trans-Mexican volcanic belt by slab rollback, asthenosphere infiltration, and variable flux melting, *Geology* 29, 507–510.

GERYA, V. T., CONNOLLY, J. A. D., YUEN, A. D., GORCZYK, W., and CAPEL, A. M. (2006), Seismic implications of mantle wedge plumes. *Phys. Earth Planet. Interiors* 156, 59–74.

GÓMEZ-TUENA, A., OROZCO-ESQUIVEL, MA. T., and FERRARI, L. (2007), Igneous petrogenesis of the trans-mexican volcanic belt, In: Alaniz-Álvarez, S. A., and Nieto-Samaniego, Á. F. (eds.) Geology of México: celebrating the centenary of the Geological Society of México: *Geological Society of America Special Paper* 422, 129–181. doi:10.1130/2007.2422(05).

GORBATOV, A. and FUKAO, Y. (2005), Tomographic search for missing link between the remnant Farallon slab and present Cocos subduction. *Geophys. J. Int.* 160, 849–854.

GREEN, D. H., FALLOON, T. J., EGGINS, S. M., and YAXLEY, G. M. (2001), Primary magmas and mantle temperatures, *Eur. J. Mineral.* 13, 437–451.

GUNG, Y. and ROMANOWICZ, B. (2004), Q tomography of the upper mantle using three component long period waveforms, *Geophys. J. Int.* 157, 813–830.

GUTSCHER, M. A., SPAKMAN, W., BIJWAARD, H., ENGDAHL, E. R. (2000), Geodynamics of flat subduction: seismicity and tomographic constraints from the Andean margin. *Tectonics* 19, 814–833.

HERZBERG, C., and O'HARA, M. J. (1998), Phase equilibrium constraints on the origin of basalts, picrites, and komatiites. *Earth Sci. Rev.* 44, 39–79.

HYNDMAN, R. D. and WANG, K. (1993), Thermal constraints on the zone of major thrust earthquake failure: the Cascadia subduction zone. *J. Geophys. Res.* B98, 2039–2060.

JÖDICKE, H., JORDING, A., FERRARI, L., ARZATE, J., MEZGER, K., and RUPKE, L. (2006), Fluid release from the subducted Cocos plate and partial melting of the crust deduced from magnetotelluric studies in southern Mexico: Implications for the generation of volcanism and subduction dynamics. *J. Geophys. Res.* 111, B08102. doi:10.1029/2005JB003739.

JOHNSON, E. R., WALLACE, P. J., DELGADO GRANADOS, H., MANEA, V. C., KENT, A. J. R., BINDEMAN, I. N., and DONEGAN C. S. (2009), Subduction-related volatile recycling and magma generation beneath Central Mexico: insights from melt inclusions, oxygen isotopes and geodynamic models. *J. Petrol.* 50(9), 1729–1764.

JONES, A. G. (1999), Imaging the continental upper mantle using electromagnetic methods. *Lithos* 48, 57–80.

JULIAN, B. (2002), Seismological detection of slab metamorphism, *Science* 296, 1625–1626.

KARATO, S. I. (1993), Importance of anelasticity in the interpretation of seismic tomography. *Geophys. Res. Lett.* 20, 1623–1626.

KODAIRA, S., IIDAKA, T., KATO, A., PARK, J.-O., IWASAKI, T., KANEDA, Y. (2004), High pore fluid pressure may cause silent slip in the Nankai Trough. *Science* 304, 1295–1298.

KOSTOGLODOV, V., SINGH, S. K., SANTIAGO, J. A., FRANCO, S. I., LARSON, K. M., LOWRY, A. R., and BILHAM, R. (2003), A large silent earthquake in the Guerrero seismic gap, Mexico., *Geophys. Res. Lett.* 30(15), 1807. doi:10.1029/2003GL017219.

MANEA, V. C. and GURNIS, M. (2007), Subduction zone evolution and low viscosity wedges and channels, *Earth Planet. Sci. Lett.* 264, 1–2, pp 22–45.

MANEA, M., MANEA, V. C., KOSTOGLODOV, V. (2003), Sediment fill of the Middle America Trench Inferred from the gravity anomalies. *Geofis. Int.* 42 (4), 603–612.

MANEA, V. C., MANEA, M., KOSTOGLODOV, V., CURRIE, C.A. and SEWELL, G. (2004), Thermal structure, coupling and metamorphism in the Mexican subduction zone beneath Guerrero, *Geophys. J. Int.* 158, 775–784. doi:10.1111/j.1365-246X.2004.02325.x.

MANEA, V. C., MANEA, M., KOSTOGLODOV, V., and SEWELL, G. (2005a), Thermo-mechanical model of the mantle wedge in Central Mexican subduction zone and a blob tracing approach for the magma transport. *Phys Earth Planet Interiors* 149, 165–186. doi:10.1016/j.pepi.2004.08.024.

MANEA, V. C., MANEA, M., KOSTOGLODOV, V., and SEWELL, G. (2005b), Thermal models, magma transport, and velocity estimation beneath southern Kamchatka. In: Foulger, G. R., Natland, J. H., Presnell, D. C., and Anderson, D. L. (eds.), *GSA Special paper: plates, plumes and paradigms* 388–31, pp 517–536.

McGEARY, S., NUR, A., BEN-AVRAHAM, Z. (1985), Spacial gaps in arc volcanism: the effect of collision or subduction of oceanic plateaus. *Tectonophysics* 119, 195–221.

McKENZIE, D., and BICKLE, M. J. (1988), The volume and composition of melt generated by extension of the lithosphere, *J. Petrol.* 29, 625–679.

MIYAZAWA, M., BRODSKY, E. E., and MORI, J. (2008), Learning from dynamic triggering of low-frequency tremor in subduction zones. *Earth Planets Space* 60, e17–e20.

NUGRAHA, A. D. and MORI, J. (2006), 3-D velocity structure in the Bungo channel and Shikoku Area, Japan, and its relationship to low-frequency earthquakes, *Geophys. Res. Lett.* 33, L24307.

OBARA, K. (2002), Nonvolcanic Deep Tremor Associated with Subduction in Southwest Japan. *Science* 296, no. 5573, pp 1679–1681. doi:10.1126/science.1070378.

OGAWA, Y., MISHINA, M., GOTO, T., SATOH, H., OSHIMAN, N., KASAYA, T., TAKAHASHI, Y., NISITANI, T., SAKANAKA, S., UYESHIMA, M., TAKAHASHI, Y., HONKURA, Y., and MATSUSHIMA, M. (2001), Magnetotelluric imaging of fluids in intraplate earthquakes zones, NE Japan back arc, *Geophys. Res. Lett.* 28, 3741–3744.

ONO, S. (1998), Stability limits of hydrous minerals in sediment and mid-ocean ridge basalt compositions: implications for water transport in subduction zones. *J. Geophys. Res.* 103 (B8), 18253–18267.

PARDO, M. and SUAREZ, G. (1995), Shape of the subducted Rivera and Cocos plates in southern Mexico: seismic and tectonic implications, *J. Geophys. Res.* 100, 12357–12373.

PAYERO, J., KOSTOGLODOV, V., SHAPIRO, N., MIKUMO, T., IGLESIAS, A., PEREZ-CAMPOS, X., and CLAYTON, R. (2008), Non-volcanic tremor observed in the Mexican subduction zone. *Geophys. Res. Lett.* 35, L07305.

PEREZ-CAMPOS, X., KIM, Y. H., HUSKER, A., DAVIS, P. M., CLAYTON, R. W., IGLESIAS, A., PACHECO, J. F., SINGH, S. K., MANEA, V. C., and GURNIS, M. (2008), Horizontal subduction and truncation of the Cocos plate beneath central Mexico, *Geophys. Res. Lett.* 35, L18303. doi:10.1029/2008GL035127.

PEREZ-GUSSINYÉ, M., LOWRY, A., PHIPPS MORGAN, J., and TASSARA, A. (2008), Effective elastic thickness variations along the Andean margin and their relationship to subduction geometry. *Geochem. Geophys. Geosyst.* 9, 2.

PRESNALL, D. C., GUDFINNSON, G. H., and WALTER, M. J. (2002), Generation of mid-ocean ridge basalts at pressures from 1 to 7 GPa, *Geochim. Cosmochim. Acta* 66, 2073–2090.

PUTIRKA, K. (1999), CPX + Liquid equilibria. *Contrib Mineral Petrol.* 135, 151–163.

PUTIRKA, K. D. (2005), Mantle potential temperatures at Hawaii, Iceland, and the mid-ocean ridge system, as inferred from olivine phenocrysts: evidence for thermally driven mantle plumes, *Geochem. Geophys. Geosyst.* 6, Q05L08. doi:10.1029/2005GC000915.

PUTIRKA, K. D., PERFIT, M., RYERSON, F. J., and JACKSON, M. G. (2007), Ambient and excess mantle temperatures, olivine thermometry, and active vs. passive upwelling. *Chem. Geol.* 24, 177–206.

ROBERGE, J., DELGADO-GRANADOS, H., and WALLACE, P. J. (2009), Mafic magma recharge supplies high CO_2 and SO_2 gas fluxes from Popocatépetl volcano, Mexico. *Geology*, vol 37, no. 2, p 107–110. doi:10.1130/G25242A.1.

RUBINSTEIN, J. L., SHELLY, D. R., and ELLSWORTH, W. L. (2010), Non-volcanic tremor: a window into the roots of fault zones. In: Cloetingh, S., Negendank, J. (eds.), *New frontiers in integrated solid earth sciences*. International year of planet earth., Springer Science + Business Media B.V. doi:10.1007/978-90-481-2737-5_8.

RÜPKE, L. H., MORGAN, J. P., HORT, M., CONNOLLY, J. A. D. (2004), Serpentine and the subduction zone water cycle. *Earth Planet. Sci. Lett.* 223, 17–34.

RYBIN, A., SPICHAK, V., BATALEV, V., SCHELOCHKOV, G., BATALEVA, E., and SAFRONOV, I. (2004), Magnetotelluric investigations of an active thrust faults in the Northern Tien Shan, Kyrgyzstan, Central Asia. IAGA WG 1.2 on electromagnetic induction in the earth. Available at *Proceedings of the 17th Workshop Hyderabad*, India, October 18–23.

SCHILLING, F. R., PARTZSCH, G. M., BRASSE, H., and SCHWARZ, G. (1997), Partial melting below the magmatic arc in the Central Andes deduced from geoelectromagnetic field experiments and laboratory data. *Phys. Earth Planet. Inter.* 103, pp 17–31.

SHELLY, D. R., BEROZA, G. C., IDE, S., and NAKAMULA, S. (2006), Low-frequency earthquakes in Shikoku, Japan, and their relationship to episodic tremor and slip, *Nature* 442, 188–191.

SONG, T. R. A., HELMBERGER, D. V., BRUDZINSKI, M. R., CLAYTON, R. W., DAVIS, P., PEREZ-CAMPOS, X., and SINGH, S. K. (2009), Subducting slab ultra-slow velocity layer coincident with silent earthquakes in southern Mexico, *Science* 324, 502–506.

SOYER, W. and UNSWORTH, M. (2006), Deep electrical structure of the northern Cascadia (British Columbia, Canada) subduction zone: Implications for the distribution of fluids. *Geology*, vol 34, no. 1, p 53–56. doi:10.1130/G21951.1.

SUAREZ, G., MONFRET, T., WITTLINGER, G., and DAVID, C. (1990), Geometry of subduction and depth of the seismogenic zone in the Guerrero gap, Mexico. *Nature* 345, 336–338.

TAMURA, Y., TATSUMI, Y., ZHAO, D. P., KIDO, Y., and SHUKUNO, H. (2002), Hot fingers in the mantle wedge: new insights into magma genesis in subduction zones. *Earth Planet. Sci. Lett.* 197, pp 105–116.

UNSWORTH, M. J., MALIN, P. E., EGBERT, G. D., and BOOKER, J. R. (1997), Internal structure of the San Andreas Fault Zone at Parkfield, California. *Geology* 25, 359–362.

YANG, T., GRAND, S. P., WILSON, D., GUZMAN-SPEZIALE, M., GOMEZ-GONZALEZ, J. M., DOMINGUEZ-REYES, T., and NI, J. (2009), Seismic structure beneath the Rivera subduction zone from finite-frequency seismic tomography. *J. Geophys. Res.* 114, B1. doi: 10.1029/2008JB005830.

(Received February 18, 2010, revised September 29, 2010, accepted September 30, 2010, Published online November 10, 2010)

Pure Appl. Geophys. 168 (2011), 1489–1499
© 2010 Springer Basel AG
DOI 10.1007/s00024-010-0238-2

Curie Point Depth Estimates and Correlation with Subduction in Mexico

MARINA MANEA[1] and VLAD C. MANEA[1]

Abstract—We investigate the regional thermal structure of the crust in Mexico using Curie Point Depth (CPD) estimates. The top and bottom of the magnetized crust were calculated using the power-density spectra of the total magnetic field from the freely available "Magnetic Anomaly Map of North America". We applied this method to estimate the regional crustal thermal structure in overlapping square windows of $2° \times 2°$. The CPD estimates range between 10 and 40 km and show several regions of relatively shallow and deep magnetic sources, with a general inverse correlation with measured heat flow. A deep CPD region (20–30 km) is located in the fore-arc area where the subducting Cocos plate has a flat-slab geometry. This deep region is bound to the NW and SE by shallow CPD areas beneath the states of Michoacan (CPD = 12–16 km) and Oaxaca (CPD = ~16 km), respectively. There is a good spatial correlation between this deep CPD area and two main fracture zones located on the incoming Cocos plate (Orozco and O'Gorman fracture zones), suggesting that subduction plays an important role in setting apart different CPD provinces along the Mexican coast. Another deep CPD (16–32 km) area corresponds to the region where the Rivera plate subducts beneath Jalisco block. The Trans-Mexican Volcanic Belt is characterized by a decrease in Curie depths from west (16–20 km) to east (10–12 km). Finally, several deep CPD areas are situated in the back-arc region where old Mesozoic terrains are present. Our results suggest that the main control on the crust's regional thermal structure in the fore-arc and volcanic arc regions is due to the subduction of the Cocos and Rivera plates beneath Mexico.

Key words: Magnetic Anomalies, Curie-point depth, flat-slab, thermal structure, Central Mexico.

1. Introduction

The temperature distribution inside the Earth is one of the most important parameters used in earth sciences. Our knowledge about the temperature variation at depth is limited because most of the direct heat flow measurements are performed in boreholes. Although these measurements are considered to be the most reliable, the available boreholes are sparse and limited in depth to only several kilometers. Also, heat flow data from shallow boreholes can be affected by groundwater circulation (ARTEMIEVA and MOONEY, 2001). However, temperature variations in the crust can be inferred indirectly from helium isotopic concentrations of local gases measured in thermal springs (POLYAK *et al.*, 1985; TARAN *et al.*, 2002), from silica concentration in thermal waters (MARVIN, 1984), or from an analysis of the power spectrum of aeromagnetic anomalies (TANAKA and ISHIKAWA, 2005). In this study we employ the spectral analysis of the aeromagnetic anomalies where a direct relationship is assumed to exist between the maximum depth of magnetized crust and Curie temperature, which is considered to be approximately 580°C for magnetite (SCHLINGER, 1985; FROST and SHIVE, 1986). Laboratory studies show that the magnetism of rocks diminishes drastically beyond the Curie temperature, and rocks can be considered non-magnetic above this temperature. Therefore, an estimate of the bottom of magnetized crust represents a direct indicator of the Curie isotherm, and variations in the thickness of the magnetized crust can be interpreted as variations in temperature. Active subduction zones influence considerably the temperature distribution in the overriding plate, especially where the slab geometry strongly varies along strike. In these areas complex thermal structures with strong lateral and depth variations are expected. The main tectonic process in Mexico is the subduction of Rivera and Cocos plates beneath the North America Plate. The slab dip angle varies greatly along the Middle America Trench (MAT) in Mexico, from a steep slab beneath the states of Jalisco and Michoacan to flat-slab subduction under

---
[1] Computational Geodynamics Laboratory, Centro de Geociencias, Campus Juriquilla-Queretaro, Universidad Nacional Autonoma de Mexico, Queretaro 76230, Mexico. E-mail: marina@geociencias.unam.mx

the states of Guerrero and Oaxaca to the south (PARDO and SUAREZ, 1995).

In recent years, several seismic studies and experiments have focused on exploring in greater detail the subduction geometry in this area (Fig. 1). The Middle America Seismic Experiment (MASE) imaged the subducted Cocos plate beneath Guerrero (CLAYTON et al., 2007). The slab geometry shows the presence of a long flat slab segment extending inland some 300 km from the trench, which then sinks into the asthenosphere at a steep angle of $\sim 65°$ (PEREZ-CAMPOS et al., 2008; PACHECO and SINGH, 2009). The "Mapping of Rivera Subduction Zone" experiment (MARS) in Western Mexico, where the Rivera plate subducts beneath the Jalisco block, revealed that the Rivera plate is subducting more steeply than does the adjacent Cocos plate (YANG et al., 2009). Subduction zone thermal models in general, and for the Mexican subduction zone in particular, show that slab geometry significantly affects the temperature distribution in the overlying plate (CURRIE et al., 2002; MANEA et al., 2004, 2005). Therefore considerable variation of the thermal structure in the fore-arc region is expected. Adequate knowledge of the thermal structure of the continental crust is essential for geodynamic studies of subduction processes for the Mexican subduction zone. Several direct measurements of continental conductive heat flow in boreholes performed by ZIAGOS et al. (1985) show significant variations within the fore-arc, volcanic arc and back-arc regions. Low heat flow values ($20–30$ mW/m^2) are registered between the coastline and the Trans-Mexican Volcanic Belt (TMVB), but they are limited to several measurements located mainly in the state of Guerrero. In this area, heat flow estimates from helium isotope concentrations show higher values of around 60 mW/m^2 (POLYAK, 1985), which approach the global average of 65 mW/m^2 (POLLACK et al., 1993). In contrast, the TMVB is characterized by elevated heat flow values of $90–100$ mW/m^2. However, these direct measurements are sparse and not distributed uniformly, and they are not sufficient to image the 3D thermal structure of the continental lithosphere, especially in Mexico, which is characterized by intense tectonic activity and a complex subduction regime. To partially overcome these limitations, PROL-LEDESMA and

JUAREZ (1986) estimated heat-flow indirectly using an empirical relationship between silica temperature and the heat flow derived by SWANBERG and MORGAN (1979). They found elevated heat flow values of $110–140$ mW/m^2 within the TMVB. However, the fore-arc area remains poorly investigated due to scarce heat-flow measurements, and the Curie-temperature isotherm can provide a proxy temperature at depth. Shallow Curie point depths (CPDs) are associated with recent magmatic activity and thinned crust, whereas deep CPDs are related with thickened, cooled or old crust. Previous CPD estimates from aeromagnetic data in the western part of the TMVB show a shallow CPD of $7–14$ km (CAMPOS-ENRIQUEZ et al., 1990). Although these estimates were in good agreement with heat flow measurements in this area, they were limited to a series of 2D profiles.

Our goal in this paper is to estimate the CPD using a spectral analysis applied to aeromagnetic anomalies, and to provide a 3D model of the Curie point isotherm for Mexico. The same method has been applied with success to several regions, such as Nevada (BLAKELY, 1988), East and Southeast Asia (TANAKA et al., 1999), Western Argentina (RUIZ and INTROCASO, 2004), Japan (TANAKA and ISHIKAWA, 2005), Turkey (DOLMAZ et al., 2005; AYDIN et al., 2005; ATES et al., 2005; BEKTAS et al., 2007), Slovakia (ROZIMANT et al., 2009) and California (ROSS et al., 2006). Although this is a convenient indirect method for inferring the thermal structure of continental crust, it has several drawbacks. First, the depth resolution is limited to the length of the aeromagnetic profile (L), the maximum CPD depth estimation is limited to $2\pi/L$ (CAMPOS-ENRIQUEZ et al., 1990). In the case of 2D magnetic maps, blocks are used instead of profiles, in which the radial power spectrum of the magnetic anomaly is estimated. BLAKELY (1995) recommends a minimum survey dimension of 160 km for deep magnetized bodies, and TANAKA and ISHIKAWA (2005) showed that a detailed Curie isotherm map for Japan islands can be obtained using square windows of $2.125° \times 2.125°$. A second limitation of the spectral method is related to the complexity of the geological structures that can cause a scattered power spectra and can lead to significant errors in CPD estimates (OKUBO and MATSUNAGA, 1994).

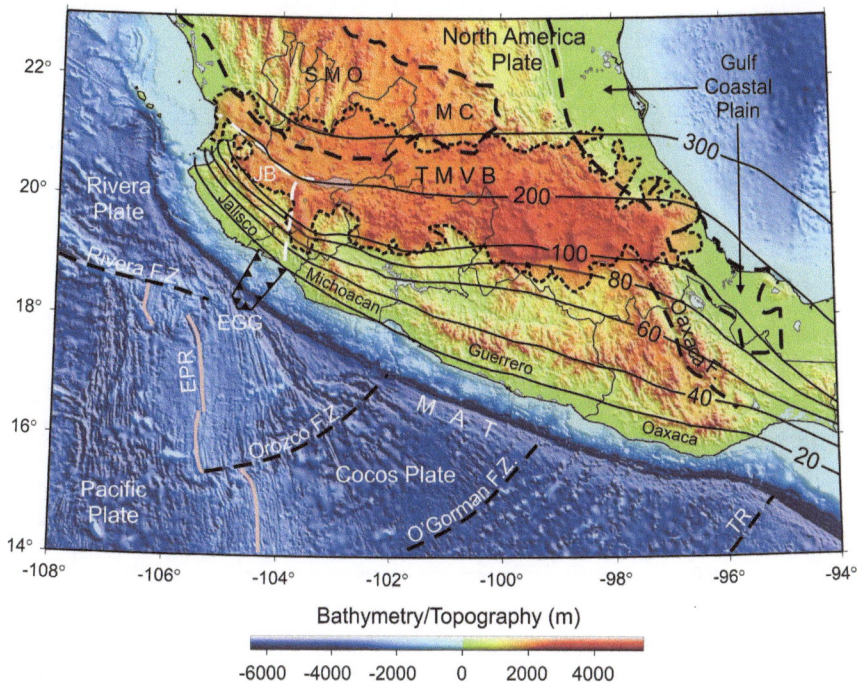

Figure 1
Simplified tectonic map of Mexico. Slab contours are from GORBATOV and FUKAO (2005). *TMVB* Trans-Mexican Volcanic Belt, *SMO* Sierra Madre Occidental, *JB* Jalisco Block, *MC* Mesa Central, *MAT* Middle America Trench, *EGG* El Gordo Graben, *TR* Tehuantepec Ridge. Bathymetric and topographic data are from SMITH and SANDWELL (1997); *EPR* East Pacific Rise, *F.Z.* fractures zones. *Continuous thin black line* coastal states (Jalisco, Michoacan, Guerrero, Oaxaca)

Our study area involves the Sierra Madre del Sur and Oriental (fore-arc), the TMVB area, the southern part of the Sierra Madre Occidental (SMO), the Jalisco Block and a small part of the Gulf Coastal Plain (Fig. 1). In these regions strong temperature contrasts are expected, controlled in part by subduction processes and in part by tectonic and lithologic limits. Using the basic relation for conductive heat transport $Q = k*\text{grad}T$, (TURCOTTE and SCHUBERT, 2002), estimations of heat-flow in the region are performed based on CPD estimates, the Curie point for magnetite of 580°C (HAGGERTY, 1978) and using an average thermal conductivity for igneous rocks of $k = 2.5$ W/m°C (STACEY, 1977).

2. Magnetic Anomalies and Spectral Analysis

The present work is based on the digital magnetic anomaly map for the North American continent, as the result of a combined effort by the Geological Survey of Canada, U.S. Geological Survey and Consejo de Recursos Minerales of Mexico (NORTH AMERICAN MAGNETIC ANOMALY GROUP 2002). The total field anomalies were calculated from measurements by subtracting the Definitive Geomagnetic Reference Field (DGRF). The study area extends from 16° to 23° N and from −95° to −105°30′ W. The aeromagnetic data grid was divided into 230 overlapping blocks, of size 2° × 2°. Each block overlaps the adjacent blocks by 30′ in all directions (Fig. 2). In each block the total field anomaly data were reduced to magnetic pole and tapered at the boundaries in order to reduce the edge effect common in FFT techniques. Using large 2° × 2° blocks, the obtained CPD is an average for the area, and it cannot resolve local thermal anomalies.

In this study we follow a technique for estimating the basal depth of magnetic sources similar to that described by TANAKA *et al.* (1999) which is based on BHATTACHARYYA and LEU (1975) and OKUBO *et al.* (1985). We calculate the radial average of the power density spectra of a magnetic anomaly in each block, and then Z_0, Z_t and Z_b, which represent the depth to

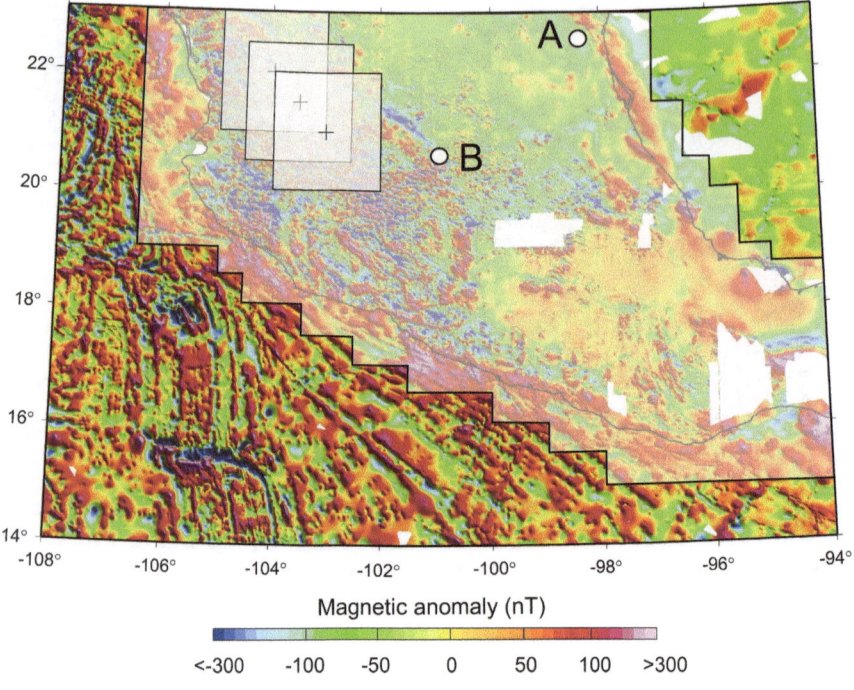

Figure 2

Magnetic anomaly map (*North American Magnetic Anomaly Group* 2002) and the study area (*transparent region*). *White areas* have no data. Each *square* ($2° \times 2°$) represents the area used to calculate Z_t and Z_0. All squares overlap $30'$. *Point A* and *B* are locations where two examples of spectra are shown (Fig. 3)

the centroid, and the top and bottom of the magnetic layer, respectively. The basal depth of the magnetic source (CPD) is computed as $Z_b = 2Z_0 - Z_t$. In Fig. 3 we show two contrasting examples of power spectrum for a shallow and deep magnetic source from different tectonic regions of Mexico (Fig. 2). Since our blocks are 210×210 km^2, the wavenumber range for the calculation of Z_b is 0.05–0.2 (rad/km). As for the Z_t estimation, we use a wavenumber range of 0.6–0.9 rad/km, similar that used in a recent study of AYDIN and OKSUM (2010). The error of the estimate is calculated for each block from the standard deviation of the power spectrum at low wave numbers (0.05–0.2 rad/km) and the straight line that constitutes the least squares fit to the spectrum in this range (OKUBO and MATSUNAGA, 1994).

3. Curie Depth Estimates

Figures 4a and b show the CPD variation and their associated errors where the uncertainties in CPD estimates are ±(2.3–3.3) km. The top bound (Z_t) ranges between 1 and 3 km, and the centroid (Z_0) varies between 5 and 20 km, thus the basal depth of the magnetic source (CPD) fluctuates from 9 (±3.3) km to 37 (±2.3) km.

In Mexico there is a strong superposition of different tectonic regimes due to complex geodynamic processes that have taken place (MORAN-ZENTENO, 1986), and the origins of geological structures are still an ongoing debate. The TMVB is characterized by Pliocene to recent andesitic to basaltic rocks, and is widely accepted to be the product of the subduction of the Cocos plate. Its unusual trend, non-parallel with the MAT, reflects the slab geometry variation along strike (Fig. 1). The SMO is composed mainly of Oligocene and Miocene volcanic rocks, which are the remnants of the earlier subduction of the Farallon plate. The Jalisco Block, located in Western Mexico, has a complex structure affected by the subduction of the Rivera plate, incipient rifting, and by recent volcanism. The principal rock units of the Jalisco Block are represented by granites, diorites, volcanic, and

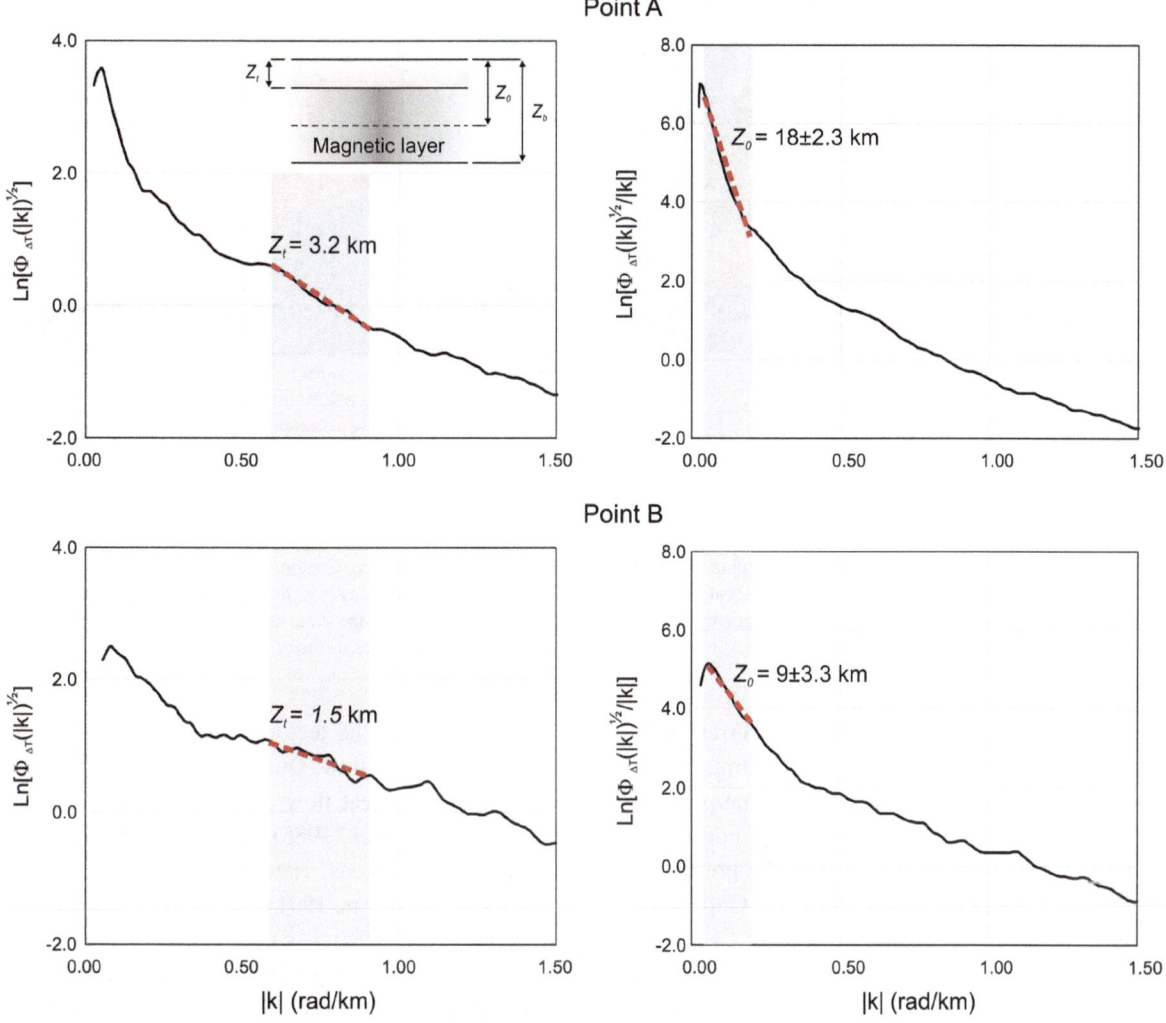

Figure 3

Radial power spectrum for two *sample points A* and *B* presented in Fig. 2. *Top* deep magnetic layer ($Z_t = 3.2$ km, $Z_0 = 18 \pm 2.3$ km; $Z_b = 32.8 \pm 2.3$ km). *Bottom* shallow magnetic layer ($Z_t = 1.5$ km, $Z_0 = 9\pm3.3$ km; $Z_b = 16.5 \pm 3.3$ km)

sedimentary rocks (FERRARI *et al.*, 2000). The oldest rocks in our study area are located in the Mesa Central and Gulf Coastal Plain (Fig. 1). They are mainly schists and gneisses of Precambrian age, and the exposed rocks are Mesozoic and Cenozoic sediments.

Shallow CPD regions correspond in general to high heat flow values (Fig. 5). In the fore-arc, two shallow CPD areas are identified, both adjacent to the flat-slab sector. To the southeast, beneath the Oaxaca state, we identified a relatively shallow CPD of 16 ± 3.3 km. Beneath the Michoacan state, our CPD estimates also show a shallow magnetic source at a

depth of 12–14 km. Shallow CPDs can be identified in several areas, such as the TMVB and the SMO. The TMVB is characterized in general by shallow CPDs, but have a consistent trend, which is shallower from the west to east, consistent with previous studies of CPDs for the TMVB (CAMPOS-ENRIQUEZ *et al.*, 1990). The western side of the TMVB is characterized by a relatively deep magnetic source of 16–20 km, and toward the east, CPD decreases to ~12 km. However, magnetic data over the eastern half of the TMVB is discontinuous and a large data gap (see Fig. 2) affects our estimates, hence CPDs in this area should be considered with caution.

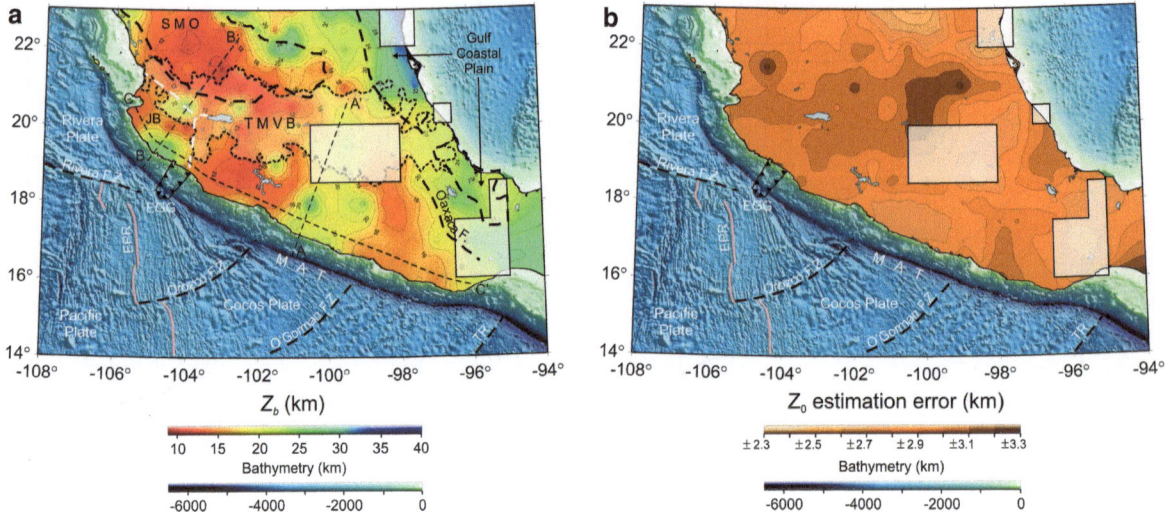

Figure 4

a The CPD map of the study area and the main morphotectonic provinces and structures in Mexico. Contours are drawn every 2 km intervals. **b** The Z_0 estimation error map of the study area. Contours are drawn every 1 km. *Transparent rectangles* represent areas where our CPD estimates are poorly constrained due to lack of magnetic measurements. Other abbreviations are the same as in Fig. 1. Location of the cross-sections in Fig. 6 (A–A′ and B–B′) and Fig. 7 (C–C′) are shown as *dashed lines*

Generally, deep CPDs (20–40 km) correspond to old terrains, as in the case of Mesa Central, tectonic blocks (i.e., the Jalisco Block), or to some particular subduction style (i.e. flat-slab). To the northeast, a deep CPD region is consistent with the presence of old Mesozoic rocks of Mesa Central and Gulf Coastal Plain. In this area we obtain the deepest CPDs, with values down to ~37 ± 2 km. The deep CPD zone located above the flat-slab is characterized by CPD in the range of 16–24 km. Also, a couple of deep CPD regions are located to the northwest, approximately matching the Jalisco block.

4. Heat Flow Estimates from CPD

Curie isotherms for magnetite are in the range of 600 ± 50°C; a common value used is 580°C (HAGGERTY, 1978). BEARDSMORE and CULL (2001) mention that different rock types have different Curie temperatures, so the Curie depth may also correspond to a compositional boundary and not necessarily represent an isotherm. Using an average thermal conductivity for igneous rocks of 2.5 W/m°C and a Curie temperature of 580°C, we obtain the heat-flow map from CPD (Fig. 5a). We compared CPD

estimations with the tectonic settings and measured or estimated heat flow. Our estimates are, in general, consistent with heat flow data from direct measurements (ZIAGOS et al., 1985), indirect estimations from silica concentrations (PROL-LEDESMA AND JUAREZ, 1986; PROL-LEDESMA, 1991) and from helium isotopes (POLYAK et al., 1985; TARAN et al., 2002).

The relatively deep CPD zone located above the flat-slab is characterized by low heat flow measurements of 16–37 mW/m² (ZIAGOS et al., 1985). Using the Curie temperature for magnetite of 580°C and an average rock conductivity for igneous rocks of 2.5 W/m°C (STACEY, 1977), we estimate heat flow values above the flat-slab region in the range of 60–80 mW/m² (Fig. 5a). These estimates are higher than the heat flow measurements performed in boreholes (13–40 mW/m² from ZIAGOS et al., 1985), but they are closer to estimates from helium isotopes measured in coastal springs (65–66 mW/m² from POLYAK et al., 1985). Farther east the CPD increases from 16 km to 28–30 km, and this transition zone fits well with the surface location of Oaxaca and Caltepec deep crustal faults (ELÍAS-HERRERA and ORTEGA-GUTIÉRREZ 2002) (Fig. 4).

Our spectral analysis revealed another deep (16–20 km) CPD area located above the subducting

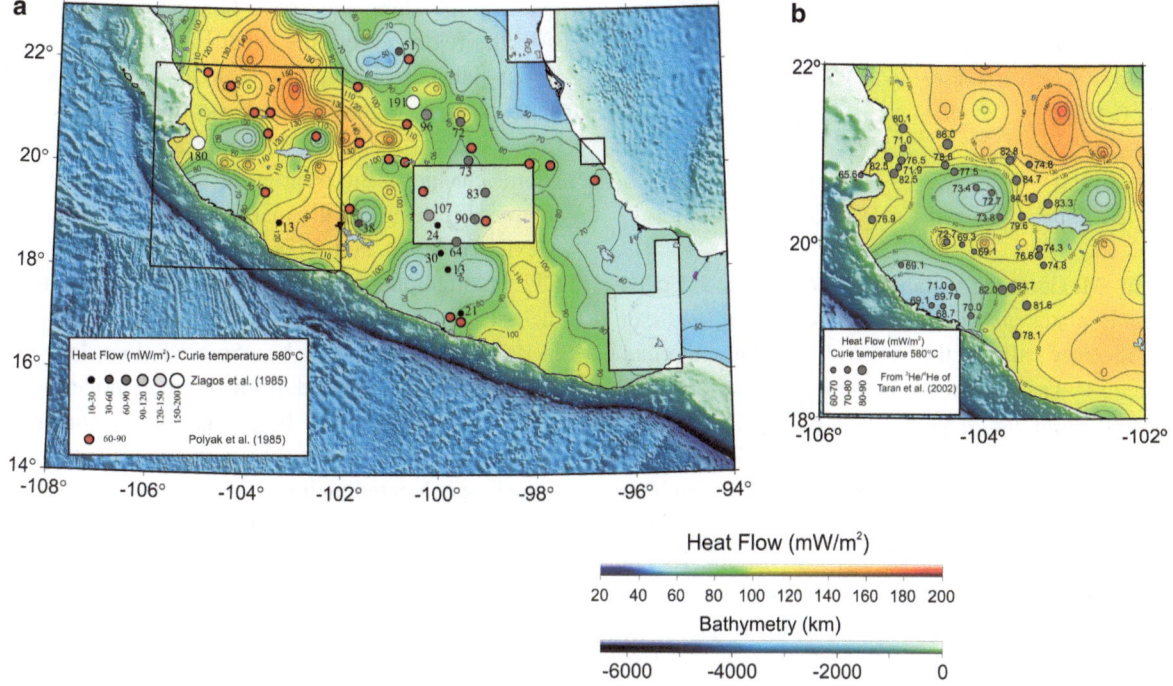

Figure 5

a Heat flow distribution from CPD estimates using a Curie temperatures of 580°C. *Gray circles* are heat flow measurements from ZIAGOS *et al.* (1985), and *red circles* represent heat flow values determined using isotope ratios (POLYAK *et al.*, 1985). **b** Heat flow map from CPD and estimation from [3]He/[4]He measurement performed by TARAN *et al.* (2002) for the Jalisco Block

Rivera plate in the Jalisco Block. Here, the borehole heat flow measurements are sparse and show quite contrasting values from 10 mW/m[2] to more than 130 mW/m[2] (ZIAGOS *et al.*, 1985), probably reflecting local effects and the effect of shallow ground water circulation. However, TARAN *et al.* (2002) provide a large set of helium isotopic composition measurements in thermal springs over the Jalisco Block. POLYAK and TOLSTIKHIN (1985) provide the empirical equation for estimating the heat flow from helium measurements, $Q = 16 \; \log(R/Ra) + 72 \; (mW/m^2)$, where R is the measured [3]He/[4]He and $Ra = 1.4 \times 10^{-6}$ is the air ratio (see also BALLENTINE and BURNARD, 2002). Using the above relation and the [3]He/[4]He ratios measured by TARAN *et al.* (2002) for the Jalisco Block, we calculated the corresponding heat flow. In Fig. 5b we present the comparison between heat flow estimates from CPD and from [3]He/[4]He. Along the coast we predict a low heat flow in the range of 55–70 mW/m[2], in remarkably good agreement with heat flow estimates from helium isotopes (60–79 mW/m[2]). The same good fit can be

observed for the second low heat flow area located farther inland (60–80 mW/m[2] from CPD compared with ~70 mW/m[2] from helium isotopes). High heat flow areas generally agree with high heat flow estimates from helium isotopes, although our predictions are higher in some regions with up to 30 mW/m[2].

For the TMVB, POLYAK *et al.* (1985) reports heat flow values estimated from helium isotopes in the range of 74–84 mW/m[2], and ZIAGOS *et al.* (1985) recorded high heat flow values (100–180 mW/m[2]) in several boreholes (Fig. 5). Our heat flow map for the TMVB shows a west-east trend with heat flow values from 60 to 150 mW/m[2] for a Curie temperature of 580°C (Fig. 5).

The SMO is characterized mainly by high heat flow estimates from silica temperature (100–116 mW/m[2]), in good agreement with our estimates from magnetic anomalies (120–150 mW/m[2]; Fig. 5).

To the northeast, a deep CPD region (20–30 km) is consistent with the presence of old Mesozoic rocks of the Mesa Central and the Gulf Costal Plain, where a relatively low heat flow (51 mW/m[2]) was recorded by ZIAGOS *et al.* (1985). In this area we also obtained low heat flow estimates from CPD in the range of 30–40 mW/m² when using a Curie temperature of 300°C (Fig. 5).

5. Subduction and Curie Depth in Mexico

In Mexico, the subducting Rivera and Cocos oceanic plates are young and therefore hot, and can significantly affect the overriding plate thermal structure. The subduction regime and style varies strongly along the MAT in Mexico (Fig. 1). The slab geometry changes from steep beneath Jalisco to flat beneath Guerrero and steep again beneath southern Mexico. The convergence rate increases steadily from NW to SE, from 3.8 cm/year for the Rivera plate to 5.8–6.1 cm/year for the Cocos plate in front of Guerrero and Oaxaca states. Also the plate age along the MAT varies from 11 Ma to the NW to 16 Ma to the SE. All this was taken into account in the thermal models which we present in Fig. 6 for two contrasting subduction geometries in Mexico, namely steep and flat slabs (CURRIE *et al.*, 2002; MANEA and MANEA, 2010).

A flat-slab creates a colder and wider fore-arc structure whereas for a steep slab the effect on the overriding plate thermal structure is more limited to the deformation front. Deep CPDs are observed in

two regions along the coast. One area corresponds with the region where the subducting Cocos plate is in flat-slab regime, and is interpreted as an effect of plate cooling. Figure 6a, b shows that lateral variations of the Curie depth does not agree well with a lateral change in the position of the 580°C isotherm inferred from recent thermal models for flat-slab in Mexico (MANEA and MANEA, 2010). Although our CPD estimations roughly reflect large variations in the crustal temperature distribution, major differences in depth estimation can be observed (Fig. 4a). The 580°C isotherm inferred from thermal modeling is located some 20–25 km below the CPD estimate in the fore-arc area. The 580°C isotherm from thermal modeling is located mainly in the oceanic lithosphere below the oceanic crust, which needs to be highly serpentinized in order to contribute to the long wavelength of the observed magnetic signal. If this is not the case, then the depth boundary for the magnetized layer is limited to the subducted oceanic crust. Indeed, we obtain a good fit, but only beneath the coast where the CPD intersects the subducting slab at ~20 km depth. Then, the misfit increases to some 15–20 km difference in depth estimates between CPD and slab surface in the fore-arc region. On the other hand, an apparently quite good fit can be observed for the volcanic area, although in this region the CPD is poorly constrained due to data gaps in the aeromagnetic anomaly. Figure 4a also shows that the

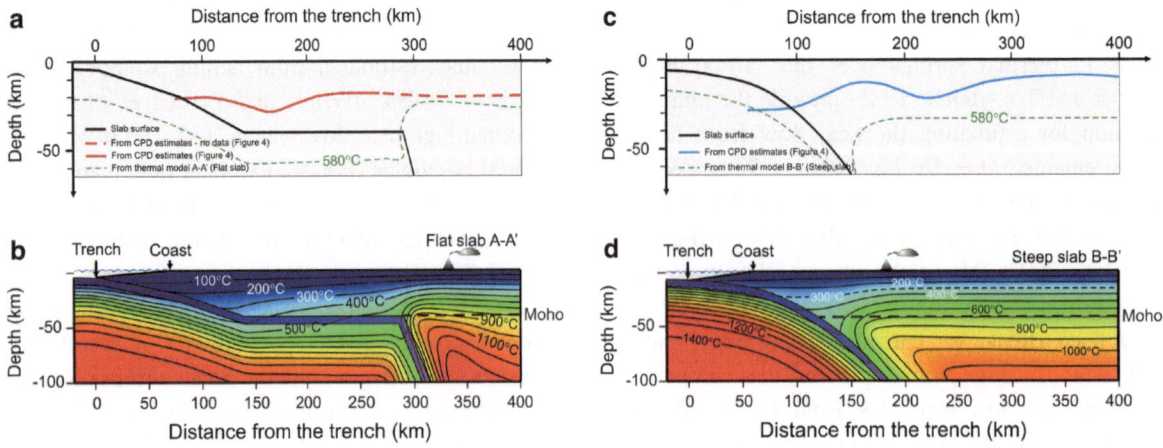

Figure 6
Thermal models and CPD variation along two profiles shown in Fig. 4. **a, b** Curie depth, the 580°C isotherm from thermal modeling of flat-slab (MANEA and MANEA, 2010), and the Cocos slab geometry. **c, d** Curie depth, the 580°C isotherm from thermal modeling (CURRIE *et al.*, 2002), and the geometry of the subducting Rivera plate

deep CPD area extends inland some 280–290 km from the trench axis, corresponding with the flat-slab segment extension from recent seismic experiments (PEREZ-CAMPOS et al., 2008). In addition, two major fracture zones on the incoming Cocos plate, the Orozco and O'Gorman fracture zones, laterally bound this deep CPD area (Fig. 4a), and can be interpreted as the lateral limits of the flat slab.

A second deep CPD area is located farther NW in the area where the Rivera plate currently subducts at a steep angle (YANG et al., 2009). In this area, Rivera plate dips into the asthenosphere at a steep angle, and this style of subduction strongly affects the regional thermal structure in the overlying plate. The isotherms tend to follow the slab trajectory and therefore the 580°C Curie isotherm is pushed down to greater depths (CURRIE et al., 2002). Thus, we consider that the steep subducting slab influences the CPD distribution in this area. However, a good fit between the 580°C isotherm and our CPD estimates is limited only the coastal area, where the oceanic crust actually intersects the CPD. Farther inland, the misfit varies between 15 and 25 km. The isoviscous mantle wedge from thermal model of CURRIE et al. (2002) could be partially responsible for this large dissimilarity. A different rheology applied to the mantle wedge (i.e. temperature-dependent) would have produced a more vigorous mantle wedge flow, a hotter lithosphere and a shallower CPD beneath the volcanic arc. A comparison between CPD estimates and thermal models shows significant differences, especially for the fore-arc area, where the 580°C isotherm is predicted to be as deep as 50 km or more. Such large differences are also observed in other subduction zones, such as Japan for example. Here, CPD inferred from aero-magnetic anomaly varies from 18 to 28 km in the fore-arc area (TANAKA and ISHIKAWA, 2005), yet thermal models for SE and NE Japan show the 580°C isotherm as deep as 50 km or more (PEACOCK and WANG, 1999). We conclude that the spectral method used cannot solve very deep magnetized bodies with sufficient precision, and probably much larger windows are necessary (MAUS et al., 1997). However, very large windows cannot provide sufficient resolution to reveal significant lateral changes in temperature that represent a key characteristic for subduction zones.

6. Conclusions

In this study we have applied spectral techniques to aeromagnetic anomalies with the aim of estimating depths to the Curie isotherm beneath Mexico, where only a few heat flow measurements are available. Then, CPD estimates are compared with the tectonic regime and main geological boundaries. Our estimates show that CPDs vary greatly according to the geological and morphotectonical context (Fig. 4). We found a correlation between the subduction regime and CPD distribution in the fore-arc region. The CPDs distribution over the flat-slab area is characterized by a deep Curie isotherm (16–24 km) that extends farther inland, some 280–290 km from the trench axis, consistent with the slab geometry inferred from recent seismic experiments. Also, the CPD estimates provide lateral constraints for the flat-slab extension along the Mexican coast. Our study revealed that the flat slab area is centered on the Guerrero seismic gap and extends ~230 km along the coast (Fig. 7). The lateral extent of the flat slab can be correlated with the Orozco and O'Gorman fracture zones located on the incoming Cocos plate. Farther northwest, the steep dipping Rivera slab tends to push Curie isotherms to greater depths, as revealed by our CPD estimations beneath Jalisco (20–32 km). However, the inland extension of the CPD here is confined only to some 100 km from the trench. The two deep CPD fore-arc regions are bounded by relatively shallow magnetic layers, with CPDs of 12–16 km beneath Michoacan and CPDs of ~16 km beneath Oaxaca (Fig. 4). Farther southeast, the Curie isotherm deepens abruptly to ~24 km, and this boundary coincides well with the Oaxaca fault.

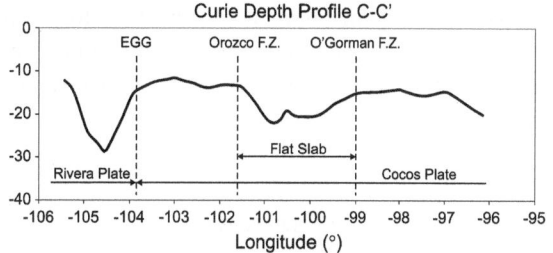

Figure 7

Along coast cross-section showing the variation of CPD. Note how principal bathymetric structures on the Cocos and Rivera plates correlate well with the main CPD changes

Shallow CPDs are generally characteristic for volcanic arcs. CPDs over the TMVB are deeper for the western side (16–20 km) and decreases to ~ 16 km for the eastern side. Also, shallow CPDs (~ 10 km) are found beneath the SMO. On the other hand, deeper CPDs (20–40 km) correspond to the Mesozoic and Gulf Coastal Plain terrains, and to the Jalisco block.

A comparison with the observed heat flow data revealed that there is a relatively good fit with the observations, especially with heat flow estimates from helium isotopes. In general, in the fore-arc region and old terrains we found a better fit between heat flow from CPD and from ^{3}He/^{4}He ratios. On the other hand, for the TMVB and SMO, our estimates are higher when compared with heat flow from ^{3}He/^{4}He ratios. This result reflects that there could be a large number of heat sources distributed in the crust, which produced the same surface heat flow but different lower crustal temperatures.

The estimated Curie depths in Mexico agree fairly well with the locations of main tectonic provinces and limits in Mexico, and also with independent estimates of heat flow, providing confidence that the spectral method used in this study is sampling the Curie-temperature isotherm and assessing average thermal conditions for the continental crust. The lateral CPD variation in the fore-arc region revealed for the first time the lateral extent along the coast (~ 230 km) of the flat slab section of the subducting Cocos plate.

Acknowledgments

The authors would like to thank Yuri Taran, A. Buyuksarac and an anonymous reviewer for their critical comments and suggestions. This study was funded by PAPIIT IN110709, PAPIIT IN115810 and CONACyT 84035.

REFERENCES

ARTEMIEVA, I.M., and MOONEY, W.D. (2001), *Thermal thickness and evolution of Precambrian lithosphere: a global study.* J. Geophys. Res. *106*(B8), 16,387–16,414, doi:10.1029/2000 JB900439.

ATES, A., BILIM, F., and BUYUKSARAC, A. (2005), *Curie-point depth investigation of Central Anatolia, Turkey.* Pure. Appl. Geophys. *162*, 357–371.

AYDIN, I., KARAT, H.I., and KOÇAK, A. (2005), *Curie-point depths map of Turkey.* Geophys. J. Int. *162*, 633–640.

AYDIN, I., and OKSUM, E. (2010), *Exponential approach to estimate the Curie-temperature depth.* J. Geophys. Eng. *7*, 113–125.

BALLENTINE, C.J., and BURNARD, P.G. (2002), *Production, release and transport of noble gases in the continental crust.* Reviews in Mineralogy and Geochemistry *47*(1), 481–538.

BEARDSMORE, G.R., and CULL, J.P. (2001), *Crustal Heat Flow. A guide to measurement and modeling.* Cambridge University Press. 324 pp.

BEKTAS, O., RAVAT, D., BUYUKSARAC, A., BILIM, F., and ATES, A. (2007), *Regional geothermal characterization of East Anatolia from aeromagnetic, heat flow and gravity data*, Pure Appl. Geophys. *164*, 975–998.

BHATTACHARYYA, B.K., and LEU, L.K. (1975), *Analysis of magnetic anomalies over Yellowstone National Park: mapping the Curie point depth isothermal surface for geothermal reconnaissance.* J. Geophys. Res. v. *80*, 4461–4465.

BLAKELY, R.J. (1988), *Curie temperature isotherm analysis and tectonic implications of aeromagnetic data from Nevada.* J. Geophys. Res. vol. *93*, no. B10, pp. 11,817–11,832.

BLAKELY, R. J. (1995), *Potential theory in gravity and magnetic applications.*, by BLAKELY, R. J. Cambridge University Press, Cambridge (UK), 1995, XIX + 441 p., ISBN 0-521-41508-X.

CAMPOS-ENRIQUEZ, J.O., ARROYO-ESQUIVEL, M.A., and URRUTIA-FUCUGAUCHI, J. (1990), *Basement, Curie isotherm and shallow-crustal structure of the Trans-Mexican Volcanic Belt, from aeromagnetic data.* Tectonophys. *172*, pp. 77–90.

CLAYTON, R.W., DAVIS, P.M., and PEREZ-CAMPOS, X. (2007), *Seismic structure of the subducted Cocos plate*, Eos Trans. AGU *88*(23), Jt. Assem. Suppl., Abstract T32A-01.

CURRIE, C.A., HYNDMAN, R.D., WANG, K., and KOSTOGLODOV, V. (2002), *Thermal models of the Mexico subduction zone: Implications for the megathrust seismogenic zone*, J. Geophys. Res. *107*(B12), 2370, doi:10.1029/2001JB000886.

DOLMAZ, M.N., USTAÖMER, T., HISARLI, Z.M., and ORBAY, N. (2005), *Curie point depth variations to infer thermal structure of the crust at the African-Eurasian convergence zone, SW Turkey.* Earth Planet Space *57*, 373–383.

ELÍAS-HERRERA, M., and F. ORTEGA-GUTIÉRREZ (2002), *Caltepec fault zone: An Early Permian dextral transpressional boundary between the Proterozoic Oaxacan and Paleozoic Acatlán complexes, southern Mexico, and regional tectonic implications*, Tectonics, *21*(3), 1013, doi:10.1029/2000TC001278.

FERRARI, L., PASQUARÉ, G., VENEGAS, S., and ROMERO, F. (2000), *Geology of the western Mexican Volcanic Belt and adjacent Sierra Madre Occidental and Jalisco block.* Geological Society of America Special Paper 334, chapter 04, p. 65–84.

FROST, B.R., and SHIVE, P.N. (1986), *Magnetic mineralogy of the lower continental crus.* J. Geophys. Res. *91*, 6513–6521.

GORBATOV, A., and FUKAO, Y. (2005), *Tomographic search for missing link between the ancient Farallon subduction and the present Cocos subduction.* Geophys. J. Int. *160*, 849–854.

HAGGERTY, S.E. (1978), *Mineralogical constraints on Curie isotherm in deep crustal magnetic anomalies.* Geophys. Res. Lett. *5*(2), pp. 105–109.

MANEA, V.C., MANEA, M., KOSTOGLODOV, V., CURRIE, C.A., and SEWELL, G. (2004), *Thermal structure, coupling and metamorphism* in the Mexican subduction zone beneath Guerrero, Geophys. J. Int., 158, 775–784, doi:10.1111/j.1365-246X.2004.02325.x.

MANEA, V.C., MANEA, M., KOSTOGLODOV, V., and SEWELL, G. (2005), Thermo-mechanical model of the mantle wedge in Central Mexican subduction zone and a blob tracing approach for the magma transport: Physics of the Earth and Planetary Interiors v. 149, p. 165–186, doi:10.1016/j.pepi.2004.08.024.

MANEA, V.C., and MANEA, M. (2010), Flat-slab thermal structure and evolution beneath central Mexico. Pure App. Geophys. This volume.

MARVIN, P.R. (1984), Regional heat flow based on the silica content of ground waters from northcentral Mexico, M.Sc. Thesis, New Mex. St. Univ., p. 107.

MAUS, S., GORDON, D., and FAIRHEAD, J.D. (1997), Curie-temperature Depth Estimation Using a Self-similar Magnetization Model, Geophys. J. Int. 129, 163–168.

MORAN-ZENTENO, D.J. (1986), Breve revision sobre la evolucion tectonica de Mexico. Geofisica Internacional, 25:9–38.

NORTH AMERICAN MAGNETIC ANOMALY GROUP (2002), Magnetic anomaly map of North America: U.S. Geological Survey Special Map, scale 1:10,000,000.

OKUBO, Y., GRAF, R.J., HANSEN, R.O., OGOWA, K., and TSU, H. (1985), Curie point depth of the island of Kyushu and surrounding areas, Japan. Geophysics, 50, 481–494.

OKUBO, Y., and MATSUNAGA, T. (1994), Curie point depth in northeast Japan and its correlation with regional thermal structure and seismicity. J. Geophys. Res. 99(B11), 22,363–22,371, doi:10.1029/94JB01336.

PACHECO, J.F., and SINGH, S.K. (2009), Seismicity and state of stress in Guerrero Segment of the Mexican subduction zone. Journal of Geophysical Research vol. 115, B01303, doi:10.1029/2009JB 006453.

PARDO, M., and SUAREZ, G. (1995), Shape of the subducted Rivera and Cocos plates in southern Mexico: Seismic and tectonic implications, J. Geophys. Res. 100, 12,357–12,373.

PEACOCK, S.M., and WANG, K. (1999), Seismic consequences of warm versus cool subduction metamorphism: Examples from southwest and northeast Japan, Science, 286, 937–939.

PEREZ-CAMPOS, X., KIM, Y.H., HUSKER, A., DAVIS, P.M., CLAYTON, R.W., IGLESIAS, A., PACHECO, J.F., SINGH, S.K., MANEA, V.C., and GURNIS, M. (2008), Horizontal subduction and truncation of the Cocos plate beneath central Mexico, Geophys. Res. Lett. 35, L18303, doi:10.1029/2008GL035127.

POLLACK, H. N., S. J. HURTER, and J. R. JOHNSON (1993), Heat flow from the Earth's interior: Analysis of the global data set, Rev. Geophys., 31(3), 267–280, doi:10.1029/93RG01249.

POLYAK, B.G., PRASOLOV, E.M., KONONOV, V.I., VERKHOVSKY, A.B., GONZALEZ, A., TEMPLOS, L.A., ESPINDOLA, J.M., ARELLIANO, J.M., and MANON, A. (1985), Isotopic composition and concentration of inert gases in Mexican hydrothermal systems. Geofisica Internacional 21, 193–227.

POLYAK, B.G., and TOLSTIKHIN, I.N. (1985), Isotopic composition of the earth's helium and the problem of the motive forces of tectogenesis. Chemical Geology: Isotope Geosciences section 52(1), 9–33.

PROL-LEDESMA, R.M., and JUAREZ, G. (1986), Geothermal map of Mexico. Journal of Volcanology and Geothermal Research, 28, 351–362.

PROL-LEDESMA, R.M. (1991), Terrestrial heat flow in Mexico. In Exploration of the deep continental crust. Terrestrial heat flow and the lithosphere structure, eds. CERMAK, V. and RYBACH L., Springer-Verlag. pp. 475–485.

ROSS, H.E., BLAKELY, R.J., and ZOBACK, M.D. (2006), Testing the use of aeromagnetic data for the determination of Curie depth in California. Geophysics, vol. 71, no. 5, L51–L59, doi:10.1190/1.2335572.

ROZIMANT, K., BUYUKSARAC, A., BEKTAS, O. (2009), Interpretation of Magnetic Anomalies and estimation of depth of magnetic crust in Slovakia. Pure. Appl. Geophys. 166, pp. 471–484.

RUIZ, F., and INTROCASO, A. (2004), Curie Point Depths beneath Precordillera Cuyana and Sierras Pampeanas obtained from spectral analysis of magnetic anomalies. Gondwana Research v.7, no. 4, pp. 1133–1142.

SCHLINGER, C.M. (1985), Magnetization of lower crust and interpretation of regional crust anomalies: example from Lofoten and Vesteralen, Norway. J. Geophys. Res. 90, 11,848–11,504.

SMITH, W.H.F., and SANDWELL, D.T. (1997), Global seafloor topography from satellite altimetry and ship depth soundings, Science v. 277, p. 1957–1962.

STACEY, F.D. (1977), Physics of the Earth 2nd edition, Wiley, New York, 414 pp.

SWANBERG, Ch.A., and MORGAN, P. (1979), The linear relation between temperatures based on the silica content of groundwater and regional heat flow: a new heat flow map of the United States. Pure. Appl. Geophys. 117, 227–241.

TANAKA, A., and ISHIKAWA, Y. (2005), Crustal thermal regime inferred from magnetic anomaly data and its relationship to seismogenic layer thickness: the Japanese islands case study. Physics of the Earth and Planetary Interiors 152, 257–266.

TANAKA, A., OKUBO, Y., and MATSUBAYASHI, O. (1999), Curie point depth based on spectrum analysis of the magnetic anomaly data in East and Southeast Asia. Tectonophysics 306, 461–470.

TARAN, Y., S. INGUAGGIATO, N. VARLEY, G. CAPASSO, and R. FAVARA (2002), Helium and carbon isotopes in thermal waters of the Jalisco block, Mexico, Geofis. Int., 41, 459–466.

TURCOTTE, D., SCHUBERT, G. (2002), Geodynamics 2nd edition, Cambridge University Press, New York.

YANG, T., GRAND, S.P., WILSON, D., GUZMAN-SPEZIALE, M., GOMEZ-GONZALEZ, J.M.T., DOMINGUEZ-REYES, and NI, J. (2009), Seismic structure beneath the Rivera subduction zone from finite-frequency seismic tomography J. Geophys. Res. 114, B1, doi: 10.1029/2008JB005830.

ZIAGOS, J.P., BALCKWELL, D.D., MOOSER, F. (1985), Heat flow in Southern Mexico and the thermal effects of subduction. J. Geophys. Res. 90, B7, pp. 5410–5420.

(Received March 7, 2010, revised October 27, 2010, accepted October 28, 2010, Published online December 14, 2010)

Pure Appl. Geophys. 168 (2011), 1501–1525
© 2010 Springer Basel AG
DOI 10.1007/s00024-010-0173-2

Evaluation of Recent Tectonomagmatic Discrimination Diagrams and their Application to the Origin of Basic Magmas in Southern Mexico and Central America

SURENDRA P. VERMA,[1] SANJEET K. VERMA,[2] KAILASA PANDARINATH,[1] and MARÍA ABDELALY RIVERA-GÓMEZ[2]

Abstract—Discrimination diagrams to decipher tectonic settings have been in use for nearly 40 years. Although old diagrams have been extensively used, the recent ones based on discriminant functions of ratio variables, with or without log-transformation, proposed during 2004–2010 for the discrimination of four tectonic settings of island arc, continental rift, ocean-island and mid-ocean ridge, were newly evaluated to show their high success rates of 57.3–100% and 58.5–100% for major-element and immobile-element based diagrams, respectively. For the continental arc of the Andes evaluated for its similarity to island arc, these four sets of diagrams showed success rates of 62.1–83.8%. These four sets of five diagrams per set were therefore used to infer tectonic setting of the Mexican Volcanic Belt (MVB), Los Tuxtlas volcanic field (LTVF), and Central American Volcanic Arc (CAVA). Using this approach, the MVB, especially its western, central and eastern parts, and the LTVF of Southern Mexico show a dominantly continental rift setting and the CAVA shows an arc setting. The west-central part of the MVB is consistent with dual tectonics of arc and rift. These results confirm the application of an unusual mantle upwelling rift-model for the Mexican on-land volcanism, whereas the conventional plate tectonic subduction model seems to be applicable for the CAVA from Guatemala to north-western Costa Rica.

Key words: Mexico, Mexican Volcanic Belt, Los Tuxtlas volcanic field, subduction, rifting, geochemistry, tectonic setting.

1. Introduction

In plate tectonics theory, it is common to consider four main tectonic settings which are related to subduction processes (island arc and continental arc), extensional processes in continents (continental rift), special melting regimes in oceans (ocean-island), and extensional processes in oceans (mid-ocean ridge). (The number of tectonic settings can be higher, for example five if the subduction setting is divided into island and continental types, or more if other subdivisions are considered.) For nearly 40 years, discrimination diagrams have constituted a widely used complementary technique to other petrological and geochemical methods for interpreting compositional data and for inferring tectonic settings of older terrenes as well as of areas with complex (or multiple) tectonic settings (e.g., PEARCE and CANN, 1971, 1973; ROLLINSON, 1993; AGRAWAL and VERMA, 2007; VERMA, 2010). Recently, VERMEESCH (2007), SHETH (2008), and VERMA (2010) have evaluated most existing discrimination diagrams.

The older bivariate and ternary discrimination diagrams are all plagued by erroneous treatment of compositional data (for correct statistical procedures, see AITCHISON, 1986; AITCHISON *et al.*, 2000; EGOZCUE *et al.* 2003; AITCHISON and EGOZCUE, 2005; BAXTER *et al.*, 2005; EGOZCUE and PAWLOWSKY-GLAHN, 2005; BUCCIANTI *et al.*, 2006; AGRAWAL and VERMA, 2007; VERMA, 2010). Some of them, such as the Zr–Zr/Y bivariate diagram of PEARCE and Norry (1979) and the MgO–FeO^t–Al$_2$O$_3$ ternary diagram of PEARCE *et al.* (1977), show very low success rates. The "success rate" of a given diagram for a tectonic setting is a statistical parameter (see VERMA, 2010) and is defined as the ratio, expressed as a percent, of the number of samples with correct tectonic discrimination and the total number of samples of that particular tectonic setting. For the evaluation of classification diagrams, the success rate is similarly defined as the correct

---

S. P. Verma was on sabbatical leave (April 2009–March 2010) at the División de Ciencias Básicas e Ingeniería, Universidad Autónoma Metropolitana-Iztapalapa, occupying the Chair (Cátedra) "Ronald Tunstall Ackroyd".

[1] Departamento de Sistemas Energéticos, Centro de Investigación en Energía, Universidad Nacional Autónoma de México, Priv. Xochicalco s/no., Col. Centro, 62580 Temixco, Morelos, Mexico. E-mail: spv@cie.unam.mx
[2] Posgrado en Ingeniería, Centro de Investigación en Energía, Universidad Nacional Autónoma de México, Priv. Xochicalco s/no., Col. Centro, 62580 Temixco, Morelos, Mexico.

rock classification for a given rock-type according to some accepted rock nomenclature (see VERMA et al., 2010).

In other discrimination diagrams, e.g., the Zr–3Y–Ti/100 ternary diagram of PEARCE and CANN (1973), the Zr/4–Y–2Nb ternary diagram of MESCHEDE (1986), and the La/10-Nb/8-Y/15 ternary diagram of CABANIS and LECOLLE (1989), most samples plot in overlap regions of two tectonic settings. However, when success rates are greater, diagrams such as Ti/Y–Zr/Y by PEARCE and GALE (1977) and the Zr–3Y–Ti/100 ternary diagram of PEARCE and CANN (1973) then discriminate only two tectonic settings under the broad names of "within-plate" and "plate margin". For some other diagrams, such as the Nb/Y–Ti/Y bivariate diagram of PEARCE (1982), the Ti/1000–V bivariate diagram of SHERVAIS (1982), the Th–Ta–Hf/3 ternary diagram of Wood (1980), the $10MnO–10P_2O_5–TiO_2$ ternary diagram of MULLEN (1983), and the Zr/4–Y–2Nb ternary diagram of MESCHEDE (1986), although the success rates for some tectonic settings may be somewhat higher, they are certainly less than those for the old discriminant function based discrimination diagrams (viz., $Score_1$–$Score_2$ diagram of BUTLER and WORONOW, 1986 and F_1– F_2 and F_2–F_3 diagrams of PEARCE, 1976) as well as for the new DF1–DF2 (2004–2008) diagrams of AGRAWAL et al. (2004, 2008) and VERMA et al. (2006). Furthermore, the older discriminant function diagrams do not comply with all the recommendations for handling of compositional data (AITCHISON, 1986; AGRAWAL and VERMA, 2007), and are also not based on representative databases.

Therefore, it appears that the newer discriminant function diagrams proposed during 2004–2008 (AGRAWAL et al., 2004, 2008; VERMA et al., 2006) obtained from linear discriminant analysis (LDA) of representative databases for four tectonic settings (island arc, continental rift, ocean island, and mid-ocean ridge) and with probability-based tectonic field boundaries (AGRAWAL, 1999) work well with success rates up to about 99% (VERMA, 2010). Still newer discrimination diagrams based on all these new concepts, viz., representative database, correct statistical treatment of compositional data, multivariate techniques of LDA, probability-based field boundaries, and fulfillment of basic assumptions of normally distributed log-ratio transformed compositional variables, have also

been proposed (VERMA and AGRAWAL, 2010) and are shown to perform with even higher success rates. The diagrams of AGRAWAL et al. (2004) and VERMA et al. (2006) are based on major-elements (the first with simple ratio variables and the second with log-transformation of ratio variables), whereas those of AGRAWAL et al. (2008) and VERMA and AGRAWAL (2010) use log-transformed ratios of immobile-elements (La, Sm, Yb, Nb, and Zr for the former and adjusted TiO_2, Nb, V, Y, and Zr for the latter).

Unfortunately, none of the existing diagrams to date is capable of successfully discriminating between the two very similar tectonic settings of island and continental arcs. Further, the continental arc setting was included in none of the four sets of the newer diagrams proposed during 2004–2010, which successfully discriminate four tectonic settings. In fact, AGRAWAL et al. (2004) noted that continental arc was missing from their diagrams and that this setting was initially included, but had to be set aside due to the significant similarities between major-element compositions of basic magmas from island and continental arcs. These authors further hypothesized that only highly differentiated rocks will give contrasting compositions for these two tectonic settings because different types of underlying crust might be involved in the genesis and evolution of magmas. To solve the problem for basic rocks, trace elements should be incorporated, although none of the more recent publications (AGRAWAL et al., 2008; VERMA and AGRAWAL, 2010) has taken these recommendations into account. Therefore, there is still a worldwide interest in discriminating between island and continental arc settings, but given the similarities of basic magmas from these two tectonic settings, as documented by AGRAWAL et al. (2004) and VERMA et al. (2006), we should further test these newer diagrams for a typical continental arc setting of the Andes in South America before their application to the main aim of the present paper.

Figure 1 illustrates the schematic representation of the main volcanic areas of Southern Mexico and Central America, constituting the Mexican Volcanic Belt (MVB), Los Tuxtlas volcanic field (LTVF) and isolated centers, such as El Chichón volcano, which merge with the Central American Volcanic Arc (CAVA). Geochemical and radiogenic isotope data for volcanic rocks from Southern Mexico, mainly from the central to eastern parts of the MVB and LTVF, were interpreted by VERMA (2002, 2004) to demonstrate a lack of relationship with the subduction process, in spite of the ongoing subduction of the Cocos plate beneath the North American plate. Such a conclusion was also drawn by VERMA (2006) for the LTVF. VERMA (2002) also showed a text-book type case for volcanic centers of the CAVA, particularly from Guatemala to north-western Costa Rica (see also, BURBACH et al., 1984; LEEMAN et al., 1994; HARRY and GREEN, 1999; PATINO et al., 2000; WALKER et al., 2001).

Nevertheless, for the MVB and LTVF, there also exist conventional subduction-related models (e.g., PARDO and SUÁREZ, 1995; NELSON et al., 1995; FERRARI et al., 2001; GÓMEZ-TUENA et al., 2007; PÉREZ-CAMPOS et al., 2008; and PACHECO and SINGH, 2010) and other proposals such as plume-related

origin (e.g., MÁRQUEZ et al., 1999a), which therefore make the tectonic setting of these provinces highly complex and controversial (e.g., FERRARI and ROSAS-ELGUERA, 1999; MÁRQUEZ et al., 1999b; SHETH et al., 2000; TORRES-ALVARDO et al., 2002). More recently, VERMA (2009) reviewed all the available geological, geochemical and geophysical evidence from the MVB, particularly for its central part, and concluded that this volcanic province should better be called the Mexican Volcanic Rift, because all the available data are more consistent with a rift rather than an arc.

These are some of the reasons that motivated us to explore more constraints for elucidating this problem of the Mexican Pacific and on-land volcanism, as well as to document the differences or similarities between the volcanism and tectonics of Southern Mexico and Central America (CAVA).

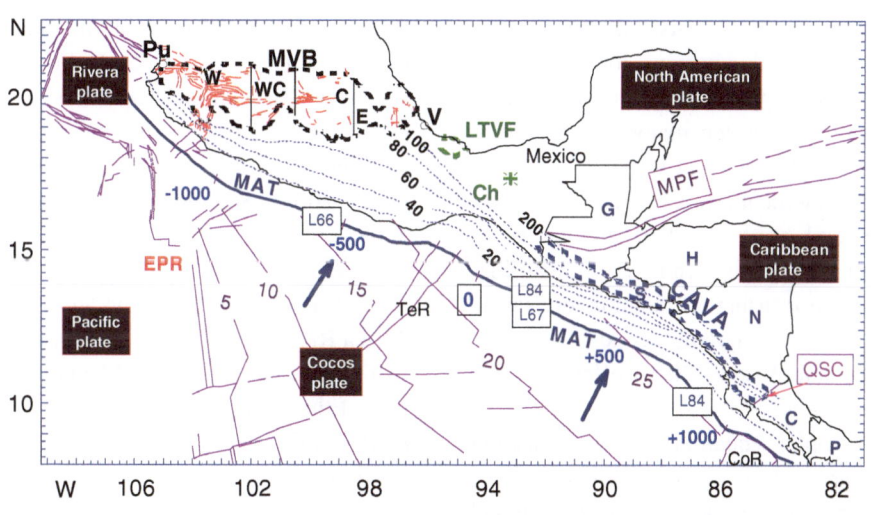

Figure 1

Simplified tectonic map of Southern Mexico and Central America (modified after VERMA, 2002, 2006). The land–ocean boundary and the subdivision of land in countries are shown by *thick solid curves*. The on-land tectonic features (fractures and faults) in the Mexican Volcanic Belt (MVB) region are from the following sources: NEGENDANK et al. (1985); JOHNSON and HARRISON (1989a, b); ALLAN et al. (1991); LYLE and NESS (1991); and SUTER et al. (1991, 1992, 1995a, b, 2001). Note the well-established triple junction in the western part of the MVB (LUHR et al., 1985; ALLAN et al., 1991). Rivera-Cocos plate diffuse boundaries are from ALLAN et al. (1991) and BANDY et al. (2000). EPR East Pacific Rise, MAT Middle America Trench, LTVF Los Tuxtlas Volcanic Field, CAVA Central American Volcanic Arc, TeR Tehuantepec Ridge, CoR Cocos Ridge, MPF Motagua-Polochic Fault, QSC Quesada Sharp Contortion, Pu Puerto Vallarta, V Veracruz, Ch El Chichón volcano, G Guatemala, S El Salvador, H Honduras; N Nicaragua, C Costa Rica, P Panamá. Legs L66, L67 and L84 are approximate Deep Sea Drilling Project Sites where drilling has recovered samples from the ocean floor. The numbers 5–25 in the Pacific Ocean give the approximate age of the ocean floor in Ma. The seismic depth contours marked 20–200 give the depth of the Benioff zone in km. Finally, the numbers 0, −500, −1,000, +500, and +1,000 are the distance in km for the transect used by VERMA (2002) to understand the similarities and differences between Southern Mexico and Central America, as inferred along the Pacific coast (MAT). The abbreviations W, WC, C, and E within the MVB refer to the approximate subdivision of the MVB into western, west-central, central, and eastern parts, which was used in the presentation and interpretation of compiled geochemical data (see Tables 5, 6, 7, 8 for more details). L66, L67, and L84 refer to the Deep Sea Drilling Project Legs and Sites, from which subducting sediments and underlying basalt samples were recovered, whose geochemical data provided constraints on the on-land volcanism (see VERMA, 2000a for more details)

Although the new discriminant function based discrimination diagrams for magmatic rocks have been tested by their respective authors using training sets by randomly dividing the databases into training and testing sets, and by Sheth (2008) using an independent database and by Verma (2010) using a much more extensive database, it would still be worthwhile to test them further using totally independent worldwide databases. If the diagrams perform satisfactorily for the discrimination of the four main tectonic settings and the continental arc of the Andes, we can then apply them to the study of fresh rocks from the subduction-related CAVA for further ascertaining the applicability of these new diagrams for a continental arc. Finally, we could determine the dominant tectonic setting(s) of controversial and complex volcanic areas of Southern Mexico (MVB and LTVF). Thus, a comprehensive database from the MVB and LTVF could be used to infer their tectonic settings from these diagrams provided, of course, they were also shown to work well for this purpose for the CAVA. The novelty of our approach lies in the multi-dimensional solution (note ten major-element ratios used by Agrawal et al., 2004 and Verma et al., 2006; and four immobile-element ratios by Agrawal et al., 2008 and Verma and Agrawal, 2010) and the use of correct statistical methodology (note log-transformation of element ratios using a common divisor in three of the four sets of diagrams; see also Agrawal and Verma, 2007 and Verma, 2010) offered by the newer sets of diagrams.

The highly complex equations used to construct these diagrams are not reproduced here to reduce journal space; they can be consulted in the original papers (Agrawal et al., 2004, 2008; Verma et al., 2006; Verma and Agrawal, 2010), or else in Verma (2010) or Verma and Rivera-Gómez (2010). Both of the latter papers can be freely downloaded from respective journal's website.

2. Databases

2.1. World Databases

For the statistical evaluation of the new discrimination diagrams, four extensive databases of basic and ultrabasic rocks of (1) island arc (IAB); (2) continental rift (CRB); (3) ocean-island (OIB); and (4) mid-ocean ridge (MORB), were separated from the combined database of all rock types. This was done from papers that were not included in any of the previous studies, in which the diagrams to be evaluated were proposed, i.e., not included in the databases used by Agrawal et al. (2004, 2008), Verma et al. (2006), and Verma and Agrawal (2010). Nevertheless, the same strict selection conditions as posed by these authors were maintained as follows: (a) age between late Miocene to Recent; (b) tectonic setting of the study area known explicitly without ambiguity; (c) rocks with adjusted silica $(SiO_2)_{adj} \leq 52\%$ after SINCLAS processing (Verma et al., 2002, 2003), i.e., after Fe-oxidation ratio adjustment (Middlemost, 1989) and on an anhydrous basis (Le Bas et al., 1986; Le Bas, 2000).

The island arc basic rocks (IAB) were compiled from the following areas and sources: Aleutian (Finney et al., 2008); Bicol and Bataan, Philippines (McDermott et al., 2005; DuFrane et al., 2006); Indonesia (Elburg and Kamenetsky, 2007; Sendjaja et al., 2009); Isu Bonin (Tamura et al., 2007); Lesser Antilles (Turner et al., 1996); Northern Honsu (Ohba et al., 2009); Solomon, SW Pacific (Schuth et al., 2009).

The continental rift basic rocks (CRB) were compiled from the following areas and sources: East Africa Rift system (Chakrabarti et al., 2009); Kenya Rift (Rogers et al., 2006; Macdonald et al., 2008); Main Ethiopian Rift (Peccerillo et al., 2007; Rooney et al., 2007; Ronga et al., 2009); and Mt. Etna, Italy (Ferlito et al., 2009).

The ocean-island basic rocks (OIB) were compiled from the following areas and sources: French Polynesia (Takamasa et al., 2009) and Hawaiian Islands (Xu et al., 2007; Dixon et al., 2008; Marske et al., 2008; Ireland et al., 2009).

The mid-ocean ridge basic rocks (MORB) were compiled from the following areas and sources: Arctic Mid-Ocean Ridge (Hellevang and Pedersen, 2008); Indian Ridge (Nakamura et al., 2007; Ray et al., 2007); and Mid-Atlantic Ridge (Debaille et al., 2006; Regelous et al., 2009).

Additionally, to further evaluate these diagrams for continental arc setting, the Andean data were compiled from the following sources listed according to the publication year: López-Escobar *et al.* (1981, 1991, 1993), Deruelle (1982); Frey *et al.* (1984), Gerlach *et al.* (1988), Hickey-Vargas *et al.* (1989), Tormey *et al.* (1991), Vergara *et al.* (2004), and Bruni *et al.* (2008).

2.2. *Central American Volcanic Arc Database*

The CAVA database was an updated version of that used by Verma *et al.* (2006), except that the information from Carr's website was not included mainly to avoid repetition. The geochemical data for CAVA basic rocks reported from different countries were compiled as follows: Guatemala (Carr, 1984; Carr *et al.*, 1990; Bardintzeff and Deniel, 1992; Walker *et al.*, 2000); El Salvador (Carr, 1984; Carr *et al.*, 1990; Agostini *et al.*, 2006); Honduras (Patino *et al.*, 1997; Walker *et al.*, 2000); Nicaragua (Carr, 1984; Hazlett, 1987; Carr *et al.*, 1990; Walker *et al.*, 1990, 2001; La Femina *et al.*, 2004); and NW Costa Rica (Reagan and Gill, 1989; Carr *et al.*, 1990; Alvarado *et al.*, 2006; Bolge *et al.*, 2006; Ryder *et al.*, 2006).

2.3. *Mexican Databases*

Databases were prepared for basic and ultrabasic rocks from the MVB and LTVF. The MVB database was an updated version of those used by Verma (2000a, 2002, 2004, 2009), Velasco-Tapia and Verma (2001), Torres-Alvarado *et al.* (2002), Verma and Hasenaka (2004), Verma *et al.* (2006), and Verma and Luhr (2010). Similarly, the LTVF database was updated from Verma (2006) and Verma *et al.* (2006).

Thus, the MVB data were compiled from the following sources listed according to the publication year: Williams (1950), Gunn and Mooser (1971), Negendank (1972), Robin (1976), Robin and Tournon (1978), Gastil *et al.* (1979), Pérez *et al.* (1979), Demant (1981), Luhr and Carmichael (1981, 1985), Verma and López (1982), Verma (1983, 2000a, 2001a, b, c, 2002, 2003), Allan and Carmichael (1984), Robin *et al.* (1984), Boudal (1985), Gilbert *et al.* (1985), Negendank *et al.* (1985), Allan (1986), Nelson (1986), Nelson and Livieres (1986),

Cathelineau *et al.* (1987), Ferriz and Mahood (1987), Hasenaka and Carmichael (1987), Martin del Pozzo *et al.* (1987), Silva Mora (1988), Luhr *et al.* (1989), Martin del Pozzo (1989), Swinamer (1989), Verma and Nelson (1989a, b), Wallace and Carmichael (1989, 1992, 1999), Robin *et al.* (1990), Allan *et al.* (1991), Lange and Carmichael (1991), Hasenaka (1992), Righter and Carmichael (1992), Robin and Potrel (1993), Ferrari *et al.* (1994, 2000), Moore *et al.* (1994), Besch *et al.* (1995), Orozco-Esquivel (1995), Righter *et al.* (1995), Carmichael *et al.* (1996, 2006), Luhr (1997), Delgado *et al.* (1998), Blatter *et al.* (2001, 2007), Blatter and Hammersley, 2009), Righter and Rosas-Elguera (2001), Velasco-Tapia and Verma (2001), Chesley *et al.* (2002), García-Palomo *et al.* (2002), Siebert and Carrasco-Núñez (2002), Gómez-Tuena *et al.* (2003), Petrone *et al.* (2003), Siebe *et al.* (2004), Verma and Hasenaka (2004), Carrasco-Núñez *et al.* (2005), Lewis-Kenedi *et al.* (2005), Schaaf *et al.* (2005), Rossotti *et al.* (2006), Frey *et al.* (2007), Orozco-Esquivel *et al.* (2007), Torres-Alvarado *et al.* (2007), Arce *et al.* (2008), Maria and Luhr (2008), Meriggi *et al.* (2008), Straub *et al.* (2008), Vigouroux *et al.* (2008), Mori *et al.* (2009), Rodríguez *et al.* (2009), and Verma and Luhr (2010).

The LTVF database included data from the following sources: Nelson and Gonzalez-Caver (1992), Verma *et al.* (1993), Nelson *et al.* (1995), Verma (2006), and Espíndola *et al.* (2009).

3. *Statistical Evaluation of Discriminant Function Discrimination Diagrams*

We used five world databases to evaluate all four sets of new discrimination diagrams (Agrawal *et al.*, 2004, 2008; Verma *et al.*, 2006; Verma and Agrawal, 2010). For each set of diagrams, five different plots were prepared and the samples in different fields counted. Then, the success rate statistics were calculated and reported. In this way, 20 diagrams were thus obtained for the world data from each tectonic setting, amounting to a total of 100 diagrams for the five databases. To conserve space, these diagrams are not included here. The results are summarized in Tables 1, 2, 3 and 4.

Table 1

Statistical evaluation information of the set of five major-element based discriminant function DF1–DF2 discrimination diagrams (AGRAWAL et al., 2004) for basic rocks from island arc (IAB), continental rift (CRB), ocean-island (OIB) and mid-ocean ridge (MORB)

Tectonic setting (figure #)	Total # samples (%)	Number of discriminated samples (%)			
		IAB	CRB	OIB	MORB
IAB–CRB–OIB–MORB					
Island arc	64 (100)	**58 (90.6)**	0 (0.0)	0 (0.0)	6 (9.4)
Continental rift	110 (100)	15 (13.6)	**63 (57.3)**	17 (15.5)	15 (13.6)
Ocean-island	233 (100)	7 (3.0)	18 (7.7)	**200 (85.8)**	8 (3.5)
Mid-ocean ridge	71 (100)	2 (2.8)	0 (0.0)	0 (0.0)	**69 (97.2)**
Andes	57 (100)	**46 (80.7)**	11 (19.3)	0 (0.0)	0 (0.0)
IAB–CRB–OIB					
Island arc	64 (100)	**64 (100)**	0 (0.0)	0 (0.0)	–
Continental rift	110 (100)	21 (19.1)	**77 (70.0)**	12 (10.9)	–
Ocean-island	233 (100)	12 (5.2)	21 (9.0)	**200 (85.8)**	–
Andes	57 (100)	**46 (80.7)**	11 (19.3)	0 (0.0)	–
IAB–CRB–MORB					
Island arc	64 (100)	**59 (92.2)**	0 (0.0)	–	5 (7.8)
Continental rift	110 (100)	14 (12.7)	**83 (75.5)**	–	13 (11.8)
Mid-ocean ridge	71 (100)	2 (2.8)	0 (0.0)	–	**69 (97.2)**
Andes	57 (100)	**46 (80.7)**	10 (17.5)	–	1 (1.8)
IAB–OIB–MORB					
Island arc	64 (100)	**59 (92.2)**	–	0 (0.0)	5 (7.8)
Ocean-island	233 (100)	1 (0.4)	–	**212 (91.0)**	20 (8.6)
Mid-ocean ridge	71 (100)	2 (2.8)	–	0 (0.0)	**69 (97.2)**
Andes	57 (100)	**44 (77.2)**	–	6 (10.5)	7 (12.3)
CRB–OIB–MORB					
Continental rift	110 (100)	–	**88 (80.0)**	8 (7.3)	14 (12.7)
Ocean-island	233 (100)	–	14 (6.0)	**205 (88.0)**	14 (6.0)
Mid-ocean ridge	71 (100)	–	0 (0.0)	0 (0.0)	**71 (100)**

Boldface italic font shows the correct (expected) tectonic setting. The results of inapplicable diagrams (diagrams from which the expected tectonic setting is absent for a given dataset) are not included in this table, e.g., Island arc is missing from the CRB–OIB–MORB combination (i.e., from the diagram that would correspond to this combination), and therefore, this setting is missing from the final part of table. Andean results are also not shown for this combination because it does not contain the expected IAB setting

The first set of major-element based discrimination diagrams (Table 1; AGRAWAL et al., 2004) works well for all tectonic settings of island arc (success rates of 90.6–100%), continental rift (57.3–80.0%), ocean-island (85.8–91.0%), mid-ocean ridge (97.2–100%), and continental arc of the Andes evaluated for its similarity to island arc (77.2–80.7%). All success rates are statistically significant because they are $\gg 33.3\%$ (being the simple "by chance" probability).

The second set of major-element based diagrams (Table 2; VERMA et al., 2006) using exactly the same samples as for the first set, were evaluated with the following success rates: island arc 96.9–100%, continental rift 70.0–96.4%, ocean-island 70.0–92.7%, mid-ocean ridge 97.2–100%, and the Andes evaluated for its similarity to island arc (73.7–82.5%).

The first set of immobile-element based diagrams (Table 3; AGRAWAL et al., 2008) was also evaluated for IAB, CRB, OIB, and MORB (three settings at a time) settings using five world databases. The success rates from these databases of island arc, continental rift, ocean-island, mid-ocean ridge, and the Andes were, respectively, 71.7–84.9%, 66.7–100%, 69.7–83.5%, 96.6–98.3%, and identical values of 83.8%.

Finally, the second set of immobile-element based diagrams (Table 4; VERMA and AGRAWAL, 2010) also complies with the requirement of normal distribution of the log-ratio variables used to construct these diagrams. This was achieved by the use of DODESYS software (VERMA and DÍAZ-GONZÁLEZ, unpublished), which uses new precise and accurate critical values for discordancy tests (BARNETT and LEWIS, 1994;

Table 2

Statistical evaluation information of the set of five discrimination diagrams based on natural logarithm transformation of major-element ratios discriminant functions DF1–DF2 (VERMA et al., 2006) for basic rocks from island arc (IAB), continental rift (CRB), ocean-island (OIB) and mid-ocean ridge (MORB)

Tectonic setting (figure #)	Total # samples (%)	Number of discriminated samples (%)			
		IAB	CRB	OIB	MORB
IAB–CRB–OIB–MORB					
Island arc	64 (100)	*64 (100)*	0 (0.0)	0 (0.0)	0 (0.0)
Continental rift	110 (100)	2 (1.8)	*87 (79.1)*	18 (16.4)	3 (2.7)
Ocean-island	233 (100)	1 (0.4)	38 (16.3)	*163 (70.0)*	30 (13.3)
Mid-ocean ridge	71 (100)	2 (2.8)	0 (0.0)	0 (0.0)	*69 (97.2)*
Andes	57 (100)	*45 (78.9)*	12 (21.1)	0 (0.0)	0 (0.0)
IAB–CRB–OIB					
Island arc	64 (100)	*64 (100)*	0 (0.0)	0(0.0)	–
Continental rift	110 (100)	0 (0.0)	*77 (70.0)*	33 (30.0)	–
Ocean-island	233 (100)	6 (2.6)	22 (9.4)	*205 (88.0)*	–
Andes	57 (100)	*42 (73.7)*	15 (26.3)	0 (0.0)	–
IAB–CRB–MORB					
Island arc	64 (100)	*64 (100)*	0 (0.0)	–	0 (0.0)
Continental rift	110 (100)	0 (0.0)	*106 (96.4)*	–	4 (3.6)
Mid-ocean ridge	71 (100)	2 (2.8)	0 (0.0)	–	*69 (97.2)*
Andes	57 (100)	*45 (78.9)*	12 (21.1)	–	0 (0.0)
IAB–OIB–MORB					
Island arc	64 (100)	*62 (96.9)*	–	2 (3.1)	0 (0.0)
Ocean-island	233 (100)	15 (6.4)	–	*216 (92.7)*	2 (0.9)
Mid-ocean ridge	71 (100)	2 (2.8)	–	0 (0.0)	*69 (97.2)*
Andes	57 (100)	*47 (82.5)*	–	8 (14.0)	2 (3.5)
CRB–OIB–MORB					
Continental rift	110 (100)	–	*81 (73.6)*	25 (22.7)	4 (3.6)
Ocean-island	233 (100)	–	30 (12.9)	*202 (86.7)*	1 (0.4)
Mid-ocean ridge	71 (100)	–	0 (0.0)	0 (0.0)	*71 (100)*

Boldface italic font shows the correct (expected) tectonic setting. The results of inapplicable diagrams (diagrams from which the expected tectonic setting is absent for a given dataset) are not included in this table, e.g., Island arc is missing from the CRB–OIB–MORB combination (i.e., from the diagram that would correspond to this combination), and therefore, this setting is missing from the final part of table. Andean results are also not shown for this combination because it does not contain the expected IAB setting

VERMA and QUIROZ-RUIZ, 2006a, b, 2008; VERMA et al., 2008). The success rates for the five world databases were as follows: island arc 92.6–100%, continental rift 58.5–100%, ocean-island identical values of 100%, mid-ocean ridge 83.1–100%, and the Andes 62.1–79.3%. We also calculated the success rates (not-tabulated) of these five entire databases before the application of DODESYS, i.e., without ascertaining that the log-transformed variables are normally distributed. A net gain of success rates was observed for the combined CRB + OIB (8.0%), OIB (16.7%) and MORB (5.4%) when discordant outlier-free data were used as compared to the entire dataset. On the contrary, a net loss was obtained for CRB (about 3.8%), IAB (about 0.1%), and the Andes (about 3.0%).

4. Use of Discriminant Function based Discrimination Diagrams for Southern Mexico and Central America

The results of the application of the four sets of new diagrams to basic rocks from Southern Mexico (MVB, arbitrarily divided into four parts: W–western, WC–west-central, C–central and E–eastern, and LTVF; see Fig. 1 for locations) and Central America (CAVA) are summarized in Tables 5, 6, 7 and 8. In these tables, success rates are calculated when the number of samples was at least 25 (an arbitrarily set limit), and the "inapplicable" results are also indicated. Although a total of 120 diagrams were prepared, we present only four sets of discrimination diagrams as follows: AGRAWAL et al. (2004) diagrams

Table 3

Statistical evaluation of the set of five discrimination diagrams based on natural logarithm transformation of trace-element ratios discriminant functions DF1–DF2 (AGRAWAL et al., 2008) for basic rocks from island arc (IAB), continental rift (CRB), ocean-island (OIB) and mid-ocean ridge (MORB)

Tectonic setting (figure #)	Total # samples (%)	Number of discriminated samples (%)				
		IAB	Within-plate			MORB
			CRB + OIB	CRB	OIB	
IAB–CRB–OIB–MORB						
Island arc	53 (100)	*45 (84.9)*	4 (7.5)	–	–	4 (7.5)
Continental rift	84 (100)	0 (0.0)	**84 (100)**	–	–	0 (0.0)
Ocean-island	109 (100)	0 (0.0)	96 (88.1)	–	–	13 (11.9)
Mid-ocean ridge	59 (100)	2 (3.4)	0 (0.0)	–	–	*57 (96.6)*
Andes	37 (100)	*31 (83.8)*	6 (26.2)	–	–	0 (0.0)
IAB–CRB–OIB						
Island arc	53 (100)	*38 (71.7)*	–	7 (13.2)	8 (15.1)	–
Continental rift	84 (100)	0 (0.0)	–	*74 (88.1)*	10 (11.9)	–
Ocean-island	109 (100)	0 (0.0)	–	33 (30.3)	*76 (69.7)*	–
Andes	37 (100)	*31 (83.8)*	–	5 (13.5)	1 (2.7)	–
IAB–CRB–MORB						
Island arc	53 (100)	*42 (79.2)*	–	7 (13.2)	–	4 (7.6)
Continental rift	84 (100)	0 (0.0)	–	84 (*100*)	–	0 (0.0)
Mid-ocean ridge	59 (100)	1 (1.7)	–	0 (0.0)	–	*58 (98.3)*
Andes	37 (100)	*31 (83.8)*	–	6 (16.2)	–	0 (0.0)
IAB–OIB–MORB						
Island arc	53 (100)	*41 (77.4)*	–	–	8 (15.1)	4 (7.5)
Ocean-island	109 (100)	0 (0.0)	–	–	*91 (83.5)*	18 (16.5)
Mid-ocean ridge	59 (100)	2 (3.4)	–	–	0 (0.0)	*57 (96.6)*
Andes	37 (100)	*31 (83.8)*	–	–	6 (16.2)	0 (0.0)
CRB–OIB–MORB						
Continental rift	84 (100)	–	–	*56 (66.7)*	28 (33.3)	0 (0.0)
Ocean-island	109 (100)	–	–	20 (18.3)	*78 (71.6)*	11 (10.1)
Mid-ocean ridge	59 (100)	–	–	1 (1.7)	0 (0.0)	*58 (98.3)*

Boldface italic font shows the correct (expected) tectonic setting. The results of inapplicable diagrams (diagrams from which the expected tectonic setting is absent for a given dataset) are not included in this table, e.g., Island arc is missing from the CRB–OIB–MORB combination (i.e., from the diagram that would correspond to this combination), and therefore, this setting is missing from the final part of table. Andean results are also not shown for this combination because it does not contain the expected IAB setting

for W-MVB and WC-MVB (Fig. 2); VERMA *et al.* (2006) diagrams for C-MVB (Fig. 3); AGRAWAL *et al.* (2008) diagrams for E-MVB (Fig. 4); and VERMA and AGRAWAL (2010) diagrams for LTVF and CAVA (Fig. 5).

We describe in detail only one set of diagrams (Fig. 2; AGRAWAL *et al.*, 2004) to illustrate their use for inferring the tectonic setting. The compiled W-MVB data (227 analyses, Table 5) plotted in Fig. 2a were counted to calculate success rates for all four tectonic settings of IAB, CRB, OIB, and MORB. The highest success rate obtained for CRB is about 78.9%, because 179 out of 227 samples plot in this field. For other tectonic settings, the success rates varied as follows: IAB 18.1% (41 out of 227 samples

plot in this field), OIB 0.9% (2 out of 227 samples), and MORB 2.2% (5 out of 227 samples). These results show that for W-MVB the expected tectonic setting from the first four-field diagram (Fig. 2a) is a continental rift setting. However, the results are still not considered definitive, because all four remaining diagrams should be examined (Fig. 2b–e). Figures 2b and c confirm the results of Fig. 2a, because 177 and 174 samples, respectively (out of 227; with respective success rate of 78.0 and 76.7%, Table 5) plot in CRB on IAB-CRB-OIB and IAB-CRB-MORB plots. In Figure 2d, corresponding to IAB-OIB-MORB, the expected CRB field is missing. Therefore, this diagram (Fig. 2d) should be eliminated from any interpretations of W-MVB data, and the results

Table 4

Statistical evaluation of the set of five discrimination diagrams based on natural logarithm transformation of trace-element ratios discriminant functions DF1–DF2 (VERMA and AGRAWAL, 2010) for basic rocks from island arc (IAB), continental rift (CRB), ocean-island (OIB) and mid-ocean ridge (MORB)

Tectonic setting (figure #)	Total # samples (%)	Number of discriminated samples (%)				
		IAB	Within-plate			MORB
			CRB + OIB	CRB	OIB	
IAB–CRB–OIB–MORB						
Island arc	27 (100)	***27 (100)***	0 (0.0)	–	–	0 (0.0)
Continental rift	65 (100)	0 (0.0)	***65 (100)***	–	–	0 (0.0)
Ocean-island	85 (100)	0 (0.0)	**85 (100)**	–	–	0 (0.0)
Mid-ocean ridge	65 (100)	10 (15.4)	0 (0.0)	–	–	***55 (84.6)***
Andes	29 (100)	***18 (62.1)***	6 (20.7)	–	–	5 (17.2)
IAB–CRB–OIB						
Island arc	27 (100)	***27 (100)***	–	0 (0.0)	0 (0.0)	–
Continental rift	65 (100)	0 (0.0)	–	***43 (66.2)***	22 (33.8)	–
Ocean-island	85 (100)	0 (0.0)	–	0 (0.0)	***85 (100)***	–
Andes	29 (100)	***23 (79.3)***	–	5 (17.2)	1 (3.4)	–
IAB–CRB–MORB						
Island arc	27 (100)	***27 (100)***	–	0 (0.0)	–	0 (0.0)
Continental rift	65 (100)	0 (0.0)	–	***65 (100)***	–	0 (0.0)
Mid-ocean ridge	65 (100)	8 (12.3)	–	0 (0.0)	–	***57 (87.7)***
Andes	29 (100)	***18 (62.1)***	–	6 (20.7)	–	5 (17.2)
IAB–OIB–MORB						
Island arc	27 (100)	***25 (92.6)***	–	–	0 (0.0)	2 (7.4)
Ocean-island	85 (100)	0 (0.0)	–	–	***85 (100)***	0 (0.0)
Mid-ocean ridge	65 (100)	11 (16.9)	–	–	0 (0.0)	***54 (83.1)***
Andes	29 (100)	***20 (69.0)***	–	–	6 (20.7)	3 (10.3)
CRB–OIB–MORB						
Continental rift	65 (100)	–	–	***38 (58.5)***	27 (41.5)	0 (0.0)
Ocean-island	85 (100)	–	–	0 (5.8)	***85 (100)***	0 (0.0)
Mid-ocean ridge	65 (100)	–	–	0 (0.0)	0 (0.0)	***65 (100)***

Boldface italic font shows the correct (expected) tectonic setting. The results of inapplicable diagrams (diagrams from which the expected tectonic setting is absent for a given dataset) are not included in this table, e.g., Island arc is missing from the CRB–OIB–MORB combination (i.e., from the diagram that would correspond to this combination), and therefore, this setting is missing from the final part of table. Andean results are also not shown for this combination because it does not contain the expected IAB setting

summarised in Table 5 ignored. The final diagram (Fig. 2e) for CRB-OIB-MORB further confirms the results of earlier diagrams (Fig. 2a–c) because 200 out of 227 samples (88.1%) plot in CRB field. Thus, for W-MVB consistent results are obtained from Fig. 2a–e.

For WC-MVB, on the other hand, this set of diagrams (Fig. 2a–e) does not provide a consistent result (Table 5). The samples plot in two distinct fields in Figs. 2a–c (IAB and CRB, with somewhat higher success rates for IAB; Table 5). Figures 2d and e do not have these two tectonic settings together, therefore high success rates are obtained for individual tectonic setting in these diagrams (61.0% for IAB in Fig. 2d and 53.7% for CRB in Fig. 2e). We

may conclude that, for WC-MVB, a dual or transitional tectonic setting between arc and rift is being indicated.

Once the functioning of the set of five diagrams (AGRAWAL *et al.*, 2004) is fully understood from the above discussion (Fig. 2, Table 5), it will not be necessary to present in detail the functioning of other diagrams (Figs. 3, 4 5, Tables 6, 7, 8; VERMA *et al.*, 2006; AGRAWAL *et al.* 2008; VERMA and AGRAWAL, 2010), and it would suffice to simply point out the most important results. The following discussion is based on these diagrams (Figs. 3, 4, 5) as well as Tables 6, 7, 8. Therefore, we will not always explicitly refer to them, because the results are summarized in these tables.

Table 5

Application of the set of five major-element based discriminant function DF1–DF2 discrimination diagrams (AGRAWAL et al., 2004) for basic rocks from Southern Mexico and Central America

Tectonic setting (figure #)	Total # samples (%)	Number of discriminated samples (%)			
		IAB	CRB	OIB	MORB
IAB–CRB–OIB–MORB					
W-MVB (Fig. 2a)	227 (100)	41 (18.1)	*179 (78.9)*	2 (0.9)	5 (2.2)
WC-MVB (Fig. 2a)	82 (100)	*42 (51.2)*	**29 (35.4)**	0 (0.0)	11 (13.4)
C-MVB	79 (100)	6 (7.6)	*49 (62.0)*	2 (2.5)	22 (27.9)
E-MVB	197 (100)	23 (11.7)	*120 (60.9)*	1 (0.5)	53 (26.9)
LTVF	75 (100)	7 (9.3)	*62 (82.7)*	0 (0.0)	6 (8.0)
CAVA	119 (100)	*89 (74.8)*	2 (1.7)	0 (0.0)	28 (23.5)
IAB–CRB–OIB					
W-MVB (Fig. 2b)	227 (100)	48 (21.1)	*177 (78.0)*	2 (0.9)	–
WC-MVB (Fig. 2b)	82 (100)	*48 (58.5)*	**34 (41.5)**	0 (0.0)	–
C-MVB	79 (100)	11 (13.9)	*66 (83.6)*	2 (2.5)	–
E-MVB	197 (100)	33 (16.8)	*164 (83.2)*	0 (0.0)	–
LTVF	75 (100)	8 (10.7)	*67 (89.3)*	0 (0.0)	–
CAVA	119 (100)	*118 (99.2)*	1 (0.8)	0 (0.0)	–
IAB–CRB–MORB					
W-MVB (Fig. 2c)	227 (100)	46 (20.2)	*174 (76.7)*	–	7 (3.1)
WC-MVB (Fig. 2c)	82 (100)	*45 (54.9)*	23 (28.0)	–	14 (17.1)
C-MVB	79 (100)	9 (11.4)	*37 (46.8)*	–	**33 (41.8)**
E-MVB	197 (100)	23 (11.7)	*131 (66.5)*	–	43 (21.8)
LTVF	75 (100)	5 (6.7)	*67 (89.3)*	–	3 (4.0)
CAVA	119 (100)	*92 (77.3)*	2 (1.7)	–	25 (21.0)
IAB–OIB–MORB					
W-MVB (Fig. 2d)[a]	227 (100)	108 (47.6)	–	82 (36.1)	37 (16.3)
WC-MVB (Fig. 2d)	82 (100)	*50 (61.0)*	–	17 (20.7)	15 (18.3)
C-MVB[a]	79 (100)	19 (24.1)	–	11 (13.9)	49 (62.0)
E-MVB[a]	197 (100)	45 (22.8)	–	55 (27.9)	97 (49.2)
LTVF[a]	75 (100)	11 (14.7)	–	8 (10.6)	56 (74.7)
CAVA	119 (100)	*91 (76.5)*	–	1 (0.8)	27 (22.7)
CRB–OIB–MORB					
W-MVB (Fig. 2e)	227 (100)	–	*200 (88.1)*	2 (0.9)	25 (11.0)
WC-MVB (Fig. 2e)	82 (100)	–	*44 (53.7)*	0 (0.0)	**38 (46.3)**
C-MVB	79 (100)	–	*62 (78.5)*	1 (1.3)	16 (20.3)
E-MVB	197 (100)	–	*147 (74.6)*	2 (1.0)	48 (24.4)
LTVF	75 (100)	–	*66 (88.0)*	2 (2.7)	7 (9.3)
CAVA[a]	119 (100)	–	6 (5.0)	0 (0.0)	113 (95.0)

MVB Mexican Volcanic Belt, *W* western, *WC* west-central, *C* central, *E* eastern, *LTVF* Los Tuxtlas volcanic field, *CAVA* Central American Volcanic Arc

[a] Inapplicable results and diagrams; boldface italic font shows the expected tectonic setting; whenever this setting is followed by a significant (>33.3%) success rate for another setting, it is shown by simple boldface number for success rate

For W-MVB, a continental rift setting indicated by the first set of major-element based diagrams (AGRAWAL *et al.*, 2004) is confirmed by the second set of major-element based diagrams (VERMA *et al.*, 2006), with even greater success rates of 84.6–93.8% (Table 6). The first set of immobile trace-element based diagrams (AGRAWAL *et al.*, 2008) indicates a dual arc and rift setting for the W-MVB, with success rates of 49.3 and 79.1% for arc and 59.7 and 68.7 for rift (Table 7). The other set of immobile element based diagrams (VERMA and AGRAWAL, 2010) indicates an arc setting with success rates of 64.1–67.0% (Table 8). Thus, for the W-MVB it appears that both subduction (e.g., BANDY *et al.*, 2000; BANDY and HILDE, 2000; YANG *et al.*, 2009) and rifting (e.g., LUHR *et al.*, 1985) processes play a significant role in controlling the magma compositions, although major-elements favor a rift setting. The transition between

Table 6

Application of the set of five discrimination diagrams based on natural logarithm transformation of major-element ratios discriminant functions DF1–DF2 (VERMA et al., 2006) for basic rocks from Southern Mexico and Central America

Tectonic setting (figure #)	Total # samples (%)	Number of discriminated samples (%)			
		IAB	CRB	OIB	MORB
IAB–CRB–OIB–MORB					
W-MVB	227 (100)	22 (9.7)	*192 (84.6)*	9 (4.0)	4 (1.8)
WC-MVB	82 (100)	**35 (42.7)**	*42 (51.2)*	0 (0.0)	5 (6.1)
C-MVB (Fig. 3a)	79 (100)	7 (8.9)	*65 (82.3)*	1 (1.2)	6 (7.6)
E-MVB	197 (100)	25 (12.7)	*164 (83.2)*	5 (2.5)	3 (1.5)
LTVF	75 (100)	6 (8.0)	*69 (92.0)*	0 (0.0)	0 (0.0)
CAVA	119 (100)	*96 (80.7)*	4 (3.4)	0 (0.0)	19 (16.0)
IAB–CRB–OIB					
W-MVB	227 (100)	19 (8.4)	*201 (88.5)*	7 (3.1)	–
WC-MVB	82 (100)	22 (26.8)	*60 (73.2)*	0 (0.0)	–
C-MVB (Fig. 3b)	79 (100)	4 (5.1)	*73 (92.4)*	2 (2.5)	–
E-MVB	197 (100)	18 (9.1)	*177 (89.8)*	2 (1.0)	–
LTVF	75 (100)	8 (10.7)	*67 (89.3)*	0 (0.0)	–
CAVA	119 (100)	*91 (76.5)*	17 (14.3)	11 (9.2)	–
IAB–CRB–MORB					
W-MVB	227 (100)	17 (7.5)	*205 (90.3)*	–	5 (2.2)
WC-MVB	82 (100)	27 (32.9)	*48 (58.5)*	–	7 (8.5)
C-MVB (Fig. 3c)	79 (100)	4 (5.1)	*69 (87.3)*	–	6 (7.6)
E-MVB	197 (100)	19 (9.6)	*175 (88.8)*	–	3 (1.5)
LTVF	75 (100)	6 (8.0)	*69 (92.0)*	–	0 (0.0)
CAVA	119 (100)	*96 (80.7)*	5 (4.2)	–	18 (15.1)
IAB–OIB–MORB					
W-MVB[a]	227 (100)	72 (31.7)	–	115 (50.7)	40 (17.6)
WC-MVB	82 (100)	**59 (72.0)**	–	11 (13.4)	12 (14.6)
C-MVB (Fig. 3d)[a]	79 (100)	19 (24.0)	–	10 (12.7)	50 (63.3)
E-MVB[a]	197 (100)	44 (22.3)	–	81 (41.1)	72 (36.5)
LTVF[a]	75 (100)	26 (34.7)	–	42 (56.0)	7 (9.3)
CAVA	119 (100)	*98 (82.4)*	–	2 (1.7)	19 (16.0)
CRB–OIB–MORB					
W-MVB	227 (100)	–	*213 (93.8)*	9 (4.0)	5 (2.2)
WC-MVB	82 (100)	–	*70 (85.4)*	0 (0.0)	12 (14.6)
C-MVB (Fig. 3e)	79 (100)	–	*71 (89.9)*	1 (1.2)	7 (8.9)
E-MVB	197 (100)	–	*186 (94.4)*	5 (2.5)	6 (3.0)
LTVF	75 (100)	–	*74 (98.7)*	0 (0.0)	1 (1.3)
CAVA[a]	119 (100)	–	86 (72.3)	0 (0.0)	33 (27.7)

MVB Mexican Volcanic Belt, *W* western, *WC* west-central, *C* central, *E* eastern, *LTVF* Los Tuxtlas volcanic field, *CAVA* Central American Volcanic Arc

[a] Inapplicable results and diagrams; boldface italic font shows the expected tectonic setting; whenever this setting is followed by a significant (>33.3%) success rate for another setting, it is shown by simple boldface number for success rate

arc and rift settings and the significant role of both of them in the genesis of magmas in the W-MVB is being pointed out for the first time on the basis of these new discrimination diagrams, which may also imply the proximity of the termination of subduction of Rivera plate and the increasingly more significant rifting of the Jalisco block from on-land Mexico.

The contrasting behavior of major- and trace-element based diagrams may be related to higher analytical errors for trace-element determinations, such as Nb, Yb, Sm, La, and Y, as compared to those for the major-elements. For the evaluation of tectonic settings of fresh rocks, the major-element based diagrams (Figs. 2, 3, Tables 5, 6; AGRAWAL et al., 2004; VERMA et al., 2006) should be preferred in comparison to the trace-element based diagrams (Figs. 4, 5, Tables 7, 8; AGRAWAL et al., 2008; VERMA and AGRAWAL, 2010).

Table 7

Application of the set of five discrimination diagrams based on natural logarithm transformation of trace-element ratios discriminant functions DF1–DF2 (AGRAWAL et al., 2008) for basic rocks from Southern Mexico and Central America

Tectonic setting (figure #)	Total # samples (%)	Number of discriminated samples (%)				
		IAB	Within-plate			MORB
			CRB + OIB	CRB	OIB	
IAB–CRB–OIB–MORB						
W-MVB	67 (100)	*33 (49.3)*	*33 (49.3)*	–	–	1 (1.5)
WC-MVB	34 (100)	*15 (44.1)*	7 (20.6)	–	–	**12 (35.3)**
C-MVB	12	0	*11*	–	–	1
E-MVB (Fig. 4a)	90 (100)	*7 (7.8)*	*71 (78.9)*	–	–	12 (13.3)
LTVF	19	6	*13*	–	–	0
CAVA	42 (100)	*22 (52.4)*	0 (0.0)	–	–	**20 (47.6)**
IAB–CRB–OIB						
W-MVB	67 (100)	*53 (79.1)*	–	7 (10.4)	7 (10.4)	–
WC-MVB	34 (100)	*12 (35.3)*	–	**18 (52.9)**	4 (11.8)	–
C-MVB	12	0	–	*11*	1	–
E-MVB (Fig. 4b)	90 (100)	18 (20.0)	–	*60 (66.7)*	12 (13.3)	–
LTVF	19	*16*	–	3	0	–
CAVA	42 (100)	*17 (40.5)*	–	14 (33.3)	11 (26.2)	–
IAB–CRB–MORB						
W-MVB	67 (100)	20 (29.9)	–	*46 (68.7)*	–	1 (1.5)
WC-MVB	34 (100)	*15 (44.1)*	–	8 (23.5)	–	11 (32.4)
C-MVB	12	0	–	*12*	–	0
E-MVB (Fig. 4c)	90 (100)	7 (7.8)	–	*71 (78.9)*	–	12 (13.3)
LTVF	19	4	–	*15*	–	0
CAVA	42 (100)	*22 (52.4)*	–	0 (0.0)	–	**20 (47.6)**
IAB–OIB–MORB						
W-MVB	67 (100)	19 (28.4)	–	–	*47 (70.1)*	1 (1.5)
WC-MVB	34 (100)	*15 (44.1)*	–	–	5 (14.7)	**14 (41.2)**
C-MVB[a]	12	0	–	–	4	8
E-MVB (Fig. 4d)[a]	90 (100)	8 (8.9)	–	–	68 (75.6)	14 (15.5)
LTVF[a]	19	6	–	–	13	0
CAVA	42 (100)	*22 (52.4)*	–	–	0 (0.0)	**20 (47.6)**
CRB–OIB–MORB						
W-MVB	67 (100)	–	–	*40 (59.7)*	22 (32.8)	5 (7.5)
WC-MVB	34 (100)	–	–	**12 (35.3)**	0 (0.0)	**22 (64.7)**
C-MVB	12	–	–	*11*	1	0
E-MVB (Fig. 4e)	90 (100)	–	–	*72 (80.0)*	3 (3.3)	15 (16.7)
LTVF	19	–	–	*19*	0	0
CAVA	42 (100)	–	–	1 (2.4)	0 (0.0)	**41 (97.6)**

MVB Mexican Volcanic Belt, *W* western; *WC* west-central; *C* central; *E* eastern; *LTVF* Los Tuxtlas volcanic field; *CAVA* Central American Volcanic Arc

[a] Inapplicable results and diagrams; boldface italic font shows the expected tectonic setting; whenever this setting is followed by a significant (>33.3%) success rate for another setting, it is shown by simple boldface number for success rate

For WC-MVB, the first set of major-element based diagrams (Table 5) indicates a dual arc and rift setting (51.2–61.0% for arc as compared to 35.4–53.7% for rift). The second set of major-element based diagrams, however, indicates a rift setting for the WC-MVB, with success rates from 51.2 to 85.4% (Table 6). Because this second set of diagrams is based on correct statistical treatment of compositional data (AITCHISON, 1986; AGRAWAL and VERMA, 2007; VERMA, 2010), the results of this set should be considered more reliable. For WC-MVB the first set of immobile trace-element based diagrams also indicates a dual arc and rift setting although success rates for rift are somewhat greater (35.3–52.9% for

Table 8

Application of the set of five discrimination diagrams based on natural logarithm transformation of trace-element ratios discriminant functions DF1–DF2 (VERMA and AGRAWAL, 2010) for basic rocks from Southern Mexico and Central America

Tectonic setting (figure #)	Total # samples (%)	Number of discriminated samples (%)				
		IAB	Within-plate			MORB
			CRB + OIB	CRB	OIB	
IAB–CRB–OIB–MORB						
W-MVB	103 (100)	***66 (64.1)***	27 (26.2)	–	–	10 (9.7)
WC-MVB	64 (100)	20 (31.2)	19 (29.7)	–	–	***25 (39.1)***
C-MVB	45 (100)	5 (11.1)	***20 (44.4)***	–	–	**20 (44.4)**
E-MVB	49 (100)	**2 (4.1)**	***30 (61.2)***	–	–	**17 (34.7)**
LTVF (Fig. 5a)	47 (100)	***22 (46.8)***	***17 (36.2)***	–	–	8 (17.0)
CAVA (Fig. 5a)	43 (100)	***29 (67.4)***	1 (2.3)	–	–	13 (30.3)
IAB–CRB–OIB						
W-MVB	103 (100)	***69 (67.0)***	–	27 (26.2)	7 (6.8)	–
WC-MVB	64 (100)	39 (60.9)	–	14 (21.9)	11 (17.2)	–
C-MVB	45 (100)	7 (15.6)	–	***31 (68.9)***	7 (15.6)	–
E-MVB	49 (100)	6 (12.2)	–	***32 (65.3)***	11 (22.4)	–
LTVF (Fig. 5b)	47 (100)	***23 (48.9)***	–	***17 (36.2)***	7 (14.9)	–
CAVA (Fig. 5b)	43 (100)	***38 (88.4)***	–	5 (11.6)	0 (0.0)	–
IAB–CRB–MORB						
W-MVB	103 (100)	***66 (64.1)***	–	29 (28.2)	–	8 (7.8)
WC-MVB	64 (100)	18 (28.1)	–	**22 (34.4)**	–	***24 (37.5)***
C-MVB	45 (100)	5 (11.1)	–	***23 (51.1)***	–	**17 (37.8)**
E-MVB	49 (100)	1 (2.0)	–	***34 (69.4)***	–	14 (28.6)
LTVF (Fig. 5c)	47 (100)	***19 (40.4)***	–	***19 (40.4)***	–	9 (19.1)
CAVA (Fig. 5c)	43 (100)	***29 (67.4)***	–	1 (2.3)	–	13 (30.2)
IAB–OIB–MORB						
W-MVB	103 (100)	***66 (64.1)***	–	–	25 (24.3)	12 (11.7)
WC-MVB	64 (100)	**22 (34.4)**	–	–	11 (17.2)	***31 (48.4)***
C-MVB	45 (100)	5 (11.1)	–	–	10 (22.2)	***30 (66.7)***
E-MVB	49 (100)	2 (4.1)	–	–	**20 (40.8)**	***27 (55.1)***
LTVF (Fig. 5d)	47 (100)	***26 (55.3)***	–	–	11 (23.4)	10 (21.3)
CAVA (Fig. 5d)	43 (100)	***27 (62.8)***	–	–	1 (2.3)	**15 (34.9)**
CRB–OIB–MORB						
W-MVB[a]	103 (100)	–	–	26 (25.2)	66 (64.1)	11 (10.7)
WC-MVB	64 (100)	–	–	20 (31.2)	13 (20.3)	***31 (48.4)***
C-MVB	45 (100)	–	–	***20 (44.4)***	8 (17.8)	**17 (37.8)**
E-MVB	49 (100)	–	–	***23 (46.9)***	14 (28.6)	12 (24.5)
LTVF (Fig. 5e)	47 (100)	–	–	***32 (68.1)***	8 (17.0)	7 (14.9)
CAVA (Fig. 5e)[a]	43 (100)	–	–	2 (4.7)	0 (0.0)	41 (95.3)

MVB Mexican Volcanic Belt, *W* western, *WC* west-central, *C* central, *E* eastern, *LTVF* Los Tuxtlas volcanic field, *CAVA* Central American Volcanic Arc

[a] Inapplicable results and diagrams; boldface italic font shows the expected tectonic setting; whenever this setting is followed by a significant (>33.3%) success rate for another setting, it is shown by simple boldface number for success rate

rift as compared to 35.3–44.1% for arc; Table 7). Finally, arc setting is indicated from the final set of immobile element based diagrams, with success rates of 34.4–60.9% (Table 8; 34.4% for rift in one diagram only). Thus, for WC-MVB also, both tectonics of arc and rift play significant role in the genesis of mafic magmas (Tables 5, 6, 7, 8).

For C-MVB, both sets of major-element based diagrams undoubtedly show a continental rift setting, with success rates of 46.8–83.6% for the first set (Table 5) and 82.3–92.4% for the second set of diagrams (Table 6); the latter set is shown to be statistically correct. For the first set of immobile element based diagrams, although the number of

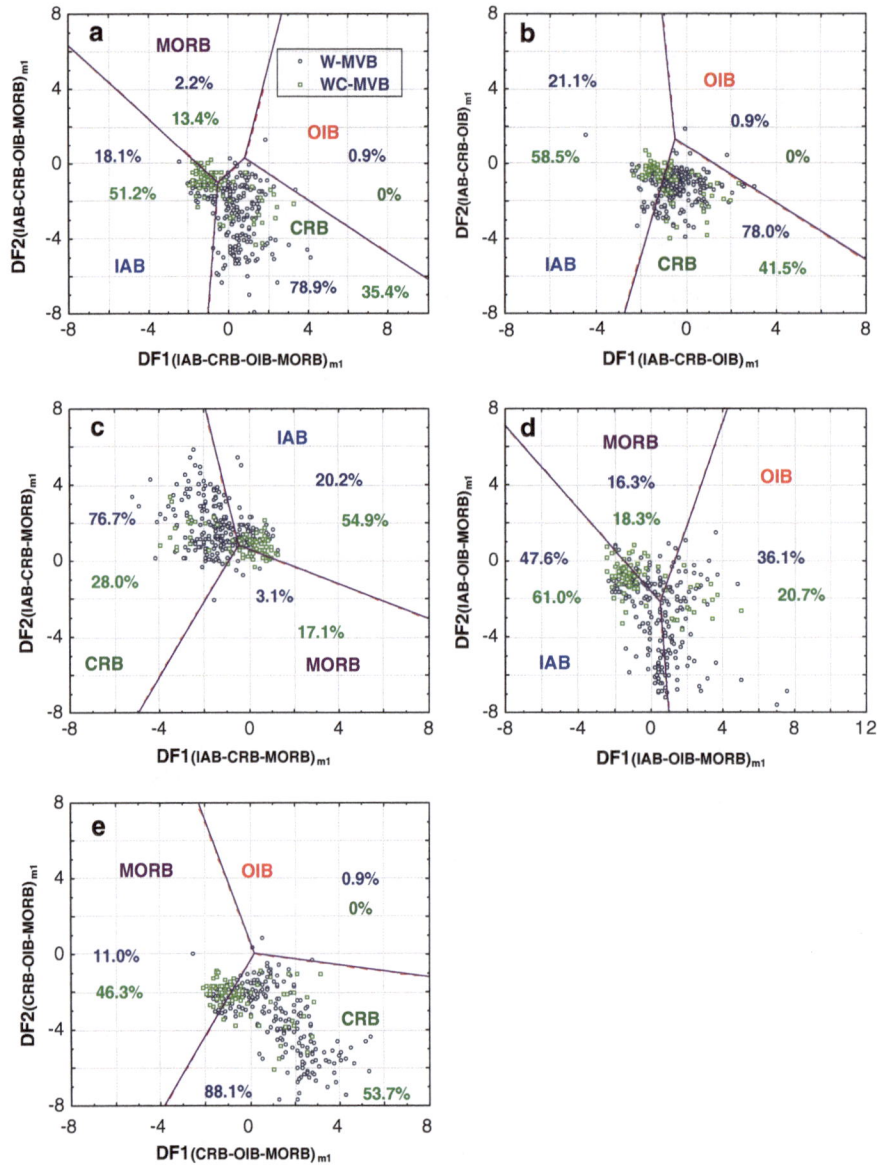

Figure 2
Application of the set of five major-element based discriminant function DF1–DF2 discrimination diagrams (see the subscript ₘ₁ in all these diagrams; AGRAWAL et al., 2004) for basic rocks from western Mexican Volcanic Belt (W-MVB) and west-central MVB (WC-MVB) compiled in this work. The percent values are given for each tectonic setting of island arc (IAB), continental rift (CRB), ocean-island (OIB), and mid-ocean ridge (MORB). The *symbols* are shown as inset in diagram (**a**). **a** Four tectonic settings IAB–CRB–OIB–MORB; **b** three tectonic settings IAB–CRB–OIB; **c** three tectonic settings IAB–CRB–MORB; **d** three tectonic settings IAB–OIB–MORB; and **e** three tectonic settings CRB–OIB–MORB

samples from the C-MVB is small (only 12 analyses; Table 7), the rift setting is more likely. Finally, from the second set of immobile element based diagrams a dual setting of rift and mid-ocean ridge (and not rift and arc) is inferred, with success rates for rift varying between 44.4 and 68.9% and for mid-ocean ridge

between 37.8 and 66.7% (Table 8). No indications of arc setting are obtained, which is consistent with the absence of deep earthquakes beneath the C-MVB in spite of a very dense seismic network (PACHECO and SINGH, 2010). These results would also be consistent with the recent suggestion by VERMA (2009) that

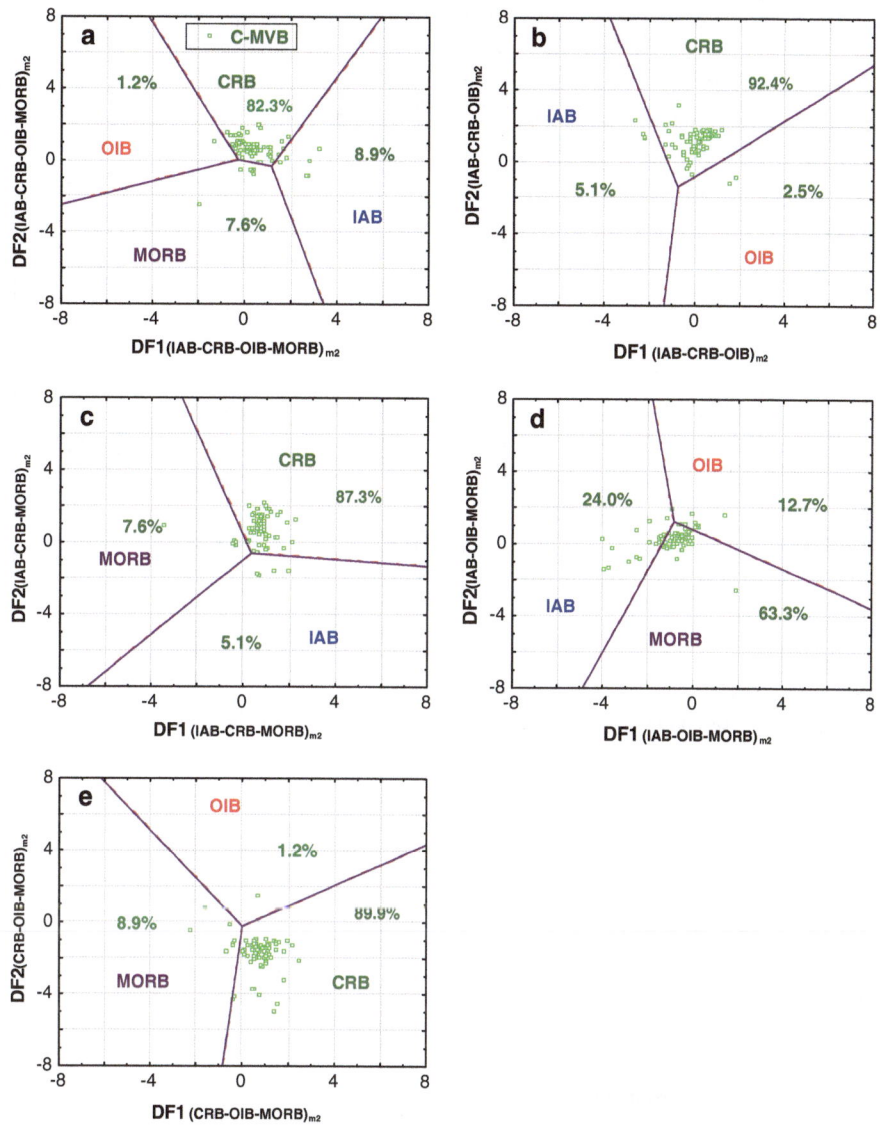

Figure 3
Application of the set of five major-element based discriminant function DF1–DF2 discrimination diagrams (see the subscript $_{m2}$ in all these diagrams; VERMA *et al.*, 2006) for basic rocks from central Mexican Volcanic Belt (C-MVB) compiled in this work. The percent values are given for tectonic setting of island arc (IAB), continental rift (CRB), ocean-island (OIB), and mid-ocean ridge (MORB). The *symbols* are shown as inset in diagram (**a**). **a** Four tectonic settings IAB–CRB–OIB–MORB; **b** three tectonic settings IAB–CRB–OIB; **c** three tectonic settings IAB–CRB–MORB; **d** three tectonic settings IAB–OIB–MORB; and **e** three tectonic settings CRB–OIB–MORB

magmas in the C-MVB are the result of continental rifting rather than the subduction of the Cocos plate beneath Mexico. The duality of tectonic setting of rift with mid-ocean ridge, and not with arc, is an interesting result at this stage, which should be confirmed by later studies.

For the final part of the MVB–E-MVB, both sets of major-element based diagrams clearly show a continental rift setting, with very high success rates of 60.9–83.2% for the first set and even higher of 83.2–94.4% for the second set. Similarly, both sets of immobile-element based diagrams favor a rift setting for the E-MVB, with success rates varying between 66.7 and 80.0% for the first set and between 46.9 and 69.4% for the second.

For LTVF, both sets of major-element based diagrams clearly indicate a rift setting, with very high

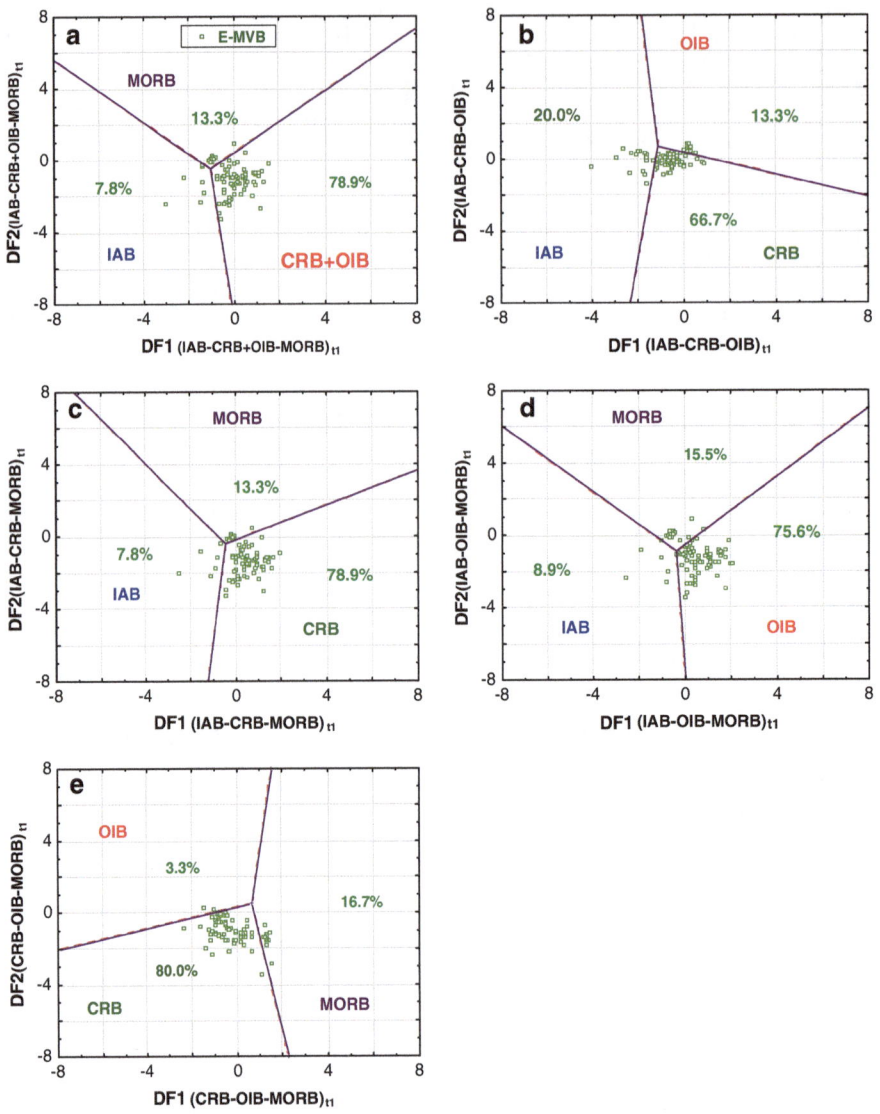

Figure 4

Application of the set of five major-element based discriminant function DF1–DF2 discrimination diagrams (see the subscript ₜ₁ in all these diagrams; AGRAWAL *et al.*, 2008) for basic rocks from eastern Mexican Volcanic Belt (E-MVB) compiled in this work. The percent values are given for each tectonic setting of island arc (IAB), combined tectonic setting of continental rift and ocean-island (CRB + OIB), continental rift (CRB), ocean-island (OIB), and mid-ocean ridge (MORB). The *symbols* are shown as inset in diagram (**a**). **a** Three tectonic settings IAB–CRB + OIB–MORB; **b** three tectonic settings IAB–CRB–OIB; **c** three tectonic settings IAB–CRB–MORB; **d** three tectonic settings IAB–OIB–MORB; and **e** three tectonic settings CRB–OIB–MORB

success rates of 82.7–89.3 and 89.3–98.7% for the first and second sets, respectively (Tables 5, 6). For the first set of immobile-element diagrams, insufficient samples (only 19 analyses) were available in our database, but the indications are mostly in favor of a rift setting (Table 7). The final set of immobile-element based diagrams shows a dual setting of arc and rift, with success rates of 40.4–55.3% and 36.2–

68.1%, respectively (Table 8). As stated above for W-MVB, the major-element based diagrams should be given higher weight when the results of these four sets of diagrams are mutually inconsistent, which would make the conclusion of a rift setting for the LTVF more likely.

For CAVA, both sets of major-element based diagrams clearly provide an arc setting, with high

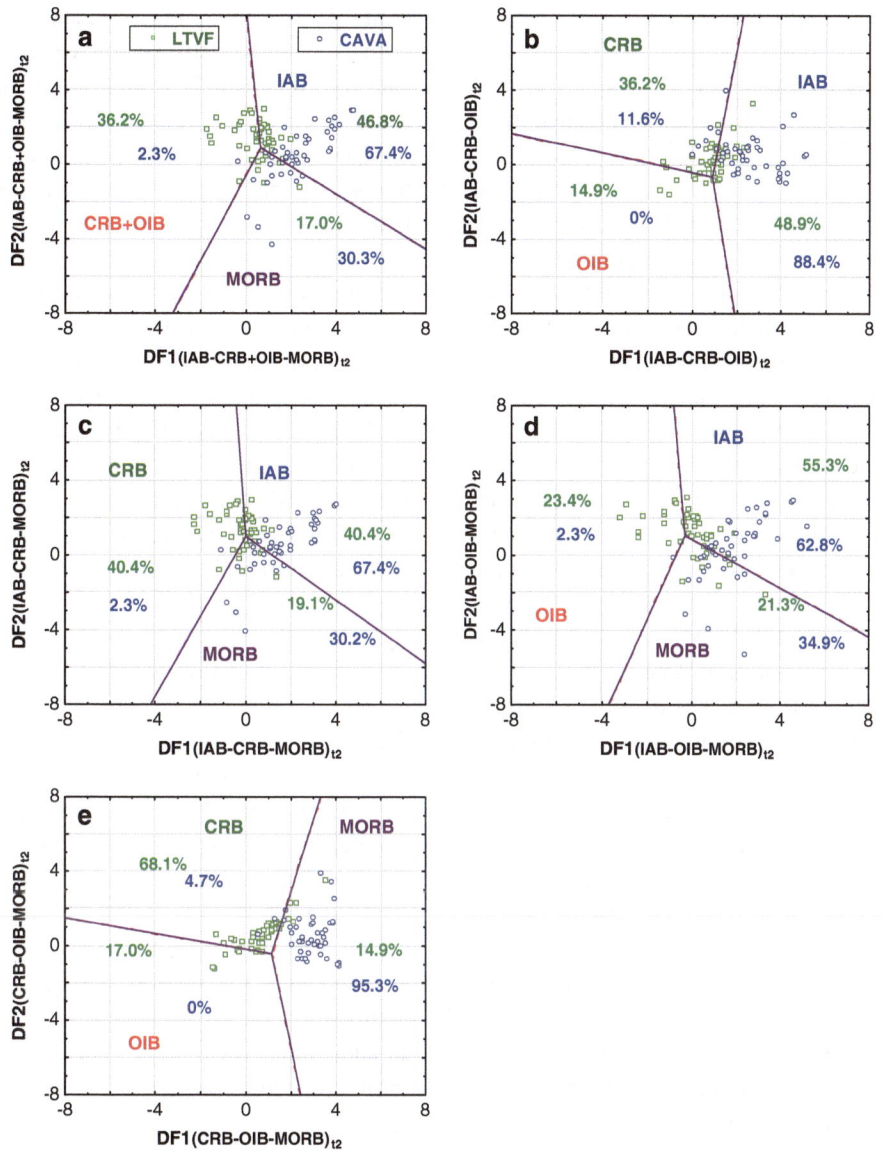

Figure 5
Application of the set of five major-element based discriminant function DF1–DF2 discrimination diagrams (see the subscript $_{t2}$ in all these diagrams; VERMA and AGRAWAL, 2010) for basic rocks from Los Tuxtlas volcanic field (LTVF) and Central America Volcanic Arc (CAVA) compiled in this work. The percent values are given for each tectonic setting of island arc (IAB), combined tectonic setting of continental rift and ocean-island (CRB + OIB), continental rift (CRB), ocean-island (OIB), and mid-ocean ridge (MORB). The *symbols* are shown as inset in diagram (**a**). **a** Three tectonic settings IAB–CRB + OIB–MORB; **b** three tectonic settings IAB-CRB-OIB; **c** three tectonic settings IAB–CRB–MORB; **d** three tectonic settings IAB–OIB–MORB; and **e** three tectonic settings CRB–OIB–MORB

success rates of 74.8–99.2% (Table 5) and 76.5–82.4% (Table 6). The first set of immobile-element diagrams show a dual setting of arc and mid-ocean ridge (with no indications of a rift setting; Table 7) for this volcanic province, with high success rates for mid-ocean ridge (47.6–97.6%) compared to arc

setting (40.5–52.4%). However, the other set of immobile-element based diagrams clearly indicates an arc setting for CAVA, with success rates of 62.8–88.4% (Table 8); the last diagram CRB-OIB-MORB should be considered as inapplicable because the expected arc (IAB) setting is absent from this case.

5. Final Considerations

All evaluations of the four tectonic settings of island arc, continental rift, ocean-island, and mid-ocean ridge in several studies (AGRAWAL *et al.*, 2004, 2008; VERMA *et al.*, 2006; AGRAWAL and VERMA, 2007; SHETH, 2008; VERMA, 2010; this work) have shown good functioning of these diagrams for samples from these four known tectonic settings. The continental arc of the Andes was also successfully evaluated for its similarities with island arc.

For their application to the MVB, LTVF, and CAVA, it is not clear why the major-elements, especially in the second set of statistically correct diagrams, clearly indicate a single tectonic setting, whereas the trace-elements, in some cases, favor a dual tectonic setting. Nevertheless, the single tectonic setting for both MVB and LTVF is shown to be a continental rift, and for CAVA an arc, which confirms the earlier conclusions by VERMA (2002, 2004, 2006) from other constraints. The duality of tectonic settings in only a few (not all) diagrams was observed for W-MVB (arc–rift), WC-MVB (arc–rift) and LTVF (arc–rift), C-MVB (rift–mid-ocean ridge), and CAVA (arc–mid-ocean ridge). Do these volcanic areas represent a true transition between two main tectonic settings, or might some other factor explain them?

This question is clearly open for future work. One reason may be related to the data quality for trace-elements, especially Nb, Yb, Sm, La, and Y, involved in these trace-element-based diagrams. Therefore, for fresh rocks compiled from Southern Mexico and Central America, the conclusions from the major-element based diagrams could be given more weight as compared to the trace-element based ones. The other source of such discrepancies may be the fact that the diagrams are strictly applicable to an island arc instead of a continental arc setting.

Nevertheless, a continental rift setting (with less or negligible influence of the subduction process) and a mantle upwelling rift-model can be inferred for the MVB and LTVF, whereas an arc setting (subduction-related model) is confirmed for the CAVA.

Finally, the clear continental rift setting for both eastern and central parts of the MVB (E-MVB and C-MVB) as well as the dominance of this tectonic setting for both western and west-central parts (W-MVB and WC-MVB) indicate that this volcanic province could be called the Mexican Volcanic Rift, provided the dominance of the rift setting is also confirmed in future for intermediate and felsic magmas. Some indications, based on large ion lithophile, rare-earth, and high field strength elements, were recently documented by VERMA (2009) for the C-MVB. For evaluating the tectonic significance from intermediate and felsic magmas, new discrimination diagrams will have to be proposed using correct statistical methodologies, as pointed out in the present work. A new computer program, TecD by VERMA and RIVERA-GÓMEZ (2010), will be available for efficient use of these four sets of discriminant function diagrams for basic and ultrabasic magmas.

Acknowledgments

We are grateful to one of the guest editors, Bill Bandy, for extending an invitation to contribute to this special issue. Two of us (SKV and MARG) are grateful to Secretaría de Relaciones Exteriores and Conacyt, respectively, for fellowships towards respective Ph.D. and Master's studies. The first author (SPV) thanks the Universidad Autónoma Metropolitana for support during the development of this work. We are grateful to Samuele Agostini and an anonymous reviewer for helpful comments on an earlier version of this paper.

REFERENCES

AGOSTINI, S., CORTI, G., DOGLIONI, C., CARMINATI, E., INNOCENTI, F., TONARINI, S., MANETTI, P., DI VINCENZO, G., and MONTANARI, D. (2006), *Tectonic and magmatic evolution of the active volcanic front in El Salvador: insight into the Berlín and Ahuachapán geothermal areas*, Geothermics 35, 368–408.

AGRAWAL, S. (1999), *Geochemical discrimination diagrams: a simple way of replacing eye-fitted boundaries with probability based classifier surfaces*, J. Geol. Soc. India 54, 335–346.

AGRAWAL, S. GUEVARA, M., and VERMA, S.P. (2004), *Discriminant analysis applied to establish major-element field boundaries for tectonic varieties of basic rocks*, Int. Geol. Rev. 46, 575–594.

AGRAWAL, S. GUEVARA, M., and VERMA, S.P. (2008), *Tectonic discrimination of basic and ultrabasic rocks through log-transformed ratios of immobile trace elements*, Int. Geol. Rev. 50, 1057–1079.

AGRAWAL, S. and VERMA, S.P. (2007), *Comment on "Tectonic classification of basalts with classification trees" by Pieter Vermeesch (2006)*, Geochim. Cosmochim. Acta *71*, 3388–3390.

AITCHISON, J., The statistical analysis of compositional data. In: *The statistical analysis of compositional data* (Chapman and Hall, London and New York, pp. 416, 1986).

AITCHISON, J., BARCELÓ-VIDAL, C., MARTÍN-FERNÁNDEZ, J.A., and PAWLOWSKY-GLAHN, V. (2000), *Logratio analysis and compositional distance*, Math. Geol. *32*, 271–275.

AITCHISON, J. and EGOZCUE, J.J. (2005), *Compositional data analysis: where are we and where should we be heading?*, Math. Geol. *37*, 829–850.

ALLAN, J.F. (1986), *Geology of the northern Colima and Zacoalco grabens, southwest Mexico: Late Cenozoic rifting in the Mexican Volcanic Belt*, Geol. Soc. Am. Bull. *97*, 473–485.

ALLAN, J.F. and CARMICHAEL, I.S.E. (1984), *Lamprophyric lavas in the Colima graben, SW Mexico*, Contrib. Mineral. Petrol. *88*, 203–216.

ALLAN, J.F., NELSON, S.A., LUHR, J.F., CARMICHAEL, I.S.E., WOPAT, M., and WALLACE, P.J., Pliocene-Holocene rifting and associated volcanism in southwest Mexico: an exotic terrane in the making. In *The Gulf and Peninsular province of the Californias* (eds. DAUPHIN, J.P. and SIMONEIT, B.R.T.) (The American Association of Petroleum Geologists, Tulsa, Oklahoma, pp. 425–445, 1991).

ALVARADO, G.E., CARR, M.J., TURRIN, B.D., SCHMINCKE, H.-U., and HUDNUT, K.W., Recent volcanic history of Irazú volcano, Costa Rica: alternation and mixing of two magma batches, and pervasive mixing. In: *Volcanic hazards in Central America* (eds. ROSE, W.I., BLUTH, G.J.S., CARR, M.J., EWERT, J., PATINO, L.C., and VALLANCE, J.) (Geological Society of America, pp. 259–276, 2006).

ARCE, J.L., MACÍAS, R., GARCÍA PALOMO, A., CAPRA, L., MACÍAS, J.L., LAYER, P., and RUEDA, H. (2008), *Late Pleistocene flank collapse of Zempoala volcano (central Mexico) and the role of fault reactivation*, J. Volcanol. Geotherm. Res. *177*, 944–958.

BANDY, W.L. and HILDE, T.W.C., Morphology and recent history of the ridge propagator system located at 18°N, 106°W. In: *Cenozoic tectonics and volcanism of Mexico* (eds. Delgado-Granados, H., Aguirre-Díaz, G., and Stock, J.M.) (Geological Society of America, pp. 29–40, 2000).

BANDY, W.L., HILDE, T.W.C., and YAN, C.-Y., The Rivera-Cocos plate boundary: implications for Rivera-Cocos relative motion and plate fragmentation. In: *Cenozoic tectonics and volcanism of Mexico* (eds. DELGADO-GRANADOS, H., AGUIRRE-DÍAZ, G., and STOCK, J.M.) (Geological Society of America, pp. 1–28, 2000).

BARDINTZEFF, J.M. and DENIEL, C. (1992), *Magmatic evolution of Pacaya and Cerro Chiquito volcanological complex, Guatemala*, Bull. Volcanol. *54*, 267–283.

BARNETT, V. and LEWIS, T., *Outliers in statistical data* (Wiley, Chichester 1994).

BAXTER, M.J., BEARDAH, C.C., COOL, H.E.M., and JACKSON, C.M. (2005), *Compositional data analysis of some alkaline glasses*, Math. Geol. *37*, 183–196.

BESCH, T., VERMA, S.P., KRAMMER, U., NEGENDANK, J.F.W., TOBSCHALL, H.J., and EMMERMANN, R. (1995), *Assimilation of sialic crustal material by volcanics of the easternmost extension of the Trans-Mexican Volcanic Belt- Evidence from Sr and Nd isotopes*, Geofís. Int. *34*, 263–281.

BLATTER, D.L., CARMICHAEL, I.S.E., DEINO, A.L., and RENNE, P.R. (2001), *Neogene volcanism at the front of the central Mexican volcanic belt: basaltic andesites to dacites, with contemporaneous shoshonites and high-TiO$_2$ lava*, Geol. Soc. Am. Bull. *113*, 1324–1342.

BLATTER, D.L., FARMER, G.L., and CARMICHAEL, I.S.E. (2007), *A north-south transect across the Central Mexican Volcanic Belt at ~100°W: spatial distribution, petrological, geochemical, and isotopic characteristics of Quaternary volcanism*, J. Petrol. *48*, 901–950.

BLATTER, D.L. and HAMMERSLEY, L. (2009), *Impact of the Orozco Fracture Zone on the Central Mexican Volcanic Belt*, J. Volcanol. Geotherm. Res. (in press).

BOLGE, L.L., CARR, M.J., FEIGENSON, M.D., and ALVARADO, G.E. (2006), *Geochemical stratigraphy and magmatic evolution at Arenal volcano, Costa Rica*, J. Volcanol. Geotherm. Res. *157*, 34–48.

BOUDAL, C. (1985), Pétrologie d'un grand volcan andésitique mexicain: le Popocatepetl: Thèse de Doctorat, Univ. Clermont-Ferrand II, 140 p.

BRUNI, S., D'ORAZIO, M., HALLER, M.J., INNOCENTI, F., MANETTI, P., PÉCSKAY, Z., and TONARINI, S. (2008), *Time-evolution of magma sources in a continental back-arc setting: the Cenozoic basalts from Sierra de San bernardo (Patagonia, Chubut, Argentina)*, Geol. Mag. *145*, 714–732.

BUCCIANTI, A., MATEAU-FIGUERAS, G., and PAWLOWSKY-GLAHN, V., Compositional data analysis in the geosciences: from theory to practice (Geological Society Special Publication No. 262, London 2006).

BURBACH, G.V., FROHLICH, C., PENNINGTON, W.D., and MATUMOTO, T. (1984), *Seismicity and tectonics of the subducted Cocos plate*, J. Geophys. Res. *89*, 7719–7735.

BUTLER, J.C. and WORONOW, A. (1986), *Discrimination among tectonic settings using trace element abundances of basalts*, J. Geophys. Res. *91*, 10289–10300.

CABANIS, B. and LECOLLE, M. (1989), *Le diagramme La/10–Y/15–Nb/8: un outil pour la discrimination des séries volcaniques et la mise en évidence des processus de mélange et/ou de contamination crustale*, C.R. Acad. Sci. Paris, *309*, 2023–2029.

CARMICHAEL, I.S.E., FREY, H.M., LANGE, R.A., and HILL, C.M. (2006), *The Pleistocene cinder cones surrounding Volcán Colima, Mexico re-visited: eruption ages and volumes, oxidation states, and sulfur content*, Bull. Volcanol. *68*, 407–419.

CARMICHAEL, I.S.E., LANGE, R.A., and LUHR, J.F. (1996), *Quaternary minettes and associated volcanic rocks of Mascota, western Mexico: a consequence of plate extension above a subduction modified mantle wedge*, Contrib. Mineral. Petrol. *124*, 302–333.

CARR, M.J. (1984), *Symmetrical and segmented variation of physical and geochemical characteristics of the Central American volcanic front*, J. Volcanol. Geotherm. Res. *20*, 231–252.

CARR, M.J., FEIGENSON, M.D., and BENNETT, E.A. (1990), *Incompatible element and isotopic evidence for tectonic control of source mixing and melt extraction along the Central American arc*, Contrib. Mineral. Petrol. *105*, 369–380.

CARRASCO-NÚÑEZ, G., RIGHTER, K., CHESLEY, J., SIEBERT, L., and ARANDA-GÓMEZ, J.J. (2005), *Contemporaneous eruption of calc-alkaline and alkaline lavas in a continental arc (Eastern Mexican Volcanic Belt): chemically heterogeneous but isotopically homogeneous source*, Contrib. Mineral. Petrol. *150*, 423–440.

CATHELINEAU, M., OLIVER, R., and NIEVA, D. (1987), *Geochemistry of volcanic series of the Los Azufres geothermal field (Mexico)*, Geofís. Int. *26*, 273–290.

CHAKRABARTI, R., BASU, A.R., SANTO, A.P., TEDESCO, D., and VASELLI, O. (2009), *Isotopic and geochemical evidence for a* hetrogeneous mantle plume origin of the Virunga volcanics, Western rift, East African Rift system, Chem. Geol. 259, 273–289.

CHESLEY, J., RUIZ, J., RIGHTER, K., FERRARI, L., and GOMEZ-TUENA, A. (2002), Source contamination versus assimilation: an example from the Trans-Mexican Volcanic Arc, Earth Planet. Sci. Lett. 195, 211–221.

DEBAILLE, V., BLICHERT-TOFT, J., AGRANIER, A., DOUCELANCE, R., SCHIANO, P., and ALBAREDE, F. (2006), Geochemical component relationship in MORB from the Mid-Atlantic Ridge, 22–35 N, Earth Planet. Sci. Lett. 241, 844–662.

DELGADO, H., MOLINERO, R., CERVANTES, P., NIETO-OBREGÓN, J., LOZANO-SANTA CRUZ, R., MACÍAS-GONZÁLEZ, H.L., MENDOZA-ROSALES, C., and SILVA-ROMO, G. (1998), Geology of Xitle volcano in southern Mexico City—a 2000-year-old monogenetic volcano in an urban area, Rev. Mex. Cienc. Geol. 15, 115–131.

DEMANT, A. (1981), L'axe néo-volcanique transmexicain, étude volcanologique et pétrographique, signification géodynamique Faculté des Sciences et Techniques de St. Jérome (Université de Droit, d'Economie et des Sciences d'Aix-Marseille).

DERUELLE, B. (1982), Petrology of the Plio-Quaternary volcanism of the south-central and meridional Andes. J. Volcanol. Geotherm. Res. 14, 77–124.

DIXON, J., CLAGUE, D.A., COUSENS, B., MONSALVE, M.L., and UHL, J. (2008), Carbonatite and silicate melt metasomatism of the mantle surrounding the Hawaiian plume: Evidence from volatiles, trace elements, and radiogenic isotopes in rejuvenated-stage lavas from Niihau, Hawaii., G3 9, 1–34.

DUFRANE, S.A., ASMEROM, Y., MUKASA, S.B., MORRIS, J.D., and DREYER, B.M. (2006), Subduction and melting processes inferred from U-series, Sr-Nd-Pb isotope, and trace element data, Bicol and Bataan arcs, Philippines., Geochemical et Cosmochimica Acta 70, 3401–3420.

EGOZCUE, J.J. and PAWLOWSKY-GLAHN, V. (2005), Groups of parts and their balances in compositional data analysis, Math. Geol. 37, 795–828.

EGOZCUE, J.J., PAWLOWSKY-GLAHN, V., MATEU-FIGUERAS, G., and BARCELÓ-VIDAL, C. (2003), Isometric logratio transformations for compositional data analysis, Math. Geol. 35, 279–300.

ELBURG, M.A. and KAMENETSKY, V.S. (2007), The origin of medium-K ankaramitic arc magmas from Lombok (Sunda arc, Indonesia): Mineral and melt inclusion evidence, Chem. Geol. 240, 260–279.

ESPÍNDOLA, J.M., ZAMORA-CAMACHO, A., GODINEZ, M.L., SCHAAF, P., and RODRÍGUEZ, S.R. (2009), The 1793 eruption of San Martín Tuxtla volcano, Veracruz, Mexico, J. Volcanol. Geotherm. Res. (in press).

FERLITO, C., COLTORTI, M., CRISTOFOLINI, R., and GIACOMONI, P.P. (2009), The contemporaneous emission of low-K and high-K trachybasalts and the role of the NE Rift during the 2002 eruptive event, Mt. Etna, Italy, Bull. Volcanol. 71, 575–587.

FERRARI, L., CONTICELLI, S., VAGGELLI, G., PETRONE, C.M., and MANETTI, P. (2000), Late Miocene volcanism and intra-arc-tectonics during the early development of the Trans-Mexican Volcanic Belt, Tectonophysics 318, 161–185.

FERRARI, L., GARDUÑO, V.H., INNOCENTI, F., MANETTI, P., PASQUARE, G., and VAGGELLI, G. (1994), A widespread mafic volcanic unit at the base of the Mexican Volcanic Belt between Guadalajara and Querétaro, Geofís. Int. 33, 107–123.

FERRARI, L., PETRONE, C.M., and FRANCALANCI, L. (2001), Generation of oceanic-island basalt-type volcanism in the western

Trans-Mexican volcanic belt by slab rollback, asthenosphere infiltration, and variable flux melting, Geology 29, 507–510.

FERRARI, L., PETRONE, C.M., and FRANCALANCI, L. (2002), Reply: "Generation of oceanic-island basalt type volcanism in the western Trans-Mexican volcanic belt by slab rollback, asthenosphere infiltration, and variable flux melting", Geology 114, 858–859.

FERRARI, L. and ROSAS-ELGUERA, J. (1999), Alkalic (ocean-island basalt type) and calc-alkaline volcanism in the Mexican volcanic belt: a case for plume-related magmatism and propagating rifting at an active margin? Comment and reply, Geology 27, 1055–1056.

FERRIZ, H. and MAHOOD, G.A. (1987), Strong compositional zonation in a silicic magmatic system: Los Humeros, Mexican Neovolcanic Belt, J. Petrol. 28, 171–209.

FINNEY, B., TURNER, S., HAWKESWORTH, C., LARSEN, J., NYE, C., GEORGE, R., BINDEMAN, I., and EICHELBERGER, J. (2008), Magmatic differentiation at an Island -arc Caldera: Okmok Volcano, Aleutian Islands, Alaska, J. Petrol. 49, 857–884.

FREY, F.A., GERLACH, D.C., HICKEY, R.L., LOPEZ-ESCOBAR, L., And MUNIZAGA-VILLAVICENCIO, F. (1984), Petrogenesis of the Laguna del Maule volcanic complex, Chile (36°S), Contrib. Mineral. Petrol. 88, 133–149.

FREY, H.M., LANGE, R.A., HALL, C.M., DELGADO-GRANADOS, H., and CARMICHAEL, I.S.E. (2007), A Pliocene ignimbrite flare-up along the Tepic-Zacoalco rift: evidence for the initial stages of rifting between the Jalisco block (Mexico) and North America, Geol. Soc. Am. Bull. 119, 49–64.

GARCÍA-PALOMO, A., MACÍAS, J.L., TOLSON, G., VALDEZ, G., and MORA, J.C. (2002), Volcanic stratigraphy and geological evolution of the Apan region, east-central sector of the Trans-Mexican volcanic belt, Geofís. Int. 41, 133–150.

GASTIL, G., KRUMMENACHER, D., and MINCH, J. (1979), The record of Cenozoic volcanism around the Gulf of California, Geol. Soc. Am. Bull. 90, 839–857.

GERLACH, D.C., FREY, F.A., MORENO-ROA, H., and LOPEZ-ESCOBAR, L. (1988), Recent volcanism in the Puyehue-Cordon Caulle region, southern Andes, Chile (40.5°S): petrogenesis of evolved lavas, J. Petrol. 29, 333–382.

GILBERT, C.M., MAHOOD, G.A., and CARMICHEL, I.S.E. (1985), Volcanic stratigraphy of the Guadalajara area, Mexico, Geofís. Int. 24, 169–191.

GÓMEZ-TUENA, A., LAGATTA, A.B., LANGMUIR, C.H., GOLDSTEIN, S.L., ORTEGA-GUTIÉRREZ, F., and CARRASCO-NUÑEZ, G. (2003), Temporal control of subduction magmatism in the eastern Trans-Mexican volcanic belt: mantle sources, slab contributions, and crustal contamination, Geochem. Geophys, Geosyst. 4, 8912. doi:10.1029/2003GC000524.

GÓMEZ-TUENA, A., LANGMUIR, C.H., GOLDSTEIN, S.L., STRAUB, S.M., and ORTEGA-GUTIÉRREZ, F. (2007), Geochemical evidence for slab melting in the Trans-Mexican Volcanic Belt, J. Petrol. 48, 537–562.

GUNN, B.M. and MOOSER, F. (1971), Geochemistry of the volcanics of central Mexico, Bull. Volcanol. 34, 577–613.

HARRY, D.L. and GREEN, N.L. (1999), Slab dehydration and basalt petrogenesis in subduction systems involving very young oceanic lithosphere, Chem. Geol. 160, 309–333.

HASENAKA, T., Chemical compositions of selected samples. In Subduction volcanism and tectonics of western Mexican Volcanic Belt. International Scientific Research Program (No. 03041014) Japan-Mexico Co-operative Research (ed. AOKI, K.) (The

Faculty of Science, Tohoku University, Sendai, Japan, pp. 238–247, 1992).

HASENAKA, T. and CARMICHAEL, I.S.E. (1987), *The cinder cones of Michoacán-Guanajuato, Central Mexico: petrology and chemistry*, J. Petrol. *28*, 241–269.

HAZLETT, R.W. (1987), *Geology of San Cristobal volcanic complex, Nicaragua*, J. Volcanol. Geotherm. Res. *33*, 223–230.

HELLEVANG, B. and PEDERSEN, R.B. (2008), *Magma ascent and crustal accretion at ultraslow-spreading ridges: constraints from plagioclase ultraphyric basalts from the Arctic Mid-Ocean Ridge*, J. Petrol. *49*, 267–294.

HICKEY-VARGAS, R., MORENO ROA, H., LOPEZ ESCOBAR, L., AND FREY, F.A. (1989), *Geochemical variations in Andean basaltic and silicic lavas from the Villarrica-Lanin volcanic chain (39.5°S): an evaluation of source heterogeneity, fractional crystallization and crustal assimilation*, Contrib. Mineral. Petrol. *103*, 361–386.

IRELAND, T.J., WALKER, R.J., and GARCIA, M.O. (2009), *Highly siderophile elements and 187Os isotope systematics of Hawaiian picrites: Implications for parental melt composition and source heterogeneity*, Chem. Geol. *260*, 112–128.

JOHNSON, C.A. and HARRISON, C.G.A. (1989a), *Thematic mapper studies of volcanism and tectonism in central Mexico*, Adv. Space Res. *9*, 85–88.

JOHNSON, C.A. and HARRISON, C.G.A. (1989b), *Tectonics and volcanism in central Mexico: a Landsat thematic mapper perspective*, Remote Sens. Environ. *28*, 273–286.

LA FEMINA, P.C., CONNOR, C.B., HILL, B.E., STRAUCH, W., and SABALLOS, A. (2004), *Magma-tectonic interactions in Nicaragua: the 1999 seismic swarm and eruption of Cerro Negro volcano*, J. Volcanol. Geotherm. Res. *137*, 187–199.

LANGE, R.A. and CARMICHAEL, I.S.E. (1991), *A potassic volcanic front in western Mexico: the lamprophyric and related lavas of San Sabastian*, Geol. Soc. Am. Bull. *103*, 928–940.

LE BAS, M.J. (2000), *IUGS reclassification of the high-Mg and picritic volcanic rocks*, J. Petrol. *41*, 1467–1470.

LE BAS, M.J., LE MAITRE, R.W., STRECKEISEN, A., and ZANETTIN, B. (1986), *A chemical classification of volcanic rocks based on the total alkali-silica diagram*, J. Petrol. *27*, 745–750.

LEEMAN, W.P., CARR, M.J., and MORRIS, J.D. (1994), *Boron geochemistry of the Central American Volcanic Arc: Constraints on the genesis of subduction-related magmas*, Geochim. Cosmochim. Acta *58*, 149–168.

LEWIS-KENEDI, C.B., LANGE, R.A., HALL, C.M., and DELGADO GRANADOS, H. (2005), *The eruptive history of the Tequila volcanic field, western Mexico: ages, volumes, and relative proportions of lava types*, Bull. Volcanol. *67*, 391–414.

LÓPEZ-ESCOBAR, L., VERGARA, M., and FREY, F.A. (1981), *Petrology and geochemistry of lavas from Antuco volcano, a basaltic volcano of the southern Andes (37°25′)*, J. Volcanol. Geotherm. Res. *11*, 329–352.

LÓPEZ-ESCOBAR, L., TAGIRI, M., and VERGARA, M. (1991), *Geochemical features of southern Andes Quaternary volcanics between 41°5′ and 43°00′S.*, Geol. Soc. Am. Special Pap. *265*, 45–56.

LÓPEZ-ESCOBAR, L., KILIAN, R., KEMPTON, P.D., and TAGIRI, M. (1993), *Petrography and geochemistry of Quaternary rocks from the southern volcanic zone of the Andes between 41°30′ and 46°00′S, Chile*, Rev. Geol. Chile *20*, 33–35.

LUHR, J.F. (1997), *Extensional tectonics and the diverse primitive volcanic rocks in the western Mexican Volcanic Belt*, Can. Min. *35*, 473–500.

LUHR, J.F., ALLAN, J.F., CARMICHAEL, I.S.E., NELSON, S.A., and HASENAKA, T. (1989), *Primitive calc-alkaline and alkaline rock types from the western Mexican Volcanic Belt*, J. Geophys. Res. *94*, 4515–4530.

LUHR, J.F. and CARMICHAEL, I.S.E. (1981), *The Colima volcanic complex, Mexico: Part II. Late-Quaternary cinder cones*, Contrib. Mineral. Petrol. *76*, 127–147.

LUHR, J.F. and CARMICHAEL, I.S.E. (1985), *Jorullo Volcano, Michoacán, México (1759–1774): The earliest stages of fractionation in calc-alkaline magmas*, Contrib. Mineral. Petrol. *90*, 142–161.

LUHR, J.F., NELSON, S.A., ALLAN, J.F., and CARMICHAEL, I.S.E. (1985), *Active rifting in southwestern Mexico: Manifestations of an incipient eastward spreading-ridge jump*, Geology *13*, 54–57.

LYLE, M. and NESS, G.E., *The opening of the southern Gulf of California*. In *The Gulf and Peninsular province of the Californias* (eds. DAUPHIN, J.P. and SIMONEIT, B.R.T.) (The American Association of Petroleum Geologists, Tulsa, Oklahoma, pp. 403–423, 1991).

MACDONALD, R., BELKIN, H.E., FITTON, J.G., ROGERS, N.W., NEJBERT, K., TINDLE, A.G., and MARSHALL, A.S. (2008), *The role of fractional crystellization, magma mixing, crystal mush remobilization and volatile-melt interactions in the genesis of young basalt-peralkaline rhyolite suite, the Greater Olkaria Volcanic Complex, Kenya Rift Valley*, J. Petrol. *49*, 1515–1547.

MARIA, A.H. and LUHR, J.F. (2008), *Lamprophyres, basanites, and basalts of the western Mexican Volcanic Belt: volatile contents and a vein-wall rock melting relationship*, J. Petrol. *49*, 2123–2156.

MÁRQUEZ, A., OYARZUN, R., DOBLAS, M., and VERMA, S.P. (1999a), *Alkalic (ocean-island basalt type) and calc-alkalic volcanism in the Mexican Volcanic Belt: a case for plume-related magmatism and propagating rifting at an active margin?*, Geology *27*, 51–54.

MÁRQUEZ, A., OYARZUN, R., DOBLAS, M., and VERMA, S.P. (1999b), *Reply (to Comment by L. Ferrari and J. Rosas Elguera on "Alkalic (ocean basalt type) and calc-alkalic volcanism in the Mexican volcanic belt: a case of plume-related magmatism and propagating rift at an active margin?" Comment and Reply*, Geology *27*, 1055–1056.

MARSKE, J.P., GARCIA, M.O., PIETRUSZKA, A.J., RHODES, J.M., and NORMAN, M.D. (2008), *Geochemical variations during Kilauea's pu'u 'O'o eruption reveal a fine-scale mixture of mantle heterogeneities within the Hawaiian plume*, J. Petrol. *49*, 1297–1318.

MARTIN DEL POZZO, A.L., Geoquímica y paleomagnetismo de la Sierra Chichinautzin *Facultad de Ciencias* (U.N.A.M., Mexico, D.F., pp. 148, 1989).

MARTIN DEL POZZO, A.L., ROMERO, V.H., and RUIZ KITCHER, R.E. (1987), *Los flujos piroclásticos del volcán de Colima, México*, Geofís. Int. *26*, 291–307.

MCDERMOTT, F., DELFIN JR, F.G., DEFANT, M.J., TURNER, S., and MAURY, R. (2005), *The Petrogenesis of volcanics from Mt. Bulusan and Mt. Mayon in the Bicol arc, the Philippines*, Contrib. Mineral. Petrol. *150*, 652–670.

MERIGGI, L., MACÍAS, J.L., TOMMASINI, S., CAPRA, L., and CONTICELLI, S. (2008), *Heterogeneous magmas of the Quaternary Sierra Chichinautzin volcanic field (central Mexico): the role of an amphibole-bearing mantle and magmatic evolution processes*, Rev. Mex. Cienc. Geol. *25*, 197–216.

MESCHEDE, M. (1986), *A method of discriminating between different types of mid-ocean ridge basalts and continental tholeiites with the Nb–Zr–Y diagram*, Chem. Geol. *56*, 207–218.

MIDDLEMOST, E.A.K. (1989), *Iron oxidation ratios, norms and the classification of volcanic rocks*, Chem. Geol. *77*, 19–26.

MOORE, G., MARONE, C., CARMICHAEL, I.S.E., and RENNE, P. (1994), *Basaltic volcanism and extension near the intersection of the Sierra Madre volcanic province and the Mexican Volcanic Belt*, Geol. Soc. Am. Bull. *106*, 383–394.

MORI, L., GÓMEZ-TUENA, A., SCHAAF, P., GOLDSTEIN, S.J., PÉREZ-ARVIZU, O., and SOLÍS-PICHARDO, G. (2009), *Lithospheric removal as a trigger for flood basalt magmatism in the Trans-Mexican Volcanic Belt*, J. Petrol. *50*, 2157–2186.

MULLEN, E.D. (1983), *MnO/TiO₂/P₂O₅: a minor element discrimination for basaltic rocks of oceanic environments and its implications for petrogenesis*, Earth Planet. Sci. Lett. *62*, 53–62.

NAKAMURA, K., KATO, Y., TAMAKI, K., and ISHII, T. (2007), *Geochemistry of hydrothermally altered basaltic rocks from the Southwest Indian Ridge near Rodriguez Triple Junction*, Marine Geol. *239*, 125–141.

NEGENDANK, J.F.W. (1972), *Geochemical aspects of volcanic rocks of the Valley of Mexico*, Geofís. Int. *11*, 267–278.

NEGENDANK, J.F.W., EMMERMANN, R., KRAWCZYK, R., MOOSER, F., TOBSCHALL, H., and WERLE, D. (1985), *Geological and geochemical investigations on the eastern Trans Mexican Volcanic Belt*, Geofís. Int. *24*, 477–575.

NELSON, S.A. (1986), *Geología del Volcán Ceboruco, Nayrit, con una estimación de riesgos de erupciones futuras*, Rev. Inst. Geol. UNAM *6*, 243–258.

NELSON, S.A. and GONZALEZ-CAVER, E. (1992), *Geology and K-Ar dating of the Tuxtla volcanic field, Veracruz, Mexico*, Bull. Volcanol. *55*, 85–96.

NELSON, S.A., GONZALEZ-CAVER, E., and KYSER, T.K. (1995), *Constraints on the origin of alkaline and calc-alkaline magmas from the Tuxtla Volcanic Field, Veracruz, Mexico*, Contrib. Mineral. Petrol. *122*, 191–211.

NELSON, S.A. and LIVIERES, R.A. (1986), *Contemporaneous calc-alkaline and alkaline volcanism at Sanganguey volcano, Nayarit, Mexico*, Geol. Soc. Am. Bull. *97*, 798–808.

OHBA, T., MATSUOKA, K., KIMURA, Y., ISHIKAWA, H., and FUJIMAKI, H. (2009), *Deep crystallization differentiation of arc tholeiite basalt magmas from Northern Honshu Arc, Japan*, J. Petrol. *50*, 1025–1046.

OROZCO-ESQUIVEL, M.T., Zur Petrologie des Vulkangebietes von Palma-Sola, Mexiko. Ein Beispiel fuer den Uebergang von anorogenem zu orogenem Vulkanismus. *Institut für Petrographie und Geochemie* (Universitaet Karlsruhe, Karlsruhe, Germany, 167 p, 1995).

OROZCO-ESQUIVEL, T., PETRONE, C.M., FERRARI, L., TAGAMI, T., and MANETTI, P. (2007), *Geochemical and isotopic variability in lavas from the eastern Trans-Mexican Volcanic Belt: slab detachment in a subduction zone with varying dip*, Lithos *93*, 149–174.

PACHECO, J.F. and SINGH, S.K. (2010), *Seismicity and state of stress in Guerrero segment of the Mexican subduction zone*, J. Geophys. Res. *115*. doi:10.1029/2009JB006453.

PARDO, M. and SUÁREZ, G. (1995), *Shape of the subducted Rivera and Cocos plates in southern Mexico: Seismic and tectonic implications*, J. Geophys. Res. *100*, 12357–12373.

PATINO, L.C., CARR, M.J., and FEIGENSON, M.D. (1997), *Cross-arc geochemical variations in volcanic fields in Honduras C.A.: progressive changes in source with distance from the volcanic front*, Contrib. Mineral. Petrol. *129*, 341–351.

PATINO, L.C., CARR, M.J., and FEIGENSON, M.D. (2000), *Local and regional variations in Central American arc lavas controled by variations in subducted sediment input*, Contrib. Mineral. Petrol. *138*, 265–283.

PEARCE, J.A. (1976), *Statistical analysis of major element patterns in basalts*, J. Petrol. *17*, 15–43.

PEARCE, J.A., Trace element characteristics of lavas from destructive plate boundaries. In: *Andesites* (ed. Thorpe, R.S.) (Wiley, Chichester, pp. 525–548, 1982).

PEARCE, J.A. and CANN, J.R. (1971), *Ophiolite origin investigated by discriminant analysis using Ti, Zr and Y*, Earth Planet. Sci. Lett. *12*, 339–349.

PEARCE, J.A. and CANN, J.R. (1973), *Tectonic setting of basic volcanic rocks determined using trace element analyses*, Earth Planet. Sci. Lett. *19*, 290–300.

PEARCE, J.A. and GALE, G.H. (1977), *Identification of ore-deposition environment from trace-element geochemistry of associated igneous host rocks*, Geol. Soc. London Spec. Publ. *7*, 14–24.

PEARCE, J.A. and NORRY, M.J. (1979), *Petrogenetic implications of Ti, Zr, Y, and Nb variations in volcanic rocks*, Contrib. Mineral. Petrol. *69*, 33–47.

PEARCE, T.H., GORMAN, B.E., and BIRKETT, T.C. (1977), *The relationship between major element chemistry and tectonic environment of basic and intermediate volcanic rocks*, Earth Planet. Sci. Lett. *36*, 121–132.

PECCERILLO, A., DONATI, C., SANTO, A.P., ORLANDO, A., YIRGU, G., and AYALEW, D. (2007), *Petrogenesis of silicic peralkaline rocks in the Ethiopian Rift: geochemical evidence and volcanological implications*, J. Afr. Earth Sci. *48*, 161–173.

PÉREZ R., J., PAL, S., TERRELL, D.J., URRUTIA F., J., and LÓPEZ M., M. (1979), *Preliminary report on the analysis of some "in-house" geochemical reference samples from Mexico*, Geofís. Int. *18*, 197–209.

PÉREZ-CAMPOS, X., KIM, Y., HUSKER, A., DAVIS, P.M., CLAYTON, R.W., IGLESIAS, A., PACHECO, J.F., SINGH, S.K., MANEA, V.C., and GURNIS, M. (2008), *Horizontal subduction and truncation of the Cocos plate beneath central Mexico*, Geophys. Res. Lett. *35*, L18303.

PETRONE, C.M., FRANCALANCI, L., CARLSON, R.W., FERRARI, L., and CONTICELLI, S. (2003), *Unusual coexistence of subduction-related and intraplate-type magmatism: Sr, Nd and Pb isotope and trace element data from the magmatism of the San Pedro-Ceboruco graben (Nayarit, Mexico)*, Chem. Geol. *193*, 1–24.

RAY, D., IYER, S.D., BANEREJE, R., MISRA, S., and WIDDOWSON, M. (2007), *A petrogenetic model of basalts from the northern Central Indian Ridge: 3–11°S*, Acta Geol. Sinica *81*, 99–112.

REAGAN, M.K. and GILL, J.B. (1989), *Coexisting calcalkaline and high-niobium basalts from Turrialba volcano, Costa Rica: implications for residual titanates in arc magma sources*, J. Geophys. Res. *94*, 4619–4633.

REGELOUS, M., NIU, Y., ABOUCHAMI, W., and CASTILLO, P.R. (2009), *Shallow origin for South Atlantic Dupal anomaly from lower continental crust: geochemical evidence from the Mid-Atlantic Ridge at 26 S*, Lithos *112*, 57–72.

RIGHTER, K. and CARMICHAEL, I.S.E. (1992), *Hawaiite and related lavas in the Atenguillo graben, western Mexican Volcanic Belt*, Geol. Soc. Am. Bull. *104*, 1592–1607.

RIGHTER, K., CARMICHAEL, I.S.E., BECKER, T.A., and RENNE, P.R. (1995), *Pliocene-Quaternary volcanism and faulting at the intersection of the Gulf of California and the Mexican Volcanic Belt*, Geol. Soc. Am. Bull. *107*, 612–626.

RIGHTER, K. and ROSAS-ELGUERA, J. (2001), *Alkaline lavas in the volcanic front of the western Mexican Volcanic Belt: geology and petrology of the Ayutla and Tapalpa volcanic fields*, J. Petrol. *42*, 2333–2361.

ROBIN, C. (1976), *Présence simultanée de magmatismes de significations tectoniques opposées dans l'Est du Mexique*, Bull. Soc. Geol. Fr. *18*, 1637–1645.

ROBIN, C., CAMUS, G., CANTAGREL, J.M., GOURGAUD, A., MOSSAND, P., VINCENT, P.M., AUBERT, M., DOREL, J., and MURRAY, J.M. (1984), *Les volcanes de Colima (Mexique)*, Bull. P.I.R.P.S.E.V. *87*, 98 p.

ROBIN, C., KOMOROWSKI, J.-C., BOUDAL, C., and MOSSAND, P. (1990), *Mixed-magma pyroclastic surge deposits associated with debris avalanche deposits at Colima volcanoes, Mexico*, Bull. Volcanol. *52*, 391–403.

ROBIN, C. and POTREL, A. (1993), *Multi-stage magma mixing in the pre-caldera series of Fuego de Colima volcano*, Geofís. Int. *32*, 605–615.

ROBIN, C. and TOURNON, J. (1978), *Spatial relations of andesitic and alkaline provinces of Mexico and Central America*, Can. J. Earth Sci. *15*, 1633–1641.

RODRÍGUEZ, S.R., MORALES-BARRERA, W., LAYER, P., and GONZÁLEZ-MERCADO, E. (2009), *A Quaternary monogenetic volcanic field in the Xalapa region, eastern Trans-Mexican Volcanic Belt: geology, distribution and morphology of the volcanic event*, J. Volcanol. Geotherm. Res. (in press).

ROGERS, N.W., THOMAS, L.E., MACDONALD, R., HAWKESWORTH, C.J., and MOKADEM, F. (2006), *238U-230Th disequilibrium in recent basalts and dynamic melting beneath the Kenya Rift*, Chem. Geol. *234*, 148–168.

ROLLINSON, H.R. (1993), *Using geochemical data: evaluation, presentation, interpretation* (Essex, Longman Scientific Technical).

RONGA, F., LUSTRINO, M., MARZOLI, A., and MELLUSO, L. (2009), *Petrogenesis of a basalt-comendite-pantellerite rock suite: the Boseti Volcanic Complex (Main Ethiopian Rift)*, Miner. Petrol. doi:10.1007/s00710-009-0064-3.

ROONEY, T., FURMAN, T., BASTOW, I., AYALEW, D., and YIRGU, G. (2007), *Lithospheric modification during crustal extension in the Main Ethiopian Rift*, J. Geophys. Res. *112*, B10201. doi: 10.1029/2006JB004916.

ROSSOTTI, A., CARRASCO-NUÑEZ, G., ROSI, M., and DI MURO, A. (2006), *Eruptive dynamics of the "Citlaltépetl pumice" at Citlaltépetl volcano, eastern Mexico*, J. Volcanol. Geotherm. Res. *158*, 401–429.

RYDER, C.H., GILL, J.B., TEPLEY III, F., RAMOS, F., and REAGAN, M. (2006), *Closed- to open-system differentiation at Arenal volcano (1968–2003)*, J. Volcanol. Geotherm. Res. *157*, 75–93.

SCHAAF, P., STIMAC, J., SIEBE, C., and MACÍAS, J.L. (2005), *Geochemical evidence for mantle origin and crustal processes in volcanic rocks from Popocatépetl and surrounding monogenetic volcanoes, central Mexico*, J. Petrol. *46*, 1243–1282.

SCHUTH, S., MUNKER, C., KONIG, S., QOPOTO, C., BASI, S., SCHONBERG, D.G., and BALLHAUS, C. (2009), *Petrogenesis of lavas along the Solomon Island Arc, SW Pacific: coupling of compositional variations and subduction zone geometry*, J. Petrol. *50*, 781–811.

SENDJAJA, Y.A., KIMURA, J., and SUNARDI, E. (2009), *Across-arc geochemical variation of Quaternary lavas in West Java, Indonesia: Mass-balance elucidation using arc basalt simulator model*, Island Arc *18*, 201–224.

SHERVAIS, J.W. (1982), *Ti-V plots and the petrogenesis of modern and ophiolitic lavas*, Earth Planet. Sci. Lett. *59*, 101–118.

SHETH, H.C. (2008), *Do major oxide tectonic discrimination diagrams work? Evaluating new log-ratio and discriminant-analysis-based diagrams with Indian Ocean mafic volcanics and Asian ophiolites*, Terra Nova *20*, 229–236.

SHETH, H.C., TORRES-ALVARADO, I.S., and VERMA, S.P. (2000), *Beyond subduction and plumes: a unified tectonic-petrogenetic model for the Mexican Volcanic Belt*, Int. Geol. Rev. *42*, 1116–1132.

SIEBE, C., RODRÍGUEZ-LARA, V., SCHAAF, P., and ABRAMS, M. (2004), *Geochemistry, Sr-Nd isotope composition, and tectonic setting of Holocene Pelado, Guespalapa and Chichinautzin scoria cones, south of Mexico City*, J. Volcanol. Geotherm. Res. *130*, 197–226.

SIEBERT, L. and CARRASCO-NUÑEZ, G. (2002), *Late-Pleistocene to precolumbian behind-the-arc mafic volcanism in the eastern Mexican Volcanic Belt; implications for future hazards*, J. Volcanol. Geotherm. Res. *115*, 179–205.

SILVA MORA, L. (1988), *Algunos aspectos de los basaltos y andesitas cuaternarias de Michoacán Oriental*, Rev. Inst. Geol. UNAM *7*, 89–96.

STRAUB, S.M., LAGATTA, A.B., POZZO, A.L.M., and LANGMUIR, C.H. (2008), *Evidence from high-Ni olivines for a hybridized peridotite/pyroxenite source for orogenic andesites from the central Mexican Volcanic Belt*, Geochem. Geophys. Geosys. *9*, 7 March 2008.

SUTER, M., AGUIRRE, G., SIEBE, C., QUINTERO, O., and KOMOROWSKI, J.C. (1991), *Volcanism and active faulting in the central part of the Trans-Mexican Volcanic Belt, Mexico*. In *Geological excursions in southern California and Mexico. Guidebook 1991 Annual Meeting Geological Society of America* (eds. WALAWENDER, M.J. and HANAN, B.B.) (Geological Society of America, San Diego, pp. 224–243, 1991).

SUTER, M., QUINTERO, O., and JOHNSON, C.A. (1992), *Active faults and state of stress in the central part of the Trans-Mexican Volcanic Belt, Mexico. 1. The Venta de Bravo fault*, J. Geophys. Res. *97*, 11983–11993.

SUTER, M., CARRILLO MARTÍNEZ, M., LÓPEZ MARTÍNEZ, M., AND FARRAR, E. (1995a), *The Aljibes half-graben—active extension at the boundary between the Trans-Mexican Volcanic Belt and the basin and range province, Mexico*, Geol. Soc. Am. Bull. *107*, 627–641.

SUTER, M., QUINTERO-LEGORRETA., O., LÓPEZ-MARTÍNEZ, M., AGUIRRE-DÍAZ, G., and FARRAR, E. (1995b), *The Acambay graben: active intraarc extension in the Trans-Mexican Volcanic Belt, Mexico*, Tectonics *14*, 1245–1262.

SUTER, M., LÓPEZ MARTÍNEZ, M., QUINTERO LEGORRETA, O., AND CARRILLO MARTÍNEZ, M. (2001), *Quaternary intra-arc extension in the central Trans-Mexican volcanic belt*, Geol. Soc. Am. Bull. *113*, 693–703.

SWINAMER, R.T., *The geomorphology, petrography, geochemistry and petrogenesis of the volcanic rocks in the Sierra del Chichinautzin, Mexico Department of Geological Sciences* (Kingston, Queen's University, 212 p, 1989).

TAKAMASA, A., NAKAI, S., SAHOO, Y., HANYUM, T., and TATSUMI, Y. (2009), *W isotope compositions of oceanic islands basalts from French Polynesia and their meaning for core-mantle interaction*, Chem. Geol. *260*, 37–46.

TAMURA, Y., TANI, K., CHANG, Q., SHUKUNO, H., KAWABATA, H., ISHIZUKA, O., AND FISKE, R.S. (2007), *Wet and dry basalt magma evolution at Torishima Volcano, Izu-Bonin Arc, Japan: the* possible role of phengite in the downgoing slab, J. Petrol. 48, 1999–2031.

TORMEY, D.R., HICKEY-VARGAS, R., FREY, F.A., and LÓPEZ-ESCOBAR, L., Recent lavas from the Andean volcanic front (33 to 42°S); interpretations of along-arc compositional variations. In Andean magmatism and its tectonic setting, Geological Society of America Special Paper (eds. HARMON, R.S. and RAPELA, C.W.) (Boulder, Colorado, Geological Society of America, pp. 57–77, 1991).

TORRES-ALVARADO, I.S., PANDARINATH, K., VERMA, S.P., and DULSKI, P. (2007), Mineralogical and geochemical effects due to hydrothermal alteration in the Los Azufres geothermal field, Mexico, Rev. Mex. Cienc. Geol. 24, 15–24.

TORRES-ALVARADO, I.S., VERMA, S.P., and VELASCO-TAPIA, F. (2002), Comment and reply to "Generation of oceanic-island basalt type volcanism in the western Trans-Mexican volcanic belt by slab rollback, asthenosphere infiltration, and variable flux melting", Geology 30, 857–858.

TURNER, S., HAWKESWORTH, C.J., CALSTEREN, P.V., HEATH, E., MACDONALD, R., and BLACK, S. (1996), U-series isotopes and destructive plate margin magma genesis in the Lesser Antilles, Earth Planet. Sci. Lett. 142, 191–207.

VELASCO-TAPIA, F. and VERMA, S.P. (2001), First partial melting inversion model for a rift-related origin of the Sierra de Chichinautzin volcanic field, central Mexican Volcanic Belt, Int. Geol. Rev. 43, 788–817.

VERGARA, M., LÓPEZ-ESCOBAR, L., PALMA, J.L., HICKEY-VARGAS, R., and ROESCHMANN, C. (2004), Late Tertiary volcanic episodes in the area of the city of Santiago de Chile: new geochronological and geochemical data, J. South. Am. Earth. Sci. 17, 227–238.

VERMA, S.P. (1983), Magma genesis and chamber processes at Los Humeros caldera, Mexico—Nd and Sr isotope data, Nature 301, 52–55.

VERMA, S.P., Geochemistry of the subducting Cocos plate and the origin of subduction-unrelated mafic volcanism at the volcanic front of the central Mexican Volcanic Belt. In Cenozoic tectonics and volcanism of Mexico (eds. DELGADO-GRANADOS, H., AGUIRRE-DÍAZ, G., and STOCK, J.M.) (Geological Society of America, pp. 195–222, 2000a).

VERMA, S.P. (2000b), Geochemical evidence for a lithospheric source for magmas from Los Humeros caldera, Puebla, Mexico, Chem. Geol. 164, 35–60.

VERMA, S.P. (2001a), Geochemical evidence for a lithospheric source for magmas from Acoculco caldera, eastern Mexican Volcanic Belt, Int. Geol. Rev. 43, 31–51.

VERMA, S.P. (2001b), Geochemical evidence for a rift-related origin of bimodal volcanism at Meseta Río San Juan, North-Central Mexican volcanic belt, Int. Geol. Rev. 43, 475–493.

VERMA, S.P. (2001c), Geochemical and Sr–Nd–Pb isotopic evidence for a combined assimilation and fractional crystallisation process for volcanic rocks from the Huichapan caldera, Hidalgo, Mexico, Lithos 56, 141–164.

VERMA, S.P. (2002), Absence of Cocos plate subduction-related basic volcanism in southern Mexico: a unique case on Earth?, Geology 30, 1095–1098.

VERMA, S.P. (2003), Geochemical and Sr-Nd isotopic evidence for a rift-related origin of magmas in Tizayuca volcanic field, Central Mexican Volcanic Belt, J. Geol. Soc. India 61, 257–276.

VERMA, S.P. (2004), Solely extension-related origin of the eastern to west-central Mexican Volcanic Belt (Mexico) from partial melting inversion model, Curr. Sci. 86, 713–719.

VERMA, S.P. (2006), Extension related origin of magmas from a garnet-bearing source in the Los Tuxtlas volcanic field, Mexico, Int. J. Earth Sci. 95, 871–901.

VERMA, S.P. (2009), Continental rift setting for the central part of the Mexican Volcanic Belt: a statistical approach, Open Geol. J. 3, 8–29.

VERMA, S.P. (2010), Statistical evaluation of bivariate, ternary and discriminant function tectonomagmatic discrimination diagrams, Turk. J. Earth Sci. 19, 185–238.

VERMA, S.P. and AGRAWAL, S. (2010), Discriminant function discrimination diagrams for basic and ultrabasic volcanic rocks through log-transformed ratios of high field strength elements and implications for petrogenetic processes, Rev. Mex. Cienc. Geol. (in press).

VERMA, S.P. and HASENAKA, T. (2004), Sr, Nd, and Pb isotopic and trace element geochemical contraints for a veined-mantle source of magmas in the Michoacán-Guanajuato volcanic field, west-central Mexican Volcanic Belt, Geochem. J. 38, 43–65.

VERMA, S.P. and LOPEZ M., (1982), Geochemistry of Los Humeros caldera, Puebla, Mexico, Bull. Volcanol. 45, 63–79.

VERMA, S.P. and LUHR, J.F. (2010), Sr, Nd, and Pb isotopic evidence for the origin and evolution of the Cántaro-Colima Volcanic Chain, Western Mexican Volcanic Belt, J. Volcanol. Geotherm. Res. (submitted).

VERMA, S.P. and NELSON, S.A. (1989a), Isotopic and trace element constraints on the origin and evolution of alkaline and calc-alkaline magmas in the northwestern Mexican Volcanic Belt, J. Geophys. Res. 94, 4531–4544.

VERMA, S.P. and NELSON, S.E. (1989b), Correction to "Isotopic and trace element constraints on the origin and evolution of alkaline and calc-alkaline magmas in the northwestern Mexican Volcanic Belt" by Surendra P. Verma and Stephen A. Nelson, J. Geophys. Res. 94, 7679–7681.

VERMA, S.P. and QUIROZ-RUIZ, A. (2006a), Critical values for six Dixon tests for outliers in normal samples up to sizes 100, and applications in science and engineering, Rev. Mex. Cienc. Geol. 23, 133–161.

VERMA, S.P. and QUIROZ-RUIZ, A. (2006b), Critical values for 22 discordancy test variants for outliers in normal samples up to sizes 100, and applications in science and engineering, Rev. Mex. Cienc. Geol. 23, 302–319.

VERMA, S.P. and QUIROZ-RUIZ, A. (2008), Critical values for 33 discordancy test variants for outliers in normal samples for very large sizes of 1,000 to 30,000, Rev. Mex. Cienc. Geol. 25, 369–381.

VERMA, S.P. and RIVERA-GÓMEZ, M.A. (2010), TecD: A new computer program for tectonomagmatic discrimination from discriminant function diagrams for basic and ultrabasic magmas, Rev. Mex. Cienc. Geol. (submitted).

VERMA, S.P., SALAZAR-V., A., NEGENDANK, J.F.W., MILÁN, M., NAVARRO-L., I., and BESCH, T. (1993), Características petrográficas y geoquímicas de elementos mayores del campo volcánico de Los Tuxtlas, Veracruz, México, Geofís. Int. 32, 237–248.

VERMA, S.P., TORRES-ALVARADO, I.S., and SOTELO-RODRÍGUEZ, Z.T. (2002), SINCLAS: standard igneous norm and volcanic rock classification system, Comput. Geosci. 28, 711–715.

VERMA, S.P., TORRES-ALVARADO, I.S., and VELASCO-TAPIA, F. (2003), A revised CIPW norm, Schweiz. Miner. Petrog. Mitteil. 83, 197–216.

VERMA, S.P., GUEVARA, M., and AGRAWAL, S. (2006), *Discriminating four tectonic settings: five new geochemical diagrams for basic and ultrabasic volcanic rocks based on log-ratio transformation of major-element data*, J. Earth Syst. Sci. *115*, 485–528.

VERMA, S.P., QUIROZ-RUIZ, A., and DÍAZ-GONZÁLEZ, L. (2008), *Critical values for 33 discordancy test variants for outliers in normal samples up to sizes 1000, and applications in quality control in Earth Sciences*, Rev. Mex. Cienc. Geol. *25*, 82–96.

VERMA, S.P., RODRÍGUEZ-RÍOS, R., and GONZÀLEZ-RAMÌREZ, R. (2010), *Statistical evaluation of classification diagrams for altered igneous rocks*, Turk. J. Earth Sci. *19*, 239–265.

VERMEESCH, P. (2007), *Tectonic discrimination diagrams revisited*, G3 7. doi:10.1029/2005GC001092.

VIGOUROUX, N., WALLACE, P.J., AND KENT, A.J.R. (2008), *Volatiles in high-K magmas from the western Trans-Mexican volcanic belt: evidence for fluid fluxing and extreme enrichment of the mantle wedge by subduction processes*, J. Petrol. *49*, 1589–1618.

WALKER, J.A., CARR, M.J., FEIGENSON, M.D., and KALAMARIDES, R.I. (1990), *The petrogenetic significance of interstratified high- and low-Ti basalts in central Nicaragua*, J. Petrol. *31*, 1141–1164.

WALKER, J.A., PATINO, L.C., CAMERON, B.I., and CARR, M.J. (2000), *Petrogenetic insights provided by compositional transects across the Central American arc: southeastern Guatemala and Honduras*, J. Geophys. Res. *105*, 18949–18963.

WALKER, J.A., PATINO, L.C., CARR, M.J., AND FEIGENSON, M.D. (2001), *Slab control over HFSE depletions in central Nicaragua*, Earth Planet. Sci. Lett. *192*, 533–543.

WALLACE, P. and CARMICHAEL, I.S.E. (1989), *Minette lavas and associated leucitites from the western front of the Mexican Volcanic Belt: petrology, chemistry and origin*, Contrib. Mineral. Petrol. *103*, 470–492.

WALLACE, P. and CARMICHAEL, I.S.E. (1992), *Alkaline and calc-alkaline lavas near Los Volcanes, Jalisco, Mexico: geochemical diversity and its significance in volcanic arcs*, Contrib. Mineral. Petrol. *111*, 423–439.

WALLACE, P.J. and CARMICHAEL, I.S.E. (1999), *Quaternary volcanism near the Valley of Mexico: implications for subduction zone magmatism and the effects of crustal thickness variations on primitive magma compositions*, Contrib. Mineral. Petrol. *135*, 291–314.

WILLIAMS, H. (1950), *Volcanoes of the Parícutin region, Mexico*, U.S. Geol. Surv. Bull. *965-B*, 1–279.

WOOD, D.A. (1980), *The application of a Th-Hf-Ta diagram to problems of tectonomagmatic classification and to establishing the nature of crustal contamination of basaltic lavas of the British Tertiary volcanic province*, Earth Planet. Sci. Lett. *50*, 11–30.

XU, G., FREY, F.A., CLAGUE, D.A., ABOUCHAMI, W., BLICHERT-TOFT, J., COUSENS, B., AND WEISLER, M. (2007), *Geochemical characteristics of West Molokai shield- and postshield-stage lavas: Constraints on Hawaiian plume models*, G3 8, 1–40.

YANG, T., GRAND, S.P., WILSON, D.S., GUZMAN-SPEZIALE, M., GOMEZ-GONZALEZ, J.M., DOMINGUEZ-REYES, T., AND NI, J. (2009), *Seismic structure beneath the Rivera subduction zone from finite-frequency seismic tomography*, J. Geophys. Res. *114*, doi: 10.1029/2008JB005830.

(Received February 15, 2010, revised May 11, 2010, accepted May 25, 2010, Published online June 22, 2010)